A BASIC/INTERMEDIATE COURSE
FOR WATER SYSTEM OPERATORS

Reference Handbook

Basic Science Concepts and Applications

PRINCIPLES and PRACTICES of
WATER SUPPLY OPERATIONS

American Water Works Association
6666 W. Quincy Ave., Denver, Colorado 80235

ISBN 0-89867-202-3

Foreword

Basic Science Concepts and Applications has been developed as a supplementary reference book to accompany Volumes 1 through 4 of the "Principles and Practices of Water Supply Operations" course of study. The sections of the reference book—Mathematics, Hydraulics, Chemistry, and Electricity—provide extended discussions of the principles and operational calculations related to selected topics in the four volumes.

Student volumes in the series are

Volume 1	*Introduction to Water Sources and Transmission*
Volume 2	*Introduction to Water Treatment*
Volume 3	*Introduction to Water Distribution*
Volume 4	*Introduction to Water Quality Analyses*

An instructor guide and solutions manual is available for each of the student volumes. The guides contain lesson plans and information about required materials for class demonstrations and discussions, suggestions for "before class activities," and examination questions and answers for each module in the volume. Included in the guide are a Selected List of Library References, Source Information for Supplementary Audio-Visual Materials, and solutions to certain Study Problems and Exercises presented in the student's text.

It is the hope of the American Water Works Association that the materials comprising this course of study will provide an adequate basis of training for basic- to intermediate-grade water treatment and water distribution operators. This training project was made possible by funding assistance from the US Environmental Protection Agency, Office of Drinking Water, under Grant No. T900632-01 awarded to the American Water Works Association.

Acknowledgments

Publication of this volume was made possible through a grant from the US Environmental Protection Agency, Office of Drinking Water, under Grant No. T900632-01 in recognition of the need for a comprehensive instructional approach to operator training in the water supply industry. John B. Mannion, special assistant for communications and training, represented the Environmental Protection Agency, Office of Drinking Water, as project officer, and Bill D. Haskins, director of education, served as project manager for the American Water Works Association.

Joanne Kirkpatrick and Benton C. Price, under contract with VTN Colorado, Inc., were the principal developers for this volume.

Special thanks is extended to the many individuals who gave liberally of their time and expertise in the technical review of all or portions of the manuscript. The following are credited for their participation on the review committee or as an independent reviewer:

Charles R. Beer, Superintendent of Water Treatment, Water Department, Denver, Colorado

Lucille M. Black, Head, Technical Review Group, Geology Division, Bendix Field Engineering Corporation, Grand Junction, Colorado

James O. Bryant Jr., Director, Environmental Resources Training Center, Southern Illinois University at Edwardsville

Susan A. Castle, Program Director for Water Pollution Training, Environmental Resources Training Center, Southern Illinois University at Edwardsville

Jack C. Dice, Quality Control Engineer, Water Department, Denver, Colorado

James T. Harvey, Superintendent of Production, Water Works, Little Rock, Arkansas

William R. Hill, Director of Technical Services, Floyd G. Browne & Associates, Ltd., Marion, Ohio

Jack W. Hoffbuhr, Chief, Drinking Water Branch, US Environmental Protection Agency, Region VIII, Denver, Colorado

Kenneth D. Kerri, Professor of Civil Engineering, School of Engineering, California State University at Sacramento

John L. Krantz, Superintendent of Wastewater Training Systems, Wastewater Treatment Plant Training Section, Detroit, Michigan

Ralph W. Leidholdt, Special Projects Engineer for Training, American Water Works Association

L.H. Lockhart, Coordinator, Manpower Planning, Development & Training, Administrative Division, South Carolina Department of Health & Environmental Control

Andrew J. Piatek Jr., Chief Operator & Superintendent, Borough of Sayreville, New Jersey

John F. Rieman, Technical Publications Editor, American Water Works Association

Tom F. Staible, Laboratory Certification Coordinator, US Environmental Protection Agency, Region VIII, Denver, Colorado

Robert L. Wubbena, Vice President, Economic and Engineering Services, Inc., Olympia, Washington

The Electricity Section was reprinted with minor changes from AWWA Manual M2, *Automation and Instrumentation,* which was developed by the AWWA Distribution Division Committee on Automation and Instrumentation. Members of the committee during the time of manuscript preparation included:

Chairman:
James R. Daneker, BIF/A Unit of General Signal, West Warwick, Rhode Island

Members:
Robert C. Freeston, American Water Works Service Co., Inc., Haddon Heights, New Jersey

Harold D. Gilman, Greeley & Hanson, Philadelphia, Pennsylvania

Karl E. Jasper, ACCO Briston, Waterbury, Connecticut

Robert Knight, St. Louis County Water Co., St. Louis, Missouri

C.N. McGehee, Department of Water & Power, Los Angeles, California

Wellman E. Nasbaum, Black & Veatch, Kansas City, Missouri

E.H. Pitman Jr., Black, Crow & Eidsness, Inc., Gainesville, Florida

Joseph V. Radziul, Water Department, Philadelphia, Pennsylvania

Richard Thompson, Fischer & Porter Co., Warminster, Pennsylvania

Richard G. Toler, City Water Board, San Antonio, Texas

Introduction for the Student

Basic Science Concepts and Applications has been prepared as a reference handbook to be used with the four-volume course "Principles and Practices of Water Supply Operations." The volumes in the series are

Volume 1   *Introduction to Water Sources and Transmission*
Volume 2   *Introduction to Water Treatment*
Volume 3   *Introduction to Water Distribution*
Volume 4   *Introduction to Water Quality Analyses*

Within each of the volumes, footnotes refer to this reference handbook wherever a more extensive explanation of a concept would aid the reader's understanding or where a detailed discussion of the application of concepts to operational situations is required.

Subject matter in this reference book is organized into four main sections— Mathematics, Hydraulics, Chemistry, and Electricity—supporting a broad range of topics in the other four volumes. The chapters in each main section need not be read in the order in which they are presented. Where topics in other sections of this work are related and considered helpful to the reader, a footnote reference is given to the related discussion.

Important terms used in the book are set in SMALL CAPITAL LETTERS where they first appear. These terms are defined fully in the glossary at the end of the book.

In general, calculations and answers are carried to two decimal places where the number is less than 999.99 (example: 27.12). Final zeros after the decimal point are not printed (example: 39.7, not 39.70). Numbers greater than 999.99 are given only to the nearest whole number (example: 1718). Tables and graphs are read with whatever precision possible. These rules are adopted for consistency; they do not reflect the accuracy of the data used in the calculations. In practice, the operator should realize that the accuracy of a final answer can never be better than the accuracy of the data used.

Table of Contents

Mathematics Section

Hydraulics Section

Chemistry Section

Electricity Section

Appendices

**Basic Science
Concepts and
Applications**

Mathematics

Mathematics 1

Powers and Scientific Notation

M1-1. Powers Notation

The most basic form of powers notation is merely a shorthand method of writing multiplication. For example, 4×4 can be written as

$$4^2$$

This is referred to as *4 to the second power* or *4 squared*. The small 2 is the EXPONENT or POWER. It tells you how many 4's are to be multiplied together: two. In expanded form,

$$4^2 = (4)(4)$$

The expression *4 to the third power* (usually called *4 cubed*) is written as

$$4^3$$

In expanded form, this notation means

$$4^3 = (4)(4)(4)$$

The following examples further illustrate this concept of powers notation.

Example 1
How is the term 2^3 written in expanded form?
The power (or exponent) of 3 means that the number is multiplied by itself three times:

$$2^3 = (2)(2)(2)$$

Example 2
How is the term ft^2 written in expanded form?
The power or exponent of 2 means that the term is multiplied by itself two times:

$$ft^2 = (ft)(ft)$$

3

Example 3
How is the term 10^5 written in expanded form?
The exponent of 5 indicates that 10 is multiplied by itself five times:

$$10^5 = (10)(10)(10)(10)(10)$$

Example 4
How is the term $(2/3)^2$ written in expanded form?
When parentheses are used, the exponent refers to the entire term within the parentheses. Therefore, in this problem, $(2/3)^2$ means

$$\left(\frac{2}{3}\right)^2 = \left(\frac{2}{3}\right)\left(\frac{2}{3}\right)$$

Sometimes a negative exponent is used with a number or term. A number with a negative exponent can be re-expressed using a positive exponent:

$$3^{-2} = \frac{1}{3^2}$$

Using another example

$$12^{-3} = \frac{1}{12^3}$$

Example 5
How is the term 7^{-2} written in expanded form?

$$7^{-2} = \frac{1}{7^2} = \frac{1}{(7)(7)}$$

Example 6
How is the term 10^{-4} written in expanded form?

$$10^{-4} = \frac{1}{10^4} = \frac{1}{(10)(10)(10)(10)}$$

If a term is given in expanded form, you should be able to determine how it would be written in exponential (or power) form. For example,

$$(ft)(ft) = ft^2$$

or

$$(3)(3)(3) = 3^3$$

Example 7
Given the following expanded term, how would the term be rewritten in exponential form?

$$(yd)(yd)(yd)$$

Since the term is multiplied by itself three times, it would be written in exponential form as

$$yd^3$$

Example 8

Write the following term in exponential form:

$$\frac{(2)(2)}{(3)(3)(3)}$$

The exponent for the numerator of the fraction is *2* and the exponent for the denominator is *3*. Therefore, the term would be written as

$$\frac{2^2}{3^3}$$

Since the exponents are not the same, no parentheses can be used to express this term.

Example 9

Write the following term in exponential form:

$$\frac{(ft)(ft)}{(in.)(in.)}$$

The exponent of both the numerator and denominator is *2*:

$$\frac{ft^2}{in.^2}$$

Since the exponents are the same, parentheses can be used to express this term, if desired:

$$\left(\frac{ft}{in.}\right)^2$$

Perhaps the two most common situations in which you may see powers used with a number or term are in denoting area or volume units ($in.^2$, ft^2, $in.^3$, ft^3) and in scientific notation.

M1-2.   Scientific Notation

SCIENTIFIC NOTATION is a method by which any number can be expressed as a number between 1 and 9 multiplied by a power of 10. Examples of numbers written in scientific notation are given below:

$$5.4 \times 10^1$$

$$1.2 \times 10^3$$

$$9.789 \times 10^4$$

$$3.62 \times 10^{-2}$$

The numbers can be taken out of scientific notation by performing the indicated multiplication. For example,

$$5.4 \times 10^1 = (5.4)(10)$$
$$= 54$$
$$1.2 \times 10^3 = (1.2)(10)(10)(10)$$
$$= 1200$$
$$9.789 \times 10^4 = (9.789)(10)(10)(10)(10)$$
$$= 97,890$$
$$= (3.62) \frac{1}{10^2}$$
$$3.62 \times 10^{-2} = (3.62) \frac{(1)}{(10)(10)}$$
$$= \frac{3.62}{(10)(10)}$$
$$= 0.0362$$

An easier way to take a number out of scientific notation is by moving the decimal point the number of places indicated by the exponent.

Rule 1
When a number is taken *out* of scientific notation, a *positive* exponent indicates a decimal point move to the *right,* and a *negative* exponent indicates a decimal point move to the *left.*

Let's look again at the examples above, using the decimal point move rather than the multiplication method.

$$5.4 \times 10^1$$

The *positive* exponent of 1 indicates that the decimal point in 5.4 should be moved one place to the *right*:

$$5.4 = 54$$

The next example is

$$1.2 \times 10^3$$

The *positive* exponent of 3 indicates that the decimal point in 1.2 should be moved three places to the *right*:

$$1.200 = 1200$$

The next example is

$$9.789 \times 10^4$$

The *positive* exponent of 4 indicates that the decimal point should be moved four places to the *right*:

$$9.7890 = 97,890$$

The final example is

$$3.62 \times 10^{-2}$$

The *negative* exponent of 2 indicates that the decimal point should be moved two places to the *left*:

$$03.62 = 0.0362$$

Example 10
Take the following number out of scientific notation:

$$7.992 \times 10^5$$

The *positive* exponent of 5 indicates that the decimal point should be moved five places to the *right*:

$$7.99200 = 799,200$$

Example 11
Take the following number out of scientific notation:

$$2.119 \times 10^{-3}$$

The *negative* exponent of 3 indicates that the decimal point should be moved three places to the *left*:

$$002.119 = 0.002119$$

Although there will be very few instances in which you will need to put a number *into* scientific notation, the method is discussed below.

To put a number into scientific notation, the decimal point is moved the number of places necessary to result in a number between 1 and 9. This number is multiplied by a power of 10, with the exponent equal to the number of places that the decimal point was moved. (Remember that if no decimal point is shown in the number to be converted, it is assumed to be at the end of the number.)

Rule 2
When a number is put *into* scientific notation, a decimal point move to the *left* indicates a *positive* exponent; a decimal point move to the *right* indicates a *negative* exponent.

Now let's try converting a few numbers into scientific notation, using the same numbers as in the previous examples.

First, let's convert

$$54$$

To obtain a number between 1 and 9, the decimal point should be moved one place to the left. This place move of 1 is the exponent of the power of 10, and the move to the *left* means that the exponent is *positive*:

$$54 = 5.4 \times 10^1$$

The next number to be put into scientific notation is

$$1200$$

To obtain a number between 1 and 9, the decimal point should be moved three places to the left. The number of place moves (3) becomes the exponent of the power of 10, and the move to the *left* indicates a *positive* exponent:

$$1200 = 1.2 \times 10^3$$

The next example is

$$97,890$$

To obtain a number between 1 and 9, the decimal point should be moved four places to the *left*, resulting in a *positive* exponent of 4:

$$97,890 = 9.789 \times 10^4$$

The final example is

$$0.0362$$

To obtain a number between 1 and 9, the decimal point must be moved two places to the right. This indicates an exponent of 2, and the move to the *right* requires a *negative* exponent:

$$0.0362 = 3.62 \times 10^{-2}$$

Example 12
Put the following number into scientific notation:

$$4,573,000$$

To obtain a number between 1 and 9, the decimal point should be moved six places to the *left*, resulting in a *positive* exponent of 6:

$$4,573,000 = 4.573 \times 10^6$$

Example 13
Convert the following decimal to scientific notation:

$$0.000375$$

To obtain a number between 1 and 9, the decimal point should be moved four places to the *right*, resulting in a *negative* exponent of 4:

$$0.000375 = 3.75 \times 10^{-4}$$

Review Questions

1. How is the term cm^2 written in expanded form?

2. How is the term 20^3 written in expanded form?

3. How is the term $(4/5)^2$ written in expanded form?

4. Write the term 10^{-3} in expanded form.

5. How would the following terms be written in exponential form?
    (a) (ft)(ft)
    (b) (2)(2)(2)(2)
    (c) $\dfrac{(in.)(in.)}{min}$
    (d) $\dfrac{1}{(10)(10)(10)}$

6. Take the following numbers out of scientific notation:
    (a) 1.921×10^2
    (b) 4.7×10^{-3}
    (c) 7.889×10^6
    (d) 2.76×10^{-1}

7. Put the following numbers into scientific notation:
    (a) 7,960
    (b) 0.0875
    (c) 30
    (d) 0.00042

Summary Answers

1. (cm)(cm)

2. (20)(20)(20)

3. $\left(\dfrac{4}{5}\right)\left(\dfrac{4}{5}\right)$

4. $\dfrac{1}{(10)(10)(10)}$

5. (a) ft^2

   (b) 2^4

   (c) $\dfrac{in.^2}{min}$

   (d) $\dfrac{1}{10^3}$ or 10^{-3}

6. (a) 192.1

   (b) 0.0047

   (c) 7,889,000

   (d) 0.276

7. (a) 7.96×10^3

   (b) 8.75×10^{-2}

   (c) 3.0×10^1

   (d) 4.2×10^{-4}

Detailed Answers

1. The exponent of 2 means the term is multiplied by itself two times
$$cm^2 = (cm)(cm)$$

2. The exponent of 3 indicates the number is multiplied by itself three times
$$20^3 = (20)(20)(20)$$

3. The exponent of 2 indicates the entire term in the parentheses is multiplied by itself two times:
$$\left(\frac{4}{5}\right)^2 = \left(\frac{4}{5}\right)\left(\frac{4}{5}\right)$$

4. The negative exponent indicates
$$10^{-3} = \frac{1}{10^3}$$

This term can then be expanded to
$$\frac{1}{10^3} = \frac{1}{(10)(10)(10)}$$

5. (a) Since the term is multiplied by itself two times, the exponent is 2:
$$(ft)(ft) = ft^2$$

  (b) The number is multiplied by itself four times. Therefore the exponent is 4:
$$(2)(2)(2)(2) = 2^4$$

  (c) The term in the numerator is multiplied by itself twice; therefore the exponent is
$$\frac{(in.)(in.)}{min} = \frac{in.^2}{min}$$

  (d) The number in the denominator is multiplied by itself three times; therefore the exponent is 3:
$$\frac{1}{(10)(10)(10)} = \frac{1}{10^3}$$

  Since the fraction is 1 over a number, it may also be written as a negative exponent:
$$\frac{1}{10^3} = 10^{-3}$$

6. (a) With the decimal point method, the *positive* exponent of *2* indicates a decimal point move of two places to the *right:*

$$1.921 = 192.1$$

(b) The *negative* exponent of *3* indicates decimal point move of three places to the *left:*

$$004.7 = 0.0047$$

(c) The *positive* exponent of *6* indicates a decimal point move of six places to the *right:*

$$7.889000 = 7,889,000$$

(d) A *negative* exponent of *1* indicates a decimal point move of one place to the *left:*

$$2.76 = 0.276$$

7. (a) To obtain a number between 1 and 9 multiplied by a power of 10, the decimal point should be moved three places to the *left*, indicating a *positive* exponent of *3:*

$$7,960 = 7.96 \times 10^3$$

(b) To obtain a number between 1 and 9, the decimal point should be moved two places to the *right*, resulting in a *negative* exponent of *2:*

$$0.0875 = 8.75 \times 10^{-2}$$

(c) To obtain a number between 1 and 9, the decimal point should be moved one place to the *left*, resulting in a *positive* exponent of *1:*

$$30 = 3.0 \times 10^1$$

(d) To obtain a number between 1 and 9, the decimal point should be moved four places to the *right*, resulting in a *negative* exponent of *4:*

$$0.00042 = 4.2 \times 10^{-4}$$

Mathematics 2

Dimensional Analysis

Dimensional analysis is a tool that you can use to determine whether you have set up a problem correctly. In checking a math setup using dimensional analysis, you work only with the dimensions or units of measure and not with the numbers themselves.

To use the dimensional analysis method, you must know three things:

- How to express a horizontal fraction (such as gal/cu ft) as a vertical fraction (such as $\dfrac{\text{gal}}{\text{cu ft}}$),
- How to divide by a fraction, and
- How to divide out or cancel terms in the numerator and denominator of a fraction.

These techniques are reviewed briefly below.

When you are using dimensional analysis to check a problem, it is often desirable to write any horizontal fractions as vertical fractions, thus:

$$\text{cu ft/min} = \frac{\text{cu ft}}{\text{min}}$$

$$\text{sec/min} = \frac{\text{sec}}{\text{min}}$$

$$\frac{\text{gal/min}}{\text{gal/cu ft}} = \frac{\dfrac{\text{gal}}{\text{min}}}{\dfrac{\text{gal}}{\text{cu ft}}}$$

When a problem involves division by a fraction, the rule is to invert (or turn over) the terms in the denominator and then multiply. For example,

$$\frac{\dfrac{\text{gal}}{\text{min}}}{\dfrac{\text{gal}}{\text{cu ft}}} = \frac{\text{gal}}{\text{min}} \times \frac{\text{cu ft}}{\text{gal}}$$

and

$$\frac{\text{lb/day}}{\text{min/day}} = \frac{\dfrac{\text{lb}}{\text{day}}}{\dfrac{\text{min}}{\text{day}}} = \frac{\text{lb}}{\text{day}} \times \frac{\text{day}}{\text{min}}$$

or

$$\frac{\text{sq in.}}{\text{sq in./sq ft}} = \frac{\text{sq in.}}{\dfrac{\text{sq in.}}{\text{sq ft}}} = \text{sq in.} \times \frac{\text{sq ft}}{\text{sq in.}}$$

Once the fractions in a problem, if any, have been rewritten in the vertical form, and division by a fraction has been re-expressed as multiplication as shown above, then the terms can be divided out or cancelled. In cancelling terms, for every term cancelled in the numerator of a fraction, a similar term must be cancelled in the denominator, and vice versa, as shown below:

$$\frac{\cancel{\text{gal}}}{\text{min}} \times \frac{\text{cu ft}}{\cancel{\text{gal}}} = \frac{\text{cu ft}}{\text{min}}$$

$$\frac{\text{lb}}{\cancel{\text{day}}} \times \frac{\cancel{\text{day}}}{\text{min}} = \frac{\text{lb}}{\text{min}}$$

$$\cancel{\text{sq in.}} \times \frac{\text{sq ft}}{\cancel{\text{sq in.}}} = \text{sq ft}$$

$$\frac{\text{cu ft}}{\cancel{\text{sec}}} \times \frac{\text{gal}}{\cancel{\text{cu ft}}} \times \frac{\cancel{\text{sec}}}{\cancel{\text{min}}} \times \frac{\cancel{\text{min}}}{\text{day}} = \frac{\text{gal}}{\text{day}}$$

You may wish to review the concept of powers[1] before continuing with the following examples in dimensional analysis.

Suppose you wish to convert 1200 cu ft volume to gallons and suppose that you know you will use 7.48 gal/cu ft in the conversion but that you don't know whether to multiply or divide by 7.48. Let's look at both possible ways and see how dimensional analysis can be used to choose the

---
[1] Mathematics Section, Powers and Scientific Notation.

correct way. *Only the dimensions* will be used to determine if the math setup is correct.

First, try multiplying the dimensions:

$$(\text{cu ft})(\text{gal/cu ft}) = (\text{cu ft})\left(\frac{\text{gal}}{\text{cu ft}}\right)$$

Or re-expressed as

$$= (\text{ft}^3)\left(\frac{\text{gal}}{\text{ft}^3}\right)$$

Then multiply the numerators and denominators

$$= \frac{(\text{ft}^3)(\text{gal})}{\text{ft}^3}$$

And cancel common terms:

$$= \frac{(\cancel{\text{ft}^3})(\text{gal})}{\cancel{\text{ft}^3}}$$

$$= \text{gal}$$

So, by dimensional analysis you know that if you *multiply* the two dimensions (cu ft and gal/cu ft), the answer you get will be in *gal,* which is what you want. Therefore, since the math setup is correct you would then multiply the numbers to obtain gal:

$$(1200 \text{ cu ft})(7.48 \text{ gal/cu ft}) = 8976 \text{ gal}$$

What would have happened if you had divided the dimensions instead of multiplying?

$$\frac{\text{cu ft}}{\text{gal/cu ft}} = \frac{\text{cu ft}}{\dfrac{\text{gal}}{\text{cu ft}}}$$

$$= (\text{cu ft})\left(\frac{\text{cu ft}}{\text{gal}}\right)$$

This can be re-expressed as

$$= (\text{ft}^3)\left(\frac{\text{ft}^3}{\text{gal}}\right)$$

Then multiply the numerators and denominators of the fraction:

$$= \frac{\text{ft}^6}{\text{gal}}$$

So had you *divided* the two dimensions (cu ft and gal/cu ft), the units of the answer would have been ft^6/gal, *not* gal. Clearly you do not want to divide in making this conversion.

Example 1

Suppose you have two terms—3 ft/sec and 6 sq ft—and you wish to obtain an answer in *cu ft/sec*. Is multiplying the two terms the correct math setup?

$$(ft/sec)(sq\ ft) = \frac{ft}{sec} \times sq\ ft$$

This can be re-expressed as

$$= \frac{ft}{sec} \times ft^2$$

Then multiply the numerators and denominators of the fraction:

$$= \frac{(ft)(ft^2)}{sec}$$

$$= \frac{ft^3}{sec}$$

The math setup is correct since the dimensions of the answer are ft³/sec. So if you multiply the numbers, just as you did the dimensions, you will get the correct answer:

$$(3\ ft/sec)(6\ sq\ ft) = 18\ cu\ ft/sec$$

Example 2

You wish to obtain an answer in square feet. If you are given the two terms—80 cu ft/sec and 3.5 ft/sec—is the following math setup correct?

$$(80\ cu\ ft/sec)(3.5\ ft/sec)$$

First, only the dimensions are used to determine if the math setup is correct. By multiplying the two dimensions you get

$$(cu\ ft/sec)(ft/sec) = \left(\frac{cu\ ft}{sec}\right)\left(\frac{ft}{sec}\right)$$

Or re-expressed as

$$= \left(\frac{ft^3}{sec}\right)\left(\frac{ft}{sec}\right)$$

Then multiply the terms in the numerators and denominators of the fraction:

$$= \frac{(ft^3)(ft)}{(sec)(sec)}$$

$$= \frac{ft^4}{sec^2}$$

The math setup is wrong since the dimensions of the answer are not sq ft. Therefore, if you multiply the numbers just as you did the dimensions, the answer will be wrong.

Let's try division of the two dimensions:

$$\frac{\text{cu ft}/\text{sec}}{\text{ft}/\text{sec}} = \frac{\dfrac{\text{cu ft}}{\text{sec}}}{\dfrac{\text{ft}}{\text{sec}}}$$

This can be re-expressed as

$$= \frac{\dfrac{\text{ft}^3}{\text{sec}}}{\dfrac{\text{ft}}{\text{sec}}}$$

Then invert the denominator and multiply:

$$= \left(\frac{\text{ft}^3}{\text{sec}}\right)\left(\frac{\text{sec}}{\text{ft}}\right)$$

$$= \frac{(\text{ft})(\text{ft})(\text{ft})(\text{sec})}{(\text{sec})(\text{ft})}$$

$$= \frac{(\text{ft})(\text{ft})(\cancel{\text{ft}})(\cancel{\text{sec}})}{(\cancel{\text{sec}})(\cancel{\text{ft}})}$$

$$= \text{ft}^2$$

This math setup is correct since the dimensions of the answer are sq ft. Therefore, if you *divide* the numbers as you did the units, the answer will also be correct:

$$\frac{80 \text{ cu ft}/\text{sec}}{3.5 \text{ ft}/\text{sec}} = 22.86 \text{ sq ft}$$

Example 3

Suppose you have been given the following problem: "The flow rate in a water line is 2.3 cfs. What is the flow rate expressed as gpm?" You then set up the math problem as shown below. Use dimensional analysis to determine if this math setup is correct.

$$(2.3 \text{ cfs})(7.48 \text{ gal}/\text{cu ft})(60 \text{ sec}/\text{min})$$

First, the flow rate is rewritten as *cu ft/sec*. Then dimensional analysis is used to check the math setup:

$$(\text{cu ft}/\text{sec})(\text{gal}/\text{cu ft})(\text{sec}/\text{min}) = \left(\frac{\text{cu ft}}{\text{sec}}\right)\left(\frac{\text{gal}}{\text{cu ft}}\right)\left(\frac{\text{sec}}{\text{min}}\right)$$

$$= \left(\frac{\text{ft}^3}{\text{sec}}\right)\left(\frac{\text{gal}}{\text{ft}^3}\right)\left(\frac{\text{sec}}{\text{min}}\right)$$

$$= \left(\frac{\cancel{\text{ft}^3}}{\cancel{\text{sec}}}\right)\left(\frac{\text{gal}}{\cancel{\text{ft}^3}}\right)\left(\frac{\cancel{\text{sec}}}{\text{min}}\right)$$

$$= \frac{\text{gal}}{\text{min}}$$

This analysis indicates that the math setup is correct as shown above.

Example 4

You have been given the following problem: "A channel is 3 ft wide with water flowing to a depth of 2 ft. The velocity in the channel is found to be 1.8 fps. What is the cubic feet per second flow rate in the channel?" You then set up the math problem as shown below. Use dimensional analysis to determine if this math setup is correct.

$$\frac{(3 \text{ ft})(2 \text{ ft})}{1.8 \text{ fps}}$$

First, the velocity is rewritten as *ft/sec*. Then dimensional analysis is used to check the math setup:

$$\frac{(\text{ft})(\text{ft})}{\text{ft}/\text{sec}} = \frac{(\text{ft})(\text{ft})}{\dfrac{\text{ft}}{\text{sec}}}$$

$$= (\text{ft})(\text{ft})\frac{\text{sec}}{\text{ft}}$$

$$= \frac{(\text{ft})(\text{ft})(\text{sec})}{\text{ft}}$$

$$= \frac{(\cancel{\text{ft}})(\text{ft})(\text{sec})}{\cancel{\text{ft}}}$$

$$= \text{ft sec}$$

Since the dimensions of the answer are incorrect, the math setup shown above is also incorrect. Had the math setup been (3 ft)(2 ft)(1.8 fps), the dimensions of the answer would have been correct.

Review Questions

1. Suppose you have the terms—2,500,000 gal/day and 1440 min/day. If you wish to obtain gallons per minute, would the following math setup be correct?

$$\frac{2,500,000 \ \ gal/day}{1440 \ \ min/day}$$

2. Suppose you had set up the following equation. If you wanted to obtain square yards, would this setup be correct?

$$(500 \ sq \ in.)(144 \ sq \ in./sq \ ft)(3 \ sq \ yd/sq \ ft)$$

3. If you wish to obtain an answer in pounds, is the following setup correct?

$$\frac{(20,000 \ cu \ ft)(7.48 \ gal/cu \ ft)}{8.34 \ lb/gal}$$

Summary Answers

1. The setup is correct.

2. The setup is not correct.

3. The setup is not correct.

Detailed Answers

1. The fraction should first be rewritten as

$$\frac{\text{gal}/\text{day}}{\text{min}/\text{day}} = \frac{\dfrac{\text{gal}}{\text{day}}}{\dfrac{\text{min}}{\text{day}}}$$

When dividing by a fraction, invert the denominator and then multiply:

$$= \left(\frac{\text{gal}}{\text{day}}\right)\left(\frac{\text{day}}{\text{min}}\right)$$

$$= \frac{\text{gal}}{\text{min}}$$

The setup is correct and therefore dividing the numbers would also be correct.

2. The fraction should first be rewritten as

$$(\text{sq in.})\left(\frac{\text{sq in.}}{\text{sq ft}}\right)\left(\frac{\text{sq yd}}{\text{sq ft}}\right) = \frac{(\text{sq in.})^2(\text{sq yd})}{(\text{sq ft})^2}$$

Therefore the math setup is not correct. Since the first term was square inches, the next term had to have square inches in the denominator:

$$(\text{sq in.})\left(\frac{\text{sq ft}}{\text{sq in.}}\right)$$

And the final term had to have square feet in the denominator if it were to divide out:

$$(\cancel{sq\ in.}) \left(\frac{\cancel{sq\ ft}}{\cancel{sq\ in.}}\right) \left(\frac{sq\ yd}{\cancel{sq\ ft}}\right) = sq\ yd$$

3. The fraction should first be re-expressed as

$$\frac{(cu\ ft) \left(\dfrac{gal}{cu\ ft}\right)}{\dfrac{lb}{gal}}$$

When dividing by a fraction, invert the denominator and then multiply:

$$= (\cancel{cu\ ft}) \left(\frac{gal}{\cancel{cu\ ft}}\right) \left(\frac{gal}{lb}\right)$$

$$= \frac{gal^2}{lb}$$

According to dimensional analysis, the math setup was *not* correct. If the lb/gal term had been multiplied rather than divided, the units would have been

$$(\cancel{cu\ ft}) \left(\frac{\cancel{gal}}{\cancel{cu\ ft}}\right) \left(\frac{lb}{\cancel{gal}}\right) = lb$$

Mathematics 3

Rounding and Estimating

The practice of *estimating the size of the answer* to a calculation helps you to avoid reporting an answer as 5000, for example, when it should be 50,000, or 40,000 when it should be 4,000,000. Suppose when you are solving a problem you forget to multiply or divide by a certain number; or suppose you punch a wrong button on the calculator. If you have an idea of what the approximate answer should be, then you can recognize an incorrect answer and recheck your arithmetic.

In general, the process of estimating the size of an answer involves two steps:
• Rounding the numbers, and
• Completing the calculation using the rounded numbers.

Rounding means replacing the final digits of a number with zeros, thus expressing the number as tens, hundreds, thousands, or tenths, hundredths, thousandths, etc. (for example, expressing 498 as 500; 0.19 as 0.2; or 5,754,193 as 6,000,000). Equations using rounded numbers are often written with an approximate sign ($\doteq$ or $\approx$) instead of an equals sign (=) to show that the answer is an estimate.

M3-1. Rounding

The technique of rounding numbers is based on a particular *place value* in the decimal system. The various place values are reviewed below to emphasize the importance of understanding this concept before proceeding further.

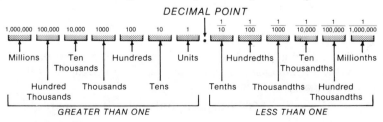

Suppose you want to round the number 2427 to the nearest hundred. Rounding this number depends only on the *size of the digit just to the right* of the hundreds place:

Hundreds place

↓

2427

↑

Digit that determines rounding

Similarly, if you want to round the number 323,772 to the nearest thousand, this rounding depends only on the *size of the digit just to the right* of the thousands place:

Thousands place ("rounding place" in this example)

↓

323,772

↑

Digit that determines rounding

Now consider the rules for *rounding whole numbers*. The procedure depends on whether the digit just to the right of the rounding place is less than 5, is 5, or is greater than 5.

Rule 1

If the digit is less than 5:   When rounding to any desired place, if the digit to the right of that place is less than 5, replace all digits to the right of the rounding place with zeros.

Rule 2

If the digit is 5 or greater than 5:   When rounding to any desired place, if the digit to the right of that place is 5 or greater than 5, increase the digit in the rounding place by 1 and replace all digits to the right of the increase with zeros.

The best way to understand rounding rules is to look at a few examples of rounding.

Example 1

Round 37, 926 to the nearest hundred.

The procedure used in this rounding depends on the digit just to the right of the hundreds place:

37,926

↑

Hundreds place

Since the digit to the right of the hundreds place is *less than 5*, the 9 is not changed and all the digits to the right of the 9 are replaced with zeros:

37,926 ≈ 37,900 (rounded to the nearest hundred)

Example 2

Round 248,722 to the nearest thousand.

The procedure used depends on the digit just to the right of the thousands place:

248,722

↑

Thousands place

Since the digit to the right of the thousands place is *greater than 5*, the 8 in the thousands place is increased by 1 and all the digits to the right of the 8 are replaced with zeros:

248,722 ≈ 249,000 (rounded to the nearest thousand)

Example 3

Round 25,675 to the nearest ten thousand.

The digit just to the right of the ten-thousands place determines how the number is rounded:

25,675

↑

Ten-thousands place

Since the digit to the right of the ten-thousands place is 5, the 2 is increased by 1 and all digits to the right of the 2 are replaced by zeros:

25,675 ≈ 30,000 (rounded to the nearest ten thousand)

Example 4

Round 14,974 to the nearest hundred.

The digit to the right of the hundreds place determines how the number is rounded:

14,974

↑

Hundreds place

Since the digit to the right of the hundreds place is 7, the 9 should be increased by 1 and all digits to the right replaced by zeros. But notice that increasing the 9 by 1 changes it to 10. The zero in the 10 replaces the 9, and the 1 is carried and added to the 4 in the thousands place:

14,974 ≈ 15,000 (rounded to the nearest hundred)

The rules for rounding decimal numbers are basically the same as those for rounding whole numbers, with the following modification:

Rule 3

When rounding decimal numbers to a place to the right of the decimal point, instead of replacing the rounded digits with zeros, drop the rounded digits.

The following examples illustrate this procedure.

Example 5

Round 37.18236 to the nearest thousandth.

As in rounding whole numbers, the procedure used depends on the size of the digit just to the right of the rounding place:

37.18236
    ↑
Thousandths place

According to the basic rounding rules, since the digit to the right of the thousandths place is *less than 5,* the 2 remains unchanged and all digits to the right of the 2 are dropped:

37.18236 ≈ 37.182 (rounded to the nearest thousandth)

Example 6

Round 5.654 to the nearest tenth.

5.654
  ↑
Tenths place

The digit to the right of the tenths place is 5. Therefore the 6 is increased by 1 and all digits to the right are dropped:

5.654 ≈ 5.7 (rounded to the nearest tenth)

Example 7

Round 483.16 to units.

483.16
   ↑
Units place

Even though the decimal point falls between the rounding place and the digit just to the right, the procedure still follows the basic rounding rules. The digit to the right of the units place is less than 5; therefore it (and all the digits after it) should be replaced with zeros. But, rounded digits on the right side of the decimal point are dropped. Therefore, the decimal number is rounded to a whole number:

483.16 ≈ 483 (rounded to the nearest unit)

M3-2. Estimating

When estimating the answer to a problem, you will find it helpful to round each number in the calculation so that only one digit remains, with the rest of the digits in the number either changed to zeros or dropped in accordance with the rounding rules. In this way the estimation can often be done in your head, or at least with a minimum of computing with pencil or calculator.

However, you should be aware that the more rounding you do in an estimation, the greater the difference may be between the DIGITS of your estimated answer and those of the actual answer. Keep this principle in mind:

> Rounding and estimating indicate the approximate size (place value) of a calculated answer but do not necessarily indicate the numerical value of the answer.

For instance, if the estimate to a problem is 40,000, you can expect the calculated answer to be in the tens of thousands (not hundreds or millions); but the value of the calculated answer could fall between 10,000 and 90,000.

Also consider what the place value of your answer could be if the digit in the estimated answer is 1 or 9: With an estimate of 900, the actual answer will probably be in the hundreds (somewhere between 100 and 900)—but it could also be a little more than 900 (such as 1020). On the other hand, if the estimate is 1000, then the calculated answer could be in the thousands, or a little less than 1000 (in the high hundreds).

Now let's look at the mechanics of estimating. Suppose you have rounded two numbers so that your estimate of the answer is 20 times 30. In making this estimate, first multiply the two digits (2 and 3):

$$(20)(30)$$
$$2 \times 3 = 6$$

Then count all the zeros in the calculation and put that amount of zeros after the 6:

$$(20)(30) = 600$$

2 zeros   2 zeros

The estimated answer is 600, so the actual answer should be in the hundreds. Let's look at a few more examples of this procedure.

Example 8

A calculation has been rounded so that it can be estimated. What is the estimated answer for the calculation?

$$(600)(3000)$$

First the 6 and the 3 are multiplied, then the total number of zeros are added:

$$(600)(3000) = \underbrace{18}\ \underbrace{00000}$$

6 × 3   5 zeros

Therefore the estimated answer for the calculation is 1,800,000; the actual answer should be in the millions or high hundred thousands.

Example 9

A calculation has been rounded to the numbers shown below. What is the estimated answer for the calculation?

$$(400)(70,000)$$

First the 4 and 7 are multiplied, then the total number of zeros are added:

$$(400)(70,000) = \underset{4 \times 7}{28,} \quad \underset{6 \text{ zeros}}{000000,}$$

The estimated answer is therefore 28,000,000; the actual answer should be in the ten millions.

Example 10

What is the estimated answer for the following rounded calculation?

$$(20)(300)(40)$$

As in the two preceding examples, the 2, 3, and 4 are multiplied and then the total number of zeros are added:

$$(20)(300)(40) = \underset{2 \times 3 \times 4}{24,} \quad \underset{4 \text{ zeros}}{0000,}$$

The estimated answer is therefore 240,000; the actual answer should be in the hundred thousands.

When three or more digits must be multiplied, rounding is sometimes required during the multiplication process. In the last example the three digits could be multiplied easily, so no rounding was needed. Suppose, however, that you wanted to estimate the answer to

$$(800)(7000)(800)$$

In this case the multiplication of the digits is not quite so obvious. Therefore, rounding during multiplication is permissible:

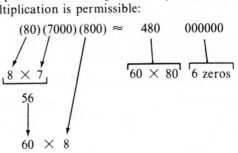

The following two examples illustrate the technique of rounding during multiplication of digits.

Example 11

The estimate of a calculation is shown below. What is the estimated answer?

$$(50)(50)(3000)$$

As in the other examples, the 5, 5, and 3 are multiplied; then the zeros are added:

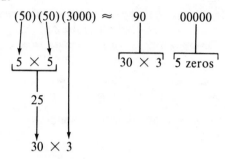

The estimated answer is therefore 9,000,000; the actual answer should be in the millions or ten millions.

Example 12

What is the estimated answer to the following calculation?

$$(9)(700)(60)(70)$$

First the 9, 7, 6, and 7 are multiplied; then the zeros are added. This problem requires more rounding during the multiplication step than was required in the previous examples:

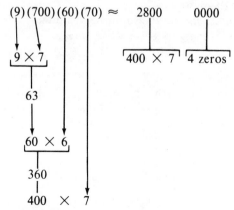

The estimated answer is 28,000,000; so the actual answer should be in the ten millions.

Sometimes the calculation being estimated involves division. When this occurs, often the estimate can be simplified by dividing out (or crossing off) final zeros; that is, zeros at the end of a WHOLE NUMBER. The next few examples illustrate this technique.

Example 13

What is the estimated answer to the following calculation?

$$\frac{40,000}{200}$$

In this problem the estimate may first be simplified. When a final zero is crossed off in the denominator (bottom) of the fraction, a final zero must be crossed off in the numerator (top) of the fraction. The simplified estimate is therefore

$$\frac{40,0\cancel{00}}{2\cancel{00}} = \frac{400}{2}$$

Then the estimate can be completed:

$$\frac{400}{2} = 200$$

Next, let's consider problems where the *estimated answer is a decimal less than 1.* For estimating purposes it is necessary to obtain only the first non-zero digit to the right of the decimal point. If an estimated answer is 0.01, for example, the actual answer could be 0.014, or 0.02, or even 0.009. The estimate of 0.01 merely shows that the answer will probably be in the *low* hundredths (although it could be in the high thousandths). The point is that the estimate should give you a "ball-park" answer, but *you must take particular care in determining where to place the decimal point* in the calculated answer.

Example 14

Complete the following estimate:

$$\frac{700}{6000}$$

First the estimate is simplified by crossing out final zeros, and then the problem is completed by division. As previously stated, for *estimated decimal answers less than 1, it is necessary to obtain only the first non-zero digit to the right of the decimal point.*

$$\frac{7\cancel{00}}{60\cancel{00}} = \frac{7}{60}$$

$$\begin{array}{r} .1 \\ 60\,)\overline{7.0} \\ \underline{6\;0} \end{array}$$

Therefore the estimated answer is 0.1; the actual answer should be in the tenths (0.1 to 0.9) or high hundredths (such as 0.09 or 0.08).

Example 15
   The estimate of a calculation is shown below. What is the estimated answer?

$$\frac{2000}{7,000,000}$$

This estimate is first simplified by crossing out final zeros:

$$\frac{2\cancel{000}}{7,000,\cancel{000}} = \frac{2}{7000}$$

Then the problem is completed by division. As in the last example, it is necessary to obtain *only the first non-zero digit* (*after the required number of zeros*) *to the right of the decimal point.*

$$
\begin{array}{r}
.0002 \\
7000) \overline{2.0000} \\
1\ 4000
\end{array}
$$

Therefore the answer to this estimate is 0.0002; the actual answer should be in the ten thousandths.

   In many practical applications involving rounding and estimating, the problem requires both multiplication and division. When this is the case, first simplify as much as possible by crossing out zeros. Next, multiply the numbers in the numerator, then those in the denominator. Finally, complete the problem by division. The next two examples illustrate this technique.

Example 16
   Calculate the estimated answer to

$$\frac{(20)(400)}{(50)(80)}$$

First, the problem is simplified by crossing out final zeros where possible:

$$\frac{(20)(4\cancel{00})}{(5\cancel{0})(8\cancel{0})} = \frac{(20)(4)}{(5)(8)}$$

The numerator and denominator each is multiplied:

$$\frac{(20)(4)}{(5)(8)} = \frac{80}{40}$$

Additional zeros may be crossed out and the problem completed by division:

$$\frac{8\cancel{0}}{4\cancel{0}} = \frac{8}{4}$$

$$= 2$$

The estimated answer is 2, so the actual answer should be in the units.

Example 17

Calculate the answer to the following estimated problem:

$$\frac{(4000)(40)}{(60)(3)}$$

First the problem is simplified by crossing out zeros:

$$\frac{(400\cancel{0})(40)}{(6\cancel{0})(3)} = \frac{(400)(40)}{(6)(3)}$$

The numerator and denominator each is multiplied:

$$\frac{(400)(40)}{(6)(3)} = \frac{16,000}{18}$$

Before the problem is completed by division, the numbers in the fraction can be rounded and additional zeros crossed out:

$$\frac{16,000}{18} \approx \frac{20,00\cancel{0}}{2\cancel{0}}$$

$$\approx \frac{2000}{2}$$

$$\approx 1000$$

The estimated answer is therefore 1000, and the actual answer should be in the thousands or high hundreds.

In the examples given previously, the numbers were already in rounded form. In practical applications, however, you must round the numbers in the problem before continuing with the estimate. And as noted at the beginning of this section, the more rounding you do, the greater the difference may be between the digits of your estimated answer and the actual answer. The estimated answer will show you what the approximate place value of the calculated answer should be, but the numerical values of the two answers may vary considerably.

Let's look at a few examples that require you to round the numbers before estimating. In each case the estimate is calculated according to the rules given in this section. Then the actual answer, using the numbers before rounding, is given to show the comparison between the estimated and actual answers.

Example 18

Estimate the answer to the following problem:

$$\frac{(29)(244)}{(12)(32)}$$

First, each term in the calculation is rounded:

$$\frac{(29)(244)}{(12)(32)} \approx \frac{(30)(200)}{(10)(30)}$$

The estimate can now be completed as in the previous examples. The problem is first simplified by crossing out zeros; then the numerator and denominator each is multiplied, and finally the estimated answer is obtained by division:

$$\frac{(3\cancel{0})(20\cancel{0})}{(1\cancel{0})(3\cancel{0})} = \frac{60}{3}$$

$$= 20$$

The estimated answer is 20; the actual answer is 18.43.

Example 19

Estimate the answer to the following problem:

$$\frac{(430)(61,702)}{(65)(65)}$$

Each term in the calculation is first rounded; then the estimation is completed as in previous examples.

$$\frac{(430)(61,702)}{(65)(65)} \approx \frac{(4\cancel{00})(60,000)}{(7\cancel{0})(7\cancel{0})}$$

$$\approx \frac{240,000}{49}$$

Then round the terms before division, cancel zeros, and complete the estimate:

$$\approx \frac{200,00\cancel{0}}{5\cancel{0}}$$

$$\approx 4000$$

The estimated answer is 4000; the actual answer is 6280.

Example 20

Estimate the answer to the following problem:

$$\frac{(291)(34)(419)}{(21)(169)}$$

First, each term is rounded and the problem is simplified by crossing out final zeros. Then the numerator and denominator are multiplied out; finally, the answer is obtained by division.

$$\frac{(291)(34)(419)}{(21)(169)} \approx \frac{(300)(30)(400)}{(20)(200)}$$

$$\approx \frac{(3)(3)(400)}{(2)(2)}$$

$$\approx \frac{3600}{4}$$

$$\approx 900$$

The estimated answer is 900; the actual answer is 1168. This problem shows how the place value of the estimated and actual answers can vary.

Up to this point we have been considering estimates using primarily rounded *whole* numbers. In some practical problems, however, you will be working with decimals less than one or with combinations of whole numbers and decimals less than one. The rounding rules for decimals already have been illustrated. But in performing the necessary multiplication and division steps in an estimate, you may have to do some of the arithmetic in long-hand in order to determine the position of the decimal point (and therefore the place value) of the answer.

Suppose, for example, you must round and estimate the following calculation:

$$\frac{(2.375)(4.75)(247)}{(61)(92)(0.785)}$$

First, round each term to one digit (plus the necessary number of zeros); then cancel the final zeros where possible:

$$\frac{(2.375)(4.75)(247)}{(61)(92)(0.785)} \approx \frac{(2)(5)(200)}{(60)(90)(0.8)}$$

Next, multiply the terms in the numerator, then those in the denominator. Round and cancel final zeros again.

$$\frac{(2)(5)(2)}{(6)(9)(0.8)} = \frac{20}{(54)(0.8)}$$

$$\approx \frac{20}{(50)(0.8)}$$

$$\approx \frac{2}{4}$$

Finally, divide to obtain the estimated answer:

$$\begin{array}{r} .5 \\ 4\overline{)\,2.0} \end{array}$$

The estimated answer is 0.5; the actual answer is 0.63.

Now consider another example:

$$\frac{(271)(32)(0.06725)}{(725)(0.0024)}$$

Round each term to one digit plus the required number of zeros:

$$\frac{(300)(30)(0.07)}{(700)(0.002)}$$

Now cancel the final zeros of whole numbers, as possible. Note that *the only zeros that can be cancelled are the final zeros of whole numbers.*

$$\frac{(3\cancel{00})(30)(0.07)}{(7\cancel{00})(0.002)} = \frac{(3)(30)(0.07)}{(7)(0.002)}$$

Next, multiply the numerator, then the denominator. You will probably want to do the arithmetic in longhand to place the decimal points properly.

$$= \frac{(90)(0.07)}{(7)(0.002)}$$

$$= \frac{6.3}{0.014}$$

Round each term again, then divide to complete the estimate:

$$\frac{6.3}{0.014} \approx \frac{6}{0.01}$$

$$\begin{array}{r} 600. \\ 0.01\overline{)\,6.00} \end{array}$$

The estimated answer is 600; the actual answer is 335.17.

As you can see from these two examples, estimates involving decimal numbers less than one can be time-consuming, and if you misplace the decimal point your answer could be wrong by one or more place values. You need to be particularly careful in estimating this type of problem because the possibility of error greatly increases.

An easier way to estimate problems having decimal numbers less than 1 is to put all the rounded numbers of the estimate into scientific notation. The rules for converting numbers to and from scientific notation are given in another section,[2] and are not discussed here. However, there are some basic rules for multiplying and dividing numbers in scientific notation that you must know.

---

[2]Mathematics Section, Powers and Scientific Notation.

Rule 1
When you *multiply* numbers in scientific notation, multiply the numbers but *add the exponents.*

Rule 2
When you *divide* in scientific notation, divide the numbers but *subtract the exponent* in the denominator from the exponent in the numerator.

First let's look at some examples that demonstrate the multiplication rule.

Example 21
Calculate the following:

$$(20)(400)$$

You can see immediately that the answer is 8000. But, for practice, put the calculation into scientific notation:

$$(20)(400) = (2 \times 10^1)(4 \times 10^2)$$

Now multiply the numbers, then add the exponents:

$$(2 \times 10^1)(4 \times 10^2) = (2)(4) \times 10^{1+2}$$
$$= 8 \times 10^3$$

Finally, take the answer out of scientific notation by moving the decimal point. The *positive* exponent of 3 indicates that the decimal point should be moved 3 places to the *right.*

$$8.000_\curvearrowright = 8000$$

Example 22
Multiply the following whole number and decimal, using scientific notation.

$$(0.0003)(300)$$

In scientific notation this becomes

$$(0.0003)(300) = (3 \times 10^{-4})(3 \times 10^2)$$

Now multiply the numbers and add the exponents:

$$(3 \times 10^{-4})(3 \times 10^2) = (3)(3) \times 10^{-4+2}$$
$$= 9 \times 10^{-2}$$

Finally, take the answer out of scientific notation by moving the decimal point. The *negative* exponent of 2 indicates a decimal point move two places to the *left.*

$$_\curvearrowleft09. = 0.09$$

The next example shows how to multiply three decimals and not lose track of the decimal point.

Example 23

Calculate the following, using scientific notation:

$$(0.003)(0.2)(0.0006)$$

First, convert to scientific notation; then multiply the numbers and add the exponents:

$$(0.003)(0.2)(0.0006) = (3 \times 10^{-3})(2 \times 10^{-1})(6 \times 10^{-4})$$
$$= (3)(2)(6) \times 10^{-3-1-4}$$
$$= 36 \times 10^{-8}$$

Finally, take the answer out of scientific notation by moving the decimal point. The *negative* exponent of 8 indicates a decimal point move of 8 places to the *left*.

$$_{\underleftarrow{\quad}}00000036. = 0.00000036$$

For estimating purposes, a *whole* number between 1 and 9 need not be converted to scientific notation. It would merely be the same as multiplying the number by 10^0. Consider how this can happen:

2×10^2 means $(2)(10)(10)$ or 200
2×10^1 means $(2)(10)$ or 20
2×10^0 means $(2)[not\ multiplied\ by\ 10]$ or   2

However, in using scientific notation to perform a calculation, you may sometimes get an answer where the positive and negative exponents cancel each other and the answer becomes a number multiplied by 10^0. How do you take such a number out of scientific notation? Since 10^0 means that the number is *not* raised to a power of 10, the answer is, in effect, already out of scientific notation, and the number stands "as is." Another way to state this is that $10^0 = 1$, so any number times 10^0 equals the number itself.

Let's consider an example:

$$(0.002)(3000)$$

First, convert to scientific notation; then multiply the numbers and add the exponents:

$$(0.002)(3000) = (2 \times 10^{-3})(3 \times 10^3)$$
$$= (2)(3) \times 10^{-3+3}$$
$$= 6 \times 10^0$$
$$= 6 \times 1$$
$$= 6$$

Now let's look at some examples using the division rule for scientific notation; that is, divide the numbers and subtract the exponent in the denominator from the exponent in the numerator.

Example 24

Calculate the following:

$$\frac{800}{20}$$

You know immediately that the answer is 40. But let's put it into scientific notation:

$$\frac{800}{20} = \frac{8 \times 10^2}{2 \times 10^1}$$

Now divide the numbers, but subtract the denominator exponent from the numerator exponent:

$$\frac{8 \times 10^2}{2 \times 10^1} = \left(\frac{8}{2}\right) \times 10^{2-1}$$

$$= 4 \times 10^1$$

Finally, take the number out of scientific notation by moving the decimal point. The *positive* exponent of 1 indicates a decimal point move of one place to the *right*.

$$4.0 = 40$$

Example 25

Calculate the following, using scientific notation:

$$\frac{800}{0.2} = \frac{8 \times 10^2}{2 \times 10^{-1}}$$

Divide the numbers and subtract the denominator exponent from the numerator exponent:

$$\frac{8 \times 10^2}{2 \times 10^{-1}} = \left(\frac{8}{2}\right) \times 10^{2-(-1)}$$

Here, the two minus signs become a plus sign (two negatives make a positive).

$$= \left(\frac{8}{2}\right) \times 10^{2+1}$$

So, the answer becomes

$$= 4 \times 10^3$$

And taking the answer out of scientific notation,

$$4 \times 10^3 = 4000$$

The next example shows how you can divide decimals and keep track of the decimal point for the answer.

Example 26

Make the following calculation, using scientific notation:

$$\frac{0.006}{0.3} = \frac{6 \times 10^{-3}}{3 \times 10^{-1}}$$

Divide the numbers, and subtract the exponent in the denominator from the exponent in the numerator:

$$\frac{6 \times 10^{-3}}{3 \times 10^{-1}} = \left(\frac{6}{3}\right) \times 10^{(-3)-(-1)}$$

Again, the two minus signs become a plus:

$$= \left(\frac{6}{3}\right) \times 10^{-3+1}$$

$$= 2 \times 10^{-2}$$

And taking the answer out of scientific notation

$$2 \times 10^{-2} = 0.02$$

Now consider the estimating problem we used previously, in which there are whole numbers and decimals less than 1 in both the numerator and denominator:

$$\frac{(271)(32)(0.06725)}{(725)(0.0024)}$$

First, round each term to one digit plus the required number of zeros; then cancel the final zeros of whole numbers where possible:

$$\frac{(271)(32)(0.06725)}{(725)(0.0024)} \approx \frac{(300)(30)(0.07)}{(700)(0.002)}$$

$$\approx \frac{(3)(30)(0.07)}{(7)(0.002)}$$

Next, convert the numbers of both the numerator and denominator to scientific notation. (Remember that a whole number between 1 and 9 need not be converted.)

$$\frac{(3)(30)(0.07)}{(7)(0.002)} = \frac{(3)(3 \times 10^{1})(7 \times 10^{-2})}{(7)(2 \times 10^{-3})}$$

Multiply the numerator, then the denominator, following the rule for multiplication of numbers in scientific notation. (Multiply the numbers but add the exponents.)

$$= \frac{(3)(3)(7) \times 10^{1-2}}{(7)(2) \times 10^{-3}}$$

$$= \frac{63 \times 10^{-1}}{14 \times 10^{-3}}$$

Round the numbers in the numerator and denominator, and cancel zeros where possible. Then, perform the division according to the division rule for scientific notation. (Divide the numbers, but subtract the denominator exponent from the numerator exponent.)

$$\frac{63 \times 10^{-1}}{14 \times 10^{-3}} \approx \frac{6\!\!\!\diagup\!\!0 \times 10^{-1}}{1\!\!\!\diagup\!\!4 \times 10^{-3}}$$

$$\approx \left(\frac{6}{1}\right) \times 10^{(-1)-(-3)}$$

Again, two minus signs become a plus:

$$\approx 6 \times 10^{-1+3}$$

$$\approx 6 \times 10^{2}$$

Finally, take the number out of scientific notation. A *positive* exponent of 2 indicates a decimal point move two places to the *right*.

$$6.00_{} = 600$$

The estimated answer is 600, so the calculated answer should be in the hundreds. (The actual answer is 355.17.)

Review Questions

1. Round 498,270 to the nearest thousand.

2. Round 2255 to the nearest hundred.

3. Round 338.23 to the nearest tenth.

4. Several calculations were rounded for estimating purposes. What are the estimated answers to the problems shown below?

   (a) $(70)(300)$

   (b) $(90)(40)$

   (c) $\dfrac{60,000}{800}$

   (d) $\dfrac{200}{70,000}$

   (e) $\dfrac{(30)(50)(600)}{(90)(20)}$

   (f) $\dfrac{(70,000)(200)}{(40)(90)(20)}$

5. Estimate the answers to the following calculations:

   (a) $\dfrac{(16)(25)}{290}$

   (b) $(0.785)(90)(90)$

   (c) $\dfrac{(730)(55)}{(42)(7.48)}$

6. Estimate the answers to the following calculations:

   (a) $\dfrac{(6.93)(3.215)(15)}{(0.785)(34)(34)}$

   (b) $\dfrac{(16.7)(790)(0.0093)}{(0.037)(57)}$

Summary Answers

1. 498,000

2. 2300

3. 338.2

4. (a) 21,000
   (b) 3600
   (c) 75
   (d) 0.002
   (e) 500
   (f) 125

5. (a) 2
   (b) 6400
   (c) 133

6. (a) 0.5
   (b) 50

Detailed Answers

1. The procedure used in rounding to the nearest thousand depends on the size of the number just to the right of the thousands place.

   498,270
   ↑
   Thousands place

   Since the digit to the right of the thousands place is *less than 5*, the 8 remains unchanged and all the digits to the right of the thousands place are replaced with zeros:

   498,270 ≈ 498,000 (rounded to the nearest thousand)

2. This problem requires rounding to the nearest hundred:

   2255
   ↑
   Hundreds place

   The digit to the right of the hundreds place is *5*. Therefore, the 2 is increased by 1 and all the digits to the right of the increase are replaced with zeros.

$$2255 \approx 2300 \text{ (rounded to the nearest hundred)}$$

3. In this problem the number is to be rounded to the nearest tenth:

338.23

↑

Tenths place

Since the digit to the right of the tenths place is *less than 5*, the 2 remains unchanged and all the digits to the right of the tenths place are dropped:

$338.23 \approx 338.2$ (rounded to the nearest tenth)

4. (a) First the 7 and 3 are multiplied, then the zeros are added to the end of the answer:

$$(70)(300) = \underbrace{21}_{7 \times 3} \quad \underbrace{000}_{3 \text{ zeros}}$$

The estimated answer is therefore 21,000; the actual answer should be in the ten thousands.

(b) The 9 and 4 are first multiplied, then the zeros are added:

$$(90)(40) = \underbrace{36}_{9 \times 4} \quad \underbrace{00}_{2 \text{ zeros}}$$

The estimated answer is therefore 3600; the actual answer should be in the thousands

(c) The problem is first simplified by crossing out zeros:

$$\frac{60,0\cancel{00}}{8\cancel{00}} = \frac{600}{8}$$

And it is completed by division:

```
        75.
  8 ) 600.
      56
      ‾‾
      40
      40
      ‾‾
```

The estimated answer is therefore 75. The actual answer should be in the tens.

(d) The problem is first simplified by crossing out zeros:

$$\frac{200}{70,000} = \frac{2}{700}$$

And division completes the estimate:

$$\begin{array}{r} .002 \\ 700\overline{)2.000} \\ \underline{00} \end{array}$$

The estimated answer is therefore 0.002. The actual answer should be in the thousandths.

(e) Simplify the problem by crossing out zeros:

$$\frac{(30)(50)(600)}{(90)(20)} = \frac{(3)(5)(600)}{(9)(2)}$$

The estimate can then be completed:

$$20 \times 600$$

$$15$$

$$\frac{(3)(5)(600)}{(9)(2)} \approx \frac{12,000}{18}$$

$$\text{or rounded to} \approx \frac{10,000}{20}$$

And dividing out the fraction:

$$\begin{array}{r} 500 \\ 2\overline{)1000} \end{array}$$

Therefore, the estimated answer is 500. The actual answer should be in the hundreds.

(f) The zeros are first crossed out where possible:

$$\frac{(70,000)(200)}{(40)(90)(20)} = \frac{(7000)(2)}{(4)(9)(2)}$$

The numerator and denominator are then multiplied out:

$$\frac{(7000)(2)}{(4)(9)(2)} \approx \frac{14,000}{80}$$

$$36$$

$$40 \times 2$$

And before division, the numbers in the fraction are further rounded with additional zeros crossed out:

$$\frac{14,000}{80} \approx \frac{10,0\cancel{0}\cancel{0}}{8\cancel{0}}$$

$$\approx \frac{1000}{8}$$

The estimate is then completed by division:

$$
\begin{array}{r}
125 \\
8 \overline{) 1000} \\
8 \\
\hline
20 \\
16 \\
\hline
40
\end{array}
$$

The estimated answer is 125; the actual answer should be in the hundreds.

5. (a) First, each term in the calculation is rounded:

$$\frac{(16)(25)}{290} \approx \frac{(20)(30)}{300}$$

And then the estimate is completed:

$$\approx \frac{(2\cancel{0})(3\cancel{0})}{3\cancel{0}\cancel{0}}$$

$$\approx \frac{(2)(3)}{(3)}$$

$$\approx \frac{6}{3}$$

$$\approx 2$$

The estimated answer is 2. The actual answer is 1.38.

(b) Each term in the calculation is first rounded:

$$(0.785)(90)(90) \approx (0.8)(90)(90)$$

Then the estimate is completed:

$$\approx (0.8)(8100)$$

$$\approx (0.8)(8000)$$

$$\approx 6400$$

The estimated answer is 6400. The actual answer is 6359.

(c) Each term in the calculation is first rounded:

$$\frac{(730)(55)}{(42)(7.48)} \approx \frac{(700)(60)}{(40)(7)}$$

And the estimate is then completed:

$$\frac{(700)(6\cancel{0})}{(4\cancel{0})(7)} = \frac{(700)(6)}{(4)(7)}$$

$$= \frac{4200}{28}$$

$$\approx \frac{400\cancel{0}}{3\cancel{0}}$$

Then dividing the fraction out,

$$
\begin{array}{r}
133 \\
3\overline{)400} \\
\underline{3} \\
10 \\
\underline{9} \\
10 \\
\underline{9} \\
\end{array}
$$

The estimated answer is therefore 133. The actual answer is 127.8.

6. (a) First, round each term to one digit (plus the necessary number of zeros); then cancel zeros where possible:

$$\frac{(6.93)(3.215)(15)}{(0.785)(34)(34)} \approx \frac{(7)(3)(2\cancel{0})}{(0.8)(30)(3\cancel{0})}$$

Next, multiply the terms in the numerator, then those in the denominator. Round out and cancel zeros again, as possible.

$$\frac{(7)(3)(2)}{(0.8)(30)(3)} = \frac{42}{72}$$

$$\approx \frac{4\cancel{0}}{7\cancel{0}}$$

$$\approx \frac{4}{7}$$

Finally, divide to obtain the estimated answer:

$$
\begin{array}{r}
.5 \\
7\overline{)4.0}
\end{array}
$$

The estimated answer is 0.5; the actual answer is 0.37.

(b) First, round each term to one digit plus the required number of zeros; then cancel final zeros where possible:

$$\frac{(16.7)(790)(0.0093)}{(0.037)(57)} \approx \frac{(2\cancel{0})(800)(0.009)}{(0.04)(6\cancel{0})}$$

Next, convert the numbers of both the numerator and denominator to scientific notation. (Remember that a whole number between 1 and 9 need not be converted.)

$$\frac{(2)(800)(0.009)}{(0.04)(6)} = \frac{(2)(8 \times 10^2)(9 \times 10^{-3})}{(4 \times 10^{-2})(6)}$$

$$= \frac{(2)(8)(9) \times 10^{2-3}}{(4)(6) \times 10^{-2}}$$

Multiply the numerator, then the denominator, following the rule for multiplication of numbers in scientific notation. (Multiply the numbers but add the exponent.)

$$= \frac{144 \times 10^{-1}}{24 \times 10^{-2}}$$

Round the numbers in the numerator and denominator, and then cancel zeros where possible. Then perform the division according to the division rule for scientific notation. (Divide the numbers, but subtract the denominator exponent from the numerator exponent.)

$$\frac{144 \times 10^{-1}}{24 \times 10^{-2}} \approx \frac{10\cancel{0} \times 10^{-1}}{2\cancel{0} \times 10^{-2}}$$

$$\approx \left(\frac{10}{2}\right) \times 10^{(-1)-(-2)}$$

The two minus signs together become a plus:

$$\approx 5 \times 10^{-1+2}$$

$$\approx 5 \times 10^{1}$$

$$\approx 50$$

The estimated answer is 50; the actual answer is 58.18.

Mathematics 4

Solving for the Unknown Value

In treatment plant operations you may use equations for such calculations as the detention time of a tank, flow rate in a channel, filter loading rate, or chlorine dosage. To make these calculations you must first know the values for all but one of the terms of the equation involved.

For example, in making a flow rate calculation,[3] you would use the equation

$$Q = AV$$

The *terms* of the equation are Q, A, and V. In solving problems using this equation, you would need to be given values to substitute for any two of the three terms. The term for which you do not have information is called the *unknown value*, or merely the *unknown*. The unknown value is often denoted by a letter such as x, but may be any other letter such as Q, A, V, etc.

Suppose you have a problem:

$$58 = (x)(25)$$

How can you determine the value of x? The rules that allow you to solve for x are discussed in this section. The discussion includes only *what* the rules are and *how* they work. (If you are interested in *why* these rules work and how they came about, you should consult a book on elementary algebra.)

Almost all problems in water treatment plant calculations have equations that involve only multiplication (such as the one just shown) and/or division. There are occasional problems, however, that use equations involving addition and subtraction. How to solve for the unknown in each of these types of equations is discussed in this section. Equations involving all four operations of multiplication, division, addition, and subtraction are encountered only in more advanced calculations and are thus beyond the scope of this text.

---

[3] Hydraulics Section, Flow Rate Problems.

M4-1. Equations Using Multiplication and Division

To solve for the unknown value in an equation using multiplication and division, you must rearrange the terms so that x is by itself. This often involves two steps:

- Move x to the numerator (top) of the fraction, if it is not already there.
- Move any other terms away from x, to the other side of the equal sign, so that x stands alone.

To accomplish these two steps, follow this rule:

Rule
In equations using multiplication and division, to move a term from one side of the equation to the other, move from the top of one side to the bottom of the other side; or from the bottom of one side to the top of the other side.

Although you rearrange only *some* of the terms in an equation when you solve for the unknown value, any of the terms in the equation may be moved if they are moved according to the rule just stated. Examples 1 and 2 show how to rearrange terms.

Example 1
Given the equation below, move the *8* to another position.

$$\frac{(x)(2)}{8} = \frac{3}{7}$$

According to the rearrangement rule, there is only one possible way to move the *8*. It must be moved to the *top* of the right side of the equation.

Left side of equation   *Right side of equation*

$$\frac{(x)(2)}{⑧} = \frac{⃝(3)}{7}$$

After making this move, the equation is

$$(x)(2) = \frac{(8)(3)}{7}$$

Example 2
Given the following equation, move the *4* to another position.

$$\frac{(5)(3)}{6} = \frac{(13)(4)}{(7)(x)}$$

There is only one possible move for the *4*. According to the rearrangement rule, it must be moved to the *bottom* of the left side of the equation.

Left side of   *Right side of*
equation   *equation*

$$\frac{(5)(3)}{(6)\bigcirc} = \frac{(13)\circled4}{(7)(x)}$$

After making this move, the equation is

$$\frac{(5)(3)}{(6)(4)} = \frac{13}{(7)(x)}$$

Rearranging the terms *always* involves this diagonal pattern. Any other movement is not permissible. The following three examples illustrate how terms can be rearranged to solve for the unknown value.

Example 3

$$\frac{x}{7} = \frac{(2)(3)}{13}$$

To solve for the unknown x in this equation, you must first ask yourself
• Is x in the numerator?
The answer is "yes." You should then ask yourself
• Is x by itself?
The answer is "no." Since x is not by itself, some rearranging is necessary.

$$\frac{x}{\circled7} = \frac{\bigcirc(2)(3)}{13}$$

$$x = \frac{(7)(2)(3)}{13}$$

Now that x is by itself, arithmetic completes the problem.

$$x = \frac{42}{13}$$

$$x = 3.23$$

Example 4

$$\frac{(3)(B)}{4} = \frac{(6)(5)}{7}$$

Again you must consider two questions:
• First, is B in the numerator?
The answer is "yes;" so then ask
• Is B by itself?
The answer is "no." Since B is not by itself, some rearranging is necessary. The *3* and *4* must be moved away from B.

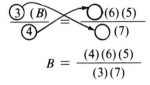

$$B = \frac{(4)(6)(5)}{(3)(7)}$$

The unknown B is now by itself, and arithmetic completes the problem.

$$B = \frac{120}{21}$$

$$B = 5.71$$

Example 5

$$\frac{2}{(3)(y)} = (7)(3)$$

Consider again the first of the two questions:
- Is the unknown y in the numerator?

The answer is "no," and so it must be moved.

Before rearranging terms, make the right side of the equation into a fraction by putting a *1* as the denominator. This does not change the value of that side of the equation and it makes the crisscross method of rearranging more apparent.

$$\frac{2}{(3)(y)} = \frac{(7)(3)}{1}$$

Now move the y according to the rearranging rule:

$$\frac{2}{(3)(y)} = \frac{(7)(3)}{1}$$

The resulting equation is

$$\frac{2}{3} = \frac{(y)(7)(3)}{1}$$

The unknown is in the numerator, so ask the second question:
- Is y by itself?

The answer is "no." Since y is not by itself, some rearranging is necessary. The *7, 3,* and *1* must be moved away from the y.

The resulting equation is

$$\frac{(2)(1)}{(3)(3)(7)} = y$$

The unknown is by itself and the problem can be completed using arithmetic:

$$\frac{2}{63} = y$$

$$0.03 = y$$

M4-2. Equations Using Addition and Subtraction

As previously mentioned, most water treatment calculations involve equations using only multiplication and division. There are some, however, such as chlorine dosage problems, that involve addition and subtraction. To solve for the unknown value in an equation using addition and subtraction, you must rearrange the terms so that x is positive and is by itself. But because in this type of equation there is neither a numerator nor a denominator, the crisscross method of rearranging *cannot be used*. Instead, the terms are moved directly from one side of the equation to the other side, according to the following rule:

Rule

In equations using addition and subtraction, when a term is moved from one side of the equation to the other side, the sign of the term must be changed.

This rule means that if a term is positive on one side of the equation, it becomes negative when moved to the other side. Conversely, if a term is negative on one side of the equation, it becomes positive when moved to the other side.

If a term does not show either a plus or minus sign in front, it is assumed to be positive. For example, in the equation

$$7-2-5+x = 19-3$$

both the *7* and *19* are considered positive although no plus sign is shown in front of them.

To solve an equation involving addition and subtraction, you must ask yourself two questions:

- Is x positive?
- Is x by itself?

The terms of the equation must be rearranged so that the answer to both questions is "yes." It is important to understand, however, that although x *as a term in the equation* should be positive, the *value* of x in the answer could be positive or negative (see Example 8).

Example 6

$$7-2-5+x = 19-3$$

In answer to the first question, x is positive. However, x is not by itself. In this problem, then, you must move the *7, 2,* and *5* to the other side of the equation. To do this, all you have to remember is to *change the sign* of the terms moved:

$$x = 19-3-7+2+5$$

The unknown has been solved for, and now arithmetic is used to complete the problem:

$$x = 16$$

Example 7

$$-18-2+5 = 7-x$$

First of all, x is not positive. To make it positive, you merely have to move it to the other side of the equation (because a negative changes to a positive in crossing to the other side of the equation):

$$x-18-2+5 = 7$$

Since x is not by itself, the *18, 2,* and *5* must be put on the other side of the equation:

$$x = 7+18+2-5$$

The unknown x has been solved for, and arithmetic is used to complete the problem:

$$x = 22$$

Example 8

$$x+4-9 = 6-23$$

In answer to the first question, the x term is positive. Since it is not by itself, however, the *4* and *9* must be moved to the other side of the equation:

$$x = 6-23-4+9$$

The unknown has been solved for, and arithmetic completes the problem:

$$x = -12$$

Here, although the x term in the equation is positive, the *value* of x is negative, because the sum of the negative numbers is larger than the sum of the positive numbers.

While the method discussed above may be used in solving for the unknown in addition and subtraction problems, most water treatment calculations you will encounter involving addition and subtraction are far simpler than the examples just given. In fact, the answer may be obvious without any rearranging of terms. To illustrate, let's look at perhaps the most common type of problem using addition and subtraction—chlorine dosage.[4] This problem is expressed by the equation

Chlorine dosage = Chlorine demand + Chlorine residual

In such problems, two of the terms are given and the third term is the unknown. As you can see in the example below, the answer can be determined without rearranging the terms:

$$7 \ mg/L = x \ mg/L + 0.2 \ mg/L$$

The answer is 6.8 mg/L, because 6.8 must be added to 0.2 to make 7.

Review Questions

In the following problems, solve for the unknown value. Rearranging the terms (if necessary) is all that is required. Do *not* complete the arithmetic once the unknown has been solved for.

1. $x = \dfrac{(7)(9)}{2}$

2. $\dfrac{(3)(x)}{2} = \dfrac{5}{3}$

3. $\dfrac{(2)(7)(5)}{(3)(3)} = \dfrac{(4)(x)}{11}$

4. $\dfrac{(7)(4)}{5} = \dfrac{3}{(x)(2)}$

5. $45 = (0.785)(5)(5)(x)$

6. $14-7+x = 3+5$

7. $-8-2 = 11-x+4$

---

[4]Chemistry Section, Chemical Dosage Calculations.

Summary Answers

1. $x = \dfrac{(7)(9)}{2}$

2. $x = \dfrac{(2)(5)}{(3)(3)}$

3. $\dfrac{(2)(7)(5)(11)}{(3)(3)(4)} = x$

4. $x = \dfrac{(5)(3)}{(7)(4)(2)}$

5. $\dfrac{45}{(0.785)(5)(5)} = x$

6. $x = 3+5-14+7$

7. $x = 11+4+8+2$

Detailed Answers

1. In this problem, x is already in the numerator and by itself. Therefore, no rearranging is necessary.

$$x = \frac{(7)(9)}{2}$$

2. The x is in the numerator in this problem, but not by itself. Therefore, the 3 and 2 must be moved to the other side of the equation.

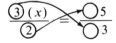

And the resulting equation is

$$x = \frac{(2)(5)}{(3)(3)}$$

3. The x is already in the numerator, but it is not by itself. Therefore, the 4 and 11 must be moved to the other side of the equation.

$$\frac{(2)(7)(5)}{(3)(3)} = \frac{(4)(x)}{(11)}$$

The resulting equation is

$$\frac{(2)(7)(5)(11)}{(3)(3)(4)} = x$$

4. The x must first be brought to the numerator:

$$\frac{(7)(4)}{5} = \frac{3}{(x)(2)}$$

After this move the equation is

$$\frac{(7)(4)(x)}{5} = \frac{3}{2}$$

Now, x must be by itself. Therefore the 7, 4, and 5 must be moved to the other side of the equation:

The resulting equation is

$$x = \frac{(5)(3)}{(7)(4)(2)}$$

5. The x is already in the numerator (it is not in the *bottom* of a fraction), but it is not by itself. Therefore, the 0.785, 5, and 5 must be moved to the other side of the equation. Before continuing with the rearrangement, however, put a 1 in the denominator of both sides of the equation to make the rearranging more apparent:

$$\frac{45}{1} = \frac{(0.785)(5)(5)(x)}{1}$$

The rearranging is

After these moves the equation is

$$\frac{(45)(1)}{(1)(0.785)(5)(5)} = x$$

or, removing the *1*'s

$$\frac{45}{(0.785)(5)(5)} = x$$

6. The x is already positive, but is not by itself. Therefore the *14* and *7* must be moved to the other side of the equation, changing their signs when they are moved:

$$x = 3+5-14+7$$

7. First x must be moved to the other side of the equation so that it will be positive:

$$x-8-2 = 11+4$$

Then the *8* and *2* must be moved away from x to the other side of the equation, changing their signs when they are moved:

$$x = 11+4+8+2$$

Mathematics 5

Ratios and Proportions

A RATIO is the relationship between two numbers and may be written using a colon (1:2, 4:7, 3:5) or as a fraction (1/2, 4/7, 3/5). When the relationship between two numbers in a ratio is the same as that between two other numbers in another ratio, the ratios are said to be in PROPORTION, or PROPORTIONATE. Stated another way, a proportion is made up of two (or more) ratios that are mathematically equal.

A method used to determine if two ratios are in proportion is CROSS MULTIPLICATION. If the answers are the same when numbers diagonally across from each other are multiplied, then the ratios are proportionate.

$$\frac{1}{2} \times \frac{2}{4}$$

| (2)(2) = 4 | (4)(1) = 4 |

The next two examples use the cross-multiplication method to determine whether ratios are proportionate.

Example 1
Are 3/7 and 42/98 proportionate?

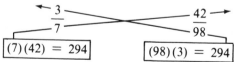

| (7)(42) = 294 | (98)(3) = 294 |

Since the answers (products) of the cross multiplication are the same (294), the ratios are proportionate.

Example 2
Are 5:13 and 22:95 proportionate?

First the ratios are written as fractions, then the cross-multiplication method is used to determine if the two ratios are proportionate:

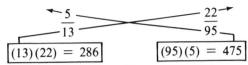

The products of cross multiplication (286 and 475) are different; therefore the ratios are not proportionate.

In examples 1 and 2, cross multiplication has been used to determine whether two ratios are proportionate. Sometimes only three of the four numbers in a proportion are known. In such cases, by solving for the unknown value[5] the fourth number can be determined.

Suppose, for example, you are given the proportion below. What is the value of the unknown number?

$$\frac{6}{13} = \frac{x}{40}$$

To determine the value of x, solve for the unknown value:

$$\frac{(40)(6)}{13} = x$$

$$18.46 = x$$

Example 3
In the following proportion, solve for the unknown value:

$$\frac{6.5}{x} = \frac{13}{20}$$

$$6.5 = \frac{(13)(x)}{20}$$

$$(20)(6.5) = (13)(x)$$

$$\frac{(20)(6.5)}{13} = x$$

$$10 = x$$

Example 4
Solve for the unknown value in the proportion:

$$\frac{x}{72} = \frac{14}{15}$$

$$x = \frac{(14)(72)}{15}$$

$$x = 67.2$$

---

[5]Mathematics Section, Solving for the Unknown Value.

In practical application problems the ratios will not be arranged as in the examples just given. You must know how to set up the proportion before you can solve for the unknown value. As previously explained, a proportion establishes a simple relation between two ratios. If any number in one of the ratios changes, then another must also change to maintain the equality. In performing various treatment plant operations you will find that a change in one operating condition produces a proportionate change in some other condition. For example, changing the speed of a metering pump changes the quantity metered; or changing the amount of chemical added to a certain volume of water changes the strength of the solution. The point is this: If you can set up a relation (ratio) for a particular operating condition, then when one of the two quantities of that ratio changes, you can calculate what the new value of the other should be.

To set up a proportion, first analyze the problem to decide what is unknown. Then decide whether you would expect the unknown value to be larger or smaller than the known value of the same unit. For example, if the unknown in a particular problem is pounds, would you expect the amount of pounds to be larger or smaller than the number of pounds given in the problem? Now write all values (including the unknown x) in fraction form, grouping values having the same units (pounds go with pounds, time with time, gallons with gallons, etc.) and putting the smaller value in the numerator (top) of each fraction and the larger value in the denominator (bottom).

Use this arrangement:

$$\underbrace{\begin{array}{c}\textit{First kind of units}\\ \textit{(for example, 1b)}\end{array}} \qquad \underbrace{\begin{array}{c}\textit{Second kind of units}\\ \textit{(for example, \$)}\end{array}}$$

$$\frac{\text{Smaller value}}{\text{Larger value}} = \frac{\text{Smaller value}}{\text{Larger value}}$$

And follow these rules:

Rule 1

If the unknown is expected to be smaller than the known value, put an x in the *numerator* of the first fraction and put the known value of the same unit in the denominator.

Rule 2

If the unknown is expected to be larger than the known value, put an x in the *denominator* of the first fraction and put the known value of the same unit in the numerator.

Rule 3

Make the two remaining values of the problem into the second fraction (the smaller value in the numerator, the larger in the denominator).

The next examples demonstrate this method of setting up proportions to solve typical problems.

Example 5

If 3 people can complete a job in 11 days, how many days will it take 5 people to complete it?

First, decide what is unknown. In this problem, a number of days is the unknown. If it takes 3 people 11 days to complete a job, then 5 people should complete the job in less time, or less than 11 days. Therefore, the unknown x should be smaller than the known (11 days). Now set up the proportion. Applying Rule 1, set up a fraction with x as the numerator and 11 days as the denominator. Then set up the other side of the proportion according to Rule 3, with 3 people as the numerator and 5 people as the denominator:

$$\frac{x \text{ days}}{11 \text{ days}} = \frac{3 \text{ people}}{5 \text{ people}}$$

Now solve for the unknown value:

$$x = \frac{(3)(11)}{5}$$

$$x = 6.6 \text{ days}$$

Example 6

If 5 lb of chemical are mixed with 2000 gal of water to obtain a desired solution, how many pounds of chemical would be mixed with 10,000 gal of water to obtain a solution of the same concentration?

Here, the unknown is some number of pounds of chemical. Analyzing the problem, you would expect that the more water you used, the more chemical would be needed to keep the same concentration. Therefore, the unknown number of pounds (x) should be greater than the known number (5 lb). Now set up the proportion using the rules given above (in this case, Rules 2 and 3):

$$\frac{5 \text{ lb}}{x \text{ lb}} = \frac{2000 \text{ gal}}{10,000 \text{ gal}}$$

And solve for the unknown value:

$$5 = \frac{(2000)(x)}{10,000}$$

$$(5)(10,000) = (2000)(x)$$

$$\frac{(5)(10,000)}{(2000)} = x$$

$$25 \text{ lb} = x$$

Example 7

If a pump will fill a tank in 13 hr at 6 gallons per minute (gpm), how long will it take a 15-gpm pump to fill the same tank?

Again, analyze the problem. Here the unknown is some number of hours. But should the answer be larger or smaller than 13 hr? If a 6-gpm pump can fill the tank in 13 hr, a larger pump (15-gpm) should be able to complete the filling in less than 13 hr. Therefore, the answer should be less than 13 hr. Now set up the proportion, following Rules 1 and 3:

$$\frac{x \text{ hr}}{13 \text{ hr}} = \frac{6 \text{ gpm}}{15 \text{ gpm}}$$

$$x = \frac{(6)(13)}{15}$$

$$x = 5.2 \text{ hr}$$

After you have worked with proportion problems for a while, you will gain an understanding that will allow you to skip some of the intermediate steps. The following examples show one "short-cut" approach that an experienced operator might use to analyze and solve proportion problems.

Example 8

To make up a certain solution, 57.2 mg of a chemical must be added to 100 gal of water. How much of the chemical should be added to 22 gal to make up the same strength solution?

To solve this problem, first decide what is unknown, and whether you expect the unknown value to be larger or smaller than the known value of the same unit. The amount of chemical to be added to 22 gal is the unknown, and you would expect this to be smaller than the 57.2 mg needed for 100 gal.

Now take the two known quantities of the same unit (22 gal and 100 gal) and make a fraction to multiply the third known quantity (57.2 mg) by. Notice that there are two possible fractions you can make with 22 and 100:

$$\frac{22}{100} \text{ or } \frac{100}{22}$$

Choose the fraction that will make the unknown number of mg less than the known (57.2 mg). Multiplying 57.2 by the fraction 22/100 would result in a number *smaller* than 57.2, since multiplying by 22/100 is roughly the same as multiplying by 1/4. Multiplying 57.2 by 100/22, however, would result in a number larger than 57.2, since multiplying by 100/22 is roughly the same as multiplying by 4.

You wish to obtain a number smaller than 57.2, so multiply by 22/100; then complete the arithmetic to solve the problem:

$$\left(\frac{22}{100}\right)(57.2) = x$$

$$\frac{(22)(57.2)}{100} = x$$

$$12.58 \text{ mg} = x$$

The key to this method is arranging the two known values of like units into a fraction that, when multiplied by the third known value, will give a result that is larger or smaller as needed.

Example 9

Operating at an 86% efficiency, a pump can fill a tank in 49 min. How long will it take the pump to fill the tank if the pump efficiency is 59%?

The unknown value is the time it will take the pump to fill the tank while operating at reduced efficiency. At the reduced pumping efficiency the unknown number of minutes should be larger than the known (49 min).

The known values of the other unit (percent) are 59% and 86%. Arrange these numbers into a fraction by which to multiply 49 min. Since the answer should be larger than 49 min, the fraction should be larger than 1 (86%/59% rather than 59%/86%).

$$\left(\frac{86\%}{59\%}\right)(49 \text{ min}) = x$$

Arithmetic completes the problem:

$$\frac{(86)(49)}{59} = x$$

$$71 \text{ min} = x$$

Using this shortcut method then, you only need to remember two things:

1. The pattern of the equation

$$\overbrace{\left(\frac{\text{Known value}}{\text{Known value}}\right)}^{\text{Similar units}} \overbrace{(\text{Known value})}^{\text{Similar units}} = \text{Unknown value}$$

2. Arrangement of the fraction so that the multiplication will result in a larger or smaller number, as expected.

Review Questions

1. Are 4/7 and 51/91 proportionate?

2. Are 108/120 and 4/5 proportionate?

3. Given the following proportion, what is the value of the unknown number?

$$\frac{7.2}{20} = \frac{x}{31}$$

4. Solve for the unknown value in the following proportion:

$$\frac{x}{14} = \frac{18}{90}$$

5. If 42 lb of chemical cost $12.20, how much will 150 lb cost?

6. A man working 8 hr each day requires 15 days to complete a job. How many days will be required to complete the job if he works 9 hr each day?

7. A pump making 32 strokes per minute discharges 600 gal of water per minute. If the speed is increased to 35 strokes per minute, what will the discharge be in gallons per minute?

8. Suppose it takes 3 gal of paint to cover 700 sq ft of a tank. How many gallons of paint would be required to cover 1150 sq ft of tank?

9. If a 4-gpm pump will fill a tank in 21 hr, what size pump would be required to complete the same job in 12 hr?

Summary Answers

1. No, they are not proportionate.

2. No, they are not proportionate.

3. 11.16

4. 2.8

5. $43.57

6. 13.33 days

7. 656.25 gpm

8. 4.93 gal

9. 7 gpm

Detailed Answers

1. Use the cross multiplication method to determine if the two ratios are proportionate:

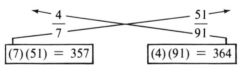

$$(7)(51) = 357 \qquad (4)(91) = 364$$

Since the products of cross multiplication are not the same, the ratios are not proportionate.

2. Use the cross multiplication method to determine if the ratios are proportionate:

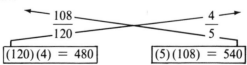

$$(120)(4) = 480 \qquad (5)(108) = 540$$

Since the products of cross multiplication are different, the ratios are not proportionate.

3. Solving for the unknown value,

$$\frac{7.2}{20} = \frac{x}{31}$$

$$\frac{(31)(7.2)}{20} = x$$

$$11.16 = x$$

4. Solving for the unknown value,

$$\frac{x}{14} = \frac{18}{90}$$

$$x = \frac{(18)(14)}{90}$$

$$x = 2.8$$

5. To solve this problem, first decide what is unknown; then decide whether the unknown should be larger or smaller than the known amount of the same unit. The cost of 150 lb of chemical is the unknown; and you might reasonably expect 150 lb of chemical to cost more than 42 lb of chemical. In other words, the unknown cost should be greater than $12.20.

Applying Rule 2, set up a fraction with x as the denominator and $12.20 as the numerator. Then set up the other side of the proportion according to Rule 3, with 42 lb as the numerator and 150 lb as the denominator:

$$\frac{\$12.20}{\$\ x} = \frac{42\ \text{lb}}{150\ \text{lb}}$$

Now solve for the unknown value:

$$12.20 = \frac{(42)(x)}{150}$$

$$(12.20)(150) = (42)(x)$$

$$\frac{(12.20)(150)}{42} = x$$

$$\$43.57 = x$$

6. First analyze the problem and decide what is unknown. The number of days is the unknown. If it takes a worker 15 days to do a job working 8 hr a day, he should be able to complete the job in fewer days working 9 hr a day.

Therefore the unknown x should be smaller than the known (15 days). Now set up the proportion, following Rules 1 and 3:

$$\frac{x \text{ days}}{15 \text{ days}} = \frac{8 \text{ hr}}{9 \text{ hr}}$$

And solve for the unknown value:

$$x = \frac{(8)(15)}{9}$$

$$x = 13.33 \text{ days}$$

7. Here, the unknown is the number of gallons per minute (gpm). Since a pump making 35 strokes per minute might well be expected to discharge more water per minute than one operating at 32 strokes, the unknown gallons per minute should be greater than 600 gpm. Now set up the proportion using Rules 2 and 3:

$$\frac{600 \text{ gpm}}{x \text{ strokes}} = \frac{32 \text{ strokes}}{35 \text{ strokes}}$$

And solve for the unknown value:

$$600 = \frac{(32)(x)}{35}$$

$$(600)(35) = (32)(x)$$

$$\frac{(600)(35)}{32} = x$$

$$656.25 \text{ gpm} = x$$

8. Here the unknown is gallons of paint needed to cover 1150 sq ft. You would expect to use more paint to cover 1150 sq ft than to cover 700 sq ft, so the unknown should be more than 3 gallons. Now set up the proportion using Rules 2 and 3:

$$\frac{3 \text{ gal}}{x \text{ gal}} = \frac{700 \text{ sq ft}}{1150 \text{ sq ft}}$$

And solve for the unknown value:

$$3 = \frac{(700)(x)}{1150}$$

$$(3)(1150) = (700)(x)$$

$$\frac{(3)(1150)}{(700)} = x$$

$$4.93 \text{ gal} = x$$

9. In this problem the unknown is pump capacity in gallons per minute. Analyzing the information given, you would expect a pump that could fill a tank in 12 hr to have a larger capacity than a pump that takes 21 hr to do the same job. Therefore, the unknown value should be larger than 4 gpm. Now set up the proportion, using Rules 2 and 3:

$$\frac{4 \text{ gpm}}{x \text{ gpm}} = \frac{12 \text{ hr}}{21 \text{ hr}}$$

And solve for the unknown value:

$$4 = \frac{(12)(x)}{21}$$

$$(4)(21) = (12)(x)$$

$$\frac{(4)(21)}{12} = x$$

$$7 \text{ gpm} = x$$

Mathematics 6

Averages

To assess the performance of a water treatment plant, much data must be collected and evaluated; and because there may be considerable variation in the information, it is often difficult to determine *trends* in performance.

The calculation of an average is a method to *group the information* so that trends in the information may be determined. When evaluating information based on averages, you must keep in mind that the "average" *reflects the general nature* of the group and does not necessarily reflect any one element of that group.

The ARITHMETIC MEAN is the most commonly used measurement of average value. It is calculated as follows:

$$\text{Average} = \frac{\text{Total of all terms}}{\text{Number of terms}}$$

Example 1

The following raw water turbidities were recorded for a week: Monday—8.2 TU; Tuesday—7.9 TU; Wednesday—6.3 TU; Thursday—6.5 TU; Friday—7.4 TU; Saturday—6.2 TU; Sunday—5.9 TU. What was the average daily turbidity?

Monday	8.2
Tuesday	7.9
Wednesday	6.3
Thursday	6.5
Friday	7.4
Saturday	6.2
Sunday	5.9
Total	48.4 TU

$$\text{Average} = \frac{\text{Total of all terms}}{\text{Number of terms}}$$

$$= \frac{48.4 \text{ TU}}{7}$$

$$= 6.91 \text{ TU}$$

Example 2

The sick days taken during the year by each of five operators were recorded as follows: Operator A—6 days; Operator B—3 days; Operator C—7 days; Operator D—0 days; Operator E—1 day. For these five men, what was the average number of days sick leave taken?

Operator A.........	6 days
Operator B.........	3 days
Operator C.........	7 days
Operator D.........	0 days
Operator E.........	1 day
Total	17 days

$$\text{Average} = \frac{\text{Total of all terms}}{\text{Number of terms}}$$

$$= \frac{17 \text{ days}}{5}$$

$$= 3.4 \text{ days sick leave}$$

Review Questions

1. The finished water turbidities for two consecutive days at a treatment plant were 0.9 TU and 0.6 TU. What was the average turbidity for that two-day period?

2. Each month the following number of pounds of alum was used. What was the average monthly requirement of alum (in pounds)?

Month	Alum (lb)	Month	Alum (lb)
January	4640	July	4810
February	4580	August	4830
March	4550	September	4760
April	4500	October	4660
May	4610	November	4610
June	4650	December	4630

3. Given the following hourly flow rates, determine the average hourly flow rate for the 24-hour period.

Time	mgd	Time	mgd
12 midnight	0.32	12 noon	1.30
1 AM	0.28	1 PM	1.25
2 AM	0.22	2 PM	1.22
3 AM	0.21	3 PM	1.23
4 AM	0.27	4 PM	1.28
5 AM	0.38	5 PM	1.37
6 AM	0.55	6 PM	1.50
7 AM	1.00	7 PM	1.69
8 AM	1.18	8 PM	1.75
9 AM	1.31	9 PM	1.67
10 AM	1.33	10 PM	0.95
11 AM	1.33	11 PM	0.50

Summary Answers

1. 0.75 TU

2. 4653 lb alum

3. 1.00 mgd

Detailed Answers

1.

$$\text{Average} = \frac{\text{Total of all terms}}{\text{Number of terms}}$$

$$= \frac{0.9 + 0.6 \text{ TU}}{2}$$

$$= \frac{1.5 \text{ TU}}{2}$$

$$= 0.75 \text{ TU}$$

2.

$$\text{Average} = \frac{\text{Total of all terms}}{\text{Number of terms}}$$

The total pounds of alum used must first be determined:

lb
4640
4580
4550
4500
4610
4650
4810
4830
4760
4660
4610
4630
55,830 lb total

The average can now be calculated:

$$\text{Average} = \frac{55,830 \text{ lb}}{12}$$

$$= 4653 \text{ lb alum}$$

3.

$$\text{Average} = \frac{\text{Total of all terms}}{\text{Number of terms}}$$

First find the total of all the hourly flow rates:

mgd	mgd
0.32	1.30
0.28	1.25
0.22	1.22
0.21	1.23
0.27	1.28
0.38	1.37
0.55	1.50
1.00	1.69
1.18	1.75
1.31	1.67
1.33	0.95
1.33	0.50
8.38	15.71

8.38 + 15.71 = 24.09 mgd total

The average can now be calculated:

$$\text{Average} = \frac{24.09 \text{ mgd}}{24}$$

$$= 1.00 \text{ mgd}$$

Mathematics 7

Percent

The word PERCENT (symbolized %) comes from the Latin words *per centum,* meaning "per one hundred." For example, if 62 percent of the voters are in favor of passing a bond issue, then 62 voters out of every 100 voters are in favor of the issue. Or if a student scores 87 percent on a test with 100 questions, he has answered 87 of the 100 questions correctly.

There is a direct relationship between percents, fractions, and decimal numbers. They are merely different ways of expressing the same mathematical proportions, as shown in the following examples.

$$20\% = \frac{20}{100} \rightarrow \text{And by dividing the fraction} = 0.20$$

$$95\% = \frac{95}{100} \rightarrow \text{And by dividing the fraction} = 0.95$$

$$5\% = \frac{5}{100} \rightarrow \text{And by dividing the fraction} = 0.05$$

$$14.5\% = \frac{14.5}{100} \rightarrow \text{And by dividing the fraction} = 0.145$$

$$0.5\% = \frac{0.5}{100} \rightarrow \text{And by dividing the fraction} = 0.005$$

Many problems involving percent require converting from a fraction to a decimal number to a percent. For example,

$$\frac{20}{100} = 0.20 = 20\%$$

$$\frac{95}{100} = 0.95 = 95\%$$

$$\frac{18}{60} = 0.30 = 30\%$$

$$\frac{4}{32} = 0.125 = 12.5\%$$

Note that to change from a decimal number to a percent, the decimal point is moved two places to the right. (The result is the same as multiplying by 100.) To convert a percent to a decimal, the decimal point is moved two places to the left (the same as dividing by 100).

A fraction is the key to calculating percent. The fraction is divided out, resulting in a decimal number which is then multiplied by 100 to be expressed as a percent. The formula for all percent problems is

$$\text{Percent} = \frac{\text{Part}}{\text{Whole}} \times 100$$

Remember that the *whole* is always the entire quantity; 100 percent of anything is the whole thing. For percentages less than 100, the *part* is less than the whole. For example, 50¢ is less than $1.00; in fact, it is 1/2 (or 50 percent) of the dollar. And $25.00 is less than $100.00; it is 1/4, or 25 percent, of the whole amount.

Examples 1—3 illustrate how to use the percent formula in solving some water treatment plant problems.

Example 1

A certain piece of equipment is having mechanical difficulties. If the equipment fails 6 times out of 25 tests, what percent failure does this represent?

The equation to be used in solving percent problems is

$$\text{Percent} = \frac{\text{Part}}{\text{Whole}} \times 100$$

In this problem, *percent* is unknown, but information is given regarding the *part* and the *whole*. Fill this information into the equation:

$$\text{Percent failure} = \frac{6 \text{ failures}}{25 \text{ tests}} \times 100$$

$$= 0.24 \times 100$$

$$= 24\% \text{ failure}$$

Example 2

A treatment plant operator has been ill during the past month and has had to miss a number of days' work. If he missed an average of 3 days out of 7, what was his percent absence?

The problem gives information regarding the *part* and the *whole*—3 days absent (part) out of a total of 7 days (whole). This information is used in the percent absence calculation as follows:

$$\text{Percent} = \frac{\text{Part}}{\text{Whole}} \times 100$$

$$\text{Percent absence} = \frac{3 \text{ days absent}}{7 \text{ days total}} \times 100$$

$$= 0.43 \times 100$$

$$= 43\% \text{ absence}$$

Example 3

The planning commission of a community voted on a bond issue concerning the construction of a new treatment plant. If 8 were in favor of the bond issue and 5 were opposed, what percent of the commission was in favor of the bond?

$$\text{Percent} = \frac{\text{Part}}{\text{Whole}} \times 100$$

$$\text{Percent in favor} = \frac{8 \text{ favor}}{13 \text{ total}} \times 100$$

$$= 0.62 \times 100$$

$$= 62\% \text{ in favor}$$

In the three examples just given, the unknown value was always *percent*. However, *any one* of the three terms (percent, part, or whole) can be the unknown value. The same general method is used in solving any percent problem. Fill the information that is given into the percent equation, and then solve for the unknown value.[6]

Example 4

The flow rate at a treatment plant is 7.2 mgd, which is 60 percent of the plant's capacity. What is the capacity of the plant?

In solving this problem, first write down the equation for percent, then determine what information is given.

$$\text{Percent} = \frac{\text{Part}}{\text{Whole}} \times 100$$

---

[6]Mathematics Section, Solving for the Unknown Value.

The 7.2 mgd is not the total capacity of the plant, so it must be the *part* in this problem. The 60 is the *percent*. Fill in the information in the problem:

$$60\% = \frac{7.2 \text{ mgd}}{x \text{ mgd}} \times 100$$

Then solve for the unknown x:

$$60 = \frac{(7.2)(100)}{x}$$

$$(60)(x) = (7.2)(100)$$

$$x = \frac{(7.2)(100)}{60}$$

$$x = 12 \text{ mgd}$$

Example 5

A tank is filled with water to 55 percent of its capacity. If the capacity of the tank is 2000 gal, how many gallons of water are in the tank?

First write down the equation for percent, then determine what information is given:

$$\text{Percent} = \frac{\text{Part}}{\text{Whole}} \times 100$$

The *percent* is 55. The question now is whether the 2000 gal is the *part* or the *whole*. Since the tank will hold only 2000 gal, that number must be the *whole*. Fill this information into the percent equation:

$$55\% = \frac{x \text{ gal}}{2000 \text{ gal}} \times 100$$

And solve for x:

$$55 = \frac{(x)(100)}{2000}$$

$$(2000)(55) = (x)(100)$$

$$\frac{(2000)(55)}{100} = x$$

$$1100 \text{ gal} = x$$

Example 6

The raw water entering a treatment plant has a turbidity of 10 TU. If the turbidity of the finished water is 0.5 TU, what is the turbidity removal efficiency of the treatment plant?

In math problems the term *efficiency* refers to percent. Therefore, another way of stating this problem is, "What percent of the turbidity has been removed?"

$$\text{Percent} = \frac{\text{Part}}{\text{Whole}} \times 100$$

Here, *percent* is the unknown and 10 TU is the *whole*. However, 0.5 TU is not the *part* removed. It is the amount of turbidity *still in* the water. The amount of turbidity removed must therefore be 10 TU − 0.5 TU, or 9.5 TU.

$$\text{Percent turbidity removed} = \frac{9.5 \text{ TU}}{10 \text{ TU}} \times 100$$

$$= 0.95 \times 100$$

$$= 95\% \text{ turbidity removal efficiency}$$

Although most percentages are less than 100 percent, in certain circumstances it is possible to have a percentage greater than 100 percent. Suppose, for example, you had $1.00 and found another dollar bill on the street. You would then have $2.00—double what you had to start, or 200 percent. If you had a dollar and found a dime, you would have $1.10, or 110 percent of what you had to start. Example 7 shows one way in which this concept of percents greater than 100 percent applies to water treatment operations.

Example 7

A treatment plant was designed to treat 15 mgd. On one particular day the plant treated 16.5 mil gal. What percent of the design capacity does this represent?

As in the other examples, first write down the equation for percent, then determine what information is given.

$$\text{Percent} = \frac{\text{Part}}{\text{Whole}} \times 100$$

In this problem the *percent* of design capacity is the unknown. The *whole* is 15 mgd, since the whole refers to the design capacity of the plant. So, what is the part? The *part* is the amount of water actually treated (16.5 mgd). And since the plant treated more water than it was designed to treat, the percent of the design capacity should be *greater than 100 percent*.

$$\text{Percent of design capacity} = \frac{\text{Part actually treated}}{\text{Total design capacity}} \times 100$$

$$= \frac{16.5 \text{ mgd}}{15 \text{ mgd}} \times 100$$

$$= 1.1 \times 100$$

$$= 110\%$$

The rule to remember is

Rule

In calculations of percents greater than 100%, the numerator of the percent equation must always be larger than the denominator.

Review Questions

1. Express the following decimal numbers as percents:
   - (a) 0.25
   - (b) 0.68
   - (c) 0.035
   - (d) 0.0005
   - (e) 1.20

2. Express the following percents as decimals:
   - (a) 15%
   - (b) 74%
   - (c) 3%
   - (d) 0.4%
   - (e) 110%

3. What percent is 1260 of 3000?

4. What is 21% of 247?

5. The number 242 is 40% of what other number?

6. If 20% of a certain number is 35, what is the number?

7. The capacity of a treatment plant is 5 mgd. If the flow at the treatment plant is 3.3 mgd, the plant is operating at what percent of its capacity?

8. From a stock of 200 machines, 82% are in use. How many machines are in use?

9. If there are 100 cu ft of water in a tank, and the tank is capable of holding 1000 gal, the tank is filled to what percent of its capacity?

10. In 1977 the average daily flow at a treatment plant was 8.5 mgd. On July 22, 1977 the plant experienced its peak day, 20 mil gal. This peak-day flow represents what percent of the average daily flow?

Summary Answers

1. (a) 25%
    (b) 68%
    (c) 3.5%
    (d) 0.05%
    (e) 120%

2. (a) 0.15
    (b) 0.74
    (c) 0.03
    (d) 0.004
    (e) 1.10

3. 42%

4. 51.87

5. 605

6. 175

7. 66%

8. 164

9. 74.8%

10. 235%

Detailed Answers

1. To convert a decimal number to a percent, move the decimal point *two places to the right* (the same as multiplying by 100).

   (a) $0.25 = 25\%$

   (b) $0.68 = 68\%$

   (c) $0.035 = 3.5\%$

   (d) $0.0005 = 0.05\%$

   (e) $1.20 = 120\%$

2. To convert a percent to a decimal, move the decimal point *two places to the left* (the same as dividing by 100).

   (a) $15\% = 0.15$

   (b) $74\% = 0.74$

   (c) $3\% = 0.03$

   (d) $0.4\% = 0.004$

   (e) $110\% = 1.10$

3.
$$\text{Percent} = \frac{\text{Part}}{\text{Whole}} \times 100$$
$$= \frac{1260}{3000} \times 100$$
$$= 0.42 \times 100$$
$$= 42\%$$

4.
$$\text{Percent} = \frac{\text{Part}}{\text{Whole}} \times 100$$
$$21\% = \frac{x}{247} \times 100$$

And solving for the unknown,
$$21 = \frac{(x)(100)}{247}$$
$$(247)(21) = (x)(100)$$
$$\frac{(247)(21)}{100} = x$$
$$51.87 = x$$

5.
$$\text{Percent} = \frac{\text{Part}}{\text{Whole}} \times 100$$

$$40\% = \frac{242}{x} \times 100$$

Then solving for x,

$$40 = \frac{(242)(100)}{x}$$

$$(x)(40) = (242)(100)$$

$$x = \frac{(242)(100)}{40}$$

$$x = 605$$

6.
$$\text{Percent} = \frac{\text{Part}}{\text{Whole}} \times 100$$

$$20\% = \frac{35}{x} \times 100$$

And solving for the unknown x,

$$20 = \frac{(35)(100)}{x}$$

$$(x)(20) = (35)(100)$$

$$x = \frac{(35)(100)}{20}$$

$$x = 175$$

7.
$$\text{Percent} = \frac{\text{Part}}{\text{Whole}} \times 100$$

$$\text{Percent of capacity} = \frac{\text{Part actually treated}}{\text{Total plant capacity}} \times 100$$

$$= \frac{3.3 \text{ mgd}}{5 \text{ mgd}} \times 100$$

$$= 0.66 \times 100$$

$$= 66\%$$

8.
$$\text{Percent} = \frac{\text{Part}}{\text{Whole}} \times 100$$

$$\text{Percent machines in use} = \frac{\text{Machines in use}}{\text{Total number of machines}} \times 100$$

$$82\% = \frac{x}{200} \times 100$$

And solving for the unknown x,

$$82 = \frac{(x)(100)}{200}$$

$$(200)(82) = (x)(100)$$

$$\frac{(200)(82)}{100} = x$$

$$164 = x$$

9. Before percent can be calculated in this problem, each of the volumes must be expressed in similar terms. Both must be expressed as cubic feet, or both must be expressed as gallons.[7] The solution is given for both options:

Option 1.   Converting gal to cu ft

$$\frac{1000 \text{ gal}}{7.48 \text{ gal/cu ft}} = 133.69 \text{ cu ft}$$

The percent can now be calculated:

$$\text{Percent} = \frac{\text{Part}}{\text{Whole}} \times 100$$

$$\text{Percent of capacity} = \frac{\text{Number of gallons in tank}}{\text{Total gallon capacity}} \times 100$$

$$= \frac{100 \text{ cu ft}}{133.69 \text{ cu ft}} \times 100$$

$$= 0.748 \times 100$$

$$= 74.8\%$$

Option 2.   Converting cu ft to gal

$$(100 \text{ cu ft})(7.48 \text{ gal/cu ft}) = 748 \text{ gal}$$

The percent can now be calculated:

$$\text{Percent} = \frac{\text{Part}}{\text{Whole}} \times 100$$

---

[7]Mathematics Section, Conversions (Cubic Feet to Gallons to Pounds).

$$\text{Percent of capacity} = \frac{\text{Number of gallons in tank}}{\text{Total gallon capacity}} \times 100$$

$$= \frac{748 \text{ gal}}{1000 \text{ gal}} \times 100$$

$$= 0.748 \times 100$$

$$= 74.8\%$$

10. In this problem the *percent* of the average daily flow is the unknown. The *whole* is 8.5 mgd, since average daily flow is the basis of comparison. The peak flow is therefore the *part*. And because the peak-day flow is *greater* than the average daily flow, the percent should be greater than 100 percent.

$$\text{Percent} = \frac{\text{Part}}{\text{Whole}} \times 100$$

$$\text{Percent of average daily flow} = \frac{\text{Peak flow}}{\text{Average daily flow}} \times 100$$

$$= \frac{20 \text{ mgd}}{8.5 \text{ mgd}} \times 100$$

$$= 2.35 \times 100$$

$$= 235\%$$

Mathematics 8

Linear Measurements

Linear measurements define the distance (or length) along a line, and can be expressed in customary or metric units. The customary units of linear measurement are inches, feet, yards, and miles; the metric units are centimetres, metres, and kilometres.

The focus of this section in on one particular linear measurement—the distance around the *outer edge* of a shape, called the PERIMETER. To determine the distance around the outer edge of an *angular shape,* such as a square, rectangle, or any other shape *with sides that are straight lines,* merely add the length of each side, as shown in the examples below.

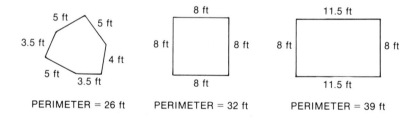

The mathematical equation for perimeter is

$$\text{Perimeter} = \text{Side}_1 \text{ Length} + \text{Side}_2 \text{ Length} + \text{Side}_3 \text{ Length} \ldots \text{etc.}$$

Use this type of calculation if, for example, you must determine the number of feet of wire needed for a fence.

Because the outside of a circle is not made up of easily measured straight lines, perimeter calculations for a circle are approached differently. The following diagram illustrates two of the linear terms associated with the circle.

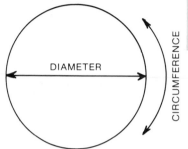

Circumference = distance *around* the circle

Diameter = distance through the center *across* the circle

As shown, the distance measured around the outside edge of a circle is called the CIRCUMFERENCE; this is just a special name for the perimeter of a circle. The DIAMETER is a straight line drawn from one side of the circle *through the center* to the other side.

If you compare the diameter of any circle to the circumference of that same circle, you notice that the circumference is just a little more than three times the length of the diameter. In mathematics, this length comparison (or ratio) of circumference to diameter is indicated by the Greek letter PI (π), and has a value of approximately 3.14.

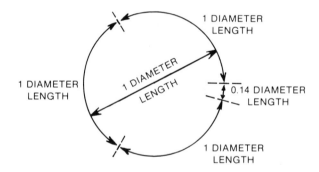

Because this relationship between the diameter length and circumference length is true for all circles, if you know the diameter of any circle you can calculate the circumference of the circle. For example, if you know the diameter of a circle is 20 ft, the distance around the circle (circumference) is roughly 3 × 20 ft, or about 60 ft. The exact circumference of this circle is:

$$(3.14)(20 \text{ ft}) = 62.8 \text{ ft}$$

The mathematical equation for circumference is

$$\text{Circumference} = (3.14)(\text{Diameter})$$

Use this type of calculation if, for example, you must determine the circumference of a circular tank.

Example 1

One of the storage areas at a treatment plant has the dimensions shown below. Determine the perimeter of this area.

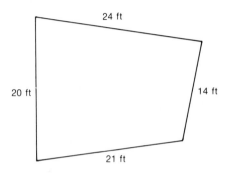

To calculate the perimeter, or distance around the area, add the lengths of the four sides:

$$\text{Perimeter} = \text{Side}_1 \text{ Length} + \text{Side}_2 \text{ Length}$$
$$+ \text{ Side}_3 \text{ Length} + \text{Side}_4 \text{ Length}$$
$$= 20 \text{ ft} + 24 \text{ ft} + 14 \text{ ft} + 21 \text{ ft}$$
$$= 79 \text{ ft}$$

Example 2

A circular clarifier has a diameter of 50 ft. What is the circumference of this tank?

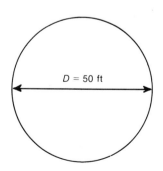

From the relationship between the diameter and circumference of a circle, you know that the circumference is approximately three times the length of the diameter. More accurately, it is

$$\text{Circumference} = (3.14)(\text{Diameter})$$
$$= (3.14)(50 \text{ ft})$$
$$= 157 \text{ ft}$$

Example 3

The circumference of a circular tank is 235.5 ft. What is the diameter of the tank?

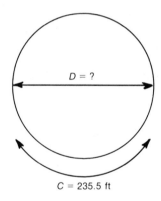

C = 235.5 ft

In this problem the circumference is known, rather than the diameter. Use the same formula for circumference and fill in the information given.

$$\text{Circumference} = (3.14)(\text{Diameter})$$
$$235.5 \text{ ft} = (3.14)(D)$$

Solve for the unknown value:[8]

$$\frac{235.5 \text{ ft}}{3.14} = D$$
$$75 \text{ ft} = D$$

---

[8] Mathematics Section, Solving for the Unknown Value.

Review Questions

1. A rectangular tank is 20 ft wide and 90 ft long. What is the perimeter of this tank?

2. A fence is going to be put around a circular area. If the diameter of the circular area is 30 ft, how many feet of fencing material will be required?

3. A circular tank has a circumference of 78.5 ft. What is the diameter of this tank?

Summary Answers

1. 220 ft
2. 94.2 ft
3. 25 ft

Detailed Answers

1. Because the tank is rectangular, its opposite sides are of equal length.

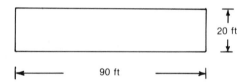

Therefore, two sides are 20 ft long and two sides are 90 ft long. The perimeter is calculated:

$$\text{Perimeter} = \text{Side}_1 + \text{Side}_2 + \text{Side}_3 + \text{Side}_4$$
$$= 20\ \text{ft} + 20\ \text{ft} + 90\ \text{ft} + 90\ \text{ft}$$
$$= 220\ \text{ft}$$

2.

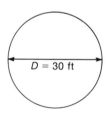

The circumference is approximately

$$3 \times 30\ \text{ft} = 90\ \text{ft}$$

For the more accurate calculation,

$$\text{Circumference} = (3.14)(\text{Diameter})$$
$$= (3.14)(30\ \text{ft})$$
$$= 94.2\ \text{ft}$$

3.

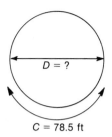

$D = ?$

$C = 78.5$ ft

Using the formula that relates circumference and diameter, fill in the given information:

$$\text{Circumference} = (3.14)(\text{Diameter})$$
$$78.5 \ \text{ft} = (3.14)(D)$$
$$\frac{78.5 \ \text{ft}}{3.14} = D$$
$$25 \ \text{ft} = D$$

Mathematics 9

Area Measurements

Area measurements define the size of the *surface* of an object. The customary units most frequently used to express this surface space are square inches, square feet, and square yards; the metric units are square millimetres, square centimetres, and square metres.

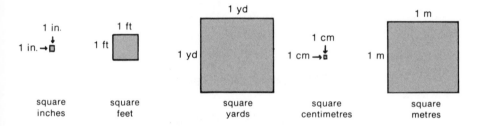

The fact that area measurements are expressed in square units does *not* mean that the surface must be a square in order to measure it. The surface of any shape can be measured. Although the shapes differ in the illustration below, the total surface area of each is the same.

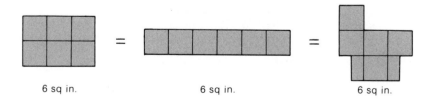

For area measurements in water treatment plant calculations, three shapes are particularly important, namely: rectangles, triangles, and circles. Most problems involve one or two combinations of these shapes. Equations (formulas) for each of the three basic shapes are presented below. Because these formulas are used so often in treatment plant calculations you should memorize them.

Surface Area Formulas

Rectangle   Area $=$ (length)(width)

$= lw$

Triangle   Area $= \dfrac{\text{(base)}\,\text{(height)}}{2}$

$= \dfrac{bh}{2}$

Circle   Area $=$ (0.785)(Diameter^2)

$=$ (0.785)(D^2)

or

$=$ (3.14)(radius^2)

$=$ (3.14)(r^2)

Use of these formulas is demonstrated in the examples that follow.

M9-1.   Area of a Rectangle

Area of a rectangle $=$ (length)(width)

$= lw$

Example 1
What is the area of the rectangle shown below?

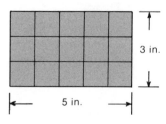

Area of rectangle $=$ (length)(width)

$=$ (5 in.)(3 in.)

$=$ 15 sq in. surface area

Example 2
What is the area of the rectangle in the diagram below?

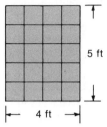

Area of rectangle = (length)(width)

= (4 ft)(5 ft)

= 20 sq ft surface area

Example 3
A tank is 10 ft long, 10 ft wide, with water to a depth of 5 ft. What is the area of the water surface of the tank?

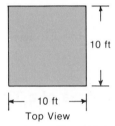

Area of rectangle = (length)(width)

= (10 ft)(10 ft)

= 100 sq ft surface area

Example 4
A sedimentation tank is 60 ft long and 20 ft wide. What is the area of the water surface in the tank?

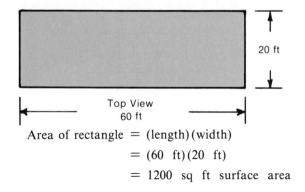

Area of rectangle = (length)(width)

= (60 ft)(20 ft)

= 1200 sq ft surface area

Example 5

A piece of land 800 ft by 1200 ft is to be purchased for expansion of water treatment facilities. How many acres is this?

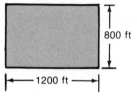

First calculate the number of sq ft in the area, then convert the sq ft to acres.[9]

$$\text{Area of rectangle} = (\text{length})(\text{width})$$
$$= (800 \ \text{ft})(1200 \ \text{ft})$$
$$= 960,000 \ \text{sq ft}$$

Since 1 acre = 43,560 sq ft, the number of acres in this area is

$$\frac{960,000 \ \text{sq ft}}{43,560 \ \text{sq ft/acre}} = 22 \ \text{acres}$$

M9-2. Area of a Triangle

$$\text{Area of a triangle} = \frac{(\text{base})(\text{height})}{2}$$

As shown in the following three examples, the *height* of the triangle *must be measured vertically from the horizontal base.*

Example 6

What is the area of the triangle in the diagram below?

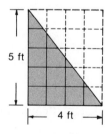

---

[9]Mathematics Section, Conversions (Area Measurement).

$$\text{Area of triangle} = \frac{(\text{base})(\text{height})}{2}$$

$$= \frac{(4\ \text{ft})(5\ \text{ft})}{2}$$

$$= 10\ \text{sq ft surface area}$$

Example 7

What is the area of the triangle shown below?

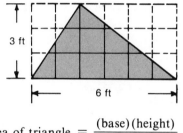

$$\text{Area of triangle} = \frac{(\text{base})(\text{height})}{2}$$

$$= \frac{(6\ \text{ft})(3\ \text{ft})}{2}$$

$$= 9\ \text{sq ft surface area}$$

Example 8

A triangular portion of the treatment plant grounds is not being used. How many square feet does this represent if the height of the triangle is 170 ft and the base is 200 ft?

$$\text{Area of triangle} = \frac{(\text{base})(\text{height})}{2}$$

$$= \frac{(200\ \text{ft})(170\ \text{ft})}{2}$$

$$= 17,000\ \text{sq ft surface area}$$

M9-3.   Area of a Circle

$$\text{Area of a circle} = (0.785)(\text{Diameter}^2)$$
$$or$$
$$= (3.14)(\text{radius}^2)$$

The more familiar formula for the area of a circle is πr^2. The r stands for the RADIUS of the circle; that is, the distance from the circle's center to its circumference. The radius of any circle is exactly *half* of the diameter.

However, since the diameter rather than the radius is generally given for circular tanks and basins, the formula using diameter is preferred for water treatment plant calculations. The relationship between the two formulas is shown in the following equations:

$$\text{Area} = \pi r^2$$

Since the radius is half the diameter, r^2 in the formula may be expressed as $(D/2)^2$:

$$\text{Area} = (\pi)\left(\frac{D}{2}\right)^2$$
$$= (\pi)\left(\frac{D}{2}\right)\left(\frac{D}{2}\right)$$

or

$$= \frac{\pi D^2}{4}$$

And, because $\pi = 3.14$,

$$\text{Area} = \left(\frac{3.14}{4}\right)(D^2)$$

The formula may be re-expressed as

$$\text{Area} = (0.785)(D^2)$$

Another advantage of the formula using diameter is that it can be better understood by diagram than the formula using radius. The diameter formula describes a surface area which can be thought of as *a square with the corners cut off.*

Let's examine the relationship between a square and a circle to better understand the formula for circle area:

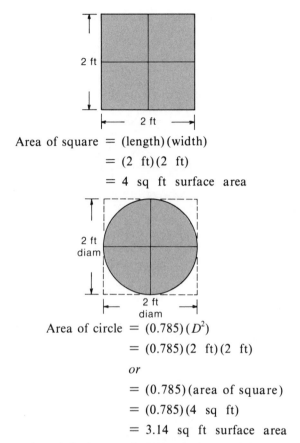

Area of square = (length)(width)

= (2 ft)(2 ft)

= 4 sq ft surface area

Area of circle = $(0.785)(D^2)$

= $(0.785)(2$ ft$)(2$ ft$)$

or

= (0.785)(area of square)

= $(0.785)(4$ sq ft$)$

= 3.14 sq ft surface area

When a circle with a 2-ft diameter is drawn inside the 2-ft square, you can see that the surface area of the circle is less than that of the square. However, it is not necessary to construct a square around each circle to be calculated. Mathematically, the D^2 (diameter squared) of the formula represents the square. Therefore, in finding the area of the circle, 0.785 is essential to the formula, because it is this factor that "cuts off the corners" of the square.

Example 9

Calculate the area of the circle shown below.

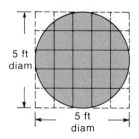

$$\text{Area of circle} = (0.785)(D)^2$$
$$= (0.785)(5 \text{ ft})(5 \text{ ft})$$
$$= 19.63 \text{ sq ft surface area}$$

Example 10

What is the area of the circle shown in the diagram below?

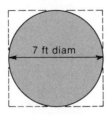

$$\text{Area of circle} = (0.785)(D)^2$$
$$= (0.785)(7 \text{ ft})(7 \text{ ft})$$
$$= 38.47 \text{ sq ft surface area}$$

Example 11

A circular clarifier has a diameter of 40 ft. What is the surface area of the clarifier?

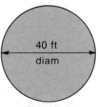

Top View

$$\text{Area of circle} = (0.785)(D)^2$$
$$= (0.785)(40 \text{ ft})(40 \text{ ft})$$
$$= 1256 \text{ sq ft surface area}$$

Review Questions

1. The top dimensions of a tank are 12 ft by 20 ft. What is the area of the water surface in the tank?

2. A piece of land 1000 ft by 2500 ft adjacent to the treatment plant is being reserved for future plant expansion. How many acres is this?

3. What is the area of a triangle that has a base of 50 ft and a height of 32 ft?

4. The bottom of a cylindrical basin has to be painted. If the basin has a diameter of 30 ft, how many square feet must be painted?

Summary Answers

1. 240 sq ft surface area

2. 57.39 acres

3. 800 sq ft surface area

4. 706.5 sq ft surface area

Detailed Answers

1.

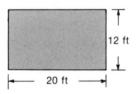

$$\text{Area of rectangle} = \text{(length)(width)}$$
$$= (20 \text{ ft})(12 \text{ ft})$$
$$= 240 \text{ sq ft surface area}$$

2. First calculate the square feet of the area, then convert the square feet to acres.

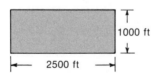

$$\text{Area of rectangle} = \text{(length)(width)}$$
$$= (2500 \text{ ft})(1000 \text{ ft})$$
$$= 2,500,000 \text{ sq ft}$$

Since 1 acre = 43,560 sq ft, determining how many acres in this area is

$$\frac{2,500,000 \text{ sq ft}}{43,560 \text{ sq ft/acre}} = 57.39 \text{ acres}$$

3.

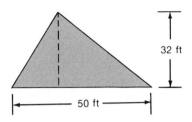

32 ft

50 ft

Area of triangle $= \dfrac{\text{(base)(height)}}{2}$

$= \dfrac{(50 \text{ ft})(32 \text{ ft})}{2}$

$= 800$ sq ft surface area

4.

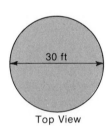

30 ft

Top View

Area of circle $= (0.785)(D^2)$

$= (0.785)(30 \text{ ft})(30 \text{ ft})$

$= 706.5$ sq ft surface area

Mathematics 10

Volume Measurements

A volume measurement defines the amount of space that an object occupies. The basis of this measurement is the *cube,* a square-sided box with all edges of equal length, as shown in the diagram below. The customary units commonly used in volume measurements are cubic inches, cubic feet, cubic yards, gallons, and acre-feet. The metric units commonly used to express volume are cubic centimetres, cubic metres, and litres.

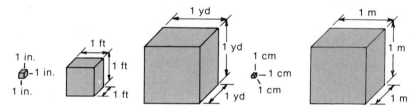

The calculations of surface area and volume are closely related. For example, to calculate the surface area of one of the cubes above, you would multiply two of the dimensions (length and width) together. To calculate the volume of that cube, however, a *third dimension* (depth) is used in the multiplication. The concept of volume can be simplified as

$$\text{Volume} = (\text{Area of surface})(\text{Third dimension})$$

The *area of surface* to be used in the volume calculation is the *representative surface area,* the side that gives the object its basic shape. For example, suppose you begin with a rectangular area as shown below. Notice the shape that would be created by stacking a number of those same rectangles one on top of the other.

Because the rectangle gives the object its basic shape in this example, it is considered the "representative area."

The same volume could have been created by stacking a number of smaller rectangles one behind the other.

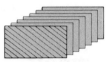

Although an object may have more than one representative surface area as illustrated in the two preceding diagrams, sometimes only one surface is the representative area. Consider, for instance, a shape like this:

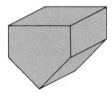

Let's compare two different sides of this shape (the top and the front) to determine if they are "representative areas." In the first case, a number of the top rectangles stacked together does not result in the same shape volume. Therefore, this rectangular area is not a "representative area." In the second case, however, a number of front shapes stacked one behind the other results in the same shape volume as the original object. Therefore, this front area may be considered a "representative surface area."

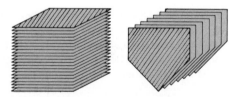

For treatment plant calculations, representative surface areas are most often rectangles, triangles, circles, or a combination of these. The following diagrams illustrate the three basic shapes for which volume calculations are made.

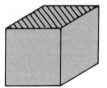

In the first diagram the rectangle defines the shape of the object; in the second diagram the triangle, rather than the rectangle, gives the trough its basic shape; in the third, the surface that defines the shape of the cylinder is the circle.

The formulas for calculating the volume of each of these three shapes are given below. Note that they are closely associated with the area formulas given previously.

Volume Formulas

$$\text{Rectangular tank} = \left(\begin{array}{c}\text{Area of}\\\text{rectangle}\end{array}\right)\left(\begin{array}{c}\text{Third}\\\text{dimension}\end{array}\right)$$

$$= (lw)\left(\begin{array}{c}\text{Third}\\\text{dimension}\end{array}\right)$$

$$\text{Trough} = \left(\begin{array}{c}\text{Area of}\\\text{triangle}\end{array}\right)\left(\begin{array}{c}\text{Third}\\\text{dimension}\end{array}\right)$$

$$= \left(\frac{bh}{2}\right)\left(\begin{array}{c}\text{Third}\\\text{dimension}\end{array}\right)$$

$$\text{Cylinder} = \left(\begin{array}{c}\text{Area of}\\\text{circle}\end{array}\right)\left(\begin{array}{c}\text{Third}\\\text{dimension}\end{array}\right)$$

$$= (0.785\ D^2)\left(\begin{array}{c}\text{Third}\\\text{dimension}\end{array}\right)$$

Use of the volume formulas is demonstrated in the examples that follow.

Example 1

Calculate the volume of water contained in the tank illustrated below if the depth of water (called SIDE WATER DEPTH, SWD) in the tank is 10 ft.

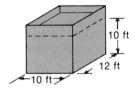

Volume = (Area of surface)(Third dimension)

In this example, the *rectangle* is the representative surface, and the dimension not used in the area calculation is the *depth*.

$$\text{Volume} = (\text{length})(\text{width})(\text{depth})$$
$$= (12\ \text{ft})(10\ \text{ft})(10\ \text{ft})$$
$$= 1200\ \text{cu ft}$$

Example 2

What is the volume (in cubic inches) of water in the trough shown below if the depth of water is 8 in.?

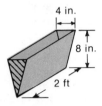

4 in.

8 in.

2 ft

First, all dimensions must be expressed in the same terms.[10] Since the answer is desired in cubic inches, the 2-ft dimension should be converted to inches:

$$(2 \text{ ft})(12 \text{ in.}/\text{ft}) = 24 \text{ in.}$$

Then

Volume = (Area of surface)(Third Dimension)

The triangle is the representative surface; the third dimension may be considered length or width—a difference in terminology.

$$\text{Volume} = \left(\frac{bh}{2}\right)(\text{length})$$

$$= \frac{(4 \text{ in.})(8 \text{ in.})}{2}(24 \text{ in.})$$

$$= \frac{(4 \text{ in.})(8 \text{ in.})(24 \text{ in.})}{2}$$

$$= 384 \text{ cu in.}$$

Example 3

What is the volume of water contained in the tank shown below if the depth of water (SWD) is 12 ft?

10 ft

12 ft

Volume = (Area of surface)(Third dimension)

---

[10] Mathematics Section, Conversions (Linear Measurement).

The circle is the representative surface, and the third dimension is depth.

$$\text{Volume} = (0.785)(D^2)(\text{depth})$$
$$= (0.785)(10 \text{ ft})^2(12 \text{ ft})$$
$$= (0.785)(10 \text{ ft})(10 \text{ ft})(12 \text{ ft})$$
$$= 942 \text{ cu ft}$$

Example 4

A tank with a cylindrical bottom has dimensions as shown below. What is the capacity of the tank? (Assume that the cross section of the bottom of the tank is a half circle.)

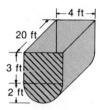

In problems involving a "representative surface area" that is a combination of shapes, it is often easier to calculate the representative surface area first, then calculate the volume:

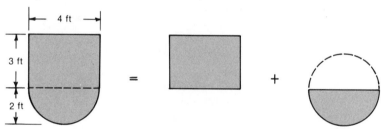

$$\begin{aligned}
\text{Representative surface area} &= \text{Area of rectangle} \\
&\quad + \text{Area of } 1/2 \text{ circle} \\
&= (4 \text{ ft})(3 \text{ ft}) + \frac{(0.785)(4 \text{ ft})(4 \text{ ft})}{2} \\
&= 12 \text{ sq ft} + 6.28 \text{ sq ft} \\
&= 18.28 \text{ sq ft}
\end{aligned}$$

And now calculate the volume of the tank:

$$\begin{aligned}
\text{Volume} &= (\text{Area of surface})(\text{Third dimension}) \\
&= (18.28 \text{ sq ft})(20 \text{ ft}) \\
&= 365.6 \text{ cu ft}
\end{aligned}$$

There are many shapes (though very few in water treatment calculations) for which the concept of a "representative surface" does not apply. The cone and sphere are notable examples of this.

Calculating the volume of a cylinder was discussed earlier. The volume of a cone represents 1/3 of that volume.

$$\text{Volume of a cone} = 1/3 \text{ (Volume of a cylinder)}$$

or

$$= \frac{(0.785)(D^2)(\text{Depth})}{3}$$

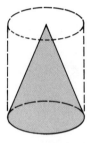

The volume of a sphere is more difficult to relate to the other calculations or even to a diagram. In this case the formula should be memorized. To express the volume of a sphere mathematically,

$$\text{Volume of a Sphere} = \left(\frac{\pi}{6}\right)(\text{Diameter})^3$$

(π is the relationship between the circumference and diameter of a circle. The number 3.14 is used for PI.)[11] The equation may be re-expressed as

$$\text{Volume of a sphere} = \left(\frac{3.14}{6}\right)(\text{Diameter})^3$$

---

[11] Mathematics Section, Linear Measurements.

Example 5

Calculate the volume of a cone with the dimensions shown on the diagram.

Volume of a cone $= 1/3$ (Volume of a cylinder)

or

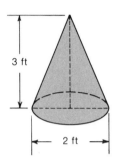

$$= \frac{(0.785)(D^2)(\text{Third dimension})}{3}$$

$$= \frac{(0.785)(2 \text{ ft})(2 \text{ ft})(3 \text{ ft})}{3}$$

$$= 3.14 \text{ cu ft}$$

3 ft

2 ft

Example 6

If a spherical tank is 30 ft in diameter, what is its capacity?

Volume of a sphere $= \left(\frac{\pi}{6}\right)(\text{Diameter})^3$

$$= \left(\frac{3.14}{6}\right)(30 \text{ ft})(30 \text{ ft})(30 \text{ ft})$$

$$= 14,130 \text{ cu ft}$$

30 ft

Occasionally it is necessary to calculate the volume of a tank that consists of two distinct shapes. In other words, there is no "representative surface area" for the entire shape. In this case, the volumes should be calculated separately, then the two volumes added. The diagrams below illustrate this method.

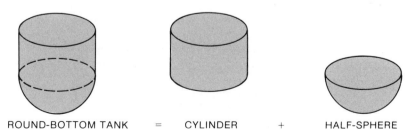

ROUND-BOTTOM TANK   =   CYLINDER   +   HALF-SPHERE

Review Questions

1. A sedimentation basin is 25 ft wide, 100 ft long, with water to a depth of 10 ft. How many cubic feet of water are in the tank?

2. Determine the cubic-foot capacity of the container shown below:

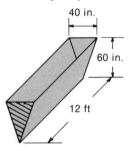

3. A circular clarifier has a diameter of 25 ft and is 11 ft deep. If the depth of the water (SWD) is 9 ft, how many cubic feet of water are in the tank?

4. One segment of a radial launder system is shown below. Given the average water depth as shown in the diagram, how many cubic feet of water are carried in this radial launder? (Assume that the bottom cross sectional area of the launder is a half circle.)

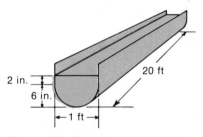

5. The circular clarifier shown below has a conical bottom. With side water depth (SWD) of 8 ft, a cone depth of 2 ft, and a diameter of 50 ft, how many cubic feet of water are in the clarifier?

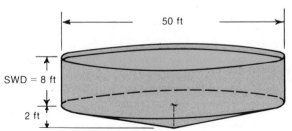

6. A tank with a sloping bottom has dimensions and a water depth as shown
   below. How many gallons of water are in this tank?

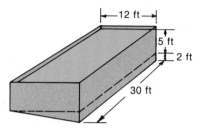

Summary Answers

1. 25,000 cu ft

2. 99.96 cu ft

3. 4416 cu ft

4. 11.2 cu ft

5. 17,008 cu ft

6. 16,157 gal

Detailed Answers

1.

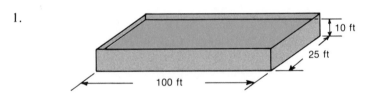

$$\begin{aligned}
\text{Volume} &= (\text{Area of surface})(\text{Third dimension}) \\
&= (lw)(\text{depth}) \\
&= (100 \text{ ft})(25 \text{ ft})(10 \text{ ft}) \\
&= 25,000 \text{ cu ft}
\end{aligned}$$

2. Before volume can be calculated, the dimensions must be expressed in similar terms. Since volume in cubic feet is desired, 40 in. and 60 in. should be converted to ft.

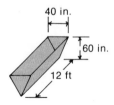

$$\frac{40 \text{ in.}}{12 \text{ in.}/\text{ft}} = 3.33 \text{ ft}$$

$$\frac{60 \text{ in.}}{12 \text{ in.}/\text{ft}} = 5 \text{ ft}$$

$$\text{Volume} = (\text{Area of surface})(\text{Third dimension})$$

$$= \left(\frac{bh}{2}\right)(\text{Third dimension})$$

$$= \frac{(3.33 \text{ ft})(5 \text{ ft})}{2}(12 \text{ ft})$$

$$= \frac{(3.33 \text{ ft})(5 \text{ ft})(12 \text{ ft})}{2}$$

$$= 99.9 \text{ cu ft}$$

3.

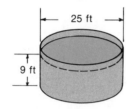

$$\text{Volume} = (\text{Area of surface})(\text{Third dimension})$$

$$= (0.785 \ D^2)(\text{Third Dimension})$$

$$= (0.785)(25 \text{ ft})(25 \text{ ft})(9 \text{ ft})$$

$$= 4416 \text{ cu ft}$$

4. The figure shows a condensed view of the launder. In this problem, the "representative surface area" is a combination of a rectangle and a half circle. First calculate the combined surface area, then the volume.
Before beginning the area calculation, convert the 2-in. dimension to feet. (It is not necessary to convert the 6-in. dimension to feet since it represents only the radius of the circle; the diameter of the circle, 1 ft, is used in the volume calculation.)

$$\frac{2 \text{ in.}}{12 \text{ in.}/\text{ft}} = 0.17 \text{ ft}$$

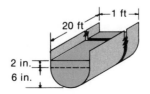

$$\text{Representative surface area} = \text{Area of rectangle} \\ + \text{Area of } 1/2 \text{ circle}$$

$$= (0.17 \text{ ft})(1 \text{ ft}) + \frac{(0.785)(1 \text{ ft})(1 \text{ ft})}{2}$$

$$= 0.17 \text{ sq ft} + 0.39 \text{ sq ft}$$

$$= 0.56 \text{ sq ft}$$

The volume of the launder can now be calculated:

$$\text{Volume} = (\text{Area of surface})(\text{Third dimension})$$

$$= (0.56 \text{ sq ft})(20 \text{ ft})$$

$$= 11.2 \text{ cu ft}$$

5. In this problem there is no "representative surface area" that is applicable for the entire tank. Therefore, the cylindrical volume is calculated first; the conical volume next, and then the two volumes are added together.

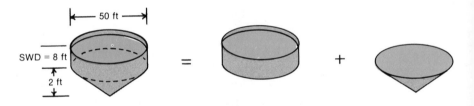

$$\text{Volume of cylinder} = (\text{Area of surface})(\text{Third dimension})$$

$$= (0.785 \; D^2)(\text{Third dimension})$$

$$= (0.785)(50 \text{ ft})(50 \text{ ft})(8 \text{ ft})$$

$$= 15,700 \text{ cu ft}$$

$$\text{Volume of cone} = 1/3 \,(\text{Volume of cylinder})$$

$$= \frac{(0.785 \; D^2)(\text{Third dimension})}{3}$$

$$= \frac{(0.785)(50 \text{ ft})(50 \text{ ft})(2 \text{ ft})}{3}$$

$$= 1308 \text{ cu ft}$$

$$\text{Total volume of tank} = \text{Volume of cylinder} \\ + \text{Volume of cone}$$

$$= 15,700 \text{ cu ft} + 1308 \text{ cu ft}$$

$$= 17,008 \text{ cu ft}$$

6. In this problem the representative surface area is a combination of a rectangle and a triangle, as shown in the diagram. Therefore the combined surface area is calculated before continuing with the volume calculation.

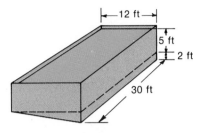

Representative = Area of rectangle
  surface area   + Area of triangle

$$= (12 \text{ ft})(5 \text{ ft}) + \frac{(12 \text{ ft})(2 \text{ ft})}{2}$$

$$= 60 \text{ sq ft} + 12 \text{ sq ft}$$

$$= 72 \text{ sq ft}$$

The volume of the tank can now be calculated. First calculate the volume in cubic feet, then convert this volume to gallons:

Volume of tank = (Area of surface)(Third dimension)

$$= (72 \text{ sq ft})(30 \text{ ft})$$

$$= 2160 \text{ cu ft}$$

Expressing this volume as gallons[12]

(2160 cu ft)(7.48 gal/cu ft) = 16,157 gal

---

[12]Mathematics Section, Conversions.

Mathematics 11

Conversions

In making the conversion from one unit to another you must know

- The number that relates the two units, and
- Whether to multiply or divide by that number.

For example, in converting from feet to inches, you must know that in 1 ft there are 12 in., and you must know whether to multiply or divide the number of feet by 12.

Although the number that relates the two units of a conversion is usually known or can be looked up, there is often confusion about whether to multiply or divide. One method to help decide whether to multiply or divide for a particular conversion is called *dimensional analysis,* discussed in a previous section.[13]

M11-1. Conversion Tables

Usually the fastest method of converting units is to use a conversion table, such as the one included in the appendix, and follow the instructions indicated by the table headings. For example, if you want to convert from feet to inches, look in the *Conversion* column of the table for *From* "feet" *To* "inches." Read across this line and perform the operation indicated by the headings of the other columns; that is, multiply the number of feet by 12 to get the number of inches.

Suppose, however, that you want to convert inches to feet. Look in the *Conversion* column for *From* "inches" *To* "feet," and read across this line. The headings tell you to multiply the number of inches by 0.08333 (which is the decimal equivalent of 1/12) to get the number of feet. *Multiplying by either 1/12 or 0.08333 is the same as dividing by 12.*

---

[13] Mathematics Section, Dimensional Analysis.

The instruction to *multiply* by certain numbers (called conversion factors) is used throughout the conversion table. There is no column headed *Divide by,* because the fractions representing division (such as 1/12) were converted to decimal numbers (such as 0.08333) when the table was prepared.

To use the conversion table, remember the following three steps:

- In the *Conversion* column, find the units you want to change *From* and *To.* (Go *From* what you have *To* what you want.)

- Multiply the *From* number you have by the conversion factor given.

- Read the answer in *To* units.

Example 1
Convert 288 in. to feet.

In the *Conversion* column of the table, find *From* "inches" *To* "feet." Reading across the line, perform the multiplication indicated; that is, multiply the number of inches (288) by 0.08333 to get the number of feet.

$$(288 \text{ in.})(0.08333) = 24 \text{ ft}$$

Example 2
A tank holds 50 gal of water. How many cubic feet of water is this, and what does it weigh?

First, convert gallons to cubic feet. Using the table, you find that to convert *From* "gallons" *To* "cubic feet," you must multiply 50 gal by 0.1337 to get the number of cubic feet:

$$(50 \text{ gal})(0.1337) = 6.69 \text{ cu ft}$$

Next, convert gallons to pounds of water. Using the table, you find that to convert *From* "gallons" *To* "pounds of water," you must multiply 50 gal by 8.34 to get the number of pounds of water:

$$(50 \text{ gal})(8.34) = 417 \text{ lb of water}$$

Notice that you could have arrived at approximately the same weight by converting 6.69 cu ft to pounds of water. Using the table,

$$(6.69 \text{ cu ft})(62.4) = 417.46 \text{ lb of water}$$

This slight difference in the two answers is due to rounding[14] numbers both when the conversion table was prepared and when the numbers are used in solving the problem. You may notice the same sort of slight difference in answers if you have to convert from one kind of units to two or three other units, depending on whether you round intermediate steps in the conversions.

---

[14]Mathematics Section, Rounding and Estimating.

M11-2. Box Method

A third method that may be used to determine whether multiplication or division is required for a particular conversion is called the *box method,* and is based on the relative size of different squares ("boxes"). The box method, used extensively throughout this section, can be used when a conversion table is not available (such as during a certification exam). This method of conversion is often slower than using the conversion table, but many people find it simpler.

Because *multiplication* is usually associated with an *increase* in size, when you move from a smaller box to a larger box, multiplication is used in the conversion:

MULTIPLICATION

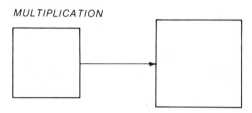

Division, on the other hand, is usually associated with a *decrease* in size. Therefore, when you move from a larger box to a smaller box, division is used in the conversion:

DIVISION

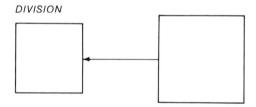

To use the box method to determine whether to multiply or divide in making a conversion, set up and label the boxes according to the following procedure:

- Write the equation that relates the two types of units involved in the conversion. (One of the two numbers in the equation must be a *one*; for example, *1* ft = 12 in., or 3 ft = *1* yd.)

- Draw a small box on the left, a large one on the right, and connect them with a line (as in the drawings at the beginning of this section).

- In the *smaller* box, write the name of the units associated with the *one* (for example, 1 *ft* = 12 in.—*ft* should be written in the smaller box). NOTE: The name of the units next to the *one* must be written in the *smaller* box, *or the box method will give incorrect results.*

- In the larger box, write the name of the remaining units. Those units will also have a number next to them, a number which is not *one*. Write that number over the line between the boxes.

Suppose, for example, that you want to make a box diagram for feet-to-inches conversions. First, write the equation that relates feet to inches:

$$1 \text{ ft } = 12 \text{ in.}$$

Next, draw the conversion boxes (smaller box on the left) and the connecting line:

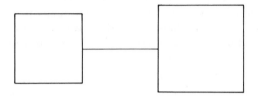

Now label the diagram. Because the number *one* is next to the units of *feet* (*1 ft*), write *ft* in the smaller box. Write the name of the other units, inches (*in.*), in the larger box. And write the number that is next to inches, *12*, over the line between the boxes.

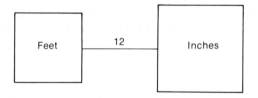

To convert from feet to inches, then, *multiply by 12* because you are moving from a smaller box to a larger box. And to convert from inches to feet, *divide by 12* because you are moving from a larger to a smaller box.

Let's look at another example of making and using the box diagram. Suppose you want to convert cubic feet to gallons. First write down the equation that relates these two units:

$$1 \text{ cu ft } = 7.48 \text{ gal}$$

Then draw the smaller and larger boxes and the connecting line; label the boxes; and write in the conversion number:

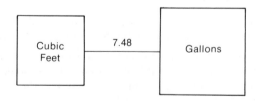

The smaller box corresponds with cubic feet and the larger box with gallons. To convert from cubic feet to gallons according to this box diagram, *multiply* by 7.48 because you are moving from a smaller to a larger box. And to convert from gallons to cubic feet, *divide* by 7.48 because you are moving from a larger to a smaller box.

The box method of conversions is used many times in the following discussions.

Cubic Feet to Gallons to Pounds Conversions

In making the conversion from cubic feet to gallons to pounds of water, you must know the following relationships:

$$1 \ \text{cu ft} = 7.48 \ \text{gal}$$

$$1 \ \text{gal} = 8.34 \ \text{lb}$$

You must also know whether to multiply or divide, and which of the above numbers are used in the conversion. The following box diagram should assist in making these decisions.

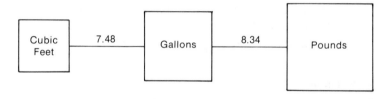

Example 3

Convert 1 cu ft to pounds.

First write down the diagram to aid in the conversion:

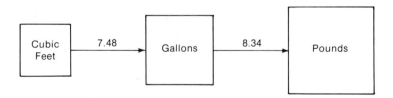

When you are converting from cubic feet to pounds, you are moving from smaller to larger boxes. Therefore *multiplication* is indicated in both conversions:

$$(1 \text{ cu ft})(7.48 \text{ gal/cu ft})(8.34 \text{ lb/gal}) = 62.38 \text{ lb}$$

$$or = 62.4 \text{ lb}$$

(rounded to the number commonly used for water treatment calculations)

Example 4

A tank has a capacity of 60,000 cu ft. What is the capacity of the tank in gallons?

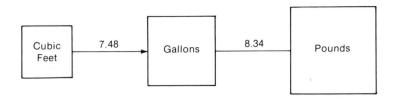

When converting from cubic feet to gallons, you are moving from a smaller to a larger box. Therefore, *multiplication* by 7.48 is indicated:

$$(60,000 \text{ cu ft})(7.48 \text{ gal/cu ft}) = 448,800 \text{ gal}$$

Example 5

If a tank will hold 1,500,000 lb of water, how many cubic feet of water will it hold?

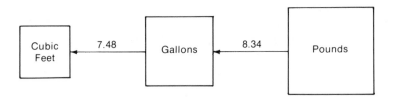

A move from larger to smaller boxes indicates *division* in both conversions:

$$\frac{1,500,000 \text{ lb}}{(8.34 \text{ lb/gal})(7.48 \text{ gal/cu ft})} = 24,045 \text{ cu ft}$$

Flow Conversions

The relationship among the various flow units is shown by the following diagram:

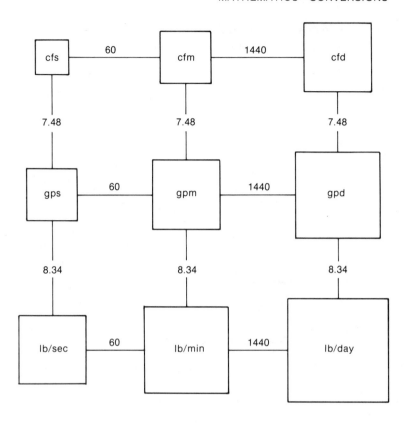

cfs	= cubic feet per second	gps	= gallons per second
cfm	= cubic feet per minute	gpm	= gallons per minute
cfd	= cubic feet per day	gpd	= gallons per day

Since flows of cubic feet per hour (cfh), gallons per hour (gph), and pounds per hour (lb/hr) are less frequently used, they have not been included in the diagram. However, some chemical feed rate calculations require converting from or to these units.[15]

The lines that connect the boxes and the numbers associated with them can be thought of as *bridges* that relate two units directly and all other units in the diagram indirectly.

The relative sizes of the boxes are an aid in deciding whether multiplication or division is appropriate for the desired conversion.

The relationship among the boxes should be understood, not merely memorized. The principle is basically the same as that described in the preceding section. For example, looking at just part of the previous diagram, notice how

---

[15]Chemistry Section, Chemical Dosage Problems (Feed Rate Conversions).

every box in a single vertical column has the same *time* units, and every box in a single horizontal row has the same *volume* units:

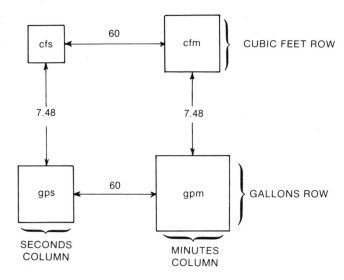

Conversion Along a Vertical Column

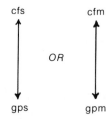

In each of these conversions the time is not changed; the conversion is from cubic feet to gallons, or vice versa.

Conversion Along a Horizontal Row

In the cfs-to-cfm conversion, cubic feet are not changed, merely the *time* associated with the cubic feet. And in the gps-to-gpm conversion, again there is only a time conversion.

Although you need not draw the nine boxes each time you make a flow conversion, it is useful to have a mental image of these boxes to make the calculations. For the examples that follow, however, the boxes are used in analyzing the conversions.

Example 6

If the flow rate in a water line is 2.3 cfs, what is this rate expressed as gallons per minute? (Assume the flow is steady and continuous.)

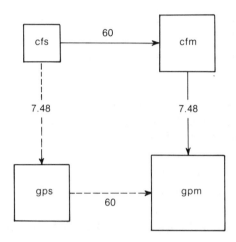

There are two possible paths from cfs to gpm. Either will give the correct answer. Notice that each path has factors of 60 and 7.48 with only a difference in order. In each case you are moving from a smaller to a larger box, and thus *multiplication* by both 60 and 7.48 is indicated:

$$(2.3 \text{ cfs})(60 \text{ sec/min})(7.48 \text{ gal/cu ft}) = 1032 \text{ gpm flow}$$

Notice that you can write both multiplication factors into the same equation—you do not need to write one equation for converting cfs to cfm and another for converting cfm to gpm.

Example 7

The flow rate to a sedimentation basin is 2,450,000 gpd. At this rate, what is the average flow in cubic feet per second?

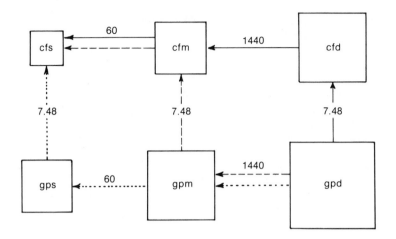

There are three possible paths from gpd to cfs. In each case you would be moving from a larger to a smaller box, thus indicating *division* by 7.48, 1440, and 60 (in any order).

$$\frac{2{,}450{,}000 \text{ gpd}}{(7.48 \text{ gal/cu ft})(1440 \text{ min/day})(60 \text{ sec/min})} = 3.79 \text{ cfs flow rate}$$

Again, the divisions are all written into one equation.

Example 8
    If a flow rate is 200,000 gpd, what is this flow expressed as pounds per minute?

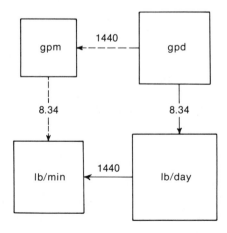

There are two possible paths from gpd to lb/min. The only difference in these paths is the order in which the numbers appear. The answer is the same in either case. In the following explanation, the solid-line path of the diagram is used.

Converting from gpd to lb/day you are moving from a smaller box to a larger box. Therefore *multiplication* by 8.34 is indicated. Then from lb/day to lb/min, you are moving from a larger to a smaller box, which indicates *division* by 1440. These multiplication and division steps are combined into one equation:

$$\frac{(200{,}000 \text{ gpd})(8.34 \text{ lb/gal})}{1440 \text{ min/day}} = 1158 \text{ lb/min flow rate}$$

Linear Measurement Conversions

Linear measurement defines the distance along a line; it is the measurement between two points. The English units of linear measurement include the inch, foot, yard, and mile. In most treatment plant calculations, however, the mile is not used. Therefore, this section discusses conversions of inches, feet, and yards only. The box diagram associated with these conversions is

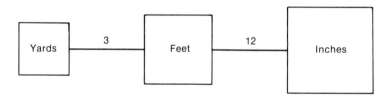

Example 9
The maximum depth of sludge drying beds is 14 in. How many feet is this?

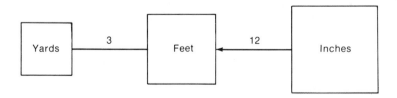

In converting from inches to feet you are moving from a larger to a smaller box. Therefore, *division* by 12 is indicated.

$$\frac{14 \text{ in.}}{12 \text{ in./ft}} = 1.17 \text{ ft depth}$$

Example 10

During backwashing, the water level drops 0.6 yd during a given time interval. How many feet has it dropped?

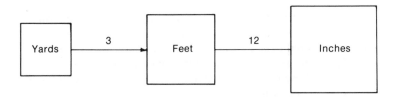

Moving from a smaller to a larger box indicates *multiplication* by 3:

$$(0.6 \text{ yd})(3 \text{ ft/yd}) = 1.8 \text{ ft drop}$$

Area Measurement Conversions

To make area conversions you work with units such as square yards, square feet, or square inches. These are derived from the multiplications:

$$(\text{yards})(\text{yards}) = \text{square yards}$$
$$(\text{feet})(\text{feet}) = \text{square feet}$$
$$(\text{inches})(\text{inches}) = \text{square inches}$$

By examining the relationship of yards, feet, and inches in linear terms, you can recognize the relationship between yards, feet, and inches in square terms. Using yards and feet for the example,

> LINEAR TERMS
> 1 yd = 3 ft

> SQUARE TERMS
> (1 yd)(1 yd) = (3 ft)(3 ft)
> 1 sq yd = 9 sq ft

This method of comparison may be used whenever you wish to compare linear terms with square terms.

Compare the diagram used for linear conversions to that used with square measurement conversions:

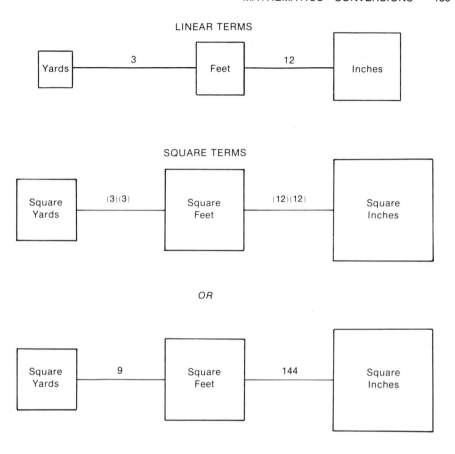

Example 11
The surface area of a sedimentation basin is 170 sq yd. How many square feet is this?

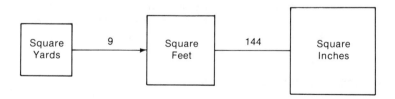

When converting from square yards to square feet, you are moving from a smaller to a larger box. Therefore, *multiplication* by 9 is indicated.

$$(170 \text{ sq yd})(9 \text{ sq ft/sq yd}) = 1530 \text{ sq ft}$$

Example 12

The cross sectional area of a pipe is 64 sq in. How many square feet is this?

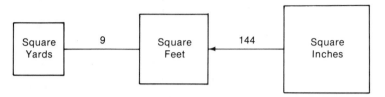

Converting from square inches to square feet, you are moving from a larger to a smaller box. *Division* by 144 is indicated.

$$\frac{64 \text{ sq in.}}{144 \text{ sq in./sq ft}} = 0.44 \text{ sq ft}$$

One other area conversion important in treatment plant calculations is that between square feet and acres. This relationship is expressed mathematically as

$$1 \text{ acre } = 43,560 \text{ sq ft}$$

A box diagram can be devised for this relationship. However, the diagram should be separate from the *sq yd – sq ft – sq in.* diagram because you usually wish to relate directly with square feet. As in the other diagrams, the relative size of the boxes is important.

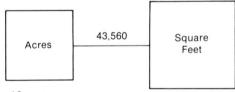

Example 13

A treatment plant requires 0.2 acre for drying beds. How many square feet are required?

Converting acres to square feet, you are moving from a smaller to a larger box. *Multiplication* by 43,560 is therefore indicated:

$$(0.2 \text{ acre})(43,560 \text{ sq ft/acre}) = 8712 \text{ sq ft}$$

Volume Measurement Conversions

To make volume conversions, you work with such units as cubic yards, cubic feet, and cubic inches. These units are derived from the multiplications:

$$(\text{yards})(\text{yards})(\text{yards}) = \text{cubic yards}$$
$$(\text{feet})(\text{feet})(\text{feet}) = \text{cubic feet}$$
$$(\text{inches})(\text{inches})(\text{inches}) = \text{cubic inches}$$

By examining the relationship of yards, feet, and inches in linear terms you can recognize the relationship between yards, feet, and inches in cubic terms. Consider the following example:

LINEAR TERMS

1 yd = 3 ft

CUBIC TERMS

(1 yd)(1 yd)(1 yd) = (3 ft)(3 ft)(3 ft)

1 cu yd = 27 cu ft

The box diagram associated with cubic conversions is

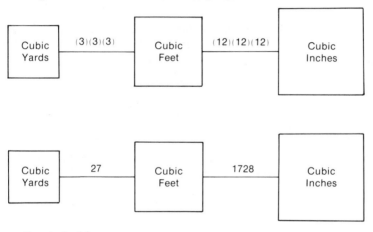

Example 14

Convert 15 cu yd to cubic inches.

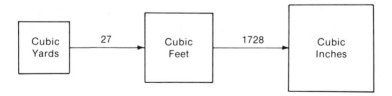

In converting from cubic yards to cubic feet and from cubic feet to cubic inches, you are moving from smaller to larger boxes. Thus, *multiplication* by 27 and 1728 is indicated:

(15 cu yd)(27 cu ft/cu yd)(1728 cu in./cu ft) = 699,840 cu in.

Example 15

The required volume for a chemical is 325 cu ft. What is this volume expressed as cubic yards?

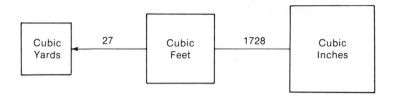

When you move from a larger to a smaller box, *division* is indicated:

$$\frac{325 \text{ cu ft}}{27 \text{ cu ft/cu yd}} = 12.04 \text{ cu yd}$$

Another volume measurement important in treatment plant calculations is that of acre-feet. A reservoir with a surface area of 1 acre and a depth of 1 ft holds exactly 1 acre-ft.

$$1 \text{ acre-ft} = 43,560 \text{ cu ft}$$

The relative sizes of the boxes are again important:

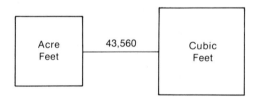

Example 16
The available capacity of a reservoir is 220,000 cu ft. What is this volume expressed in terms of acre-feet?

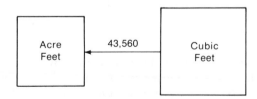

When you move from the larger to the smaller box, *division* by 43,560 is indicated:

$$\frac{220,000 \text{ cu ft}}{43,560 \text{ cu ft/acre-ft}} = 5.05 \text{ acre-ft}$$

Milligrams per Litre, Grains per Gallon, and Parts per Million Conversions

A milligram-per-litre (mg/L) concentration can also be expressed in terms of grains per gallon (gpg) or parts per million (ppm). However, of the three, the preferred unit of concentration is mg/L. To convert milligrams per litre to percent or to pounds per day (lb/day), refer to the Chemistry Section of this handbook.[16]

Milligrams per Litre to Grains per Gallon. Conversions between milligrams per litre and grains per gallon are based on the relationship:

$$1 \text{ gpg} = 17.12 \text{ mg/L}$$

As with any other conversion, often the greatest difficulty in converting from one term to another is deciding whether to multiply or divide by the number given. The box diagram below should help in making this decision.

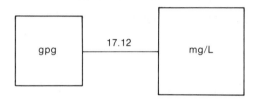

Example 17

Convert 25 mg/L to grains per gallon.

In this example you are converting from milligrams per litre to grains per gallon. Therefore, as shown below, you are moving from the larger to the smaller box:

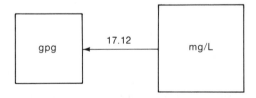

Larger to smaller indicates *division* by 17.12:

$$\frac{25 \text{ mg/L}}{17.12 \text{ mg/L/gpg}} = 1.46 \text{ gpg}$$

---

[16]Chemistry Section, Chemical Dosage Problems.

Example 18

Express a 20 gpg concentration in terms of milligrams per litre.

In this example, you are converting from grains per gallon to milligrams per litre. Therefore, you are moving from the smaller to the larger box:

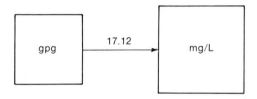

Smaller to larger indicates *multiplication* by 17.12:

(20 gpg)(17.12 mg/L/gpg) = 342.4 mg/L

Example 19

If the dosage rate of alum is 1.5 gpg, what is the dosage rate expressed in milligrams per litre?

The desired conversion is from grains per gallon to milligrams per litre:

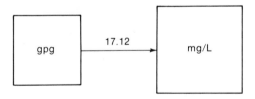

Smaller to larger indicates *multiplication* by 17.12:

(1.5 gpg)(17.12 mg/L/gpg) = 25.68 mg/L

Milligrams per Litre to Parts per Million. The concentration of impurities in water is usually so small that it is measured in milligrams per litre. This means that the impurities in a standard volume (a litre) of water are measured by weight (milligrams). Concentrations in the range of 0-2000 mg/L are roughly equivalent to concentrations expressed as the same number of parts per million (ppm). For example, "12 mg/L of calcium in water" expresses roughly the same concentration as "12 ppm calcium in water." However, milligrams per litre is the preferred unit of concentration.

M11-3.   Metric System Conversions

In order to convert from the English system of customary units to the metric system, or vice versa, you must understand how to convert within the metric system. This requires a knowledge of the common metric prefixes. *These prefixes should be learned before any conversions are attempted.*

Table 1.   Metric System Notations

Prefix	Symbol	Mathematical Value		Powers Notation
giga	G	1,000,000,000		10^9
mega	M	1,000,000		10^6
kilo	k	1,000		10^3
hecto*	h	100		10^2
deka*	da	10		10^1
(none†)	(none)	1		10^0
deci*	d	1/10	*or* 0.1	10^{-1}
centi*	c	1/100	*or* 0.01	10^{-2}
milli	m	1/1,000	*or* 0.001	10^{-3}
micro	μ	1/1,000,000	*or* 0.000 001	10^{-6}
nano	n	1/1,000,000,000	*or* 0.000 000 001	10^{-9}

*Use of these units should be avoided when possible.
†Primary units, such as metres, litres, grams.

As shown in Table 1, the metric system is based on POWERS or multiples of ten,[17] just like the decimal system is. These prefixes may be associated with positions in the place value system.

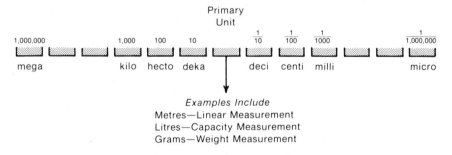

Examples Include
Metres—Linear Measurement
Litres—Capacity Measurement
Grams—Weight Measurement

Understanding the position of these prefixes in the place value system is important because the method discussed below for metric-to-metric conversions is based on this system.

Metric-to-Metric Conversions. When making conversions of linear measurement (metres), capacity measurement (litres), and weight measurement (grams), each change in prefix place value represents one decimal point move. This system of conversion is demonstrated by the following examples.

---

[17]Mathematics Section, Powers and Scientific Notation.

Example 20
Convert 1 metre to decimetres (dm).

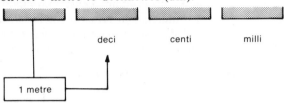

Converting from metres to decimetres requires moving one place to the right. Therefore move the decimal point from its present position *one place to the right.*

$$1.0_\lrcorner = \boxed{10 \text{ decimetres}}$$

Example 21
Convert 1 gram to (a) decigrams; (b) centigrams; and (c) milligrams.

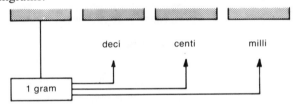

(a) Move the decimal point one place to the right.

$$1.0_\lrcorner = \boxed{10 \text{ decigrams}}$$

(b) Move the decimal point two places to the right.

$$1.00_\lrcorner = \boxed{100 \text{ centigrams}}$$

(c) Move the decimal point three places to the right.

$$1.000_\lrcorner = \boxed{1000 \text{ milligrams}}$$

This system of conversion applies whether you are converting from the primary unit, as in examples 22 and 23, or from any other unit.

Example 22
Convert 1 decilitre to millilitres.

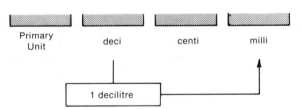

Move the decimal point two places to the right.

$$1.00 \quad = \quad \boxed{100 \text{ millilitres}}$$

This system of conversion also applies whether the number you are converting is 1.0, as in examples 20 through 22, or any other number.

Example 23

Convert 3.5 kilograms to grams.

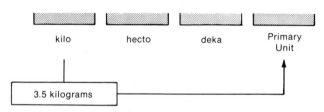

Move the decimal point three places to the right.

$$3.500 \quad = \quad \boxed{3500 \text{ grams}}$$

Example 24

Convert 0.28 centimetre to metres.

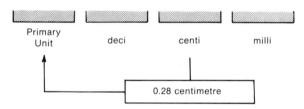

Move the decimal point two places to the left.

$$00.28 \quad = \quad \boxed{0.0028 \text{ metre}}$$

Most metric conversion errors are made in moving the decimal point to the left. You must be very careful in moving the decimal point *from its present position,* counting every number (including zeros) to the left as a decimal point move.

Example 25

Convert 1750 litres to kilolitres.

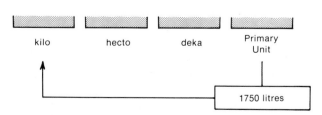

Move the decimal point three places to the left.

$$1\underset{\longleftarrow}{750.} = \boxed{1.75 \text{ kilolitres}}$$

In the examples just given, there were no conversions of square or cubic terms. However, area and volume measurements can be expressed as square and cubic metres, centimetres, kilometres, etc. The following discussion shows the special techniques needed for converting between these units.

Square metres indicates the mathematical operation:

$$(\text{metre})(\text{metre}) = \text{square metres}$$

Square metres may also be written as m^2. The exponent of *2* indicates that *metre* appears twice in the multiplication, as shown above.[18]

In conversions, the term *square* (or exponent of *2*) indicates that *each prefix place value move requires 2 decimal point moves.* All other aspects of the conversions are similar to the preceding examples 20–25.

Example 26
Convert 1 square metre to square decimetres.

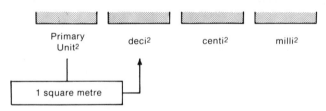

Converting from square metres to square decimetres requires moving one place value to the right. In square terms, *each prefix place move requires 2 decimal point moves.* Making the move in groups of two may be easier:

$$1.\underset{\longrightarrow}{00} = \boxed{100 \text{ square decimetres}}$$

Now check this conversion. From Example 20 you know that 1 metre = 10 decimetres. Squaring both sides of the equation,

$$(1 \text{ metre})(1 \text{ metre}) = (10 \text{ decimetres})(10 \text{ decimetres})$$

$$1 \text{ square metre} = 100 \text{ square decimetres}$$

Example 27
Convert 32,000 square metres to square kilometres.

---

[18]Mathematics Section, Powers and Scientific Notation.

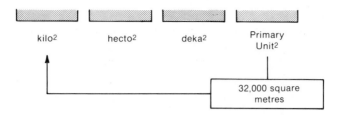

Three place value moves to the left indicate six total decimal point moves to the left:

$$032{,}000. \quad = \quad \boxed{0.032 \ \text{square} \ \text{kilometre}}$$

Cubic metres indicates the mathematical operation:

$$(\text{metres})(\text{metres})(\text{metres}) = \text{cubic} \ \text{metres}$$

Cubic metres may also be written as m^3. The exponent of *3* indicates that *metre* appears three times in the multiplication. When you are converting cubic terms, *each prefix place value move requires 3 decimal point moves.* Again, it may be easier to make the decimal point moves in groups—groups of three for cubic-term conversions as opposed to groups of two for square-term conversions.

Example 28
   Convert 1 cubic metre to cubic decimetres.

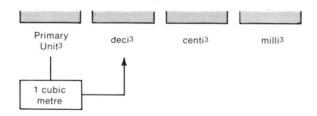

Converting from cubic metres to cubic decimetres requires moving one place value to the right. In cubic terms, *each prefix place value move requires 3 decimal point moves:*

$$1.000 \quad = \quad \boxed{1000 \ \text{cubic decimetres}}$$

Now check this conversion. From Example 20 you know that 1 metre = 10 decimetres. Cubing both sides of the equation,

$$(1 \ \text{metre})(1 \ \text{metre})(1 \ \text{metre}) = (10 \ \text{dm})(10 \ \text{dm})(10 \ \text{dm})$$
$$1 \ \text{cubic metre} = 1000 \ \text{cubic decimetres}$$

Example 29

Convert 155,000 cubic millimetres to cubic metres.

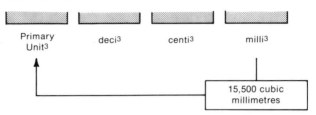

Three place value moves to the left indicate three *groups of 3* decimal point moves to the left:

$$\underset{\longleftarrow}{000115,000.} = \boxed{0.000155 \text{ cubic metre}}$$

Cross-System Conversions. For conversions from the English system to the metric system, or vice versa, the conversion table appearing in Appendix 1 and discussed at the beginning of this section is useful.

Example 30

Convert 20 gal to litres.

The factor given in the table to convert *From* "gallons" *To* "litres" is 3.785. Therefore the conversion is

$$(20 \text{ gal})(3.785) = 75.7 \text{ L}$$

Example 31

Convert 3.7 acres to square metres.

The factor given in the table for converting *From* "acres" *To* "square metres" is 4047. The conversion is therefore

$$(3.7 \text{ acre})(4047) = 14,974 \text{ m}^2$$

Example 32

Convert 0.8 m/sec to feet/minute.

The factor given in the table is 196.8. Therefore the conversion is

$$(0.8 \text{ m/sec})(196.8) = 157.44 \text{ fpm}$$

Occasionally when making cross-system conversions, you may not find the factor in the table for the two terms of interest. For example, you may wish to convert from inches to decimetres and the table only gives factors for converting from inches to centimetres or inches to millimetres. Or you may wish to convert cubic centimetres to gallons and the table only gives factors for converting cubic metres to gallons.

In situations such as these, it is usually easiest to make sure that the English system unit is in the desired form and then make any necessary changes with the metric unit. As shown in the example problems under "Metric-to-Metric Conversions," changing units in the metric system requires only a decimal point move.

The following two examples illustrate how the cross-system conversion may be made when the table does not give the precise units you need.

Example 33

The water depth in a channel is 1.2 ft. How many decimetres is this?

First check the conversion table to see if a factor is given for converting feet to decimetres. The conversion factor is given only for feet to kilometres, metres, centimetres, or millimetres.

To make the conversion from feet to decimetres then, first convert from feet to the closest metric unit to decimetres given in the table (centimetres); then convert that answer to decimetres. The conversion from centimetres to feet is

$$(1.2 \text{ ft})(30.48) = 36.58 \text{ cm}$$

Then converting from centimetres to decimetres,

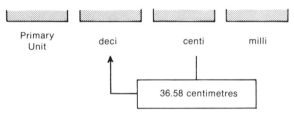

Move the decimal point one place to the left:

$$36.58 \text{ cm} = 3.658 \text{ dm}$$

Example 34

If you use 0.12 kg of a chemical to make up a particular solution, how many ounces of that chemical are used?

First, check the conversion table to determine if a factor is given for converting kilograms to ounces. No such conversion factor is given.

Since you wish the English unit (ounces) to remain in the desired form, try to find a conversion in the table from some other metric unit (such as milligrams or grams) to ounces.

The conversion from grams to ounces is given in the table. Therefore, first convert the 0.12 kg to grams, then convert grams to ounces:

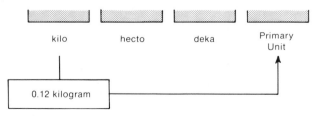

Move the decimal point three places to the right:

$$0.12 \text{ kg} = 120 \text{ g}$$

Then

$$(120 \text{ g})(0.03527) = 4.23 \text{ oz}$$

M11-4.   Temperature Conversions

The formulas used for Fahrenheit and Celsius temperature conversions are

$$°C = 5/9(°F - 32°)$$
$$°F = 9/5(°C) + 32°$$

These formulas are difficult to remember unless used frequently. There is, however, another method of conversion which is perhaps easier to remember since the same three steps are used for both Fahrenheit and Celsius conversions:

Step 1:   Add 40°

Step 2:   Multiply by 5/9 or 9/5

Step 3:   Subtract 40°

The only variable in this method is the choice of 5/9 or 9/5 in the multiplication step. To make this choice you must know something about the two scales. As shown below, on the Fahrenheit scale the freezing point of water is 32°, whereas it is 0° on the Celsius scale. The boiling point of water is 212° on the Fahrenheit scale and 100° on the Celsius scale.

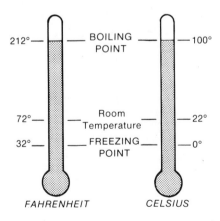

Thus for the same temperature, higher numbers are associated with the Fahrenheit scale and lower numbers with the Celsius scale. This information helps you decide whether to multiply by 5/9 or 9/5. Let's look at a few conversion problems to see how the three-step process works.

Example 35

Suppose that you wish to convert 212° F to Celsius. From the sketch of the two scales you know that the answer should be 100°C. But let's verify it using the three-step process.

The first step is to add 40°:

$$\begin{array}{r} 212° \\ +\ 40° \\ \hline 252° \end{array}$$

Next, 252° must be multiplied by either 5/9 or 9/5. Since the conversion is to the *Celsius* scale, you will be moving to a number *smaller* than 252. Multiplying by 9/5 is roughly the same as multiplying by 2, which would double 252 rather than make it smaller. On the other hand, multiplying by 5/9 is about the same as multiplying by 1/2, which would cut 252 in half.

In this problem, since you wish to move to a smaller number, you should multiply by 5/9:

$$\left(\frac{5}{9}\right)(252°) = \frac{1260°}{9}$$

$$= 140°$$

The problem can now be completed using step 3 (subtract 40°):

$$\begin{array}{r} 140° \\ -\ 40° \\ \hline 100° \end{array}$$

Therefore, 212° F = 100° C

Example 36

Convert 0°C to Fahrenheit.

The sketch of the two scales indicates that 0°C = 32°F, but this can be verified using the three-step method of conversion:

Step 1: Add 40°.

$$\begin{array}{r} 0° \\ +40° \\ \hline 40° \end{array}$$

Step 2: In this problem you are going from Celsius to Fahrenheit. Therefore you will be moving from a smaller number to a larger number and 9/5 should be used in the multiplication:

$$\left(\frac{9}{5}\right)(40°) = \frac{360°}{5}$$

$$= 72°$$

Step 3: Subtract 40°.

$$
\begin{array}{r}
72° \\
-40° \\
\hline
32°
\end{array}
$$

Thus, 0°C = 32°F

Example 37

A thermometer indicates that the water temperature is 15°C. What is this temperature expressed in degrees Fahrenheit?

Step 1: Add 40°.

$$
\begin{array}{r}
15° \\
+40° \\
\hline
55°
\end{array}
$$

Step 2: Moving from a smaller number (Celsius) to a larger number (Fahrenheit) indicates multiplication by 9/5:

$$\left(\frac{9}{5}\right)(55°) = \frac{495°}{5}$$

$$= 99°$$

Step 3: Subtract 40°.

$$
\begin{array}{r}
99° \\
-40° \\
\hline
59°
\end{array}
$$

Therefore, 15°C = 59° F.

Although it is useful to know how to make these temperature conversion calculations, in practical application you may wish to use a temperature conversion table such as the one in Appendix A. Let's look at a couple of example conversions using the table.

Example 38

Normal room temperature is considered to be 68° F. What is this expressed in degrees Celsius?

Use the Fahrenheit-to-Celsius temperature conversion table. Coming down the Fahrenheit column to 68°, you can see that 68° F = 20° C.

Example 39

Convert 90° C to degrees Fahrenheit.

Use the Celsius-to-Fahrenheit temperature conversion table. Coming down the Celsius column to 90°, you can see that 90° C = 194° F.

Review Questions

1. Cubic feet to gallons to pounds conversions:

   (a) Express 1500 cu ft in terms of gallons.

   (b) Convert 1,000,000 lb to cubic feet.

   (c) If a tank has a capacity of 25,000 gal of water, what is its capacity expressed in pounds?

2. Flow conversions:

   (a) Convert 5 cfs to gallons per day (gpd).

   (b) Convert the flow rate of 16,000,000 lb/day to cubic feet per second (cfs).

   (c) The flow in a water main is 4.5 mgd. Assuming the flow is steady and continuous, what is this flow rate expressed in cubic feet per second (cfs)?

   (d) The flow to a filter is 1700 gpm. Express this flow rate as gallons per day (gpd).

3. Linear measurement conversions:

   (a) Convert 4.7 ft to inches.

   (b) During backwashing, the water level drops 20 in. How many yards does it drop?

4. Area of measurement conversions:

   (a) Convert 20 sq yd to square inches.

   (b) The surface area of a clarifier is 800 sq ft. Express this area in terms of square yards.

   (c) If 5000 sq ft are required for a particular storage area, what is this area expressed in terms of acres?

5. Volume measurement conversions:

   (a) Convert 500 cu ft to cubic inches.

   (b) The volume of a chemical storage tank is 20 cu yd. How many cubic feet is this?

   (c) The volume of sludge drying beds at a plant is 0.5 acre-ft. What is this volume expressed in cubic feet?

6. Milligrams per litre and grains per gallon conversions:

   (a) Convert 180 mg/L to grains per gallon (gpg).

   (b) Express the concentration of 12 gpg in terms of milligrams per litre (mg/L).

   (c) At a particular treatment plant the dosage rate for alum is 30 mg/L. What is this dosage rate expressed in terms of grains per gallon (gpg)?

7. Metric system conversions:

   (a) Convert 14.5 m to millimetres.

   (b) Express 150 mL in terms of litres.

   (c) Convert 2 km^2 to square metres.

   (d) Convert 20,000 cm^3 to cubic metres.

   (e) The filter loading at a treatment plant is 4 gpm/sq ft. Using the conversion table in the appendix, express this loading rate as millimetres per second (mm/s).

   (f) The daily flow rate at a plant is 20 ML/day. Using the conversion table in the appendix, express this flow rate in million gallons per day (mgd).

8. Temperature conversions:

(a) Using the three-step conversion procedure, convert 60° F to degrees Celsius.

(b) Using the three-step conversion procedure, convert 60° C to degrees Fahrenheit.

(c) The raw water entering a treatment plant is 57° F. What is this temperature expressed in degrees Celsius? (Use the temperature conversion table given in Appendix A.)

(d) The temperature in a storeroom is 30° C. What is the temperature expressed in degrees Fahrenheit? (Use the temperature conversion table given in the appendix.)

Summary Answers

1. (a) 11,220 gal
   (b) 16,030 cu ft
   (c) 208,500 lb

2. (a) 3,231,360 gpg
   (b) 2.97 cfs
   (c) 6.96 cfs
   (d) 2,448,000 gpd

3. (a) 56.4 in.
   (b) 0.56 yd

4. (a) 25,920 sq in.
   (b) 88.89 sq yd
   (c) 0.11 acre

5. (a) 864,000 cu in.
   (b) 540 cu ft
   (c) 21,780 cu ft

6. (a) 10.51 gpg
   (b) 205.44 mg/L
   (c) 1.75 gpg

7. (a) 14,500 mm
   (b) 0.150 L
   (c) 2,000,000 m^2
   (d) 0.020 m^3
   (e) 2.72 mm/s
   (f) 5.28 mgd

8. (a) 15.56°C
   (b) 140°F
   (c) 13.9°C
   (d) 86°F

Detailed Answers

1. Cubic feet to gallons to pounds conversions:

   (a) In the conversion from cubic feet to gallons, the move is from a smaller to a larger box. Therefore *multiplication* by 7.48 is indicated:

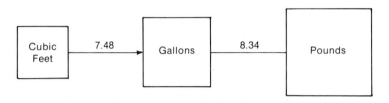

$$(1500 \text{ cu ft}) (7.48 \text{ gal/cu ft}) = 11,220 \text{ gal}$$

   (b) In the conversion from pounds to gallons and from gallons to cubic feet, the move is from a larger to a smaller box. Therefore *division* is indicated in each case:

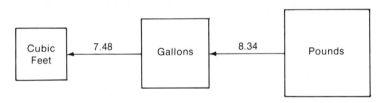

$$\frac{1,000,000 \text{ lb}}{(7.48 \text{ gal/cu ft}) (8.34 \text{ lb/gal})} = 16,030 \text{ cu ft}$$

   (c) In the conversion from gallons to pounds, the move is from a smaller to a larger box. Therefore *multiplication* by 8.34 is indicated:

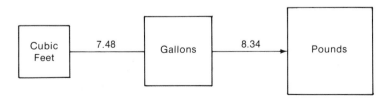

$$(25,000 \text{ gal}) (8.34 \text{ lb/gal}) = 208,500 \text{ lb}$$

2. Flow conversions:
   (a)

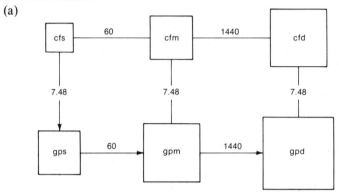

There are three possible paths from cfs to gpd. And the only difference among these paths is the *order* in which the numbers are multiplied or divided. The resulting answers will be the same. One of these paths is noted in the diagram above.

In each conversion (cfs to gps; gps to gpm; gpm to gpd) the move is from a smaller to a larger box. Therefore, *multiplication* by 7.48, 60, and 1440 is indicated:

$$(5 \text{ cfs})(7.48 \text{ gal/cu ft})(60 \text{ sec/min})(1440 \text{ min/day}) = 3,231,360 \text{ gpd}$$

(b)

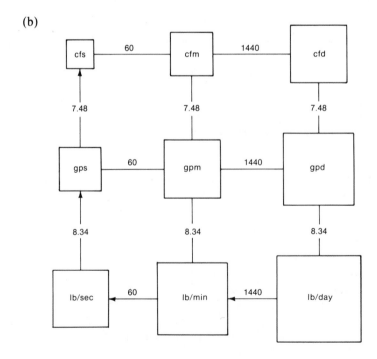

There are several possible paths from lb/day to cfs. As previously mentioned, the only difference among these paths is the order in which the numbers are multiplied or divided. The resulting answer will be the same. One of these paths is noted in the diagram above.

In each conversion on the path from lb/day to cfs, the move is from a larger to a smaller box. Therefore *division* by 1440, 60, 8.34, and 7.48 is indicated:

$$\frac{16,000,000 \text{ lb/day}}{(1440 \text{ min/day})(60 \text{ sec/min})(8.34 \text{ lb/gal})(7.48 \text{ gal/cu ft})} = 2.97 \text{ cfs}$$

(c)

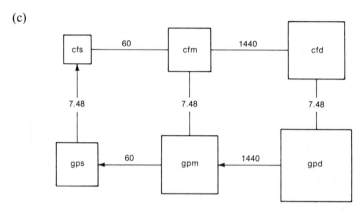

First, express 4.5 mgd as gpd:

$$4.5 \text{ mgd} = 4,500,000 \text{ gpd}$$

Only one of the three possible paths between gpd and cfs is shown in the diagram above. In making the conversion from gpd to gpm, gpm to gps, and gps to cfs, the move is from a larger to a smaller box. Thus *division* by 1440, 60, and 7.48 is indicated:

$$\frac{4,500,000 \text{ gpd}}{(1440 \text{ min/day})(60 \text{ sec/min})(7.48 \text{ gal/cu ft})} = 6.96 \text{ cfs}$$

(d)

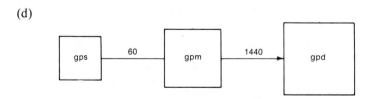

In the conversion from gpm to gpd, the move is from a smaller to a larger box, indicating *multiplication* by 1440:

$$(1700 \text{ gpm})(1440 \text{ min/day}) = 2,448,000 \text{ gpd}$$

3. Linear measurement conversions:
   (a)

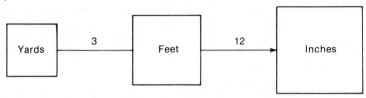

   In the conversion from feet to inches, the move is from a smaller to a larger box. This indicates *multiplication* by 12:

$$(4.7 \text{ ft})(12 \text{ in./ft}) = 56.4 \text{ in.}$$

   (b)

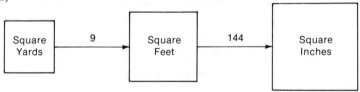

   In the conversion from inches to feet and from feet to yards, the move is from larger to smaller boxes. Therefore *division* by 12 and 3 is indicated:

$$\frac{20 \text{ in.}}{(12 \text{ in./ft})(3 \text{ ft/yd})} = 0.56 \text{ yd}$$

4. Area measurement conversions:
   (a)

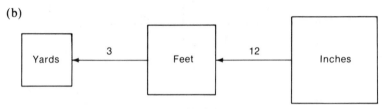

   In the move from smaller to larger boxes, *multiplication* by 9 and 144 is indicated:

$$(20 \text{ sq yd})(9 \text{ sq ft/sq yd})(144 \text{ sq in./sq ft}) = 25,920 \text{ sq in.}$$

   (b)

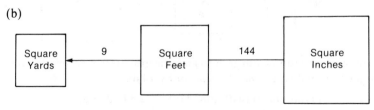

From square feet to square yards, the move is from a larger to a smaller box. *Division* by 9 is indicated:

$$\frac{800 \text{ sq ft}}{9 \text{ sq ft/sq yd}} = 88.89 \text{ sq yd}$$

(c)

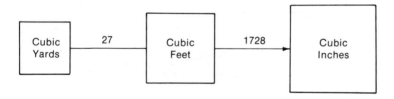

In the move from a larger to a smaller box, *division* by 43,560 is indicated:

$$\frac{5000 \text{ sq ft}}{43,560 \text{ sq ft/acre}} = 0.11 \text{ acre}$$

5. Volume measurement conversions:
   (a)

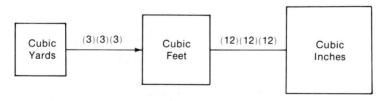

In the conversion from cubic feet to cubic inches, the move is from a smaller to a larger box. This indicates *multiplication* by 1728:

(500 cu ft)(1728 cu in./cu ft) = 864,000 cu in.

(b)

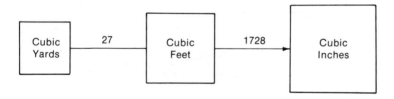

The move from a smaller to a larger box indicates *multiplication* by 27.

(20 cu yd)(27 cu ft/cu yd) = 540 cu ft

(c)

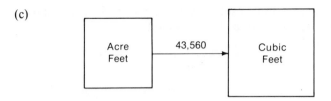

Multiplication by 43,560 is indicated since the move is from a smaller to a larger box:

$$(0.5 \text{ acre-ft})(43,560 \text{ cu ft/acre-ft}) = 21,780 \text{ cu ft}$$

6. Milligrams per litre and grains per gallon conversions:
    (a) In this problem the conversion is from milligrams per litre to grains per gallon. When you move from a larger to a smaller box as shown below, *division* by 17.12 is indicated:

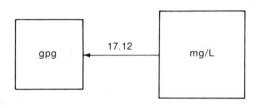

$$\frac{180 \text{ mg/L}}{17.12 \text{ mg/L/gpg}} = 10.51 \text{ gpg}$$

    (b) Since the conversion is from grains per gallon to milligrams per litre, you are moving from a smaller to a larger box. *Multiplication* by 17.12 is therefore indicated:

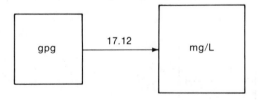

$$(12 \text{ gpg})(17.12 \text{ mg/L/gpg}) = 205.44 \text{ mg/L}$$

    (c) This problem involves a conversion from milligrams per litre to grains per gallon. The move from the larger to the smaller box indicates *division* by 17.12:

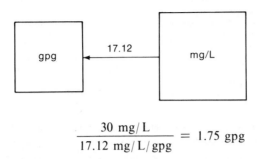

$$\frac{30 \text{ mg}/\text{L}}{17.12 \text{ mg}/\text{L}/\text{gpg}} = 1.75 \text{ gpg}$$

7. Metric system conversions:

(a)

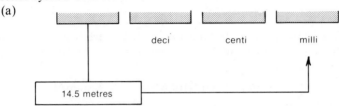

Since the conversion involves moving three places to the right in the place value system, the decimal point must be moved *three places to the right.*

$$14.500_\blacktriangle = \boxed{14,500 \text{ millimetres}}$$

(b)

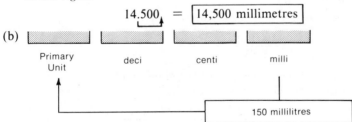

A decimal point move of *three places to the left* is required:

$$_\blacktriangle 150. = \boxed{0.150 \text{ litres}}$$

(c)

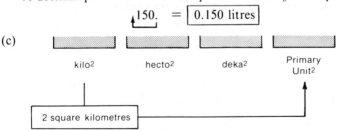

When you are converting square terms, each place value move requires two decimal point moves. In this problem then, three *place value moves* to the right indicate a total of six *decimal point* moves to the right.

$$2.000000_\blacktriangle = \boxed{2,000,000 \text{ square metres}}$$

(d)

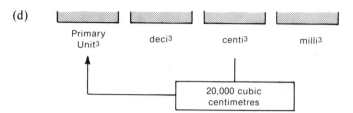

When you are converting cubic terms, each place value move requires three decimal point moves. In this problem then, two *place value moves* to the left indicate a total of six *decimal point* moves to the left.

$$020{,}000. = \boxed{.020 \text{ cubic metres}}$$

(e) The factor given in the table for converting from gallons per minute per square foot to millimetres per second is 0.6790. Therefore the conversion is

$$(4 \text{ gpm/sq ft})(0.6790) = 2.72 \text{ mm/s}$$

NOTE: Converting a measurement of flow-rate-per-area (gpm/sq ft) to a measurement of length-per-time (mm/s) may seem to be incorrect. The following analysis shows, however, that both units measure the same thing:

The units *gpm/sq ft* may be written

$$\frac{\text{gal/min}}{\text{sq ft}}$$

Gallons are a measurement of volume, and 1 gal = 0.1227 cu ft. So gallons in the expression can be converted to cubic feet:

$$\frac{\text{gal/min}}{\text{sq ft}} = \frac{0.1337 \text{ cu ft/min}}{\text{sq ft}}$$

Next, expand the square- and cubic-feet terms and cancel:

$$= \frac{0.1337 \, (\cancel{ft})(\cancel{ft})(ft)/\min}{(\cancel{ft})(\cancel{ft})}$$

$$= 0.1337 \text{ ft/min}$$

The resulting unit, feet per minute, is a measurement of length-per-time that expresses the same thing as the original measurement of flow-rate-per-area. Simply converting this to metric units gives millimetres per second. (The conversion factor used in the problem, 0.6790, combines into one number all the factors for converting gallons to cubic feet, feet to millimetres, and minutes to seconds.)

(f) The factor given in the table for converting from ML/day to mgd is 0.2642. Therefore the conversion is

$$(20 \text{ ML/d})(0.2642) = 5.28 \text{ mgd}$$

8. Temperature conversions:

(a) *Step 1:* Add 40°.

$$60°$$
$$\underline{+40°}$$
$$100°$$

Step 2: In this problem the conversion is from Fahrenheit to Celsius. Thus the conversion involves a change from a larger to a smaller number, which indicates that $5/9$ should be used in the multiplication:

$$\left(\frac{5}{9}\right)(100°) = \frac{500°}{9}$$
$$= 55.56°$$

Step 3: Subtract 40°.

$$55.56°$$
$$\underline{-40.00°}$$
$$15.56°$$

Therefore $60° F = 15.56° C$.

(b) *Step 1:* Add 40°.

$$60°$$
$$\underline{+40°}$$
$$100°$$

Step 2: In this problem the conversion is from Celsius to Fahrenheit. Since the conversion involves a change from a smaller to a larger number, $9/5$ is used in the multiplication:

$$\left(\frac{9}{5}\right)(100°) = \frac{900°}{5}$$
$$= 180°$$

Step 3: Subtract 40°.

$$180°$$
$$\underline{-40°}$$
$$140°$$

Therefore $60° C = 140° F$.

(c) According to the Fahrenheit-to-Celsius conversion table, $57° F = 13.9° C$.

(d) As shown in the Celsius-to-Fahrenheit conversion table, $30° C = 86° F$.

Mathematics 12

Graphs and Tables

Often the operator of a water treatment plant is responsible for collecting and analyzing data about many aspects of treatment. These data and their analyses are important in making operation modifications and in summarizing operating trends.

Graphs and tables are a means of sorting and organizing data so that trends in information can be identified and interpreted. Graphs and tables may also be regarded as tools for decision making. This section explains how to read the graphs and tables that you may encounter in treatment plant operation, and also how to construct some of the more simple graphs.

M12-1. Graphs

There are four types of graphs—bar, circle, line, and nomographs—frequently encountered in water treatment data presentation and analysis. The first three (bar, circle, and line graphs) are commonly used to summarize water treatment data, while nomographs are usually used to solve mathematical problems pertaining to chemical feed rates.

Before considering the distinctive features of each type of graph, you should have a general understanding of the scales used for graphs, and of how to read or interpret information using the scales.

Scales are series of intervals, usually marked along the side and bottom of a graph, that represent the range of values of the data. On a typical graph, the bottom (horizontal) scale indicates one kind of information, and the side (vertical) scale indicates another kind. For example, the horizontal scale may show the months or years during which a certain water treatment was performed, and the vertical scale of the same graph may show the number of gallons of water treated.

A scale in which the intervals (marks or lines) are equally spaced is called an

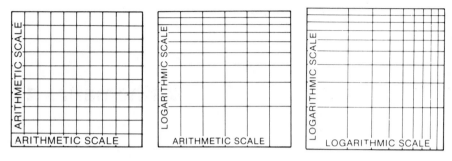

Figure 1.   **Figure 2.**   **Figure 3.**

ARITHMETIC SCALE. An example of a grid in which both the horizontal and vertical scales are arithmetic (equally spaced) is shown in Figure 1.

A LOGARITHMIC, or LOG, SCALE is one in which the intervals are varied logarithmically and therefore not equally spaced. Discussions of logarithms and of construction of a log scale are beyond the scope of this text. However, figures 2 and 3 illustrate log scales. Figure 2 is an example of a grid having an arithmetic horizontal scale and a logarithmic vertical scale; Figure 3 is an example of a grid in which both the horizontal and vertical scales are logarithmic.

The technique used to determine values that fall between the marked intervals on a scale is called INTERPOLATION. A good example of everyday interpolation is the way that you use an ordinary ruler. Although only the inches are numbered on a ruler, the spaces between the inch marks are divided into equal intervals (1/2 in., 1/4 in., 1/8 in., etc; or, on some rulers, 0.1 in., 0.2 in., 0.3 in., etc.) so that you have no difficulty measuring dimensions between the inches.

Interpolation on arithmetic scales of a graph is similar to using a ruler. For example, where would you read a value of *3* on an arithmetic scale like this?

Every other interval is numbered, and the unnumbered intervals fall halfway between the numbers, so *3* is read where the arrow indicates:

If every fourth interval is numbered, then the space between numbers is divided into four equal parts:

And the values assigned to the in-between divisions on the scale are

Suppose every fourth interval is marked thus:

Then the in-between divisions are

As the examples above illustrate, when you interpolate on an arithmetic scale you must divide the space between two known points into equal intervals. On a logarithmic scale, however, the space is not divided evenly—the divisions are larger in the first part of the space, becoming smaller in the last part of the space. Let's compare the two types of scales, using 2 and 3 as the known points on both.

Arithmetic scale (equal divisions):

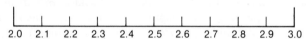

Logarithmic scale (unequal divisions):

Explaining the construction of a log scale is beyond the scope of this section, so when you need to interpolate on such a scale you will have to estimate the logarithmic divisions between two known numbers.

Let's look now at how to number the ten subdivisions between two known numbers on a log scale. Suppose the two known points are 7 and 8, as shown below. How would each of the subdivisions be numbered?

Since the two known numbers represent *units* in the decimal system, the subdivisions are numbered in *tenths*. The starting point is *7.0*. The next lines are therefore numbered *7.1, 7.2, 7.3,* and so on, up to *7.9* and *8.0:*

What would the numbering be if the two known numbers were *70* and *80?*

Since the two known numbers represent *tens* in the decimal system, the subdivisions are numbered in *units*. The starting point is *70* and the numbering continues, with *71, 72, 73,* up to *79* and *80:*

Let's look at a last example using whole numbers. Suppose the two known points this time are *700* and *800*. How would the subdivisions be numbered then?

In this example the two known numbers represent *hundreds* in the decimal system. Therefore, the subdivisions are numbered in *tens*. The starting point is *700*. The next lines are numbered *710, 720, 730,* and so on, up to *790* and *800:*

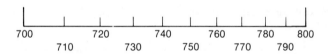

What happens in numbering the subdivisions if the two known numbers are decimal numbers less than one? For example, suppose the two known numbers are *0.07* and *0.08*.

Since the two known numbers represent *hundredths* in the decimal system, the subdivisions are numbered in *thousandths*. Therefore, the starting point is *0.070* and the succeeding lines are numbered *0.071, 0.072, 0.073,* etc. up to *0.079* and *0.080:*

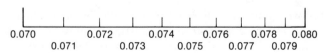

Now let's consider the four types of graphs you are most likely to use in analyzing water treatment data. (Several examples in the following discussion of graphs require interpolation to determine the answer.)

Bar Graphs

In a bar graph the data being compared are represented by bars, which may be drawn vertically or horizontally, as shown in figures 4 and 5. The information that the graph represents is usually given along the bottom and left side of the graph. In addition, information is sometimes given at the end of each bar to make reading the graph a little easier. The title of the graph indicates the general type of information being presented.

Two special types of bar graphs are the *historical* and the *divided.* On the historical bar graph, which represents data at various *times,* the bars are drawn vertically. The time factor is given on the horizontal scale, and other information is represented on the vertical scale. In Figure 6, for example, time is represented on the horizontal scale and billions of gallons on the vertical scale.

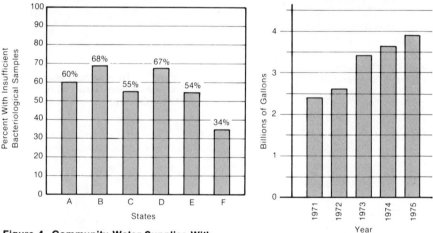

Figure 4. Community Water Supplies With Insufficient Bacteriological Monitoring (Vertical Bar Graph)

Figure 6.   Total Annual Water Production (Historical Bar Graph)

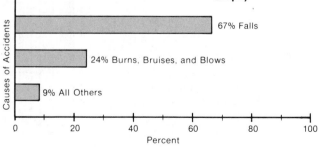

Figure 5.   Common Causes of Accidents (Horizontal Bar Graph)

The divided bar graph (Figure 7) is often used when summarizing data pertaining to percent. The sum of the percents must equal 100.

Occasionally you may see a bar graph that has a broken scale, such as that shown in Figure 8. The break, usually on the vertical scale and denoted by a zigzag line, is made to avoid unnecessarily long bars.

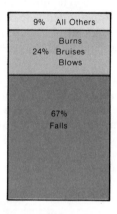

Figure 7.   Common Causes of Accidents (Divided Bar Graph)

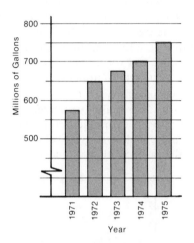

Figure 8.   Total Annual Water Production (Bar Graph With Broken Scale)

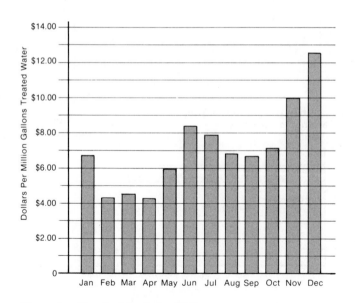

Figure 9.   Chemical Costs for 1977

Example 1

Given the bar graph in Figure 9, answer the following questions.

(a) What was the chemical cost per million gallons of treated water for the month of March?

(b) If the total water treated during July was 137 mil gal, what was the total chemical cost for that month?

(c) How much more were the chemical costs (per million gallons of treated water) for December than for February?

(a) The bar height corresponding with March indicates that the chemical costs for the month were approximately $4.50 per million gallons treated water.

(b) To calculate the total chemical costs for the month, first determine the cost per million gallons treated water. The bar height corresponding with July indicates the cost was about $7.90 for each million gallons of water treated. Since a total of 137 mil gal of water was treated during the month, the total chemical cost was

$$(\$7.90/\text{mil gal})(137 \text{ mil gal}) = \$1082.30$$

(c) The chemical costs for the month of December were approximately $12.50 per million gallons treated water, whereas the costs for February were only about $4.30 per million gallons treated water. The difference between these costs is

$12.50   December cost
−4.30   February cost
$8.20   Difference in cost

Therefore, the chemical costs for December were about $8.20 per million gallons *more* than the chemical costs for February.

Circle Graphs

The circle graph, sometimes referred to as a circle chart or pie chart, is similar to the divided bar graph mentioned above in that the sum of the different parts must equal 100. The circle graph is often used when summarizing data pertaining to percent because it is an effective way of showing the relationship of the whole to its parts, as in Figure 10.

Figure 10.
Water Use by
Category for 1977

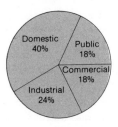

Example 2

Using the circle graph in Figure 10 as representing the water use by category for a particular water system, answer the following questions:

(a) What percent of the total water use was for domestic purposes?

(b) If the total water use for 1977 was 980 mgd, how much much water (in million gallons) was used for domestic purposes?

(a) According to the circle graph in Figure 10, 40 percent of the total water use was for domestic purposes.

(b) Since both the *percent domestic use* and the *total water used* are known, the actual amount of water used for domestic purposes can be calculated:[19]

$$\text{Percent Domestic Water} = \frac{\text{Domestic water use}}{\text{Total water use}} \times 100$$

Filling in the information,

$$40 = \frac{x \text{ mil gal domestic water use}}{980 \text{ mil gal total water use}} \times 100$$

This may also be expressed as

$$40 = \frac{(x)(100)}{980}$$

Then solve for the unknown value:[20]

$$(980)(40) = (x)(100)$$

$$\frac{(980)(40)}{100} = x$$

392 mil gal domestic use $= x$

---

[19] Mathematics Section, Percent.

[20] Mathematics Section, Solving for the Unknown Value.

Line Graphs

Line graphs are perhaps the most common type you will encounter in water treatment data presentation. As shown in figures 11 and 12, line graphs fall into two general categories: broken-line and smooth-line. Of the two, smooth-line graphs are more often found in water treatment literature.

Broken-line graphs are merely dots (data points) connected by straight line segments. Connecting the data points helps highlight the trend in the data.

When the in-between points are analyzed and found to be predictable, a smooth-line graph can be constructed, using advanced mathematical calculations to determine the proper shape of the curve.

The scales on a line graph are usually numbered along the bottom and left side of the graph, with additional numbering occasionally found along the right side and top. The scales on a line graph are most often arithmetic (equally spaced), although for some types of data logarithmic (graduated interval) scales are used.

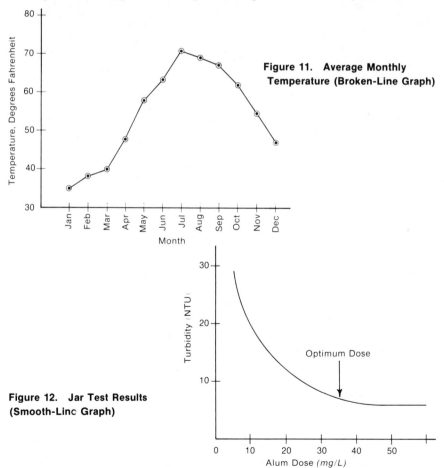

Figure 11. Average Monthly Temperature (Broken-Line Graph)

Figure 12. Jar Test Results (Smooth-Linc Graph)

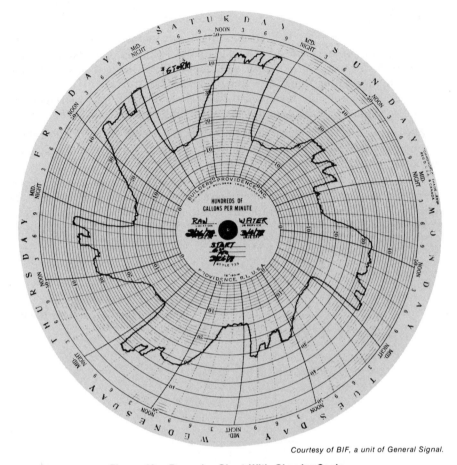

Figure 13.   Recorder Chart With Circular Scale

In Figure 13, a line graph is shown with a circular scale—somewhat like arithmetic graph paper stretched around into a circle. Reading this type of scale is not much different than reading the more conventional arithmetic scales.

The following sample problems include some line graphs with arithmetic scales and some with logarithmic scales.

Example 3

Given the graph in Figure 14, answer the following questions.

(a) What is represented by the vertical scale?

(b) What is represented by the horizontal scale?

(c) About how many waterborne outbreaks were reported in the 1951-1955 period? In the 1971-74 period?

(a) As shown on the graph, the vertical scale represents the average annual number of waterborne disease outbreaks.

(b) The horizontal scale indicates the years in which those outbreaks occurred.

(c) As shown at point *A*, there were 10 outbreaks reported during the 1951-1955 period. At point *B* it is necessary to interpolate to determine that there were about 23 outbreaks reported during the 1971-1974 period.

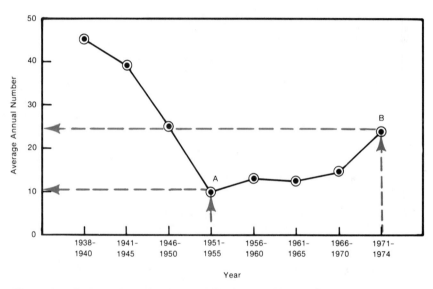

Figure 14.   Average Annual Number of Waterborne Disease Outbreaks, 1938-1974

Example 4

Given the graph in Figure 15, answer the following questions.

(a) If the staff gage on a 12-in. Parshall flume indicates a head of 18 in., what is the flow in million gallons per day?

(b) If the staff gage on a 12-in. Parshall flume indicates a head of 7 1/2 in., what is the flow in million gallons per day?

(c) If there is a head of 15 in. on a 12-in. Parshall flume, what is the flow in million gallons per day?

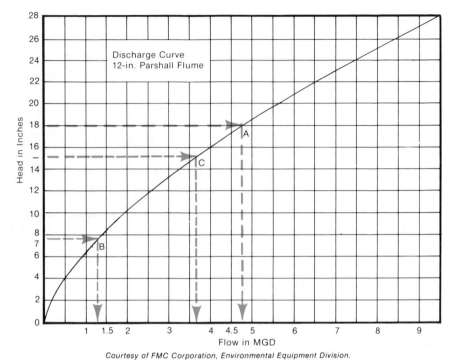

Courtesy of FMC Corporation, *Environmental Equipment Division.*

Figure 15.   Discharge Curve for 12-in. Parshall Flume

Note that, since all of the answers fall between numbers marked on the scale, interpolation is necessary to answer all of the questions.

(a) As shown by point *A* on the graph, the flow corresponding to a head of 18 in. would be halfway between *4.5* mgd and *5.0* mgd. By interpolation, the indicated flow is 4.75 mgd.

(b) Interpolation is again necessary to locate point *B* (7 1/2 in. head). At this head the flow would be 1.25 mgd.

(c) At a head of 15 in. (point *C*), the flow is somewhere between *3.5* mgd and *3.75* mgd. Let's take a closer look at the scale between *3* and *4* mgd to determine the indicated flow.

Halfway between *3* and *4* mgd is *3.5* mgd:

And *3.25* is halfway between *3.0* mgd and *3.5* mgd. Similarly, *3.75* mgd is halfway between *3.5* mgd and *4.0* mgd:

The question in this problem is, *what flow rate is indicated by the arrow?*

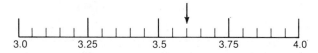

Notice that from *3.5* to *3.75*, the space has been divided into five parts, or fifths, and that the arrow lies 2/5 of the distance from *3.5* to *3.75*. The distance from *3.5* to *3.75* is

$$\begin{array}{r} 3.75 \\ -3.50 \\ \hline 0.25 \end{array}$$

And now 2/5 of the distance can be calculated:

$$\left(\frac{2}{5}\right)(0.25) = \frac{0.50}{5}$$
$$= 0.10$$

Note that 2/5 could also have been converted to a decimal and then multiplied to get the same answer:

$$(0.4)(0.25) = 0.10$$

The arrow lies *0.10* further on the scale than *3.5*:

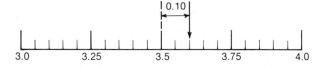

Therefore, the arrow must be at *3.5* + *0.10*, or *3.60* mgd. A head of 15 in. indicates a flow of 3.60 mgd.

Example 5

Given the graph in Figure 16, answer the following questions. (Notice that both the horizontal and vertical scales in this graph are logarithmic.)

(a) If the staff gage shows a head of 5 1/2 in. on a 90-deg **V**-notch weir, what is the flow in cubic feet per second?

(b) With a head of 0.6 ft on a 90-deg **V**-notch weir, what is the flow in cubic feet per second?

(c) Using the answer to part b, what is this flow expressed in gallons per day?

(d) If the staff gage shows a head of 1 1/2 in. on a 90-deg **V**-notch weir, what is the flow in cubic feet per second?

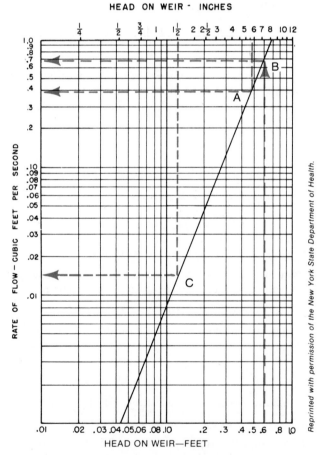

HEAD ON WEIR - INCHES

Reprinted with permission of the New York State Department of Health.

Figure 16.   Flow Rate Through 90-deg V-Notch Weir

(a) As shown by point *A*, a head of 5 1/2 in. on a 90-deg **V**-notch weir indicates a flow of 0.4 cfs.

(b) Point *B* shows that a head of 0.6 ft on a 90-deg **V**-notch weir indicates a flow of 0.7 cfs.

(c) Convert 0.7 cfs to gallons per day:[21]

$$(0.7 \text{ cfs})(60 \text{ sec/min})(1440 \text{ min/day})(7.48 \text{ gal/cu ft}) = 452{,}390 \text{ gpd}$$

(d) A head of 1 1/2 in. (point *C*) indicates the flow is somewhere between *0.01 cfs* and *0.02 cfs*. As in other example problems above, interpolation is needed to determine the indicated flow. However, interpolation in this problem is a little different.

In Example 4(c) the interpolation was on an *arithmetic scale* (the space between two known points was equally divided). This problem uses a *logarithmic scale,* so the space between two known points must be divided logarithmically. (There is a general discussion of logarithmic scales at the beginning of this section on graphs.)

Consider the part of the scale needed for this problem; that is, the section between *0.01* and *0.02* cfs:

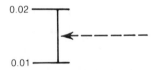

Estimate and mark off the logarithmic scale between the two known numbers:

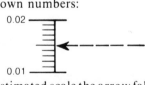

On the estimated scale the arrow falls on the halfway mark. The estimated scale begins at the point *0.01* (equal to 0.010) and is divided into ten parts, so the first subdivision line is *0.011*; the second is 0.012; and the halfway line is *0.015*. Therefore, a head of 1 1/2 in. on a 90-deg **V**-notch weir indicates a flow rate of approximately 0.015 cfs.

Sometimes a graph will have more than one line plotted on it. The lines may be interrelated or they may be unrelated lines that have been grouped on one graph for convenience. The next two problems involve such multi-lined graphs.

---

[21] Mathematics Section, Conversions (Flow Conversions).

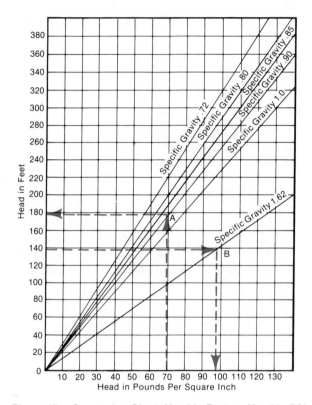

Figure 17.   Conversion Chart: Head in Feet vs. Head in PSI

Example 6

Use the graph in Figure 17 to answer the following questions:

(a) You wish to pump a particular solution which has a specific gravity[22] of 0.90. If you are pumping against a head of 70 psi, how many feet of head are you pumping against?

(b) Suppose you are pumping a liquid which has a specific gravity of 1.62. If you are pumping against a total head of 140 ft, how many pounds per square inch are you pumping against?

There are five data lines on this graph. Each is *unrelated* to the next except that each reflects data for a particular specific gravity.

---

[22] Hydraulics Section, Density and Specific Gravity.

(a) In this part of the problem you are interested only in the line depicting data for a specific gravity of 0.90. Point *A* on the graph indicates where 70 psi intersects the line labeled *Sp. Gr. 0.90*. Reading the scale directly to the left of this point indicates that 70 psi is equivalent to a head of 180 ft.

(b) For this part of the problem only the line relating data for a specific gravity of 1.62 is needed. Following the line across from *140 ft* to point *B* and from there downward indicates that a head of 140 ft is equivalent to a pressure between 90 and 100 psi. Since the arrow does not fall directly on a numbered line, interpolation must be used to determine the answer.

   The arrow falls somewhere between 95 and 100 psi, at about 98 psi. Therefore, for a liquid with a specific gravity of 1.62, a head of 140 ft is approximately equal to 98 psi.

In pump characteristics curves such as Figure 18A, four types of information are given by the scales:

- Pump capacity *Q*
- Pump head *H*
- Pump efficiency *E* (sometimes labeled η)
- Brake horsepower *P*

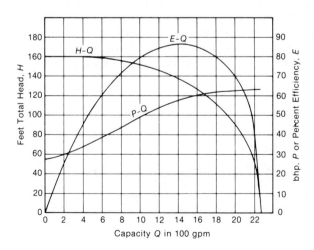

Figure 18A. Pump Characteristics Curve

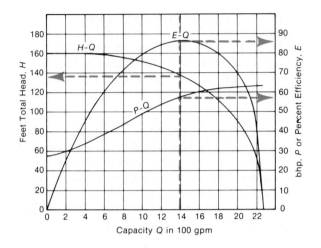

Figure 18B.   Pump Characteristics Curve, Capacity Known

The *H-Q* curve shows the relationship between head and capacity (flow rate); the *P-Q* curve shows the relationship between horsepower and capacity; and the *E-Q* curve shows the relationship between pump efficiency and capacity. These curves are referred to as

$$H\text{-}Q \text{ curve } = \text{ Head curve}$$

$$P\text{-}Q \text{ curve } = \text{ Power curve}$$

$$E\text{-}Q \text{ curve } = \text{ Efficiency curve}$$

By using the pump characteristics curves[23] you can determine *any three* characteristics when given information about the fourth. (Either the head or the capacity is normally the known characteristic.) The key to solving this kind of pump problem is to use the known characteristic to determine the location of a specific vertical line and to determine the points where it intersects all three pump curves.

Example 7

Use the graph in Figure 18A to answer the following questions:

(a) Assume the capacity of the pump to be 1400 gpm. Determine the brake horsepower (bhp), total head in feet, and efficiency of the pump at this capacity.

(b) Determine the capacity, total head, and efficiency of the pump at 50 bhp.

---

[23] Hydraulics Section, Pumping Problems (Pump Curves).

(a) In the first part of this example, the known characteristic is *capacity* (1400 gpm). How can head, efficiency, and power be determined? Of the four types of pump characteristics problems, this is probably the easiest to solve. First, draw a vertical line upward from *14* on the horizontal (capacity) scale (**Figure 18b**).

Note that the vertical line intersects each of the three curves, as indicated by the three dots. As you follow the vertical line upward, it first intersects the *P-Q* (power) curve. To determine what the power is at this point, move horizontally to the right toward the *bhp* scale. You should hit this scale at about 57 bhp.

Now look at the point where the vertical line (1400 gpm) intersects the *H-Q* (head) curve. To determine the head at this point, move horizontally to the left toward the *head* scale. You should hit this scale at about 137 ft.

Finally, look at the point where the vertical line (1400 gpm) intersects the *E-Q* (efficiency) curve. To determine the efficiency at this point, move horizontally to the right to the *efficiency* scale. The reading is about 86 percent.

Knowing just the pump capacity of 1400 gpm, you were able to determine the following characteristics:

$$\text{Brake horsepower } P = 57 \text{ bhp}$$

$$\text{Pump head } H = 137 \text{ ft}$$

$$\text{Pump efficiency } E = 86\%$$

(b) In the second part of this example, 50 bhp is the known characteristic. This problem is approached in much the same manner as part (a), above. Locate *50* on the *bhp* scale, then draw a horizontal line until you reach the *P-Q* (power) curve. This intersection point indicates where the vertical line should be drawn, as shown in Figure 18c.

The vertical line enables you to determine the three unknown characteristics. At its base on the *capacity* scale, the line falls somewhere between *10* and *12,* not quite halfway. Therefore you can estimate a capacity of about 1090 gpm.

As you move upward, the vertical line first intersects the *P-Q* curve. This point corresponds with 50 bhp, the information given in the problem.

As you move further upward, the vertical line intersects the *H-Q* curve. To determine the head value

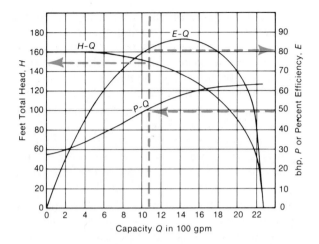

Figure 18c.   Pump Characteristics Curve, bhp Known

at this point, draw a line horizontally to the left toward the *head* scale. The indicated head is about 150 ft.

The last point on the vertical line is on the *E-Q* curve. Draw a line horizontally to the right, to the *efficiency* scale. The pump efficiency is about 82 percent.

Therefore, given a brake horsepower of 50, you were able to determine the following characteristics:

$$\text{Pump capacity } Q = 1090 \text{ gpm}$$

$$\text{Pump head } H = 150 \text{ ft}$$

$$\text{Pump efficiency } E = 82\%$$

Nomographs

A NOMOGRAPH is a graph in which three or more scales are used to solve mathematical problems. The most commonly used nomographs in the water treatment field are those which pertain to chemical feed rates. In such problems, usually the plant flow and desired chemical concentrations are known and the chemical feed rate must be determined. Let's look at a few example problems using nomographs.

Example 8

Assuming that a treatment plant flow is 600 gpm and the desired fluoride concentration is 0.9 ppm, use the nomograph in Figure 19 to answer the following questions.

   (a) How many pounds per day of sodium silico-fluoride (Na_2SiF_6) should be added to obtain the desired fluoride concentration?

(b) How many gallons per day of a 4 percent saturated solution of sodium fluoride (NaF) should be added to obtain the desired fluoride concentration?

First, locate one end of a straightedge at *600 gpm* (point *A*) on the nomograph. From there, draw a line through *0.9 ppm* (point *B*), extending the line until it intersects with the *fluoride ion* line (point *C*). Then, to obtain the pounds-per-day or gallons-per-day feed rates for each type of fluoride listed, draw a horizontal line from point *C* to the right.

(a) The indicated feed rate for Na_2SiF_6 (point *D*) is 11 lb/day.

(b) The rate at which a 4 percent saturated solution of NaF must be added to obtain the desired fluoride concentration is 44.3 gpd.

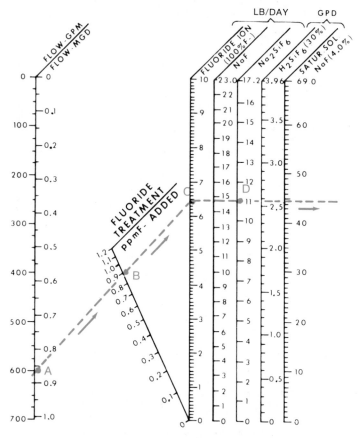

Figure 19. Fluoridation Nomograph

Example 9

A dosage of 2 grains per gallon (gpg) of alum liquor is required for treatment of 20 mgd water flow. Using the nomograph in Figure 20, determine how many gallons per day of alum liquor must be fed.

First, place one end of a straightedge at *2 gpg* on the dosage concentration scale (point *A*). From there, draw a straight line to *20 mgd* on the water flow scale (point *B*). Then from the middle scale determine the alum liquor feed rate (point *C*). In this case the feed rate is 45 gallons per *hour*. But the question asked for gallons per *day*.

Convert gallons per hour to gallons per day:

$$(45 \text{ gph})(24 \text{ hr/day}) = 1080 \text{ gpd}$$

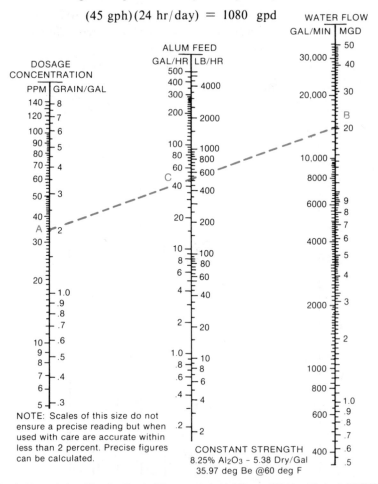

NOTE: Scales of this size do not ensure a precise reading but when used with care are accurate within less than 2 percent. Precise figures can be calculated.

Figure 20.   Alum Liquor Feed Nomograph

Constructing Simple Graphs

In summarizing water treatment data, the three types of graph most commonly used are bar, circle, and broken-line.

When constructing a bar graph, you should include the following components:

- A title
- Clearly labeled horizontal and vertical scales (except when constructing a divided bar graph).
- Descriptions of the bars (when needed for clarity)

The bars on the graph should be the same width and should be spaced an equal distance apart. Let's look at an example of constructing a bar graph.

Example 10

Given the following monthly turbidity data, construct a bar graph.

Month	Average Turbidity Value	Month	Average Turbidity Value
Jan	0.70	Jul	0.59
Feb	0.61	Aug	0.68
Mar	0.64	Sep	0.72
Apr	0.78	Oct	0.83
May	0.91	Nov	0.78
Jun	0.71	Dec	0.71

First the horizontal and vertical scales must be established. Since *time* is part of this graph, a *vertical bar graph* should be constructed with the time factor along the horizontal scale. There is no set rule for the distance between the bars. However, they should be of equal width and spaced a visually pleasing, equal distance apart.

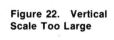

Figure 21. Vertical Scale Too Small

Figure 22. Vertical Scale Too Large

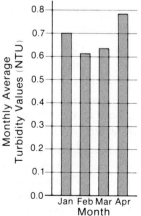

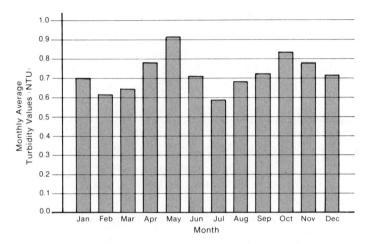

Figure 23.   Completed Bar Graph for Example 10

In this example it is especially important to give some thought to the spacing on the vertical scale. This scale will determine the lengths of the bars. As shown in Figure 21, if the scale is too small all the bars will be short, with little distinguishable difference between the bar heights. If the scale is too large, as in Figure 22, bars will be longer than necessary.

Once an appropriate vertical scale has been selected, the bar graph can be completed. As you can see from Figure 23, although some of the details of the data cannot be determined (for instance, it would be difficult to distinguish between a bar height of 0.77 and 0.78), it is much easier to see at a glance the trend in turbidity values over the 12-month period.

When constructing a circle graph, you should include the following components:

- A title
- A description of each part of the circle, including the percentage it represents

The most important thing you will need to know in constructing a circle graph is how to determine the correct size for each part. The following examples illustrate how the size of each part is determined.

Example 11

Suppose you are given information that, of all accidents, 65 percent are due to falls, 25 percent are due to burns, bruises, and blows, and 10 percent are due to all other causes. Construct a cricle graph depicting this information.

First you must determine what size to make each part. A full circle is 360 deg. Each part, therefore, will represent a percentage[24] of the 360 deg:

$(0.65)(360°) = 234°$ (represents falls)

$(0.25)(360°) = 90°$ (represents burns, bruises, and blows)

$(0.10)(360°) = 36°$ (represents all other causes)

$\overline{360°}$ (total)

Draw these angles with a protractor.

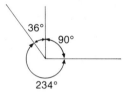

And then draw the circle and label the graph (Figure 24).

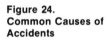

**Figure 24.
Common Causes of
Accidents**

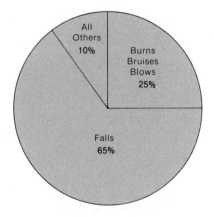

Example 12

In 1975 the national average water use was 414 billion gallons per day (bgd). Suppose that 29 bgd were for public water supplies, 140 bgd for irrigation, 5 bgd for rural uses, and 240 bgd for self-supplied industry. Report this information in terms of a circle graph.

First determine the percent of the whole that each part represents:

---

[24]Mathematics Section, Percent.

$$\frac{29 \text{ bgd public water supplies}}{414 \text{ bgd total}} \times 100 = 7\% \quad \text{(public water supplies)}$$

$$\frac{140 \text{ bgd irrigation}}{414 \text{ bgd total}} \times 100 = 34\% \quad \text{(irrigation)}$$

$$\frac{5 \text{ bgd rural uses}}{414 \text{ bgd total}} \times 100 = 1\% \quad \text{(rural uses)}$$

$$\frac{240 \text{ bgd self-supplied industry}}{414 \text{ bgd total}} \times 100 = \underline{58\%} \quad \text{(self-supplied industry)}$$
$$100\%$$

Once the percentages have been established, calculate the angle that each part represents:

$$(0.07)(360°) = 25° \quad \text{(public water supplies)}$$
$$(0.34)(360°) = 122° \quad \text{(irrigation)}$$
$$(0.01)(360°) = 4° \quad \text{(rural uses)}$$
$$(0.58)(360°) = \underline{209°} \quad \text{(self-supplied industry)}$$
$$360° \quad \text{(total)}$$

From this information draw the angles with a protractor:

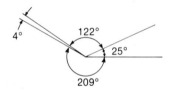

And then draw the circle and label the graph (Figure 25).

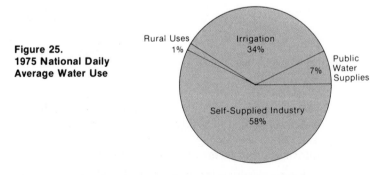

**Figure 25.
1975 National Daily
Average Water Use**

When constructing a broken-line graph, you should include the following components:

- A title
- At least one vertical and one horizontal scale, clearly labeled

Plot each data point on the graph and draw a connecting line between the points. The following example illustrates the construction of a broken-line graph.

Example 13

Given the following average monthly flows, construct a broken-line graph.

Month	Average Monthly Flow in mgd	Month	Average Monthly Flow in mgd
Jan	1.6 mgd	Jul	6.3 mgd
Feb	1.4 mgd	Aug	5.7 mgd
Mar	2.0 mgd	Sep	3.8 mgd
Apr	2.6 mgd	Oct	2.8 mgd
May	4.5 mgd	Nov	2.4 mgd
Jun	5.4 mgd	Dec	1.8 mgd

As in constructing a bar graph, you must first select a suitable scale. The scale should be one that is neither blown out of proportion nor unnecessarily compressed. The same basic principles apply as those discussed with respect to bar graph scales.

Since *time* is part of this graph, the time information should be placed along the horizontal scale. *Flow in mgd,* therefore, will go along the vertical scale. Then plot the points on the graph and draw the connecting lines (Figure 26). Notice that the small dots on the graph are circled so that data points are not obscured by the lines connecting them.

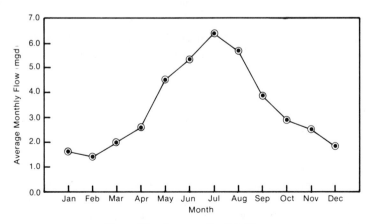

Figure 26. Average Monthly Flow

M12-2. Tables

For the most part, tables that you will use in water treatment plant operations are plainly marked and need no explanation regarding use. However, the use of a table having numerical scales for column and row headings can be clarified by example. This type of table is more complex than most others you may encounter.

Example 14
Suppose the staff gage measuring the depth of flow over a 90-deg **V**-notch weir indicates a head of 1.35 ft. Using Table 2, determine the flow rate in cubic feet per second and million gallons per day.

To read Table 2 you must know that part of the head is read on the vertical, or side, scale and part on the horizontal, or top, scale. In this problem, 1.3 ft of the head is read on the side scale, while the remaining portion, 0.05 ft is read on the top scale. These two readings pinpoint the cubic-feet-per-second and million-gallons-per-day discharge from the weir. Moving to the right of the *1.3* reading until you reach the column headed *0.05,* you can see that the indicated discharge is 5.29 cfs or 3.42 mgd.

Let's look at two more head readings on Table 2 to illustrate how the scales should be used in determining discharge. Suppose the staff gage indicated a head of 0.88 ft. Where on the top and side scales would this be read?

$$0.88 = 0.8 + 0.08$$
$$\text{side} \quad \text{top}$$
$$\text{scale} \quad \text{scale}$$

The flow for 0.88 ft head is shown on the table as 1.82 cfs, or 1.17 mgd.
How would the reading be located if the staff gage indicated a head of 1.9 ft?

$$1.9 = 1.90 = 1.9 + 0.00$$
$$\text{side} \quad \text{top}$$
$$\text{scale} \quad \text{scale}$$

(The flow for 1.9 ft head is shown as 12.4 cfs, or 8.04 mgd.)

Example 15
Assume that you have a Parshall flume with a 3-in. throat width, and a staff gage water depth reading of 0.62 ft. Using Table 3, determine the cubic-foot-per-second (cfs) and million-gallons-per-day (mgd) flow rates.

Table 2. Discharge of 90-deg V-Notch Weirs*

Head	.00		.01		.02		.03		.04		.05		.06		.07		.08		.09	
Ft.	CFS	MGD	CFS	MGD	CFS	MGD	CFS	MGD	CFS	MGD	CFS	MGD	CFS	MGD	CFS	MGD	CFS	MGD	CFS	MGD
0.1	.008	.005	.010	.006	.012	.008	.015	.010	.018	.012	.022	.014	.026	.017	.030	.019	.034	.022	.039	.025
0.2	.045	.029	.051	.033	.057	.037	.063	.041	.071	.046	.078	.050	.086	.056	.095	.061	.104	.067	.113	.073
0.3	.123	.080	.134	.086	.145	.094	.156	.101	.169	.109	.181	.117	.194	.126	.208	.135	.223	.144	.237	.153
0.4	.253	.163	.269	.174	.286	.185	.303	.196	.321	.207	.340	.219	.359	.232	.379	.245	.399	.258	.420	.272
0.5	.442	.286	.464	.300	.487	.315	.511	.330	.536	.346	.561	.362	.587	.379	.613	.396	.640	.414	.668	.432
0.6	.697	.451	.727	.470	.757	.489	.788	.509	.819	.529	.852	.550	.885	.579	.919	.594	.953	.616	.989	.639
0.7	1.02	.662	1.06	.686	1.10	.711	1.14	.736	1.18	.761	1.22	.787	1.26	.814	1.30	.841	1.34	.868	1.39	.896
0.8	1.43	.925	1.48	.954	1.52	.984	1.57	1.01	1.62	1.04	1.66	1.08	1.71	1.11	1.76	1.14	1.82	1.17	1.87	1.21
0.9	1.92	1.24	1.97	1.28	2.03	1.31	2.08	1.35	2.14	1.38	2.20	1.42	2.26	1.46	2.32	1.50	2.38	1.54	2.44	1.58
1.0	2.50	1.62	2.56	1.66	2.63	1.70	2.69	1.74	2.76	1.78	2.82	1.82	2.89	1.87	2.96	1.91	3.03	1.96	3.10	2.00
1.1	3.17	2.05	3.24	2.10	3.32	2.14	3.39	2.19	3.47	2.24	3.55	2.29	3.62	2.34	3.70	2.39	3.78	2.44	3.86	2.50
1.2	3.94	2.55	4.03	2.60	4.11	2.66	4.19	2.71	4.28	2.77	4.37	2.82	4.45	2.88	4.54	2.94	4.63	2.99	4.72	3.05
1.3	4.82	3.11	4.91	3.17	5.00	3.23	5.10	3.30	5.20	3.36	5.29	3.42	5.39	3.48	5.49	3.55	5.59	3.61	5.69	3.68
1.4	5.80	3.75	5.90	3.81	6.01	3.88	6.11	3.95	6.22	4.02	6.33	4.09	6.44	4.16	6.55	4.23	6.66	4.30	6.77	4.38
1.5	6.89	4.45	7.00	4.53	7.12	4.60	7.24	4.68	7.36	4.75	7.48	4.83	7.60	4.91	7.72	4.99	7.84	5.07	7.97	5.15
1.6	8.09	5.23	8.22	5.31	8.35	5.40	8.48	5.48	8.61	5.56	8.74	5.65	8.88	5.74	9.01	5.82	9.15	5.91	9.28	6.00
1.7	9.42	6.09	9.56	6.18	9.70	6.27	9.84	6.36	9.98	6.45	10.1	6.55	10.3	6.64	10.4	6.73	10.6	6.83	10.7	6.93
1.8	10.9	7.02	11.0	7.12	11.2	7.22	11.3	7.32	11.5	7.42	11.6	7.52	11.8	7.62	11.9	7.73	12.1	7.83	12.3	7.93
1.9	12.4	8.04	12.6	8.15	12.7	8.25	12.9	8.36	13.1	8.47	13.3	8.58	13.4	8.69	13.6	8.80	13.8	8.91	14.0	9.03
2.0	14.1	9.14	14.3	9.25	14.5	9.37	14.7	9.49	14.9	9.60	15.0	9.72	15.2	9.84	15.4	9.96	15.6	10.1	15.8	10.2

*Reprinted with permission of Leupold & Stevens, Inc., P.O. Box 688, Beaverton, OR 97005, from Stevens Water Resources Book. Formula: $CFS = 2.50\ H^{5/2}$ and $MGD = CFS \times 0.646$.

As in the last problem, this table has both vertical and horizontal scales. However, the scales here are somewhat different. The vertical scale shows head while the horizontal scale shows the Parshall flume throat width. Once these two variables are known, the corresponding flow can be determined.

In this problem the head is given as 0.62 ft and the throat width as 3 in. Move to the right of *0.62 ft* until you come under the *3-in.* column heading. The indicated discharge is 0.474 cfs, or 0.306 mgd.

Table 3.   Flow Rate Through Parshall Flumes*

Head	DISCHARGE THROUGH THROAT WIDTH, W, OF —													
	1 in.		2 in.		3 in.		6 in.		9 in.		12 in.		18 in.	
Ft.	CFS	MGD	CFS	MGD	CFS	MGD	CFS	MGD	CFS	MGD	CFS	MGD	CFS	MGD
.46	.101	.065	.203	.131	.299	.193	.61	.39	.94	.61	1.23	.79	1.82	1.18
.47	.105	.068	.210	.136	.309	.200	.63	.41	.97	.63	1.27	.82	1.88	1.22
.48	.108	.070	.217	.140	.319	.206	.65	.42	1.00	.65	1.31	.85	1.94	1.25
.49	.112	.072	.224	.145	.329	.213	.67	.43	1.03	.67	1.35	.87	2.00	1.29
.50	.115	.074	.230	.149	.339	.219	.69	.45	1.06	.69	1.39	.90	2.06	1.33
.51	.119	.077	.238	.154	.350	.226	.71	.46	1.10	.71	1.44	.93	2.13	1.38
.52	.123	.079	.245	.158	.361	.233	.73	.47	1.13	.73	1.48	.96	2.19	1.42
.53	.126	.081	.253	.164	.371	.240	.76	.49	1.16	.75	1.52	.98	2.25	1.45
.54	.130	.084	.260	.168	.382	.247	.78	.50	1.20	.78	1.57	1.01	2.32	1.50
.55	.134	.087	.268	.173	.393	.254	.80	.52	1.23	.79	1.62	1.05	2.39	1.54
.56	.138	.089	.275	.178	.404	.261	.82	.53	1.26	.81	1.66	1.07	2.45	1.58
.57	.141	.091	.283	.183	.415	.268	.85	.55	1.30	.84	1.70	1.10	2.52	1.63
.58	.145	.094	.290	.187	.427	.276	.87	.56	1.33	.86	1.75	1.13	2.59	1.67
.59	.149	.096	.298	.193	.438	.283	.89	.58	1.37	.89	1.80	1.16	2.66	1.72
.60	.153	.099	.306	.198	.450	.291	.92	.59	1.40	.90	1.84	1.19	2.73	1.76
.61	.157	.101	.314	.203	.462	.299	.94	.61	1.44	.93	1.88	1.22	2.80	1.81
.62	.161	.104	.322	.208	.474	.306	.97	.63	1.48	.96	1.93	1.25	2.87	1.85
.63	.165	.107	.330	.213	.485	.313	.99	.64	1.51	.98	1.98	1.28	2.95	1.91
.64	.169	.109	.338	.218	.497	.321	1.02	.66	1.55	1.00	2.03	1.31	3.02	1.95
.65	.173	.112	.347	.224	.509	.329	1.04	.67	1.59	1.03	2.08	1.34	3.09	2.00

*Reprinted with permission of Leupold & Stevens, Inc., P.O. Box 688, Beaverton, OR 97005, from *Stevens Water Resources Book.* Formula: $CFS = 2.50\ H^{5.2}$ and $MGD = CFS \times 0.646$.

Review Questions

1. Use the graph in Figure 27A to answer the following questions:

   (a) What is the maximum combined chlorine residual (in milligrams per litre) shown on the graph?

   (b) As shown on the graph, what happens when 0.5 to 0.8 mg/L of chlorine is added to the water?

   (c) How much chlorine must be added before free available chlorine residual begins to form?

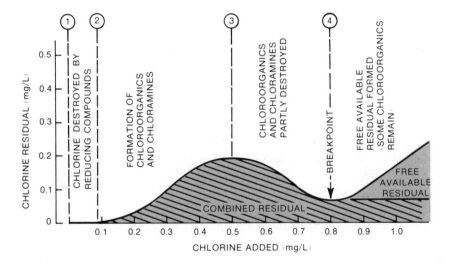

Figure 27A.   Breakpoint Chlorination

2. The graph shown in Figure 12 of the text illustrates typical jar test results when aluminum sulfate is used as a coagulant for a water supply. According to this graph, what is the relationship between turbidity and alum dose?

3. Use the 7-day circular recorder chart shown in Figure 13 of the text to determine the flow rate in gallons per minute at

   (a) 9 PM Friday

   (b) 9 AM Sunday

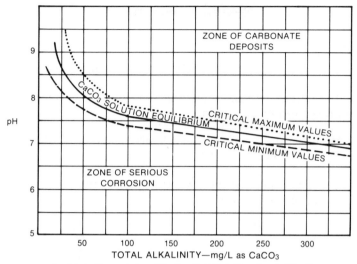

Reprinted with permission of the New York State Department of Health.

Figure 28A.   Relationship of pH vs. Alkalinity

4. The graph in Figure 28A shows the relationship between pH and alkalinity. Based on this graph, answer the following questions:

   (a) What part of the graph indicates water that would neither cause carbonate deposits nor create serious corrosion?

   (b) If a water supply were found to have 100 mg/L alkalinity and a pH of 7.5, would you expect carbonate deposits or corrosion in the distribution system?

   (c) If a water supply has an alkalinity of 140 mg/L and a pH of 6.75, would you expect this water to be a source of carbonate deposits or corrosion?

5. You are using a pump with the characteristics shown in the graph of Figure 18A in the text, and you are pumping against a 95-ft total head. What is the pump capacity, bhp, and efficiency?

6. A dosage of 12 ppm of dry aluminum sulfate is required for a treatment plant flow of 43 mgd. Use the nomograph of Figure 29A to determine the pounds per day of dry alum needed.

7. A cylindrical tank has a diameter of 30 ft and the depth of water in the tank is 5 ft. Use the nomograph of Figure 30A to determine how many gallons of water are in the tank.

8. A staff gage measuring the depth of flow over a 60-deg **V**-notch weir indicates a head of 0.84 ft. Use Table 4 to determine the flow rate in cubic feet per second and million gallons per day.

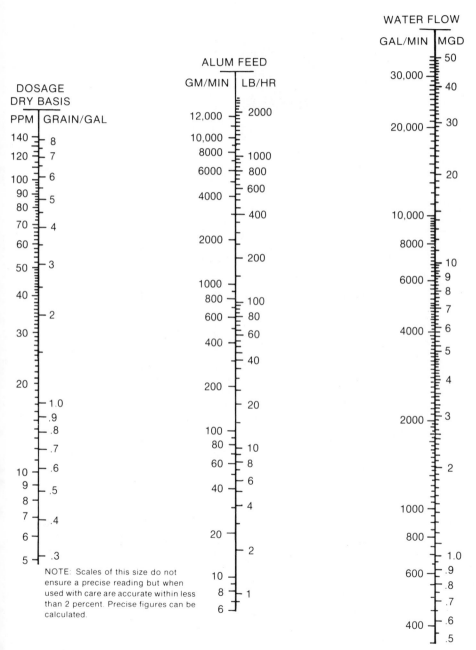

NOTE: Scales of this size do not ensure a precise reading but when used with care are accurate within less than 2 percent. Precise figures can be calculated.

Figure 29A.   Dry Alum Feed Nomograph

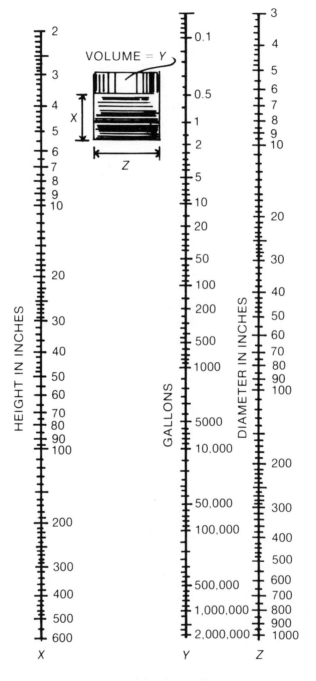

Reprinted with permission of Diesel Equipment Superintendent Magazine, 22 S. Smith St., Norwalk, CT 06856, ©Diesel Power Magazine, Nov. 1940.

Figure 30A.   Capacity of Cylindrical Tanks

Table 4. Discharge of 60-deg V-Notch Weirs*

Head	.00		.01		.02		.03		.04		.05		.06		.07		.08		.09	
Ft.	CFS	MGD	CFS	MGD	CFS	MGD	CFS	MGD	CFS	MGD	CFS	MGD	CFS	MGD	CFS	MGD	CFS	MGD	CFS	MGD
0.1	.005	.003	.006	.004	.007	.005	.009	.006	.011	.007	.013	.008	.015	.010	.017	.011	.020	.013	.023	.015
0.2	.026	.017	.029	.019	.033	.021	.037	.024	.041	.026	.045	.029	.050	.032	.055	.035	.060	.039	.065	.042
0.3	.071	.046	.077	.050	.084	.054	.090	.058	.097	.063	.105	.068	.112	.073	.120	.078	.128	.083	.137	.089
0.4	.146	.094	.155	.100	.165	.107	.175	.113	.185	.120	.196	.127	.207	.134	.219	.141	.230	.149	.243	.157
0.5	.255	.165	.268	.173	.281	.182	.295	.191	.309	.200	.324	.209	.339	.219	.354	.229	.370	.239	.386	.249
0.6	.402	.260	.419	.271	.437	.282	.455	.294	.473	.306	.492	.318	.511	.330	.530	.343	.550	.356	.571	.369
0.7	.592	.382	.613	.396	.635	.410	.657	.425	.680	.439	.703	.454	.727	.470	.751	.485	.775	.501	.801	.517
0.8	.826	.534	.852	.551	.879	.568	.906	.585	.933	.603	.961	.621	.990	.640	1.02	.658	1.05	.678	1.08	.697
0.9	1.11	.717	1.14	.737	1.17	.757	1.20	.778	1.24	.799	1.27	.820	1.30	.842	1.34	.864	1.37	.887	1.41	.910
1.0	1.44	.933	1.48	.956	1.52	.980	1.55	1.00	1.59	1.03	1.63	1.05	1.67	1.08	1.71	1.10	1.75	1.13	1.79	1.16
1.1	1.83	1.18	1.87	1.21	1.92	1.24	1.96	1.27	2.00	1.29	2.05	1.32	2.09	1.35	2.14	1.38	2.18	1.41	2.23	1.44
1.2	2.28	1.47	2.32	1.50	2.37	1.53	2.42	1.56	2.47	1.60	2.52	1.63	2.57	1.66	2.62	1.69	2.67	1.73	2.73	1.76
1.3	2.78	1.80	2.83	1.83	2.89	1.87	2.94	1.90	3.00	1.94	3.06	1.97	3.11	2.01	3.17	2.05	3.23	2.09	3.29	2.12
1.4	3.35	2.16	3.41	2.20	3.47	2.24	3.53	2.28	3.59	2.32	3.65	2.36	3.72	2.40	3.78	2.44	3.85	2.48	3.91	2.53
1.5	3.98	2.57	4.04	2.61	4.11	2.66	4.18	2.70	4.25	2.74	4.31	2.79	4.39	2.83	4.46	2.88	4.53	2.93	4.60	2.97
1.6	4.67	3.02	4.75	3.07	4.82	3.12	4.89	3.16	4.97	3.21	5.05	3.26	5.12	3.31	5.20	3.36	5.28	3.41	5.36	3.46
1.7	5.43	3.51	5.52	3.57	5.60	3.62	5.68	3.67	5.76	3.72	5.85	3.78	5.93	3.83	6.01	3.89	6.10	3.94	6.19	4.00
1.8	6.27	4.05	6.36	4.11	6.45	4.17	6.54	4.22	6.63	4.28	6.72	4.34	6.81	4.40	6.90	4.46	6.99	4.52	7.09	4.58
1.9	7.18	4.64	7.28	4.70	7.37	4.76	7.47	4.83	7.56	4.89	7.66	4.95	7.76	5.02	7.86	5.08	7.96	5.14	8.06	5.21
2.0	8.16	5.28	8.27	5.34	8.37	5.41	8.47	5.48	8.58	5.54	8.68	5.61	8.79	5.68	8.90	5.75	9.00	5.82	9.11	5.89

*Reprinted with permission of Leupold & Stevens, Inc., P.O. Box 688, Beaverton, OR 97005, from Stevens Water Resources Book. Formula: $CFS = 1.43\ H^{5/2}$ and $MGD = CFS \times 0.646$.

Table 5. Flow Through Rectangular Weirs With End Contractions*

| | | | | | LENGTH OF WEIR CREST IN FEET | | | | | | |
| | 1 | | 1½ | | 2 | | 3 | | 4 | | 5 | |
Ft.	CFS	MGD	CFS	MGD	CFS	MGD	CFS	MGD	CFS	MGD	CFS	MGD
.01	.003	.002	.005	.003	.007	.004	.010	.006	.013	.009	.016	.011
.02	.009	.006	.014	.009	.019	.012	.028	.018	.038	.024	.047	.030
.03	.017	.011	.026	.017	.034	.022	.052	.034	.069	.045	.086	.056
.04	.026	.017	.040	.026	.053	.034	.080	.052	.106	.069	.133	.086
.05	.037	.024	.055	.036	.074	.048	.111	.072	.148	.096	.186	.120
.06	.048	.031	.073	.047	.097	.063	.146	.094	.195	.126	.244	.158
.07	.061	.039	.091	.059	.123	.079	.184	.119	.246	.159	.308	.199
.08	.074	.048	.112	.072	.149	.097	.225	.145	.300	.194	.375	.243
.09	.088	.057	.133	.086	.178	.115	.268	.173	.358	.231	.448	.289
.10	.103	.067	.155	.101	.208	.135	.313	.203	.418	.271	.523	.339
.11	.119	.077	.178	.116	.241	.155	.363	.234	.485	.312	.607	.390
.12	.135	.087	.204	.132	.273	.177	.411	.266	.549	.355	.687	.445
.13	.152	.098	.230	.149	.308	.199	.464	.300	.620	.401	.776	.501
.14	.169	.110	.256	.166	.343	.222	.517	.335	.691	.447	.865	.560
.15	.188	.121	.285	.184	.382	.246	.576	.372	.770	.496	.964	.621
.16	.206	.133	.313	.202	.419	.271	.632	.408	.845	.546	1.058	.684
.17	.225	.146	.342	.221	.458	.296	.691	.447	.924	.598	1.157	.748
.18	.245	.158	.372	.240	.499	.323	.753	.487	1.007	.651	1.261	.815
.19	.266	.171	.404	.260	.542	.349	.818	.527	1.094	.706	1.370	.884
.20	.286	.185	.435	.281	.584	.377	.882	.569	1.180	.762	1.478	.954
.21	.308	.198	.468	.302	.629	.405	.950	.612	1.271	.819	1.592	1.026
.22	.329	.212	.501	.323	.673	.434	1.017	.656	1.361	.878	1.705	1.100
.23	.350	.226	.534	.345	.717	.463	1.084	.701	1.451	.938	1.818	1.175
.24	.373	.241	.569	.367	.765	.494	1.157	.746	1.549	.999	1.941	1.252
.25	.395	.255	.603	.390	.811	.524	1.227	.793	1.643	1.062	2.059	1.330
.26	.419	.270	.640	.413	.860	.555	1.303	.840	1.745	1.125	2.187	1.410
.27	.442	.285	.675	.436	.909	.587	1.376	.889	1.843	1.190	2.310	1.492
.28	.465	.301	.712	.460	.958	.619	1.451	.938	1.944	1.256	2.437	1.575
.29	.489	.316	.750	.484	1.009	.652	1.529	.988	2.049	1.324	2.569	1.659
.30	.514	.332	.788	.509	1.061	.685	1.608	1.039	2.155	1.392	2.702	1.745
.31	.539	.348	.827	.534	1.114	.719	1.689	1.090	2.264	1.461	2.839	1.832
.32	.564	.364	.866	.559	1.167	.754	1.770	1.143	2.373	1.532	2.976	1.921
.33	.589	.381	.905	.585	1.220	.788	1.851	1.196	2.482	1.603	3.113	2.011
.34	.651	.397	.945	.610	1.275	.824	1.935	1.250	2.595	1.676	3.225	2.102
.35	.642	.414	.987	.637	1.332	.859	2.022	1.304	2.712	1.750	3.402	2.195

*Reprinted with permission of Leupold & Stevens, Inc., P.O. Box 688, Beaverton, OR 97005, from *Stevens Water Resources Book*. Formula: $CFS = 0.666 (5L - H)H^{3/2}$ and $MGD = CFS \times 0.646$.

9. Suppose that contracted rectangular weirs (weirs with end contractions) are used to measure flow at your water treatment plant. The weir crest is 3 ft long and the staff gage indicates that water is flowing to a depth of 0.25 ft. Use Table 5 to determine the flow rate in million gallons per day (mgd).

10. Construct a bar graph illustrating the monthly chlorine-use data below.

Month	Chlorine Used (lb)	Month	Chlorine Used (lb)
Jan	5,200	Jul	16,500
Feb	3,400	Aug	12,600
Mar	4,500	Sep	9,200
Apr	5,500	Oct	6,100
May	7,200	Nov	5,900
Jun	15,400	Dec	4,400

11. The allocation of tax dollars for a community is shown below. Construct a circle graph depicting this information.

Schools	$13,500,000
Streets	1,687,500
Water	2,250,000
Wastewater Disposal	1,125,000
Refuse Disposal	1,125,000
Administration	2,250,000
Miscellaneous	562,500

12. The amount of polymer (rounded to the nearest 5 lb) used each month during 1977 is given below. Construct a broken-line graph depicting this information.

Month	Polymer used (lb)	Month	Polymer used (lb)
Jan	90	Jul	530
Feb	95	Aug	495
Mar	110	Sep	220
Apr	160	Oct	115
May	480	Nov	445
Jun	555	Dec	230

Summary Answers

1.   (a) The highest point combined chlorine residual reaches is 0.2 mg/L.

    (b) Some of the chloroorganics and chloramines are destroyed, resulting in a decrease in the chlorine residual level.

    (c) Free available chlorine begins to form when 0.8 mg/L of chlorine has been added to the water.

2. The addition of alum up to about 35 mg/L produces a significant reduction in turbidity. Beyond an alum dose of 35 mg/L, however, a very large increase in chemical dosage is needed to produce a small increase in turbidity removal.

3.   (a) 4000 gpm

    (b) Approximately 1650 gpm

4.   (a) The area on the graph between the dotted line (critical maximum values) and the dashed line (critical minimum values).

    (b) According to the graph, the water would not cause carbonate deposits nor result in serious corrosion.

    (c) According to the graph, the water would be a source of serious corrosion.

5. Pump capacity Q = 1950 gpm
Brake horsepower P = 63 bhp
Pump efficiency E = 74%

6. Approximately 4320 lb/day

7. Approximately 27,000 gal

8. 0.933 cfs or 0.603 mgd

9. 0.793 mgd

10. (See Detailed Answers for graph.)

11. (See Detailed Answers for graph.)

12. (See Detailed Answers for graph.)

Detailed Answers

1. (a) As indicated by point A on Figure 27B, the highest value that combined chlorine residual reaches is 0.2 mg/L.

    (b) When 0.5 to 0.8 mg/L (point 3 to point 4) of chlorine is added to the water, some of the chloroorganics and chloramines are destroyed, resulting in a decrease in the combined chlorine residual level.

    (c) Free available chlorine residual begins to form when 0.8 mg/L of chlorine has been added to the water (point 4 on the graph).

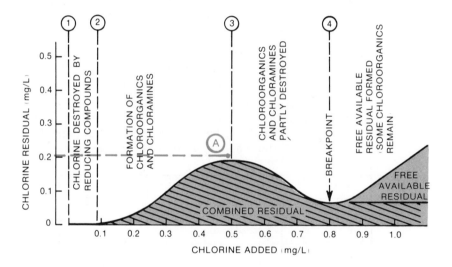

Figure 27B.   Breakpoint Chlorination

2. As shown in the graph of Figure 12, as the alum dose increases from 0 to 20 mg/L, the turbidity level drops drastically. Then from a dose of 20 to 30 mg/L alum, the turbidity continues to drop but not quite as fast as it did from 0 to 20 mg/L alum dose. Past about 35 mg/L alum dose there is very little decline in turbidity. In other words, beyond a dose of 35 mg/L, it takes a very large increase in chemical dosage to produce a small increase in turbidity removal. Therefore, 35 mg/L should be considered the optimal dose for this water.

3. (a) As shown in the graph (Figure 13 of text), the number on the circular chart recorder which corresponds with 9 PM on Friday is 40.

The center of the graph indicates that the flow chart is measured in *hundreds of gpm.* Therefore, a reading of 40 on the chart indicates a flow rate of

$$(40)(100 \text{ gpm}) = 4000 \text{ gpm}$$

(b) As seen on the graph, the number on the chart that corresponds with 9 AM on Sunday is approximately *16.5.* (The number read from the chart may vary from 16.3 to 16.6 because the precise position of the line must be estimated.)

Because the chart is measured in *hundreds of gallons,* a reading of 16.5 on the chart indicates a flow rate of

$$(16.5)(100 \text{ gpm}) = 1650 \text{ gpm}$$

4. (a) The area on the graph (Figure 28B) that falls *between the dotted line* (critical maximum values) *and the dashed line* (critical minimum values) indicates a water that does not cause carbonate deposits and is not seriously corrosive.

(b) Point *A* on the graph corresponds with an alkalinity of 100 mg/L and pH of 7.5. This water falls within the "safe" zone. Therefore the water would not cause carbonate deposits or result in serious corrosion.

(c) Since an alkalinity of 140 is not actually given on the scale, the scale divisions between *100* and *150* are first estimated. A pH of 6.75 falls halfway between 6.5 and 7.0. The two lines intersect at point *B* and indicate that this water would be a source of serious corrosion.

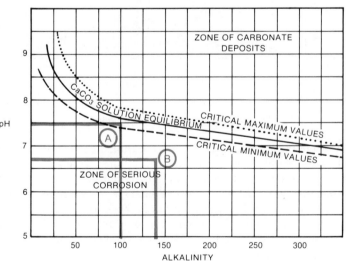

Reprinted with permission of the New York State Department of Health.

Figure 28B.   Relationship of pH vs. Alkalinity

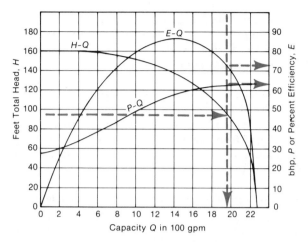

Figure 18D.   Pump Characteristics Curve

5. As shown in Figure 18D, a head of 95 ft is located on the head scale; then a horizontal line is drawn until it intersects the head (*H-Q*) curve. This point indicates where the vertical line should be drawn.

   Using this vertical line, you can determine that the pump characteristics are:

$$\text{Pump capacity } Q = 1950 \text{ gpm}$$

$$\text{Brake horsepower } P = 63 \text{ bhp}$$

$$\text{Pump efficiency } E = 74\%$$

6. First, one end of a straightedge is placed at *12 ppm* (point *A* on Figure 29B, page 206) on the dosage concentration scale. A straight line is then drawn from 12 ppm over to *43 mgd* on the water flow scale (point *B*). Now from the middle scale (point *C*), the dry alum feed rate can be determined. Since you want the answer in pounds per day, read the side of the scale that gives the feed rate in pounds: about 180 lb/hr. Then convert this feed rate to pounds per day:

$$(180 \text{ lb/hr}) (24 \text{ hr/day}) = 4320 \text{ lb/day}$$

7. Since both height and diameter scales on the nomograph (Figure 30B) are given in inches, before the nomograph can be used, the height and diameter in feet must be converted to inches:[25]

   Height:
   $$(5 \text{ ft}) (12 \text{ in./ft}) = 60 \text{ in.}$$

   Diameter:
   $$(30 \text{ ft}) (12 \text{ in./ft}) = 360 \text{ in.}$$

---

[25]Mathematics Section, Conversions (Linear Measurement).

To use the nomograph, first place one end of the straightedge at *60* on the height scale (point *A* in Figure 30B). From point *A*, draw a straight line over to *360* on the diameter scale (point *C*), determine the gallons of water contained in the tank: about 27,000 gal.

8. On Table 4, part of the head in feet (0.8) is given by the vertical or side scale and part (0.04) is given by the horizontal or top scale. These two readings are then used to pinpoint the cubic feet per second and/or million gallons per day discharge from the weir. Move to the right of the *0.8* reading until you come under the *0.04* column heading. The indicated discharge is 0.933 cfs, or 0.603 mgd.

9. In Table 5, the vertical scale shows head in feet while the horizontal scale pertains to the length of the weir crest in feet. Since in this problem these two variables are known, the corresponding flow can be determined. Move to the right of the *0.25* reading until you come under the *3* column heading. The indicated flow rate is 0.793 mgd.

10. First establish the horizontal and vertical scales. Since *time* is part of this graph, a vertical bar graph should be constructed with *time* on the horizontal scale. Then construct the bars, of equal width and spaced an equal distance apart, as shown in Figure 31.

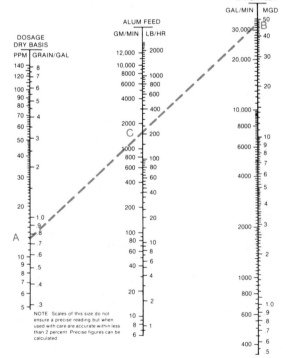

Figure 29B.
Dry Alum Feed
Nomograph

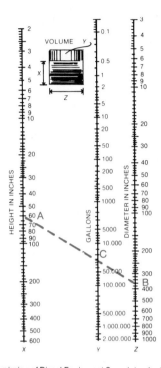

Reprinted with permission of Diesel Equipment Superintendent Magazine, 22 S. Smith St., Norwalk, CT 06856, ©Diesel Power Magazine, Nov. 1940.

Figure 30B. Capacity of Cylindrical Tanks

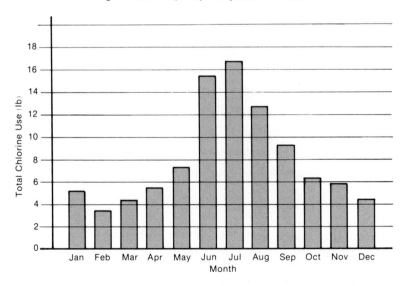

Figure 31. Chlorine Usage, by Month

11. To construct the circle graph, follow these steps:
    - Determine the percent of the whole each part represents
    - Determine the angle each part represents
    - Draw the angles and circle
    - Label the graph

First determine the percent of the whole which each part represents:

$$\frac{\$13,500,000 \text{ schools}}{\$22,500,000 \text{ total}} \times 100 = 60\% \text{ for schools}$$

$$\frac{\$1,687,500 \text{ streets}}{\$22,500,000 \text{ total}} \times 100 = 7.5\% \text{ for streets}$$

$$\frac{\$2,250,000 \text{ water}}{\$22,500,000 \text{ total}} \times 100 = 10\% \text{ for water}$$

$$\frac{\$1,125,000 \text{ wastewater disposal}}{\$22,500,000 \text{ total}} \times 100 = 5\% \text{ for wastewater disposal}$$

$$\frac{\$1,125,000 \text{ refuse disposal}}{\$22,500,000 \text{ total}} \times 100 = 5\% \text{ for refuse disposal}$$

$$\frac{\$2,250,000 \text{ administration}}{\$22,500,000 \text{ total}} \times 100 = 10\% \text{ for administration}$$

$$\frac{\$562,500 \text{ miscellaneous}}{\$22,500,000 \text{ total}} \times 100 = 2.5\% \text{ for miscellaneous}$$

Once the percents have been calculated, the angle representing each part can be determined:

$$(0.60)(360°) = 216° \text{ for schools}$$
$$(0.075)(360°) = 27° \text{ for streets}$$
$$(0.10)(360°) = 36° \text{ for water}$$
$$(0.05)(360°) = 18° \text{ for wastewater disposal}$$
$$(0.05)(360°) = 18° \text{ for refuse disposal}$$
$$(0.10)(360°) = 36° \text{ for administration}$$
$$(0.025)(360°) = 9° \text{ for miscellaneous}$$

From this information draw the angles and circle, then label the graph (Figure 32).

12. First select suitable scales—ones that are neither blown out of proportion nor unnecessarily compressed. Since *time* is part of the graph, place it along the horizontal scale. *Polymer use* is therefore placed along the vertical scale. Then plot the points on the graph and draw the connecting lines (Figure 33).

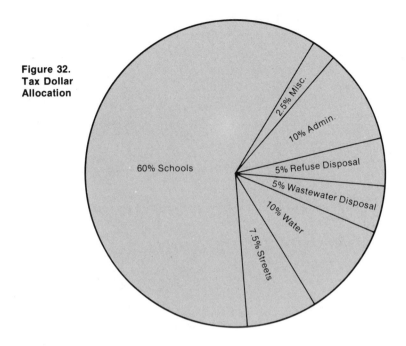

**Figure 32.
Tax Dollar
Allocation**

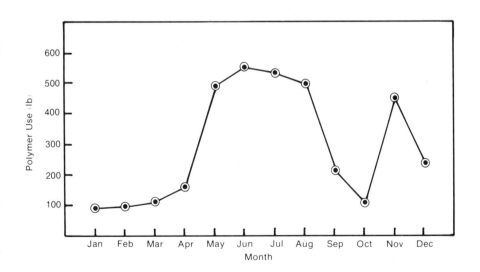

Figure 33.   1977 Polymer Use, by Month

Mathematics 13

Per Capita Water Use

The amount of water a community uses is, of course, of primary importance in water treatment. The water use establishes the amount of water that must be treated. One method of expressing the water use of a community is GALLONS PER CAPITA PER DAY (GPCD). The term *per capita* means "per person." Calculating gallons per capita per day merely determines the average *water use per person.* The mathematical formula used in calculating gallons per capita per day is

$$\text{gpcd} = \frac{\text{Water used (gpd)}}{\text{Total number of people}}$$

Note that the "total number of people" refers to the total number of people served by the water system; the water system service population is not always the same as the population of the city or town in which the system is located. The following examples illustrate the calculation of gallons per capita per day.

Example 1
The total water use on a particular day was 4,503,000 gal. If the water system served a population of 19,800, what was this water use expressed in gpcd?

$$\text{gpcd} = \frac{\text{Water used (gpd)}}{\text{Total number of people}}$$

$$= \frac{4,503,000 \text{ gpd flow}}{19,800 \text{ people}}$$

$$= 227.42 \text{ gpcd water used}$$

Example 2
During 1977, the average daily water use at a particular water system was 8.4 mgd. If the water system served a

population of 26,900, what was this average daily water use expressed in gallons per capita per day?

$$gpcd = \frac{\text{Water used (gpd)}}{\text{Total number of people}}$$

$$= \frac{8,400,000 \text{ gpd flow}}{26,900 \text{ people}}$$

$$= 312.27 \text{ gpcd}$$

Example 3

The average daily water use during 1976 for a certain water system serving a population of 10,300 was 2.7 mgd. The percent of water used in each of four categories is given below. Use this information to determine the 1976 gallons-per-capita-per-day water use for each of the four categories.

Domestic uses	46%
Commercial uses	16%
Industrial uses	25%
Public uses	13%

First, calculate the total gallons-per-capita-per-day water use for 1976:

$$gpcd = \frac{\text{Water used (gpd)}}{\text{Total number of people}}$$

$$= \frac{2,700,000 \text{ gpd flow}}{10,300 \text{ people}}$$

$$= 262.14 \text{ gpcd}$$

Next, use the percents given to calculate the water use by category:[26]

$(262.14 \text{ gpcd}) (0.46) = 120.58 \text{ gpcd for domestic uses}$

$(262.14 \text{ gpcd}) (0.16) = 41.94 \text{ gpcd for commercial uses}$

$(262.14 \text{ gpcd}) (0.25) = 65.54 \text{ gpcd for industrial uses}$

$(262.14 \text{ gpcd}) (0.13) = 34.08 \text{ gpcd for public uses}$

---

[26] Mathematics Section, Percent.

Review Questions

1. A water system serves a population of 110,800. If the average daily water use during one week was 38.8 mgd, what was the average gallons-per-capita-per-day water use?

2. A water system serves a population of 57,500. The average daily water use during July was 16.6 mgd. Using the percents listed below, determine the 1976 gpcd water use for each of the three categories:

Domestic uses	62%
Commercial uses	27%
Public uses	11%

3. The 1976 average per capita water use for a particular water system was 236 gpcd. If the water system serves a population of 45,700, what was the average daily water use (in gallons per day) for that period?

4. Suppose that the average daily flow for a water system in 1977 was 0.72 mgd. If the average per capita water use for that period was 195 gpcd, how many people did that water system serve?

Summary Answers

1. 350.18 gpcd

2. 178.99 gpcd domestic uses
   77.95 gpcd commercial uses
   31.76 gpcd public uses

3. 10,785,200 gpd

4. 3692 people

Detailed Answers

1. The average water use for each day that week was 38.8 mgd. If the water system served 110,800 people, then the gallons-per-capita-per-day water use was

$$gpcd = \frac{\text{Water used (gpd)}}{\text{Total number of people}}$$

$$= \frac{38,800,000 \text{ gpd flow}}{110,800 \text{ people}}$$

$$= 350.18 \text{ gpcd}$$

2. First, calculate the total gpcd water use:

$$gpcd = \frac{\text{Water used (gpd)}}{\text{Total number of people}}$$

$$= \frac{16,600,000 \text{ gpd flow}}{57,500 \text{ people}}$$

$$= 288.70 \text{ gpcd}$$

Now calculate the gallons-per-capita-per-day water use for each of three categories:

$$(288.70 \text{ gpcd})(0.62) = 178.99 \text{ gpcd for domestic uses}$$

$$(288.70 \text{ gpcd})(0.27) = 77.95 \text{ gpcd for commercial uses}$$

$$(288.70 \text{ gpcd})(0.11) = 31.76 \text{ gpcd for public uses}$$

3. This problem can be approached in a similar manner to other gallons-per-capita-per-day problems. First, fill in the information given:

$$\text{gpcd} = \frac{\text{Water used (gpd)}}{\text{Total number of people}}$$

$$236 \text{ gpcd} = \frac{x \text{ gpd flow}}{45,700 \text{ people}}$$

Then solve for the unknown value:[27]

$$(45,700)(236) = x \text{ gpd flow}$$
$$10,785,200 \text{ gpd} = x \text{ gpd flow}$$

4. As in Problem 3, the same gallons-per-capita-per-day formula is used. Merely fill in the information given:

$$\text{gpcd} = \frac{\text{Water used (gpd)}}{\text{Total number of people}}$$

$$195 \text{ gpcd} = \frac{720,000 \text{ gpd flow}}{x \text{ people}}$$

Then solve for the unknown value:

$$(x)(195) = 720,000$$
$$x = \frac{720,000}{195}$$
$$= 3692 \text{ people}$$

---

[27]Mathematics Section, Solving for the Unknown Value.

Mathematics 14

Domestic Water Use Based on Household Fixture Rates

Of the four principal uses of community water supplies—domestic, commercial, industrial, and public—domestic consumption often accounts for the largest portion. The approximate amounts of water required for typical household activities are tabulated in Table 6. Wasteful practices will, of course, greatly increase the quantities used.

Table 6. Household Fixture Use Rate

Use	Approximate Rate
Laundry	20 to 45 gal per load
Shower	20 to 30 gal per shower
Tub bath	30 to 40 gal per bath
Dishwashing	15 to 30 gal per load
Toilet	3.5 to 7 gal per flush
Drinking water	1–2 qt/day/person
Garbage disposal	5 gpm*
Car washing	5 gpm*
Lawn watering	7 to 43 gal per 100 ft² per wk†

*Assuming that a faucet draws 5 gpm fully open.

†The lower value applies in humid areas, the higher value in arid areas, and a midrange value in semi-arid areas.

This information, while not directly applicable to daily operation of a water treatment plant, helps to identify the factors that contribute to domestic water consumption. The following examples illustrate how to use these household fixture rates to estimate the water demand of an individual residence.

Example 1

The actual uses of water by a particular family of three are described by the list below. With this information and the approximate values given in the Household Fixture Use Rate Table, calculate the average amount of water used daily by this household. (When a range is given for the approximate rate, use the lower value; e.g., 20 gal per shower.)

```
Laundry ................... 4 loads per wk
Showers ...................... 2 per day
Tub baths..................... 1 per day
Dishwashing................ 1 load per day
Toilet.................... 10 flushes per day
Drinking water ................ 3 people
Garbage disposal............ 1 min per day
Car washing................. 5 min per wk
Lawn watering.................... 4200 ft²
```

To calculate the average amount of water used daily by this family, first determine the amount for each activity, then add these individual amounts.

Laundry: The approximate use rate for laundry is 20 gal/load. For 4 loads/wk, the total water use for the week is

$$(20 \text{ gal/load})(4 \text{ loads/wk}) = 80 \text{ gal/wk}$$

The daily water use for laundry is

$$\frac{80 \text{ gal/wk}}{7 \text{ days/wk}} = 11.43 \text{ gpd}$$

Shower: The approximate use rate for showers is 20 gal/shower. For 2 showers/day the daily water use is

$$(20 \text{ gal/shower})(2 \text{ showers/day}) = 40 \text{ gpd}$$

Tub bath: The approximate use rate is 30 gal/bath. In this problem there is 1 bath/day, so the water use is

$$(30 \text{ gal/bath})(1 \text{ bath/day}) = 30 \text{ gpd}$$

Dishwashing: The approximate use rate is 15 gal/load. At 1 load/day then,

$$(15 \text{ gal/load})(1 \text{ load/day}) = 15 \text{ gpd}$$

Toilet: The use rate for toilets is 3.5 gal/flush. At 10 flushes/day the water used is therefore

$$(3.5 \text{ gal/flush})(10 \text{ flushes/day}) = 35 \text{ gpd}$$

Drinking: At a use rate of 1 qt/day/person, the daily water use for a family of three is

$$(1 \text{ qt/person})(3 \text{ people}) = 3 \text{ qt}$$

Expressed in terms of gallons per day this is

$$\frac{3 \text{ qt}}{4 \text{ qt/gal}} = 0.75 \text{ gpd}$$

Garbage disposal: The approximate use rate is 5 gpm. In this problem the garbage disposal is used 1 min/day. Therefore the daily water use is

$$(5 \text{ gpm})(1 \text{ min/day}) = 5 \text{ gpd}$$

Car washing: In this problem, water is used for car washing 5 min/wk. At an approximate rate of 5 gpm, the water used is

$$(5 \text{ gpm})(5 \text{ min/wk}) = 25 \text{ gal/wk}$$

Converting this to gallons used per day,

$$\frac{25 \text{ gal/wk}}{7 \text{ days/wk}} = 3.57 \text{ gpd}$$

Lawn watering: The use rate is 7 gal/100 ft^2/wk. This means that for each 100 ft^2 of lawn, 7 gal of water are used each week. Therefore the water use during one week is

$$(7 \text{ gal/100 ft}^2)(42 \text{ segments of } 100 \text{ ft}^2) = 294 \text{ gal/wk}$$

and the daily water use is

$$\frac{294 \text{ gal/wk}}{7 \text{ days/wk}} = 42 \text{ gpd}$$

Now add all the daily amounts calculated above to determine the average amount of water used by this household per day:

Laundry	11.43 gpd
Shower	40 gpd
Tub bath........................	30 gpd
Dishwashing....................	15 gpd
Toilet..........................	35 gpd
Drinking water	0.75 gpd
Garbage disposal................	5 gpd
Car washing....................	3.57 gpd
Lawn watering................	+ 42 gpd
Total	182.75 gpd

Rounded to the nearest gallon, the water used daily by this family is 183 gal.

Example 2

The major water uses of a family of four are listed below. With the approximate values given in the Household Fixture Use Rate Table, determine the average amount of water used each day by this family. Express the answer in *gpcd* and identify the one activity that represents the highest water consumption per day. (When a range of values is given in the table, use the middle value.)

Laundry ....5 loads per wk
Showers .... 2 per day
Tub baths....................... 2 per day
Dishwashing................. 1 load per day
Toilet....................12 flushes per day
Drinking water .... 4 people
Garbage disposal............. 2 min per day
Car washing................. 10 min per wk
Lawn watering.................... 5000 ft²

As in Example 1, first calculate the amount of water used daily for each activity, then add these amounts.

Laundry: In this problem you were instructed to use the middle value when a range is shown for the approximate use rate. For laundry the range is 20 to 45 gal. To determine the middle value, add the two numbers and divide by two:

$$\text{Step 1:} \quad \begin{array}{r} 20 \\ +45 \\ \hline 65 \end{array}$$

$$\text{Step 2:} \quad \frac{65}{2} = 32.5 \text{ (middle value)}$$

Now continue with the water use calculation:

$$(32.5 \text{ gal/load})(5 \text{ loads/wk}) = 162.5 \text{ gal/wk}$$

Expressing this in terms of gpd,

$$\frac{162.5 \text{ gal/wk}}{7 \text{ days/wk}} = 23.21 \text{ gpd}$$

Showers: Again, first determine the middle value of the range given for shower use rate. Although the midpoint between 20 and 30 is obvious (25), verify this by the midpoint calculation method:

$$\text{Step 1:} \quad \begin{array}{r} 20 \\ +30 \\ \hline 50 \end{array}$$

Step 2: $\dfrac{50}{2} = 25$ (middle value)

Next, calculate the daily water use:

(25 gal/shower)(2 showers/day) = 50 gpd

Tub baths: A range of 30 to 40 gal/bath is given in the rate table. The middle value is 35 gal. The daily water use for tub baths is

(35 gal/bath)(2 baths/day) = 70 gpd

Dishwashing: The table indicates that the approximate water use rate for dishwashing is 15 to 30 gal/load. The midpoint of this range may not be obvious by visual inspection. Using the midpoint calculation,

Step 1:
$$\begin{array}{r} 15 \\ +30 \\ \hline 45 \end{array}$$

Step 2: $\dfrac{45}{2} = 22.5$ (middle value)

The water used daily for dishwashing can now be determined:

(22.5 gal/load)(1 load/day) = 22.5 gpd

Toilet: The range given for the approximate water use rate is 3.5 to 7 gal. Using the midpoint calculation,

Step 1:
$$\begin{array}{r} 3.5 \\ + 7.0 \\ \hline 10.5 \end{array}$$

Step 2: $\dfrac{10.5}{2} = 5.25$ (middle value)

Then calculate the water used daily:

(5.25 gal/flush)(12 flushes/day) = 63 gpd

Drinking water: Since there are four in the family, the water used daily for drinking purposes (using the middle value of the range) is

(1.5 qt/person)(4 people) = 6 qt

To be consistent with the other calculations, this water use must be expressed in gallons:

$$\frac{6 \text{ qt/day}}{4 \text{ qt/gal}} = 1.5 \text{ gpd}$$

Garbage disposal: Since no range in use rates is given, the water used daily can be calculated directly:

$$(5 \text{ gpm})(2 \text{ min/day}) = 10 \text{ gpd}$$

Car washing: No range in use is given. The water used weekly is

$$(5 \text{ gpm})(10 \text{ min/wk}) = 50 \text{ gal/wk}$$

Express this water use in gallons per day:

$$\frac{50 \text{ gal/wk}}{7 \text{ days/wk}} = 7.14 \text{ gpd}$$

Lawn watering: Calculate the midpoint of the range:

Step 1:
$$
\begin{array}{r}
7 \\
+43 \\
\hline
50
\end{array}
$$

Step 2: $\dfrac{50}{2} = 25$ (middle value)

Therefore the water used weekly is

$$(25 \text{ gal/100 ft}^2)(50 \text{ segments of } 100 \text{ ft}^2) = 1250 \text{ gal/wk}$$

Express this in terms of gallons per day:

$$\frac{1250 \text{ gal/wk}}{7 \text{ days/wk}} = 178.57 \text{ gpd}$$

Now add all the daily amounts calculated above:

Laundry	23.21	gpd
Showers	50	gpd
Tub baths	70	gpd
Dishwashing	22.5	gpd
Toilet	63	gpd
Drinking water	1.5	gpd
Garbarge disposal	10	gpd
Car washing	7.14	gpd
Lawn watering	178.57	gpd
Total	425.92	gpd

In this problem you were asked to express the water use as gpcd:[28]

---

[28] Mathematics Section, Per Capita Water Use.

$$\text{gpcd} = \frac{\text{gpd}}{\text{Total number of people}}$$

$$= \frac{425.92 \text{ gpd}}{4 \text{ people}}$$

$$= 106.48 \text{ gpcd}$$

$$= 106 \text{ gpcd (rounded to the nearest whole number)}$$

In the second part of the problem you were asked to determine which activity represents the highest water use per day. According to the calculations above, lawn watering constitutes the highest daily water use for this family.

Review Questions

1. Using the Household Fixture Use Rate Table given in the text and the actual water usage listed below for a family of five, calculate the average amount of water used by this household each day. (When a range is given in the table for the approximate rate, use the higher value.)

> Laundry . . . . 4 loads per wk
> Showers . . . . 4 per day
> Tub baths . . . . 1 per day
> Dishwashing . . . . 2 loads per day
> Toilet . . . . 15 flushes per day
> Drinking water . . . . . 5 people
> Garbage disposal . . . 2 min per day
> Lawn watering . . . . . 2100 ft^2

2. The principal water uses of a family of two are listed below. With this information and the approximate values given in the Household Fixture Use Rate Table, calculate the average amount of water used daily by this family. Express the answer in gallons per capita per day and identify the one activity that represents the highest water use per day. (When a range is given in the table for the approximate rate, use the middle value.)

> Laundry . . . . 2 loads per wk
> Showers . . . . 2 per day
> Dishwashing . . . . 4 loads per wk
> Toilet . . . . 4 flushes per day
> Drinking water . . . . . 2 people
> Garbage disposal . . . 3 min per wk
> Lawn watering . . . . . 1000 ft^2

Summary Answers

1. 492 gpd

2. 66 gpcd; showers

Detailed Answers

1. First calculate the amount used daily for each activity, then add these amounts:

   Laundry:

   $$(45 \text{ gal/load})(4 \text{ loads/wk}) = 180 \text{ gal/wk}$$

   Then

   $$\frac{180 \text{ gal/wk}}{7 \text{ days/wk}} = 25.71 \text{ gpd}$$

   Showers:

   $$(30 \text{ gal/shower})(4 \text{ showers/day}) = 120 \text{ gpd}$$

   Tub baths:

   $$(40 \text{ gal/bath})(1 \text{ bath/day}) = 40 \text{ gpd}$$

   Dishwashing:

   $$(30 \text{ gal/load})(2 \text{ loads/day}) = 60 \text{ gpd}$$

   Toilet:

   $$(7 \text{ gal/flush})(15 \text{ flushes/day}) = 105 \text{ gpd}$$

   Drinking water:

   $$(2 \text{ qt/person})(5 \text{ people}) = 10 \text{ qt}$$

   Expressing this as gallons per day:

   $$\frac{10 \text{ qt/day}}{4 \text{ qt/gal}} = 2.5 \text{ gpd}$$

   Garbage disposal:

   $$(5 \text{ gpm})(2 \text{ min/day}) = 10 \text{ gpd}$$

   Lawn watering:

   First determine the weekly water use:

   $$(43 \text{ gal/100 ft}^2)(21 \text{ segments of 100 ft}^2) = 903 \text{ gal/wk}$$

Then convert to gallons used daily:

$$\frac{903 \text{ gal/wk}}{7 \text{ days/wk}} = 129 \text{ gpd}$$

Now add the daily water uses calculated above to determine the average amount of water used daily by this family:

Laundry .... 25.71 gpd
Showers 120   gpd
Tub baths..................... 40   gpd
Dishwashing................... 60   gpd
Toilet........................ 105   gpd
Drinking water .... 2.5   gpd
Garbage disposal.............. 10   gpd
Lawn watering................. 129   gpd

                Total   492.21 gpd

Rounded to the nearest gallon, the water used daily by this family is 492 gal.

2. Before the gallons-per-capita-per-day water use of this family and the activity using the most water can be determined, the amount of water used daily for each activity must be calculated:

Laundry: Determine the middle value in the rate-of-use range.

Step 1:   20
              +45
              ———
               65

Step 2: $\dfrac{65}{2} = 32.5$ (middle value)

Now calculate the daily water use for this activity:

(32.5 gal/load)(2 loads/wk) = 65 gal/wk

Expressed as gallons per day:

$$\frac{65 \text{ gal/wk}}{7 \text{ days/wk}} = 9.29 \text{ gpd}$$

Showers: The midpoint (25 gal/shower) in the use rate range is used in the calculation:

(25 gal/shower)(2 showers/day) = 50 gpd

Dishwashing: First, calculate the midpoint in the use rate range:

Step 1:   15
              +30
              ———
               45

$$Step\ 2: \quad \frac{45}{2} = 22.5 \text{ (middle value)}$$

Using this middle value, calculate the weekly water use:

$$(22.5 \text{ gal/load})(4 \text{ loads/wk}) = 90 \text{ gal/wk}$$

Express this as gallons used per day:

$$\frac{90 \text{ gal/wk}}{7 \text{ days/wk}} = 12.86 \text{ gpd}$$

Toilet: Calculate the midpoint in the use rate range:

$$Step\ 1: \quad \begin{array}{r} 3.5 \\ +\ 7.0 \\ \hline 10.5 \end{array}$$

$$Step\ 2: \quad \frac{10.5}{2} = 5.25 \text{ (middle value)}$$

Now calculate the amount of water used daily:

$$(5.25 \text{ gal/flush})(4 \text{ flushes/day}) = 21 \text{ gpd}$$

Drinking water: The midpoint (1.5 qt/person/day) is used:

$$(1.5 \text{ qt/person})(2 \text{ people}) = 3 \text{ qt}$$

Convert this to gallons per day:

$$\frac{3 \text{ qt/day}}{4 \text{ gal/qt}} = 0.75 \text{ gpd}$$

Garbage disposal:

$$(5 \text{ gpm})(3 \text{ min/wk}) = 15 \text{ gal/wk}$$

Express this as gallons per day:

$$\frac{15 \text{ gal/wk}}{7 \text{ days/wk}} = 2.14 \text{ gpd}$$

Lawn watering: First calculate the midpoint in the use rate range:

$$Step\ 1: \quad \begin{array}{r} 7 \\ +43 \\ \hline 50 \end{array}$$

$$Step\ 2: \quad \frac{50}{2} = 25 \text{ (middle value)}$$

Now the water used weekly for lawn watering can be calculated:

$$(25 \text{ gal/100 ft}^2)(10 \text{ segments of 100 ft}^2) = 250 \text{ gal/wk}$$

Expressing this as gallons per day,

$$\frac{250 \text{ gal/wk}}{7 \text{ days/wk}} = 35.71 \text{ gpd}$$

Now add all the daily amounts calculated above:

Laundry	9.29 gpd
Showers	50 gpd
Dishwashing.....................	12.86 gpd
Toilet..........................	21 gpd
Drinking water	0.75 gpd
Garbage disposal................	2.14 gpd
Lawn watering...................	35.71 gpd
Total	131.75 gpd

As part of this problem, you were asked to express this water use in terms of gpcd:

$$\text{gpcd} = \frac{\text{Water used}}{\text{Total number of people}}$$

$$= \frac{131.75 \text{ gpd}}{2 \text{ people}}$$

$$= 65.88 \text{ gpcd}$$

$$= 66 \text{ gpcd}$$

(rounded to nearest whole number)

As the second part of the problem, you were to determine which activity represented the highest daily water use. According to the calculations above, *showers* constitute the highest daily water use for this family.

Mathematics 15

Water Use per Unit of Industrial Product Produced

Water used by industry is commonly measured in terms of the products produced. For example, the water required by a green bean cannery is typically 20,000 gal for each ton of green beans processed; for the paper industry, 47,000 gal of water is typically required for each ton of paper produced. Table 7 summarizes water requirements for some typical industries. The following examples illustrate how this information can be used in calculating water requirements for a particular industrial plant.

Example 1

A new paper mill is being planned for construction in your area. The planners have requested that water from your treatment plant be used in meeting their paper processing needs. If the paper mill is planning to produce 25 tons of paper daily at full production, how much water will the mill require daily? (Use Table 7 for water requirements per product produced.)

According to Table 7, the water requirement for each ton of paper produced is 47,000 gal. Therefore, the daily water requirement for 25 tons of paper is

$$(47,000 \text{ gal water/ton})(25 \text{ tons}) = 1,175,000 \text{ gal water required}$$

Example 2

In 1977, the total production of the sulfur manufacturing industry was 1095 tons of sulfur. On the average, how much water was required each day, assuming a 7-day work week? (Use Table 7 for water requirements per product produced.)

First calculate the amount of water required for the full year, then calculate the amount required per day. According to Table 7, the water requirement for each ton of sulfur produced is 3000 gal. Therefore, the water requirement for the year's production was

$$(3000 \text{ gal water/ton})(1095 \text{ tons}) = 3,285,000 \text{ gal water required}$$

Since sulfur was produced each day of the year, there were 365 days of sulfur production. With this information, the average amount of water required for each day can be determined:

$$\frac{3,285,000 \text{ gal water/yr}}{365 \text{ day/yr}} = 9000 \text{ gpd water required}$$

Table 7. Water Requirements in Selected Industries*

Process	Water Required (in gallons per unit of product noted)
Canneries:	
Green beans, per ton	20,000
Peaches and pears, per ton	5,300
Other fruits and vegetables, per ton	2,000–10,000
Chemical industries:	
Ammonia, per ton	37,500
Carbon dioxide, per ton	24,500
Gasoline, per 1,000 gal	7,000–34,000
Lactose, per ton	235,000
Sulfur, per ton	3,000
Food and beverage industries:	
Beer, per 1,000 gal	15,000
Bread, per ton	600–1,200
Meat packing, per ton live weight	5,000
Milk products, per ton	4,000–5,000
Whiskey, per 1,000 gal	80,000
Pulp and paper:	
Pulp, per ton	82,000–230,000
Paper, per ton	47,000
Textiles:	
Bleaching, per ton cotton	72,000–96,000
Dyeing, per ton cotton	9,500–19,000
Mineral products:	
Aluminum (electrolytic smelting), per ton	56,000 (max)
Petroleum, per barrel of crude oil	800–3,000
Steel, per ton	1,500–50,000

*Sources:
Metcalf and Eddy, *Wastewater Engineering*, McGraw-Hill Book Company, 1972.
Fair, Geyer and Okun, *Elements of Water Supply and Wastewater Disposal*, John Wiley Sons, 2nd Ed., 1971.

Review Questions

1. A new ammonia plant is scheduled to be built in your area. The peak production of this plant is expected to be 18 tons per day. If this industry expects to obtain its water requirements from your treatment plant, how much water will be required from your treatment plant daily? (Use Table 7 for water requirements per product.)

2. An industry that manufactures steel requires 17,000 gal of water per ton of steel produced. Suppose that the 1977 steel production for this industry was 18,720 tons. Assuming 312 days of production during 1977, on the average, how much water was required each day of production?

Summary Answers

1. 675,000 gpd

2. 1,020,000 gal each production day

Detailed Answers

1. According to Table 7 in the text, the water requirement for each ton of ammonia produced is 37,500 gal. Thus, the water requirement for 18 tons of ammonia per day would be:

   (37,500 gal water/ton)(18 tons/day) = 675,000 gpd water required

2. First, calculate the water required for the year:

   (17,000 gal water/ton)(18,720 tons) = 318,240,000 gal water required

   In this example, the water use was divided among 312 days. Using this information, the average amount of water used each day of production can be calculated:

   $$\frac{318,240,000 \text{ gal water/yr}}{312 \text{ days production/yr}} = 1,020,000 \text{ gpd water required}$$

Mathematics 16

Average Daily Flow

The amount of water a community uses every day can be expressed in terms of an AVERAGE DAILY FLOW (ADF); that is, the average[29] of the actual daily flows that occur within a period of time, such as a week, a month, or a year. This is expressed mathematically as

$$ADF = \frac{\text{Sum of all daily flows}}{\text{Total number of daily flows used}}$$

The average daily flow can reflect a week's data used in the averaging (weekly ADF), a month's data used in the averaging (monthly ADF), or a year's data used in the averaging (annual ADF or AADF—the most commonly calculated average).

Average daily flow is important because it is used in several treatment plant calculations. The following examples illustrate the calculation of average daily flow.

Example 1
In January 1978, a total of 68,920,000 gal of water was treated at a plant. What was the average daily flow at the treatment plant for this period?

In this problem the sum of all daily flows has already been determined—a *total* of 68,920,000 gal was treated for the month. January has 31 days; fill the information into the ADF formula:

---

[29]Mathematics Section, Averages.

$$\text{ADF} = \frac{\text{Sum of all daily flows}}{\text{Total number of daily flows used}}$$

$$= \frac{68{,}920{,}000 \text{ gal}}{31 \text{ days}}$$

$$= 2{,}223{,}226 \text{ gpd}$$

Example 2

A water treatment plant reported that the total volume of water treated for the calendar year 1977 was 152,655,000 gal. What was the annual average daily flow for 1977?

As in the previous example, the sum of all daily flows has already been determined—a total of 152,655,000 gal was treated during the year. Knowing that there are 365 days in a year, calculate the average daily flow:

$$\text{ADF} = \frac{\text{Sum of all daily flows}}{\text{Total number of daily flows used}}$$

$$= \frac{152{,}655{,}000 \text{ gal}}{365 \text{ days}}$$

$$= 418{,}233 \text{ gpd}$$

Example 3

The following daily flows (in million gallons per day) were treated during June 1978 at a water treatment plant. What was the average daily flow for this period?

June	1	6.21	June	11	7.59	June	21	6.43
	2	6.68		12	7.01		22	6.26
	3	7.31		13	6.85		23	6.87
	4	7.80		14	6.43		24	7.27
	5	6.77		15	6.52		25	7.95
	6	6.32		16	6.79		26	7.33
	7	5.96		17	6.91		27	6.72
	8	5.83		18	7.37		28	6.51
	9	6.09		19	7.02		29	5.92
	10	7.22		20	6.88		30	5.90

To calculate the average daily flow for this time period, first add all daily flows. The total of these flows is 202.72 mil gal. With this information, calculate the average daily flow for the period:

$$\text{ADF} = \frac{\text{Sum of all daily flows}}{\text{Total number of daily flows used}}$$

$$= \frac{202.72 \text{ mil gal}}{30 \text{ days}}$$

$$= 6.76 \text{ mgd}$$

In some problems you will not know *actual daily* flows for each day during a particular period, but you will know the ADF for that time period. ADF information can be used in calculating other ADF's. The following example contrasts the two methods of calculating ADF's.

Example 4

The volume of water (in million gallons) treated for each day during a 2-week period is listed below. What is the average daily flow for this 2-week period?

Week 1		Week 2	
Sunday	2.41	Sunday	2.52
Monday	3.37	Monday	3.39
Tuesday	3.44	Tuesday	3.48
Wednesday	3.61	Wednesday	3.88
Thursday	3.23	Thursday	3.19
Friday	2.86	Friday	2.82
Saturday	2.75	Saturday	2.70

Since you are given the actual flows for each day in the 2-week period, you could calculate the average daily flow in a similar way to the three example problems above. That is,

$$ADF = \frac{\text{Sum of all daily flows}}{\text{Total number of daily flows used}}$$

$$= \frac{43.65 \text{ mil gal}}{14 \text{ days}}$$

$$= 3.12 \text{ mgd}$$

Suppose, however, that in this problem you did not know the actual flow for each day in the week, but you knew the ADF for Week 1 (3.10 mgd) and the ADF for Week 2 (3.14 mgd). The ADF for the 2-week period can still be calculated, but *not* using the ADF equation given above (because in this case you do not know the *sum* of all daily flows). However, a similar formula is used:

$$ADF = \frac{\text{Sum of all weekly ADF's}}{\text{Total number of weekly ADF's used}}$$

In this case:

$$ADF = \frac{3.10 \text{ mgd} + 3.14 \text{ mgd}}{2}$$

$$= \frac{6.24 \text{ mgd}}{2}$$

$$= 3.12 \text{ mgd}$$

Notice that, although only ADF information was used, the answer was the same as when all 14 actual daily flows were used in the calculation.

In a similar manner, ADF's for each of 12 months can be used to determine the average daily flow for a year (ANNUAL AVERAGE DAILY FLOW or AADF). The equation that is used is

$$\text{AADF} = \frac{\text{Sum of all monthly ADF's}}{\text{Total number of monthly ADF's used}}$$

Example 5
The average daily flow (in million gallons per day) at a treatment plant for each month in the year is given below. Using this information, calculate the annual average daily flow.

January	10.71	July	11.96
February	9.89	August	12.24
March	10.32	September	11.88
April	10.87	October	11.53
May	11.24	November	11.36
June	11.58	December	10.98

If you knew all 365 flows for the year, the annual average daily flow (AADF) would be calculated using the formula

$$\text{AADF} = \frac{\text{Sum of all daily flows}}{\text{Total number of daily flows used}}$$

In this problem, however, average daily flows for each month of the year are given. Therefore, the average daily flow is calculated using the formula

$$\text{AADF} = \frac{\text{Sum of all monthly ADF's}}{\text{Total number of monthly ADF's used}}$$

Filling in the information given in this problem

$$\text{AADF} = \frac{134.56 \text{ mgd}}{12}$$

$$= 11.21 \text{ mgd}$$

Review Questions

1. The total water treated for the months of June, July, and August is listed below. Using this information, calculate the average daily flow for this 3-month period.

June	95,440,000 gal
July	107,913,000 gal
August	115,359,000 gal

2. A water treatment plant reported that the total water treated for the calendar year 1976 (a leap year) was 9975 million gallons. What was the annual average daily flow for that year?

3. The following total monthly flows (in million gallons) were recorded for the year. Based on this information, what was the annual average daily flow for the year?

January	172.91	July	192.55
February	174.03	August	189.41
March	179.62	September	186.07
April	181.13	October	181.92
May	181.79	November	178.66
June	186.68	December	174.32

4. The average daily flows (in million gallons per day) for each month of 1977 are listed below. Using this information, determine the annual average daily flow.

January	1.71	July	2.44
February	1.88	August	2.27
March	1.83	September	2.06
April	1.91	October	1.93
May	1.97	November	1.85
June	2.13	December	1.76

Summary Answers

1. 3,464,261 gpd ADF

2. 27.25 mgd AADF

3. 5.97 mgd AADF

4. 1.98 mgd AADF

Detailed Answers

1. First, add the flows for the 3-month period:

June	95,440,000 gal
July	107,913,000 gal
August	115,359,000 gal
Total	318,712,000 gal

When calculating the average daily flow in this problem, you also need to know the total number of days:

June	30 days
July	31 days
August	31 days
Total	92 days

With this information, calculate the average daily flow for the 3-month period:

$$ADF = \frac{\text{Sum of all daily flows}}{\text{Total number of daily flows used}}$$

$$= \frac{318,712,000 \text{ gal}}{92 \text{ days}}$$

$$= 3,464,261 \text{ gpd}$$

2. Calculate the annual average daily flow for 1976 using the formula

$$AADF = \frac{\text{Sum of all daily flows}}{\text{Total number of daily flows used}}$$

Filling in the information given in the problem:

$$AADF = \frac{9{,}975 \text{ mil gal}}{366 \text{ days (leap year)}}$$

$$= 27.25 \text{ mgd}$$

3. Given total monthly flows, you can calculate the average daily flow for the year using the formula

$$AADF = \frac{\text{Sum of all daily flows}}{\text{Total number of daily flows used}}$$

The sum of all monthly total flows equals 2,179.09 mil gal for the year. Now calculate average daily flow:

$$AADF = \frac{2{,}179.09 \text{ mil gal}}{365 \text{ days}}$$

$$= 5.97 \text{ mgd}$$

4. In this problem, the average daily flows for each month are given rather than total flows for the month. When calculating average daily flow for this problem, use the following formula:

$$AADF = \frac{\text{Sum of all monthly ADF's}}{\text{Total number of monthly ADF's used}}$$

$$= \frac{23.74 \text{ mgd}}{12}$$

$$= 1.98 \text{ mgd}$$

Mathematics 17

Surface Overflow Rate

The faster the water leaves a sedimentation tank, or clarifier, the more turbulence is created and the more suspended solids are carried out in the effluent. Overflow rate—the speed with which water leaves the sedimentation tank—is controlled by increasing or decreasing the flow rate into the tank.

The SURFACE OVERFLOW RATE measures the amount of water leaving a sedimentation tank per square foot of tank surface area. The treatment plant operator must be able to determine the surface overflow rate that produces the best quality effluent leaving the tank.

The following figure can be associated with problems involving surface overflow rate.

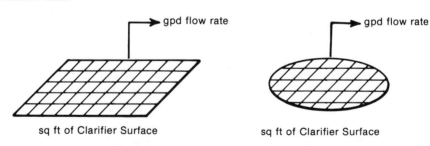

| SURFACE OVERFLOW RATE | SURFACE OVERFLOW RATE |
| for Rectangular Clarifier | for Circular Clarifier |

Since surface overflow is the gallons per day flow *up and over* each square foot of tank surface, the corresponding mathematical equation is

$$\text{Surface overflow rate} = \frac{\text{gpd flow}}{\text{sq ft tank surface}}$$

Notice that the depth of the sedimentation tank is not a consideration in the calculation of surface overflow rate.

Example 1

The flow to a treatment plant is 1.2 mgd. If the sedimentation tank is 70 ft long, 15 ft wide, and 7 ft deep, what is the surface overflow rate?

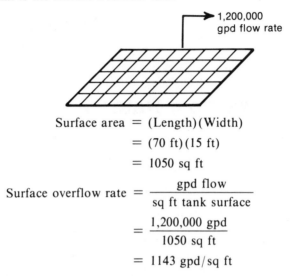

1,200,000 gpd flow rate

$$\text{Surface area} = (\text{Length})(\text{Width})$$
$$= (70 \text{ ft})(15 \text{ ft})$$
$$= 1050 \text{ sq ft}$$

$$\text{Surface overflow rate} = \frac{\text{gpd flow}}{\text{sq ft tank surface}}$$
$$= \frac{1,200,000 \text{ gpd}}{1050 \text{ sq ft}}$$
$$= 1143 \text{ gpd/sq ft}$$

Example 2

A circular sedimentation basin with a 55-ft diameter receives a flow of 2,075,000 gpd. What is the surface overflow rate?

$$\text{Surface area} = 0.785 \text{ D}^2$$
$$= (0.785)(55 \text{ ft})(55 \text{ ft})$$
$$= 2375 \text{ sq ft}$$

$$\text{Surface overflow rate} = \frac{\text{gpd}}{\text{sq ft tank surface}}$$
$$= \frac{2,075,000 \text{ gpd}}{2375 \text{ sq ft}}$$
$$= 873.68 \text{ gpd/sq ft}$$

In the previous two examples, the surface overflow rate was the unknown factor. However, *any one* of three factors (surface overflow rate, gallons per day, or surface area) can be unknown, and if the other two factors are known, the same mathematical setup can be used to solve for the unknown value.

Example 3

A 70-ft diameter tank has a surface overflow area of 730 gpd/sq ft. What is the daily flow to the tank?

$$\text{Surface area} = 0.785 \ D^2$$
$$= (0.785)(70 \text{ ft})(70 \text{ ft})$$
$$= 3847 \text{ sq ft}$$

$$\text{Surface overflow rate} = \frac{\text{gpd flow}}{\text{sq ft tank surface}}$$

$$730 \text{ gpd/sq ft} = \frac{x \text{ gpd}}{3847 \text{ sq ft}}$$

$$(3847)(730) = x$$

$$2{,}808{,}310 \text{ gpd} = x \text{ gpd flow}$$

Example 4

A sedimentation tank receives a flow of 5.4 mgd. If the surface overflow rate is 689 gpd/sq ft, what is the surface area of the sedimentation tank?

$$\text{Surface overflow rate} = \frac{\text{gpd flow}}{\text{sq ft tank surface}}$$

$$689 \text{ gpd/sq ft} = \frac{5{,}400{,}000 \text{ gpd}}{x \text{ sq ft}}$$

$$(689)(x) = 5{,}400{,}000$$

$$x = \frac{5{,}400{,}000}{689}$$

$$x = 7837 \text{ sq ft}$$

Review Questions

1. The flow to a sedimentation tank is 2.02 mgd. If the tank is 80 ft long and 20 ft wide, what is the surface overflow rate?

2. A circular clarifier has a diameter of 60 ft. If the flow to the clarifier is 2.8 mgd, what is the surface overflow rate?

3. A sedimentation tank receives a flow of 2,812,500 gpd. If the surface overflow rate is 1250 gpd/sq ft and the length of the tank is 90 ft, what is the width of the tank?

Summary Answers

1. 1263 gpd/sq ft

2. 990.8 gpd/sq ft

3. 25 ft

Detailed Answers

1.
$$\text{Surface area} = (\text{Length})(\text{Width})$$
$$= (80 \text{ ft})(20 \text{ ft})$$
$$= 1600 \text{ sq ft}$$

$$\text{Surface overflow rate} = \frac{\text{gpd flow}}{\text{sq ft tank surface}}$$
$$= \frac{2{,}020{,}000 \text{ gpd}}{1600 \text{ sq ft}}$$
$$= 1263 \text{ gpd/sq ft}$$

2.
$$\text{Surface area} = 0.785 \text{ D}^2$$
$$= (0.785)(60 \text{ ft})(60 \text{ ft})$$
$$= 2826 \text{ sq ft}$$

$$\text{Surface overflow rate} = \frac{\text{gpd flow}}{\text{sq ft tank surface}}$$
$$= \frac{2{,}800{,}000 \text{ gpd}}{2826 \text{ sq ft}}$$
$$= 990.8 \text{ sq ft}$$

3. Solve this problem in the same way as other surface overflow problems by filling in the given information:

$$\text{Surface area} = (\text{Length})(\text{Width})$$
$$= (90 \text{ ft})(x \text{ ft})$$

$$\text{Surface overflow rate} = \frac{\text{gpd flow}}{\text{sq ft tank surface}}$$
$$1250 \text{ gpd/sq ft} = \frac{2{,}812{,}500 \text{ gpd}}{(90)(x) \text{ sq ft}}$$

Then solve for the unknown value:

$$(x)(1250) = \frac{2,812,500}{90}$$

$$x = \frac{2,812,500}{(90)(1250)}$$

$$x = \frac{2,812,500}{112,500}$$

$$x \text{ ft} = 25 \text{ ft}$$

Mathematics 18

Weir Overflow Rate

The calculation of WEIR OVERFLOW RATE is important in detecting high veloci-ties near the weir, which adversely affect the efficiency of the sedimentation process. With excessively high velocities, the settling solids are pulled over the weirs and into the effluent troughs.

In calculating the weir overflow rate, you will be concerned with the *gallons per day* flowing *over each foot of weir length*. The following figures can be associated with weir overflow rate in rectangular and circular sedimentation basins.

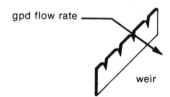

WEIR OVERFLOW RATE
Rectangular Clarifier
(Sedimentation Tank)

WEIR OVERFLOW RATE
Circular Clarifier

Since weir overflow rate is gallons-per-day flow over each foot of weir length, the corresponding mathematical equation is

$$\text{Weir overflow rate} = \frac{\text{gpd flow}}{\text{ft weir length}}$$

Example 1
If a sedimentation tank has a total of 90 ft of weir over which the water flows, what is the weir overflow rate when the flow is 1,098,000 gpd?

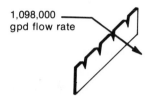

1,098,000 gpd flow rate

90-ft weir

$$\text{Weir overflow rate} = \frac{\text{gpd flow}}{\text{ft weir length}}$$

$$= \frac{1,098,000 \text{ gpd}}{90 \text{ ft}}$$

$$= 12,200 \text{ gpd/ft}$$

Example 2

A circular clarifier receives a flow of 3.6 mgd. If the diameter is 80 ft, what is the weir overflow rate?

Before you can calculate the weir overflow rate, you must know the total length of the weir. The relationship of the diameter and circumference of a circle is the key to determining this.[30]

$$\text{Circumference} = (3.14)(\text{Diameter})$$

In this problem, the diameter is 80 ft. Therefore, the length of weir (circumference) is

$$\text{Circumference} = (3.14)(80 \text{ ft})$$

$$= 251.2 \text{ ft}$$

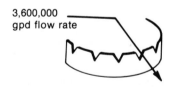

3,600,000 gpd flow rate

251.2 ft weir

Now solve for the weir overflow rate:

$$\text{Weir overflow rate} = \frac{\text{gpd flow}}{\text{ft weir length}}$$

$$= \frac{3,600,000 \text{ gpd}}{251.2 \text{ ft}}$$

$$= 14,331 \text{ gpd/ft}$$

---

[30]Mathematics Section, Linear Measurements.

Example 3

A clarifier has a diameter of 60 ft and receives a flow of 1.87 mgd. What is the weir overflow rate?

If the diameter is 60 ft, then the weir length (circumference) is

$$\text{Circumference} = (3.14)(60 \text{ ft})$$
$$= 188.4 \text{ ft}$$

Now solve for the weir overflow rate:

$$\text{Weir overflow rate} = \frac{\text{gpd flow}}{\text{ft weir length}}$$
$$= \frac{1,870,000 \text{ gpd}}{188.4 \text{ ft}}$$
$$= 9,926 \text{ gpd/ft}$$

Review Questions

1. A sedimentation tank has a total of 120 ft of weir over which the water flows. What is the weir overflow rate when the flow is 1.4 mgd?

2. The diameter of the weir in a circular clarifier is 55 ft. What is the weir overflow rate if the flow over the weir is 2.24 mgd?

3. The weir overflow rate for a particular sedimentation tank is 10,400 gpd/ft, and the tank has a total of 105 ft of weir over which the water flows. What is the gallons-per-day flow over the weir?

4. A circular clarifier receives a flow of 1,931,000 gpd. The weir overflow for the clarifier is 12,300 gpd/ft. What is the diameter of the weir?

Summary Answers

1. 11,667 gpd/ft

2. 12,970 gpd/ft

3. 1,092,000 gpd

4. 50-ft diameter

Detailed Answers

1. Calculate weir overflow rate by dividing the flow by the total weir length:

$$\text{Weir overflow rate} = \frac{\text{gpd flow}}{\text{ft weir length}}$$

$$= \frac{1,400,000 \text{ gpd}}{120 \text{ ft}}$$

$$= 11,667 \text{ gpd/ft}$$

2. Before the weir overflow rate can be calculated, the total feet of weir must be determined. Since the weir's diameter has been given, the total feet of weir around the clarifier can be calculated:

$$\text{Circumference} = (3.14)(55 \text{ ft})$$

$$= 172.7 \text{ ft}$$

Continue with the weir overflow calculation:

$$\text{Weir overflow rate} = \frac{\text{gpd flow}}{\text{ft weir length}}$$

$$= \frac{2,240,000 \text{ gpd}}{172.7 \text{ ft}}$$

$$= 12,970 \text{ gpd/ft}$$

3. This problem is approached in a similar way to previous problems by filling in the given information:

$$\text{Weir overflow rate} = \frac{\text{gpd flow}}{\text{ft weir length}}$$

$$10,400 \text{ gpd/ft} = \frac{x \text{ gpd}}{105 \text{ ft}}$$

Then solve for the unknown value:

$$(10,400)(105) = x$$

$$1,092,000 \text{ gpd flow} = x$$

4. Since this problem involves a circular clarifier, the total feet of weir is equal to:

$$\text{Circumference} = (3.14)(x \text{ ft})$$

Now fill the known information in the equation:

$$\text{Weir overflow rate} = \frac{\text{gpd flow}}{\text{ft weir length}}$$

$$12,300 \text{ gpd/ft} = \frac{1,931,100 \text{ gpd}}{(3.14)(x \text{ ft})}$$

And then solve for the unknown value:

$$(12,300)(x) = \frac{1,931,100}{3.14}$$

$$x = \frac{1,931,100}{(3.14)(12,300)}$$

$$x = \frac{1,931,100}{38,622}$$

$$x = 50 \text{ ft diameter}$$

Mathematics 19

Filter Loading Rate

The FILTER LOADING RATE is measured in gallons of water applied to each square foot of surface area. Or, stated more precisely, filter loading rate measures the amount of flowing down through each square foot of filter area. Whereas the surface overflow rate is measured in gallons per *day* per square foot, the filter loading rate is measured in gallons per *minute* per square foot.

To assist in the mathematical calculation, the following figure can be associated with problems involving filter loading rate.

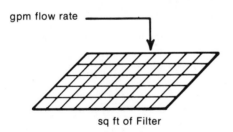

FILTER LOADING RATE

The mathematical equation associated with filter loading rate is

$$\text{Filter loading rate} = \frac{\text{gpm flow}}{\text{sq ft filter area}}$$

Sometimes the filter loading rate is measured as the fall of water (in inches) through a filter per minute. This is expressed mathematically as:

$$\text{Filter loading rate} = \frac{\text{Inches of water fall}}{\text{Minutes}}$$

Some typical loading rates for various types of filters are tabulated below.

Type of Filter	Common Loading Rate
Slow sand filters	0.016 to 0.16 gpm/sq ft
Rapid sand filters	2 gpm/sq ft
Dual media (coal/sand)	2 to 4 gpm/sq ft
Multi-media (Coal/sand/garnet or coal/sand/ilmenite)	5 to 10 gpm/sq ft

Example 1
A rapid sand filter is 10 ft wide and 15 ft long. If the flow through the filter is 450,000 gpd, what is the filter loading rate in gallons per minute per square foot?

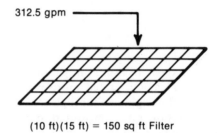

312.5 gpm

(10 ft)(15 ft) = 150 sq ft Filter

First, convert the gallons per day to gallons per minute:[31]

$$\frac{450,000 \text{ gpd}}{1440 \text{ min/day}} = 312.5 \text{ gpm}$$

Then express the filter loading rate mathematically as:

$$\text{Filter loading rate} = \frac{\text{gpm flow}}{\text{sq ft filter area}}$$

$$= \frac{312.5 \text{ gpm}}{150 \text{ sq ft}}$$

$$= 2.08 \text{ gpm/sq ft}$$

Example 2
A slow sand filter is 25 ft wide and 40 ft long. If the filter receives a flow of 173,000 gpd, what is the filter loading rate in gallons per minute per square foot?

As in the previous example, convert the flow from gallons per day to gallons per minute:

$$\frac{173,000 \text{ gpd}}{1440 \text{ min/day}} = 120.14 \text{ gpm}$$

---

[31]Mathematics Section, Conversions (Flow Conversions).

And fill this information into the equation:

$$\text{Filter loading rate} = \frac{\text{gpm flow}}{\text{sq ft filter area}}$$

$$= \frac{120.14 \text{ gpm}}{(25 \text{ ft})(40 \text{ ft})}$$

$$= 0.12 \text{ gpm/sq ft}$$

Filter loading rates, as previously noted, are sometimes expressed as the vertical movement of water in the filter in one minute; that is, inches per minute (in./min) fall in the water level.

The gallons per minute per square foot and inches per minute measurements are directly related as noted by the following equation:

$$1 \text{ gpm/sq ft} = 1.6 \text{ in./min}$$

The box method of conversion can be used when converting from one term to another.[32]

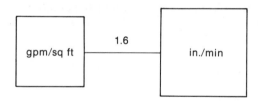

The following examples illustrate the conversion of these two terms.

Example 3

How many inches drop per minute corresponds to a filter loading rate of 2.6 gpm/sq ft?

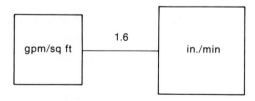

In converting from gallons per minute per square foot to inches per minute filter loading rate, you are moving from a smaller box to a larger box. Therefore, multiplication by 1.6 is indicated:

$$(2.6 \text{ gpm/sq ft})(1.6) = 4.16 \text{ in./min}$$

---

[32] Mathematics Section, Conversions.

Example 4

What is the filter loading rate in gallons per minute per square foot of a sand filter in which the water level dropped 15 in. in 3 min. after the influent valve was closed?

First, determine the inches per minute drop in water level:

$$\frac{15 \text{ in.}}{3 \text{ min}} = 5 \text{ in./min}$$

Next, convert to gallons per minute per square foot.

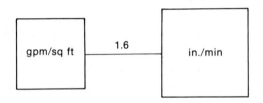

Converting from inches per minute to gallons per day per square foot, you move from a larger box to a smaller box and divide by 1.6:

$$\frac{5 \text{ in./min}}{1.6} = 3.13 \text{ gpm/sq ft}$$

Example 5

A rapid sand filter is 15 ft wide and 40 ft long. If the flow through the filter is 1.6 mgd, what is the filter loading rate in gallons per minute per square foot?

First, convert the gallons per day to gallons per minute:

$$\frac{1,600,000 \text{ gpd}}{1440 \text{ min/day}} = 1111 \text{ gpm}$$

Then express the filter loading rate mathematically as:

$$\text{Filter loading rate} = \frac{\text{gpm flow}}{\text{sq ft filter area}}$$

$$= \frac{1111 \text{ gpm}}{(15 \text{ ft})(40 \text{ ft})}$$

$$= 1.85 \text{ gpm/sq ft}$$

Review Questions

1. A rapid sand filter is 12 ft wide and 20 ft long. If the flow through the filter is 0.83 mgd, what is the filter loading rate in gallons per minute per square foot?

2. How many inches drop per minute corresponds to a filter loading rate of 1.9 gpm/sq ft?

3. What is the filter loading rate in gallons per minute per square foot of a sand filter in which the water level dropped 10 in. in 3 min after the influent valve was closed?

Summary Answers

1. 2.4 gpm/sq ft

2. 3.04 in./min

3. 2.06 gpm/sq ft

Detailed Answers

1. First, convert gallons per day to gallons per minute:

$$\frac{830{,}000 \text{ gpd}}{1440 \text{ min/day}} = 576.39 \text{ gpm}$$

Now express the problem mathematically as:

$$\text{Filter loading rate} = \frac{\text{gpm flow}}{\text{sq ft filter area}}$$

$$= \frac{576.39 \text{ gpm}}{(12 \text{ ft})(20 \text{ ft})}$$

$$= 2.4 \text{ gpm/sq ft}$$

2. Converting from gallons per minute per square foot to inches per minute, you move from a smaller box to a larger box and you multiply by 1.6:

$$(1.9 \text{ gpm/sq ft})(1.6) = 3.04 \text{ in./min}$$

3. First, determine the inches per minute drop in the water level:

$$\frac{10 \text{ in.}}{3 \text{ min}} = 3.33 \text{ in./min}$$

In converting from inches per minute to gallons per minute per square foot, you are moving from a larger box to a smaller one. Therefore, you divide by 1.6:

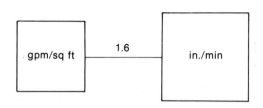

$$\frac{3.33 \text{ in./min}}{1.6} = 2.06 \text{ gpm/sq ft}$$

Mathematics 20

Filter Backwash Rate

The FILTER BACKWASH RATE is measured in gallons of water flowing upward (backward) each minute through a square foot of filter surface area. The units of measure for backwash rate are the same as those for filter loading rate; that is, gallons per *minute* per square foot of filter area.

To assist in the mathematical calculation, the following diagram can be associated with problems involving filter backwash rate.

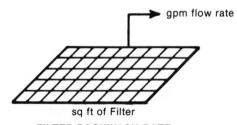

FILTER BACKWASH RATE

The mathematical equation associated with filter backwash rate is

$$\text{Filter Backwash Rate} = \frac{\text{gpm flow}}{\text{sq ft filter area}}$$

Sometimes the backwash rate is measured as the rise of water in *inches* that occurs each minute and can be expressed mathematically as

$$\text{Filter Backwash Rate} = \frac{\text{Inches of water rise}}{\text{Minutes}}$$

Typically, backwash rates will vary from 15 gpm/sq ft to 22.5 gpm/sq ft. This is equivalent to rise in the water level of from 24 to 36 in./min.

Example 1

A rapid sand filter is 10 ft wide and 16 ft long. If backwash water is flowing upward at a rate of 4,950,000 gpd, what is the backwash rate in gallons per minute per square foot?

First convert flow from gallons per day to gallons per minute:[33]

$$\frac{4,950,000 \text{ gpd}}{1440 \text{ min/day}} = 3438 \text{ gpm}$$

Therefore, there are 3438 gpm flowing upward through a filter with a surface area of 160 sq ft.

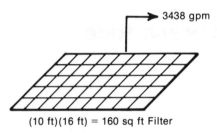

3438 gpm

(10 ft)(16 ft) = 160 sq ft Filter

This can be written mathematically as

$$\text{Filter backwash rate} = \frac{\text{gpm flow}}{\text{sq ft filter area}}$$

$$= \frac{3438 \text{ gpm}}{160 \text{ sq ft}}$$

$$= 21.49 \text{ gpm/sq ft}$$

Example 2

A mixed-media filter is 25 ft wide and 32 ft long. If the filter receives a backwash flow of 17,300,000 gpd, what is the filter backwash flow rate in gpm/sq ft?

As in the last example, first convert the backwash flow from gallons per day to gallons per minute:

$$\frac{17,300,000 \text{ gpd}}{1440 \text{ min/day}} = 12,014 \text{ gpm}$$

12,014 gpm

(25 ft)(32 ft) = 800 sq ft Filter

With this information, calculate filter backwash rate:

---

[33]Mathematics Section, Conversions (Flow Conversions).

$$\text{Filter backwash rate} = \frac{\text{gpm flow}}{\text{sq ft filter area}}$$

$$= \frac{12,014 \text{ gpm}}{(25 \text{ ft})(32 \text{ ft})}$$

$$= \frac{12,014 \text{ gpm}}{800 \text{ sq ft}}$$

$$= 15.02 \text{ gpm/sq ft}$$

Filter backwash rates, as noted earlier, are sometimes expressed in terms of vertical rise of water in a time interval measured in minutes, for example, inches per minute (in./min). The gallons per minute per square foot and inches per minute units of measure are directly related to each other as shown in Example 3.

Example 3

How many inches rise per minute corresponds to a filter backwash rate of 18 gpm/sq ft?

As in the preceding module (Filter Loading Rate), use the box method to convert from gallons per minute per square foot to inches per minute. Moving from a smaller to a larger box indicates multiplication:[34]

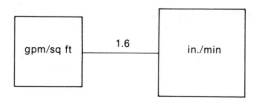

$$(18 \text{ gpm/sq ft})(1.6) = 28.8 \text{ in./min}$$

Review Questions

1. A rapid sand filter is 15 ft wide and 30 ft long. If the backwash water flow rate is 12 mgd, determine the filter backwash rate in gallons per minute per square foot.

2. What is the filter backwash rate in inches per minute corresponding to a filter backwash rate of 16 gpm/sq ft?

3. During backwashing of a rapid sand filter, the operator measured the backwash flow rate to be 27 in./min. Express this filter backwash rate in gallons per minute per square foot.

---

[34] Mathematics Section, Conversions.

Summary Answers

1. 18.52 gpm/sq ft

2. 25.6 in./min

3. 16.88 gpm/sq ft

Detailed Answers

1. First, convert the backwash flow from million gallons per day to gallons per minute:

$$\frac{12,000,000 \text{ gpd}}{1440 \text{ min/day}} = 8333 \text{ gpm}$$

This indicates that 8333 gpm are flowing upward through a filter having an area of 450 sq ft. This can be written mathematically as

$$\text{Filter backwash rate} = \frac{\text{gpm flow}}{\text{sq ft filter area}}$$

$$= \frac{8333 \text{ gpm}}{450 \text{ sq ft}}$$

$$= 18.52 \text{ gpm/sq ft}$$

2.

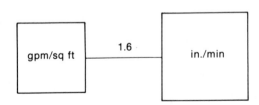

Converting from gallons per minute per square foot to inches per minute, you are moving from a smaller box to a larger box. Therefore multiplication by 1.6 is indicated:

$$(16 \text{ gpm/sq ft})(1.6) = 25.6 \text{ in./min}$$

3.

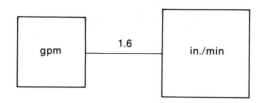

Converting from inches per minute to gallons per minute per square foot, you are moving from a larger box to a smaller box. Therefore you should divide by 1.6:

$$\frac{27 \text{ in./min}}{1.6} = 16.88 \text{ gpm/sq ft}$$

Mathematics 21

Mud Ball Calculation

Mud balls, a conglomeration of floc particles and sand, are a common problem in sand filters. The presence of mud balls is the basic cause of many filter malfunctions. Therefore, measuring and controlling the amount of accumulated mud balls are of prime importance in sand filter operation.

To measure the amount of mud balls in a filter, a tube about 3-in. in diameter and 6-in. long is pushed down into the sand of the filter and then lifted, full of sand. A sieve is used to separate the sand from any mud balls present.

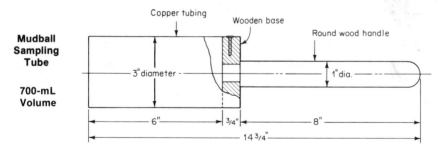

The volume of mud balls can then be determined by placing the balls in a graduated cylinder containing a known amount of water. As mud balls are put in, the water level rises. The volume of mud balls therefore is the *increase in volume* in the graduated cylinder. The percent volume of mud balls can then be determined. Mathematically, this is:[35]

$$\text{Percent volume of mud balls} = \frac{\text{Volume of mud balls}}{\text{Total volume of sample collected}} \times 100$$

---

[35]Mathematics Section, Percent.

The following example problems illustrate the calculation of percent mud balls.

Example 1
A sample of sand taken from the filter for mud ball determination has a volume of 3500 mL. If the volume of mud balls in the sample is found to be 110 mL, what is the percent volume of mud balls in the sample?

$$\text{Percent volume of mud balls} = \frac{\text{Volume of mud balls}}{\text{Total volume of sample}} \times 100$$

$$= \frac{110 \text{ mL}}{3500 \text{ mL}} \times 100$$

$$= 0.0314 \times 100$$

$$= 3.14\% \text{ mud balls}$$

Example 2
A sample of sand taken from the filter for mud ball determination has a volume of 5600 mL. If the volume of mud balls in the sample is found to be 30 mL, what is the percent volume of mud balls in the sample?

$$\text{Percent volume of mud balls} = \frac{\text{Volume of mud balls}}{\text{Total volume of sample}} \times 100$$

$$= \frac{30 \text{ mL}}{5600 \text{ mL}} \times 100$$

$$= 0.0054 \times 100$$

$$= 0.54\% \text{ mud balls}$$

Review Questions

1. A sample taken from the filter for mud ball determination has a volume of 4900 mL. If the volume of mud balls in the sample is found to be 165 mL, what is the percent volume of mud balls in the sample?

2. A mud ball test on a filter shows that there is a 88 mL volume of mud balls in the 3500 mL sample. What is the percent volume of mud balls in the sample?

Summary Answers

1. 3.4% volume mud balls

2. 2.5% volume mud balls

Detailed Answers

1. Percent volume of mud balls $= \dfrac{\text{Volume of mud balls}}{\text{Total volume of sample}} \times 100$

$= \dfrac{165 \text{ mL}}{4900 \text{ mL}} \times 100$

$= 0.034 \times 100$

$= 3.4\%$ volume mud balls

2. Percent volume of mud balls $= \dfrac{\text{Volume of mud balls}}{\text{Total volume of sample}} \times 100$

$= \dfrac{88 \text{ mL}}{3500 \text{ mL}} \times 100$

$= 0.025 \times 100$

$= 2.5\%$ volume mud balls

Mathematics 22

Detention Time

The concept of detention time is used in conjunction with many treatment plant processes including flash mixing, coagulation/flocculation, and sedimentation. DETENTION TIME refers to the length of time a drop of water or a suspended particle remains in a tank or chamber. For example, if water entering a 50-ft long tank has a flow velocity of 1 fps, an average of 50 sec would elapse from the time a drop entered the tank until it left the tank.

Detention time may also be thought of as the number of minutes or hours required for each tank emptying. The mental image might be one of the flow from the time water enters the tank until it leaves the tank completely ("plug flow"), as shown in the following figure.

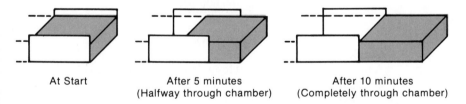

| At Start | After 5 minutes (Halfway through chamber) | After 10 minutes (Completely through chamber) |

TEN MINUTES DETENTION TIME

Typical ranges of detention times for various treatment processes are given below:

Tank	Detention Time Range
Flash mixing basin	30-60 sec
Flocculation basin	20-60 min
Sedimentation basin	1-12 hr

The equation used to calculate detention time is

$$\text{Detention time} = \frac{\text{Volume of tank (in gallons)}}{\text{Flow rate (in gallons/time)}}$$

If the flow rate used in the equation is *gallons per day*, then the detention time calculated will be expressed in *days*. If the flow rate used in the equation is *gallons per minute*, the detention time calculated will be expressed in *minutes*. The calculation method is illustrated in the following examples.

Example 1

A flash mixing basin has a capacity of 1800 gal. If the flow to the mixing basin is 37 gps, what is the detention time?

Since the flow rate is expressed in seconds, the detention time is expressed in seconds

$$\text{Detention time} = \frac{\text{Volume of tank}}{\text{Flow rate}}$$

$$= \frac{1800 \text{ gal}}{37 \text{ gps}}$$

$$= 48.65 \text{ sec}$$

Example 2

A sedimentation tank has a capacity of 60,000 gal. If the hourly flow to the clarifier is 20,800 gph, what is the detention time?

Since the flow rate is expressed in hours, the detention time calculated is also in hours:

$$\text{Detention time} = \frac{\text{Volume of tank}}{\text{Flow rate}}$$

$$= \frac{60,000 \text{ gal}}{20,800 \text{ gph}}$$

$$= 2.88 \text{ hr}$$

Example 3

A flocculation basin is 20 ft long, 10 ft wide, and has a side water depth (SWD) of 10 ft. If the flow to the basin is 520 gpm, what is the detention time?

Since the flow rate is expressed in minutes, the detention time calculated is also in minutes:

$$\text{Detention time} = \frac{\text{Volume of tank}}{\text{Flow rate}}$$

$$= \frac{(20 \text{ ft})(10 \text{ ft})(10 \text{ ft})(7.48 \text{ gal/cu ft})}{520 \text{ gpm}}$$

$$= \frac{14,960 \text{ gal}}{520 \text{ gpm}}$$

$$= 28.77 \text{ min}$$

Example 4

The flow rate through a circular clarifier is 4,752,000 gpd. If the clarifier is 65 ft in diameter and 12 ft deep, what is the clarifier detention time in hours?

To calculate a detention time in hours, first express the gallons-per-day flow rate as gallons per hour.[36]

$$\frac{4,752,000 \text{ gpd}}{24 \text{ hr/day}} = 198,000 \text{ gph}$$

Then calculate the detention time:

$$\text{Detention time} = \frac{\text{Volume of tank}}{\text{Flow rate}}$$

$$= \frac{(0.785)(65 \text{ ft})(65 \text{ ft})(12 \text{ ft})(7.48 \text{ gal/cu ft})}{198,000 \text{ gph}}$$

$$= \frac{297,700 \text{ gal}}{198,000 \text{ gph}}$$

$$= 1.5 \text{ hr}$$

Review Questions

1. A flash mixing basin has a capacity of 2400 gal. If the flow rate to the basin is 43 gps, what is the detention time?

2. A sedimentation tank is 60 ft long, 12 ft wide, and has water to a depth of 12 ft. If the flow rate to the tank is 21,600 gph, what is the detention time?

3. A flocculation basin has a capacity of 30,000 gal. If the flow rate to the basin is 1,728,000 gpd, what is the detention time in minutes?

4. A circular clarifier receives a flow of 4.2 mgd. If the detention time in the clarifier is 1.8 hr, what is the capacity of the clarifier?

---

[36]Chemistry Section, Chemical Dosage Problems (Feed Rate Problems).

Summary Answers

1. 55.81 sec

2. 2.99 hr

3. 25 min

4. 315,000 gal

Detailed Answers

1. Since the flow rate is expressed in seconds, the calculated detention time is in seconds:

$$\text{Detention time} = \frac{\text{Volume of tank}}{\text{Flow rate}}$$

$$= \frac{2400 \text{ gal}}{43 \text{ gps}}$$

$$= 55.81 \text{ sec}$$

2. Since the flow rate is expressed in hours, the calculated detention time is in hours:

$$\text{Detention time} = \frac{\text{Volume of tank}}{\text{Flow rate}}$$

$$= \frac{(60 \text{ ft})(12 \text{ ft})(12 \text{ ft})(7.48 \text{ gal/cu ft})}{21,600 \text{ gph}}$$

$$= \frac{64,627 \text{ gal}}{21,600 \text{ gph}}$$

$$= 2.99 \text{ hr}$$

3. Since the problem requires detention time to be expressed in *minutes*, the flow rate must first be converted from gallons per day to gallons per minute:[37]

$$\frac{1,728,000 \text{ gpd}}{1440 \text{ min/day}} = 1200 \text{ gpm}$$

---

[37]Mathematics Section, Conversions (Flow Conversions).

And now the detention time can be calculated:

$$\text{Detention time} = \frac{\text{Volume of tank}}{\text{Flow rate}}$$

$$= \frac{30,000 \text{ gal}}{1200 \text{ gpm}}$$

$$= 25 \text{ min}$$

4. In this problem, since the detention time is expressed in *hours,* flow rate must be in corresponding units. Therefore flow rate is expressed in terms of gallons per hour:

$$\frac{4,200,000 \text{ gpd}}{24 \text{ hr/day}} = 175,000 \text{ gph}$$

Although the unknown in this problem is different than in previous problems, the solution can be approached in the same way. Fill in the given information:

$$\text{Detention time} = \frac{\text{Volume of tank}}{\text{Flow rate}}$$

$$1.8 \text{ hr} = \frac{x \text{ gal}}{175,000 \text{ gph}}$$

And then solve for the unknown value:

$$(175,000)(1.8) = x$$

$$315,000 \text{ gal} = x \text{ gal}$$

Mathematics 23

Well Problems

The three basic calculations related to water well performance are well yield, drawdown, and specific capacity. These calculations provide information for selecting appropriate pumping equipment and for identifying any changes in the productive capacity of the well.

M23-1. Well Yield

The WELL YIELD is the volume of water that is discharged from a well during a specified time period. This discharge may be a result of pumping, as in most cases, or of free flow in the case of a flowing artesian well. Well yield is normally measured as the pumping rate in gallons per minute (gpm). This can be expressed mathematically as

$$\text{Well yield} = \frac{\text{Gallons}}{\text{Minute}}$$

The easiest method of measuring well yield is by placing a flow meter (usually an orifice meter) on the downstream, or discharge, side of the pump. However, to determine relatively small well yields, another method often used is measuring the time required to fill a container with a known volume. The following examples illustrate the second method of calculating well yield. A similar calculation is discussed elsewhere in this handbook.[38]

Example 1
During a test for well yield, the time required to fill a 55-gal barrel was 25 sec. Based on this pumping rate, what was the well yield in gallons per minute?

$$\text{Well yield} = \frac{\text{Gallons}}{\text{Minute}}$$

---

[38] Hydraulics Section, Pumping Problems (Pumping Rates).

The well yield is to be reported in gallons per *minute,* so first express the 25 sec given in the problem as minutes:

$$\frac{25 \text{ sec}}{60 \text{ sec/min}} = 0.42 \text{ min}$$

Now fill in the information given and complete the calculation:

$$\text{Well yield} = \frac{55 \text{ gal}}{0.42 \text{ min}}$$

$$= 130.95 \text{ gpm}$$

Example 2

If it takes a well pump 41 sec to fill a 55-gal barrel, what is the well yield in gallons per minute?

The equation used in calculating well yield

$$\text{Well yield} = \frac{\text{Gallons}}{\text{Minute}}$$

Before filling in the equation with the given information, the seconds must be expressed as minutes:

$$\frac{41 \text{ sec}}{60 \text{ sec/min}} = 0.68 \text{ min}$$

The well yield problem can now be solved by filling in the given information and completing the division indicated:

$$\text{Well yield} = \frac{55 \text{ gal}}{0.68 \text{ min}}$$

$$= 80.88 \text{ gpm}$$

Example 3

What is the well yield in gallons per minute if it takes the pump 3.5 min to fill a tank 5 ft square to a depth of 3 ft?

The equation used in calculating well yield is:

$$\text{Well yield} = \frac{\text{Gallons}}{\text{Minute}}$$

Before information can be filled in the equation, the gallon volume of the tank must be calculated:[39]

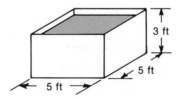

3 ft

5 ft

5 ft

---

[39]Mathematics Section, Volume Measurements.

$$\text{Volume} = (\text{Area})(\text{Third dimension})$$
$$= (5 \text{ ft})(5 \text{ ft})(3 \text{ ft})$$
$$= 75 \text{ cu ft}$$

And converting volume from cubic feet to gallons:[40]

$$(75 \text{ cu ft})(7.48 \text{ gal}/\text{cu ft}) = 561 \text{ gal}$$

Now the well yield problem can be determined by filling in the given information and completing the calculation:

$$\text{Well yield} = \frac{561 \text{ gal}}{3.5 \text{ min}}$$
$$= 160.29 \text{ gpm}$$

M23-2. Drawdown

The DRAWDOWN of a well is the amount the water level *drops* once pumping begins. As illustrated below, drawdown is the difference between the STATIC WATER LEVEL, (SWL, the level when no water is being taken from the aquifer, either by pumping or by free flow) and the PUMPING WATER LEVEL (PWL, the level when the pump is in operation).

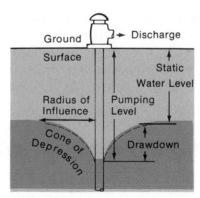

This is expressed mathematically as:

Drawdown = Pumping water level − Static water level

The following two examples illustrate how to calculate well drawdown.

Example 4

The water level in a well is 25 ft below the ground surface when the pump is not in operation. If the water level is 48 ft below the ground surface when the pump is in operation, what is the drawdown in feet?

---

[40]Mathematics Section, Conversions (Cubic Feet to Gallons to Pounds).

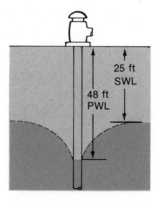

Drawdown is the measure of water level *drop* once the pump has been turned on. In this problem, the drawdown is

Drawdown = Pumping water level − Static water level

= 48 ft − 25 ft

= 23 ft drawdown

Example 5
When the pump is not in operation the water level in the well is 39 ft below ground surface. The water level drops to 57 ft when the pump is in operation. What is the drawdown in ft?

Drawdown = Pumping water level − Static water level

= 57 ft − 39 ft

= 18 ft drawdown

M23-3. Specific Capacity

The SPECIFIC CAPACITY of a well is a measure of the well yield per unit of drawdown. It is usually expressed in terms of gallons-per-minute well yield per foot of drawdown. Therefore, the equation used to calculate specific capacity is

$$\text{Specific capacity} = \frac{\text{Well yield in gpm}}{\text{Drawdown in ft}}$$

The following examples illustrate the calculation of specific capacity.

Example 6
If the well yield is 170 gpm when the drawdown for the well is 18 ft, what is the specific capacity of the well?
The equation used in calculating specific capacity is

$$\text{Specific capacity} = \frac{\text{Well yield in gpm}}{\text{Drawdown in ft}}$$

To solve the problem, fill in the information given in the equation and complete the calculation:

$$\text{Specific capacity} = \frac{170 \text{ gpm}}{18 \text{ ft}}$$

$$= 9.44 \text{ gpm/ft}$$

Example 7

It takes a well pump 0.6 min to fill a 55-gal barrel. If the drawdown while the pump is in operation is 11 ft, what is the specific capacity of the well?

To calculate the specific capacity of the well, you must know the gallons-per-minute well yield and the feet of drawdown.

$$\text{Well yield} = \frac{55 \text{ gal}}{0.6 \text{ min}}$$

$$= 91.67 \text{ gpm}$$

Using the gallons-per-minute well yield and drawdown information, calculate the specific capacity of the well:

$$\text{Specific capacity} = \frac{\text{Well yield in gpm}}{\text{Drawdown in ft}}$$

$$= \frac{91.67 \text{ gpm}}{11 \text{ ft}}$$

$$= 8.33 \text{ gpm/ft}$$

Review Questions

1. For a certain test of well yield it took 29 sec to fill a 55-gal barrel. Based on this pumping rate, what is the well yield in gallons per minute?

2. What is the well yield if it takes a pump 1 min 46 sec to fill a tank 3-ft wide and 4-ft long to a depth of 3 ft?

3. The water level in a well is 57 ft from ground surface when the pump is not in operation. If the water level is 91 ft below ground surface when the pump is in operation, what is the drawdown in feet?

4. If the well yield was 210 gpm when the drawdown for the well was 27 ft, what was the specific capacity of the well at that time?

5. It takes a pump 0.34 min to fill a 55-gal barrel. If the water level in the well is 70 ft below ground surface when the pump is not in operation, and 83 ft below ground surface when the pump is in operation, what is the specific capacity of the well?

Summary Answers

1. 114.58 gpm well yield

2. 152.14 gpm well yield

3. 34 ft drawdown

4. 7.78 gpm/ft specific capacity

5. 12.44 gpm/ft specific capacity

Detailed Answers

1. Since the well yield is desired in gallons per minute, first express 29 sec as minutes:

$$\frac{29 \text{ sec}}{60 \text{ sec/min}} = 0.48 \text{ min}$$

Now solve the well yield problem by filling in the given information and completing the calculation:

$$\text{Well yield} = \frac{\text{Gallons}}{\text{Minute}}$$

$$= \frac{55 \text{ gal}}{0.48 \text{ min}}$$

$$= 114.58 \text{ gpm well yield}$$

2. The equation used in calculating well yield in gpm is:

$$\text{Well yield} = \frac{\text{Gallons}}{\text{Minute}}$$

Before information can be filled in the equation, the gallon volume of the tank must be calculated and the time must be expressed as minutes. First, calculate the gallon volume of the tank:

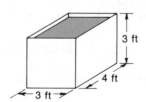

$$\text{Volume} = (\text{Area})(\text{Third dimension})$$
$$= (3 \text{ ft})(4 \text{ ft})(3 \text{ ft})$$
$$= 36 \text{ cu ft}$$

Then convert volume from cubic feet to gallons:

$$(36 \text{ cu ft})(7.48 \text{ gal/cu ft}) = 269.28 \text{ gal}$$

Now the 1 min 46 sec must be expressed as minutes. Therefore, convert the 46 sec to minutes:

$$\frac{46 \text{ sec}}{60 \text{ sec/min}} = 0.77 \text{ min}$$

Therefore, the total pumping time is 1.77 min. The well yield problem can now be solved by filling in known information and completing the calculation:

$$\text{Well yield} = \frac{269.28 \text{ gal}}{1.77 \text{ min}}$$
$$= 152.14 \text{ gpm}$$

3. Drawdown is the measure of water level *drop* once the pump has been turned on. In this problem, the water level dropped from 57 ft below the ground (pump off) to 91 ft below the ground (pump on). Therefore, the drawdown is

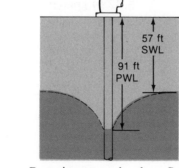

$$\text{Drawdown} = \text{Pumping water level} - \text{Static water level}$$
$$= 91 \text{ ft} - 57 \text{ ft}$$
$$= 34 \text{ ft drawdown}$$

4. Fill the information given in the problem into the specific capacity equation:

$$\text{Specific capacity} = \frac{\text{Well yield in gpm}}{\text{Drawdown in ft}}$$
$$= \frac{210 \text{ gpm}}{27 \text{ ft}}$$
$$= 7.78 \text{ gpm/ft}$$

5. To calculate the specific gravity of the well, you must know the gallons-per-minute well yield and the feet of drawdown. Using the information given in the problem, first calculate the gallons-per-minute well yield, then the feet of drawdown:

$$\text{Well yield} = \frac{\text{Gallons}}{\text{Minute}}$$

$$= \frac{55 \text{ gal}}{0.34 \text{ min}}$$

$$= 161.76 \text{ gpm}$$

$$\text{Drawdown} = \text{Pumping water level} - \text{Static water level}$$

$$= 83 \text{ ft} - 70 \text{ ft}$$

$$= 13 \text{ ft drawdown}$$

Using the gallons-per-minute well yield and drawdown information, calculate the specific capacity of the well:

$$\text{Specific capacity} = \frac{\text{Well yield in gpm}}{\text{Drawdown in ft}}$$

$$= \frac{161.76 \text{ gpm}}{13 \text{ ft}}$$

$$= 12.44 \text{ gpm/ft}$$

**Basic Science
Concepts and
Applications**

Hydraulics

Hydraulics 1

Density and Specific Gravity

H1-1. Density

In common speech, when we say that one substance is heavier than another we mean that any given volume of the substance is heavier than the same volume of the other substance. For example, any given volume of steel is heavier than the same volume of aluminum, so we say that steel is heavier than—or has greater density than—aluminum.

For scientific and technical purposes, the DENSITY of a body or material is precisely defined as *the weight per unit of volume.* In the water supply field, perhaps the most common measures of density are pounds per cubic foot (lb/cu ft or lb/ft^3) and pounds per gallon (lb/gal). For example, the density of dry materials, such as sand, activated carbon, lime, and soda ash, is usually expressed in pounds per cubic foot. The density of liquids, such as water, liquid alum, or liquid chlorine, can be expressed either as pounds per cubic foot or as pounds per gallon. The density of gases, such as air, chlorine gas, methane, or carbon dioxide, is normally expressed in pounds per cubic foot.

The density of a substance *changes slightly as the temperature of the substance changes.* This happens because substances usually increase in size (volume) as they become warmer, as illustrated in Figure 1. Because of expansion with warming, the same weight is spread over a larger volume, so the density is less when a substance is warm than when it is cold.

Similarly, a change in pressure will change the volume occupied by a substance. As a result, *density varies with pressure*, increasing as pressure increases and decreasing as pressure decreases (Figure 2).

The effects of pressure and temperature on solids and liquids are very small and are usually ignored. However, temperature and pressure have a significant effect on the density of gases and whenever the density of a gas is given, then the temperature and pressure at that density are usually also given.

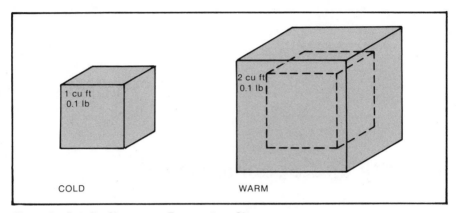

Figure 1.   Density Changes as Temperature Changes

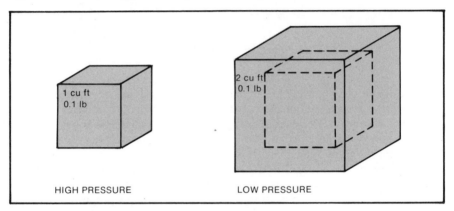

Figure 2.   Density Changes as Pressure Changes

Table 1 indicates how the density of water (usually specified as 62.4 lb/cu ft), varies with temperature. Water is unusual in that it is most dense at 39.2°F (4.0°C), and it becomes less dense below or above that temperature.

Table 2 shows some densities of typical solid, liquid, and gaseous substances. You'll notice that some of the solids given in Table 2 have density reported as BULK DENSITY. Bulk density is defined as the weight of a cubic foot of material as it would be shipped from the supplier to the treatment plant. Bulk density is much less than laboratory density because its calculation includes the volume of the air mixed in with the material; the amount of air (and therefore the bulk density) varies according to whether the material comes in rock, crystal, pellet, granular, or powder form. For example, for laboratory purposes the density of pure sodium chloride (table salt) is about 135 lb/cu ft. However, the bulk density of sodium chloride as it is shipped in rock form (rock salt) is only 50-60 lb/cu ft. This means that over half the volume of a bulk container of rock salt is occupied by air between the individual pieces.

Table 1.   Density of Water

Temperature		Density	Temperature		Density
°F	°C	lb/cu ft	°F	°C	lb/cu ft
32.0	0	62.416	90.0	32.2	62.118
35.0	1.7	62.421	100.0	37.8	61.998
39.2	4.0	62.424	120.0	48.9	61.719
40.0	4.4	62.423	140.0	60.0	61.386
50.0	10.0	62.408	160.0	71.1	61:006
60.0	15.6	62.366	180.0	82.2	60.586
70.0	21.1	62.300	200.0	93.3	60.135
80.0	26.7	62.217	212.0	100.0	59.843

Table 2.   Densities of Various Substances

Substance	Density	
	lb/cu ft	lb/gal
Solids		
Activated carbon*†	8–28 (Avg. 12)	
Lime*†	20–50	
Dry alum*†	60–75	
Aluminum (at 20°C)	168.5	
Steel (at 20°C)	486.7	
Copper (at 20°C)	555.4	
Liquids		
Propane (–44.5°C)	36.5	4.88
Gasoline†	43.7	5.84
Water (4°C)	62.4	8.34
Fluosilicic acid		
(30%, –8.1°C)	77.8–79.2	10.4–10.6
Liquid alum		
(36°Be, 15.6°C)	83.0	11.09
Liquid chlorine		
(–33.6°C)	97.3	13.01
Sulfuric acid		
(18°C)	114.2	15.3
Gases		
Methane (0°C, 14.7 psia)	0.0344	
Air (20°C, 14.7 psia)	0.075	
Oxygen (0°C, 14.7 psia)	0.089	
Hydrogen sulfide†	0.089	
Carbon Dioxide†	0.115	
Chlorine Gas (0°C, 14.7 psia)	0.187	

*Bulk density of substance.
†Temperature and/or pressure not given.

H1-2. Specific Gravity

Since density can be expressed as pounds per cubic foot, pounds per gallon, pounds per cubic inch, or even grams per cubic centimeter, it is sometimes difficult to compare the density of one substance with another. Specific gravity is one way around this problem. Although there may be many numbers that express the density of the same substance (depending on the units used) there is only one specific gravity associated with each substance (for one particular temperature and pressure). The SPECIFIC GRAVITY of a substance is the density of that substance compared to a "standard" density.

Specific Gravity of Solids and Liquids

The standard density used for solids and liquids is that of water, 62.4 lb/cu ft, or 8.34 lb/gal. Therefore, the specific gravity of a solid or liquid is the density of that solid or liquid *compared to the density of water*. It is the RATIO[1] of the density of that substance to the density of water.

For example, the density of granite rock is about 162 lb/cu ft, and the density of water is 62.4 lb/cu ft. The specific gravity of granite is found by this ratio:

$$\text{Specific Gravity} = \frac{162 \text{ lb/cu ft}}{62.4 \text{ lb/cu ft}}$$

$$= 2.60$$

In this case, the specific gravity (the ratio of the density of granite to the density of water) indicates that a cubic foot of granite weighs about 2 1/2 times as much as a cubic foot of water.

Let's look at another example. The density of SAE 30 motor oil is about 56 lb/cu ft. Its specific gravity is therefore

$$\text{Specific Gravity} = \frac{56 \text{ lb/cu ft}}{62.4 \text{ lb/cu ft}}$$

$$= 0.90$$

In other words, specific gravity in this example tells you that oil is only 9/10 as dense as water. Because a cubic foot of oil weighs less than a cubic foot of water, oil floats on the surface of water.

Table 3 lists specific gravities for various liquids and solids.

Example 1
Aluminum weighs approximately 168 lb/cu ft. What is the specific gravity of aluminum?

To calculate specific gravity, compare the weight of a cubic foot of aluminum with the weight of a cubic foot of water:

---

[1]Mathematics Section, Ratios and Proportions.

Table 3. Specific Gravities of Various Solids and Liquids

Substance	Specific Gravity
Solids	
Aluminum (20°C)	2.7
Steel (20°C)	7.8
Copper (20°C)	8.9
Activated carbon*†	0.13–0.45 (Avg. 0.19)
Lime*†	0.32–0.80
Dry alum*†	0.96–1.2
Soda ash*†	0.48–1.04
Coagulant aids (polyelectrolytes)*†	0.43–0.56
Table salt*†	0.77–1.12
Liquids	
Liquid alum (36°Be, 15.6°C)	1.33
Water (4°C)	1.00
Fluosilicic acid (30%, –8.1°C)	1.25–1.27
Sulfuric acid (18°C)	1.83
Ferric chloride (30%, 30°C)	1.34

*Bulk density used to determine specific gravity.
†Temperature and/or pressure not given.

$$\frac{\text{Specific Gravity}}{\text{of Aluminum}} = \frac{\text{aluminum lb/cu ft}}{\text{water lb/cu ft}}$$

$$= \frac{168 \text{ lb/cu ft}}{62.4 \text{ lb/cu ft}}$$

$$= 2.69$$

Example 2

If the specific gravity of a certain oil is 0.92, what is the density (in pounds per cubic foot) of that oil?

Approach this problem in the same general way as the previous example. First fill in the given information:

$$\frac{\text{Specific Gravity}}{\text{of Oil}} = \frac{\text{oil lb/cu ft}}{\text{water lb/cu ft}}$$

$$0.92 = \frac{x \text{ lb/cu ft}}{62.4 \text{ lb/cu ft}}$$

Then solve for the unknown value:[2]

$$(62.4)(0.92) = x$$

$$57.41 \text{ lb/cu ft} = x$$

Therefore the density of the oil is 57.41 lb/cu ft.

The most common use of specific gravity in water treatment operations is in gallons-to-pounds conversions. In many such cases the liquids being handled have a specific gravity of 1.00 or very nearly 1.00 (between 0.98 and 1.02), so 1.00

---

[2]Mathematics Section, Solving for the Unknown Value.

may be used in the calculations without introducing more than a 2–percent error. However, in calculations involving a liquid with a specific gravity less than 0.98 or greater than 1.02 (such as liquid alum), the conversions from gallons to pounds must take specific gravity into account. The technique is illustrated in the following two examples.

Example 3
There are 1240 gal of a certain liquid in a tank. If the specific gravity of the liquid is 0.93, how many pounds of liquid are in the tank?

Normally, in converting from gallons to pounds, the factor 8.34 lb/gal would be used.[3] However, in this example the substance has a specific gravity different than water (1.0), so the 8.34 factor must be adjusted. Multiply the 8.34 lb/gal by the specific gravity to obtain the adjusted factor:

$$(8.34 \ \text{lb/gal})(0.93) = 7.76 \ \text{lb/gal}$$

Then convert 1240 gal to pounds using the corrected factor:

$$(1240 \ \text{gal})(7.76 \ \text{lb/gal}) = 9622 \ \text{lb}$$

Example 4
Suppose you wish to pump a certain liquid at the rate of 25 gpm. How many pounds per day will you be pumping if the liquid weighs 74.9 lb/cu ft?

If the liquid being pumped were water, then you would make the following calculations:[4]

$$(25 \ \text{gpm})(8.34 \ \text{lb/gal})(1440 \ \text{min/day}) = 300,240 \ \text{lb/day}$$

However, the liquid being pumped has a greater density than water; therefore, you will have to adjust the factor of 8.34 lb/gal, as follows: first, determine the specific gravity of the liquid:

$$\begin{aligned}\frac{\text{Specific Gravity}}{\text{of the Liquid}} &= \frac{\text{lb/cu ft of the liquid}}{\text{lb/cu ft of water}} \\ &= \frac{74.9 \ \text{lb/cu ft}}{62.4 \ \text{lb/cu ft}} \\ &= 1.20\end{aligned}$$

Next, calculate the corrected lb/gal factor:

$$(8.34 \ \text{lb/gal})(1.20) = 10.01 \ \text{lb/gal}$$

Then convert the gpm flow rate to lb/day:

$$(25 \ \text{gpm})(10.01 \ \text{lb/gal})(1440 \ \text{min/day}) = 360,360 \ \text{lb/day} \ \text{pumped}$$

---

[3] Mathematics Section, Conversions (Cubic Feet to Gallons to Pounds).

[4] Mathematics Section, Conversions (Flow Rate).

Specific Gravity of Gases

The specific gravity of a gas is usually determined by comparing the density of the gas with the density of air, which is 0.075 lb/cu ft at a temperature of 20°C and a pressure of 14.7 psia (pounds per square inch absolute)—the pressure of the atmosphere at sea level.[5] For example, the density of chlorine gas is 0.187 lb/cu ft. Its specific gravity would be calculated as follows:

$$\text{Specific Gravity of Cl}_2 \text{ Gas} = \frac{\text{Cl}_2 \text{ gas lb/cu ft}}{\text{air lb/cu ft}}$$

$$= \frac{0.187 \text{ lb/cu ft}}{0.075 \text{ lb/cu ft}}$$

$$= 2.49$$

This tells you that chlorine gas is about 2 1/2 times as dense as air. Therefore, when chlorine gas is introduced into a room it will concentrate at the bottom of the room. This is important to know since chlorine is a deadly poisonous gas. Table 4 lists specific gravities for various gases.

Table 4. Specific Gravities of Various Gases

Gas	Specific Gravity	
Hydrogen (0°C; 14.7 psia)	0.07	When released in
Methane (0°C; 14.7 psia)	0.46	a room will first
Carbon Monoxide*	0.97	rise to the ceiling area.
Air (20°C; 14.7 psia)	1.00	
Nitrogen (0°C; 14.7 psia)	1.04	When released in
Oxygen (0°C; 14.7 psia)	1.19	a room will first
Hydrogen Sulfide*	1.19	settle to the
Carbon Dioxide*	1.53	floor area.
Chlorine Gas (0°C; 14.7 psia)	2.49	
Gasoline Vapor*	3.0	

*Temperature and pressure not given.

Review Questions

1. The weight of 3 cu ft of mercury is 2546 lb. (a) What is the density of mercury? (b) What is the specific gravity of mercury?

2. The specific gravity of ammonia gas is 0.60. If released, will this gas accumulate first in the ceiling area or the floor area of a room?

3. If the specific gravity of a load of copper sulfate crystals is 1.41, what is the density (pounds per cubic foot) of the copper sulfate crystals?

4. You are pumping 30 gpm of a liquid which has a specific gravity of 1.14. At this rate, how many pounds per day are you pumping?

---

[5] Hydraulics Section, Pressure and Force.

Summary Answers

1. (a) 848.67 lb/cu ft; (b) 13.6
2. Ceiling area
3. 87.98 lb/cu ft
4. 410,832 lb/day

Detailed Answers

1. (a)

$$\text{Density} = \frac{\text{lb}}{\text{cu ft}}$$

$$= \frac{2546 \text{ lb mercury}}{3 \text{ cu ft}}$$

$$= 848.67 \text{ lb/cu ft}$$

(b)

$$\frac{\text{Specific Gravity}}{\text{of Mercury}} = \frac{\text{mercury lb/cu ft}}{\text{water lb/cu ft}}$$

$$= \frac{848.67 \text{ lb/cu ft}}{62.4 \text{ lb/cu ft}}$$

$$= 13.6$$

2. The specific gravity of air is 1.00. Any gas with a smaller specific gravity than air is lighter than air; hence, if released, will first rise to the ceiling area of a room. The gas described in this problem (NH_3) has a specific gravity *less* than air; therefore it will first accumulate in the ceiling area of the room.

3. Use the equation for specific gravity to solve this problem:

$$\frac{\text{Specific Gravity of}}{\text{Copper Sulfate Crystals}} = \frac{\text{copper sulfate lb/cu ft}}{\text{water lb/cu ft}}$$

$$1.41 = \frac{x \text{ lb/cu ft}}{62.4 \text{ lb/cu ft}}$$

Solve for the unknown value:

$$(62.4)(1.41) = x \text{ lb/cu ft}$$

$$87.98 = x \text{ lb/cu ft}$$

Therefore the density of this load of copper sulfate crystals is 87.98 lb/cu ft.

4. Before making the gallons per minute conversion to pounds per day, correct the 8.34 lb/gal factor:

$$(8.34 \text{ lb/gal})(1.14) = 9.51 \text{ lb/gal}$$

Now continue with the conversion:

$$(30 \text{ gpm})(9.51 \text{ lb/gal})(1440 \text{ min/day}) = 410,832 \text{ lb/day}$$

Hydraulics 2

Pressure and Force

H2-1. Pressure

The flow of water in a system is dependent on the amount of force causing the water to move. PRESSURE is defined as the amount of force acting (pushing) on a unit area.

As shown in Figure 3, pressure may be expressed in different ways depending on the unit area selected. Normally however, the unit area of a square inch and the expression of pressures in pounds per square inch (psi) are preferred.

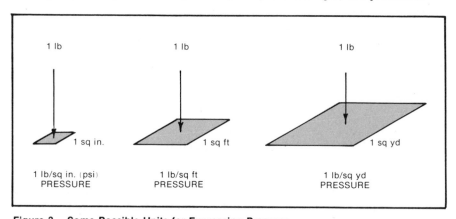

Figure 3. Some Possible Units for Expressing Pressure

In the operation of a water treatment system, you will be primarily concerned with the pressures exerted by water. Water pressures are directly related to both the height (depth) and density of water. Suppose, for example, you have a container 1 ft by 1 ft by 1 ft (a cubic-foot container) which is filled with water. What is the pressure on the square-foot bottom of the container?

Pressure is expressed in pounds per unit area. In this case, since the density of water is 62.4 lb/cu ft, the force of the water pushing down on the square foot surface area is 62.4 lb (Figure 4).

From this information, the pressure in pounds per square inch (psi) can also be determined. Convert pounds per square foot to pounds per square inch (psi):

$$62.4 \ \frac{lb}{sq \ ft} = \frac{62.4 \ lb}{(1 \ ft)(1 \ ft)}$$

$$= \frac{62.4 \ lb}{(12 \ in.)(12 \ in.)}$$

$$= \frac{62.4 \ lb}{144 \ sq \ in.}$$

$$= 0.433 \ \frac{lb}{sq \ in.}$$

$$= 0.433 \ psi$$

This means that a foot-high column of water above a square-inch surface area weighs about half a pound (0.433 lb), resulting in a pressure of 0.433 psi (Figure 5).

The factor 0.433 allows you to convert from pressure measured in feet of water to pressure measured in pounds per square inch, as shown later in this section. A conversion factor can also be developed for converting from psi to feet, as follows:

Since 1 ft is equivalent to 0.433 psi, set up a ratio[6] to determine how many feet are equivalent to 1 psi:

$$\frac{1 \ ft}{0.433 \ psi} = \frac{x \ ft}{1 \ psi}$$

Then solve for the unknown value:[7]

$$\frac{(1)(1)}{0.433} = x$$

$$2.31 \ ft = x$$

Therefore 1 psi is equivalent to the pressure created by a column of water 2.31 ft high (Figure 6).

Since the density of water is assumed to be a constant 62.4 lb/cu ft in most hydraulic calculations, the height of the water is the most important factor in determining pressures. It is the height of the water which determines the pressure over the square-inch area. Notice that, as long as the height of the water stays the same, changing the shape of the container does not change the pressure at the bottom or at any level in the container, as illustrated in Figure 7.

Each of the containers is filled with water to the same height. Therefore the

---

[6]Mathematics Section, Ratios and Proportions.
[7]Mathematics Section, Solving for the Unknown Value.

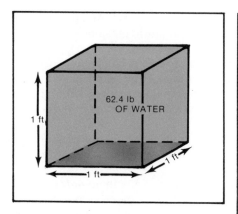

Figure 4.   Cubic Foot of Water

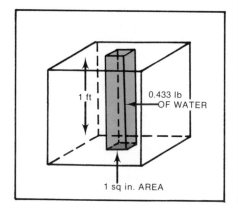

Figure 5.   1 ft Water = 0.433 psi

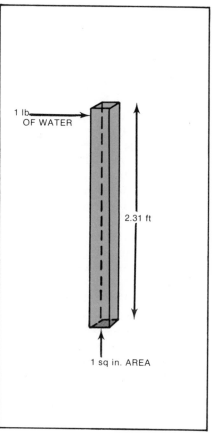

Figure 6.   1 psi = 2.31 ft Water

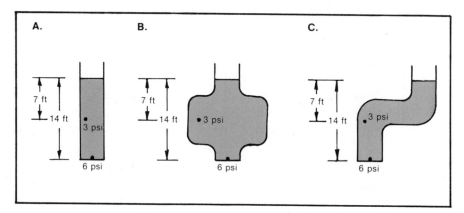

Figure 7.   Pressure Depends on Weight of Water, Not Shape of Container

pressures against the bottoms of the containers are the same. And the pressure at any depth in one container is the same as the pressure at the same depth in either of the other containers.

In water system operation, the shape of the container can help to maintain a usable volume of water at higher pressures. For example, suppose you have an elevated storage tank and a standpipe that contain equal amounts of water. When the water levels are the same, the pressures at the bottom of the tanks are the same (Figure 8A).

However, if half of the water is withdrawn from each tank, the pressure at the bottom of the elevated tank will be greater than the pressure at the bottom of the standpipe (Figure 8B).

Because of the direct relationship between the pressure in *pounds per square inch* at any point in water and the *height in feet* of water above that point, pressure can be measured either in pounds per square inch or in feet (of water), called "head."

When the water pressure in a main or in a container is measured by a gage the pressure is referred to as GAGE PRESSURE. If measured in pounds per square inch, the gage pressure is expressed as POUNDS PER SQUARE INCH GAGE (PSIG). The gage pressure is not the total pressure within the main. Gage pressure does not show the pressure of the atmosphere, which is equal to approximately 14.7 psi at sea level. Because atmospheric pressure is exerted everywhere (against the outside of the main as well as the inside, for instance), it can generally be neglected in water system calculations. However, for certain calculations the total (or absolute) pressure must be known. The ABSOLUTE PRESSURE IN POUNDS PER SQUARE INCH (PSIA) is obtained by adding the gage and atmospheric pressures. For example, in a main under 50-psi gage pressure, the absolute pressure would be 50 psig + 14.7 psi = 64.7 psia. A line under a partial vacuum, with a negative gage pressure of −2, would have an absolute pressure of (14.7 psi) + (−2 psig) = 12.7 psia. (See Figure 9.)

Pressure gages can also be calibrated in feet of head. A pressure gage reading

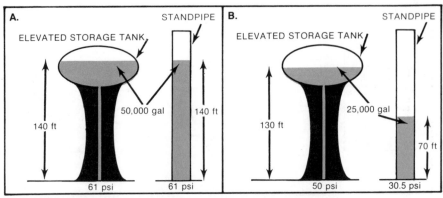

Figure 8. Shape of Container Used to Keep Water at a More Nearly Constant Height, Reducing Pressure Variations.

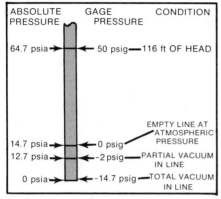

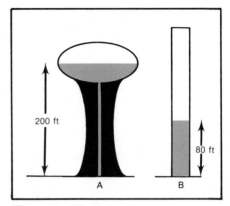

Figure 9.   Gage vs. Absolute Pressure   **Figure 10.   (Example 4.)**

of 14 ft of head, for example, means that the pressure is equivalent to the pressure exerted by a column of water 14 ft high. The equations that relate gage pressure in pounds per square inch to pressure in feet of head are given below. In this handbook, conversions from one term to another will use the first equation only.

$$1 \text{ psig} = 2.31 \text{ ft head}$$
$$1 \text{ ft head} = 0.433 \text{ psig}$$

The following example problems illustrate the conversion from feet of head to pounds per square inch gage and from pounds per square inch gage to feet of head. The method used is similar to the conversion approach (box method) discussed in the mathematics section.[8]

Example 1

Convert a pressure of 14 ft to pounds per square inch gage.

$$1 \text{ psig} = 2.31 \text{ ft head}$$

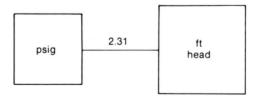

Using the diagram above, in moving from ft head to psig, you are moving from a larger box to a smaller box. Therefore, you should divide by 2.31:

$$\frac{14 \text{ ft}}{2.31 \text{ ft/psig}} = 6.06 \text{ psig}$$

---

[8] Mathematics Section, Conversions.

Example 2

A head of 250 ft of water is equivalent to what pressure in pounds per square inch gage?

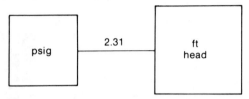

When you move from a larger box to a smaller box, division by 2.31 is indicated:

$$\frac{250 \ \text{ft}}{2.31 \ \text{ft}/\text{psig}} = 108.23 \ \text{psig}$$

Example 3

A pressure of 32 psig is equivalent to how many feet of water?

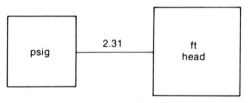

When you move from a smaller box to a larger box, multiplication by 2.31 is indicated:

$$(32 \ \text{psig})(2.31 \ \text{ft}/\text{psig}) = 73.92 \ \text{ft} \ \text{head}$$

Example 4

What would the pounds per square inch gage pressure readings be at points A and B in the diagram in Figure 10?

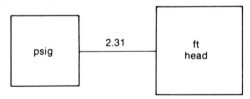

In each case the conversion is from feet of head to pounds per square inch gage pressure. Therefore, when you move from a larger box to a smaller box, division by 2.31 is indicated:

$$\text{Pressure at A} = \frac{200 \ \text{ft}}{2.31 \ \text{ft}/\text{psig}}$$

$$= 86.58 \ \text{psig}$$

$$\text{Pressure at B} = \frac{80 \text{ ft}}{2.31 \text{ ft/psig}}$$

$$= 34.63 \text{ psig}$$

Example 5

Gages are being used in the water system in Figure 11 to measure pressure at various points. The readings are in units of pounds per square inch gage. What is the equivalent pressure at each point in feet of head?

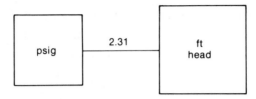

In each case the conversion is from pounds per square inch gage to feet of head. Because you are moving from the smaller box to the larger box, multiplication by 2.31 is indicated:

$$\text{Pressure at A} = (56 \text{ psig})(2.31 \text{ ft/psig})$$

$$= 129.36 \text{ ft of head}$$

$$\text{Pressure at B} = (44 \text{ psig})(2.31 \text{ ft/psig})$$

$$= 101.64 \text{ ft of head}$$

$$\text{Pressure at C} = (32 \text{ psig})(2.31 \text{ ft/psig})$$

$$= 73.92 \text{ ft of head}$$

$$\text{Pressure at D} = (20 \text{ psig})(2.31 \text{ ft/psig})$$

$$= 46.2 \text{ ft of head}$$

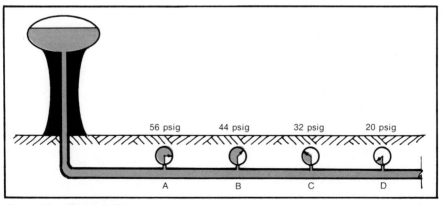

Figure 11.   (Example 5.)

H2-2. Force

The pressure exerted on a surface can also be expressed in terms of force applied to that surface. For example, if a pressure of 5 psig is exerted on a surface 2 in. by 3 in. then a force of 5 lb is pressing down on each square inch of surface area. The total force exerted on the surface would be 30 lb (Figure 12).

The general equation that is used in calculating force due to pressure

$$\frac{\text{Force}}{\text{(lb)}} = \frac{\text{Pressure}}{\text{(psig)}} \times \frac{\text{Area}}{\text{(sq in.)}}$$

Using the previous example, the force would have been calculated as[9]

$$\text{Force} = (5 \text{ psig})(2 \text{ in.})(3 \text{ in.})$$
$$= (5 \text{ psig})(6 \text{ sq in.})$$
$$= 30 \text{ lb force}$$

It is a basic principle of hydraulics that, if two containers of static (unmoving) fluid are connected with a pipeline, then the fluid will exert the same pressure everywhere within the system. This principle and the force/pressure equation can be used to explain the operation of a hydraulic jack.

The jack in Figure 13 has an operating piston with a surface area of 5 sq in. and a lifting piston with a surface area of a 100 sq in. A force of 150 lb is applied to the operating piston. What pressure is created within the hydraulic system of the jack, and what is the resulting force exerted by the lifting piston?

The pressure created within the jack by the operating piston is calculated as follows, using the force/pressure equation:

$$\frac{\text{Force}}{\text{(lb)}} = \frac{\text{Pressure}}{\text{(psig)}} \times \frac{\text{Area}}{\text{(sq in.)}}$$

$$\begin{array}{c}\text{Force on Operating} \\ \text{Piston}\end{array} = \left(\begin{array}{c}\text{Pressure within} \\ \text{jack}\end{array}\right)\left(\begin{array}{c}\text{Area of Operating} \\ \text{Piston}\end{array}\right)$$

Fill in the known information:

$$150 \text{ lb} = (x \text{ psig})(5 \text{ sq in.})$$

Then solve for the unknown value:

$$150 = (x)(5)$$
$$\frac{150}{5} = x$$
$$30 \text{ psig} = x$$

---

[9]Mathematics Section, Area Measurement.

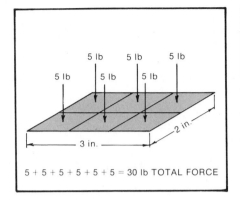

Figure 12. 5 psig Applied to
6 sq in. Surface

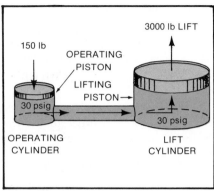

Figure 13. Hydraulic Jack

Therefore the pressure exerted by the operating piston is 30 psig. And because the pressure is the same everywhere within the closed hydraulic system of the jack, the pressure on the lift piston is also 30 psig. Use the force-pressure equation again to determine the total lifting force exerted by the lift piston:

$$\frac{\text{Force}}{\text{(lb)}} = \frac{\text{Pressure}}{\text{(psig)}} \times \frac{\text{Area}}{\text{(sq in.)}}$$

$$\text{Force on Lifting Cylinder} = \binom{\text{Pressure within}}{\text{jack}}\binom{\text{Area of Lifting}}{\text{Cylinder}}$$

$$= (30 \text{ psig})(100 \text{ sq in.})$$

$$\text{Force on Lifting Cylinder} = 3000 \text{ lb}$$

Therefore, the jack illustrated increases the force applied to the operating cylinder about 20 times (150 lb force compared to 3000 lb force).

Example 6

The pressure on a particular surface is 12 psig. If the surface area is 120 sq in., what is the force exerted on that surface?

$$\frac{\text{Force}}{\text{(lb)}} = \frac{\text{Pressure}}{\text{(psig)}} \times \frac{\text{Area}}{\text{(sq in.)}}$$

$$= (12 \text{ psig})(120 \text{ sq in.})$$

$$= 1440 \text{ lb total force}$$

Example 7

If there is a pressure of 40 psig against a surface 2 ft by 1 ft, what is the force against the surface?

The area of the surface is

$$(2 \text{ ft})(1 \text{ ft}) = 2 \text{ sq ft}$$

However, since the area must be expressed in *square inches* in order to use the equation, square feet must be converted to square inches:[10]

$$(2 \text{ sq ft})(144 \text{ sq in./sq ft}) = 288 \text{ sq in.}$$

Now calculate force:

$$\frac{\text{Force}}{\text{(lb)}} = \frac{\text{Pressure}}{\text{(psig)}} \times \frac{\text{Area}}{\text{(sq in.)}}$$
$$= (40.0 \text{ psig})(288 \text{ sq in.})$$
$$= 11{,}520 \text{ lb force}$$

Review Questions

1. Convert a pressure of 148 ft to pounds per square inch gage.

2. A pressure of 240 psig is equivalent to how many feet of water?

3. If the water in a tank is 32 ft deep, what is the gage reading (pounds per square inch gage) at the bottom of the tank? (See Figure 14.)

4. A pressure gage at the ground level of a water tower reads 105 psig. What is the height from the ground to the water surface in the tower? (See Figure 15.)

5. The pressure on a surface is 37 psig. If the surface area is 260 sq in., what is the force exerted on the surface?

6. If there is a pressure of 160 psig pressing against a surface 10 ft by 12 ft, what is the force exerted against the surface?

---
[10] Mathematics Section, Conversions (Areas).

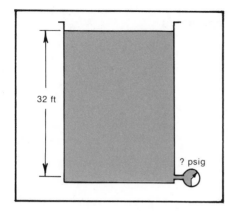

32 ft

? psig

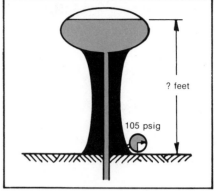

? feet

105 psig

Figure 14.   (Review Question 3.)   **Figure 15.   (Review Question 4.)**

Summary Answers

1. 64.07 psig

2. 554.4 ft of head

3. 13.85 psig

4. 242.55 ft of head

5. 9620 lb total force

6. 2,764,800 lb total force

Detailed Answers

1.

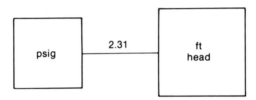

Moving from the larger box (ft) to the smaller box (psig), you should divide by 2.31:

$$\frac{148 \text{ ft}}{2.31 \text{ ft/psi}} = 64.07 \text{ psig}$$

2.

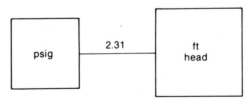

Moving from the smaller box (psig) to the larger box (ft), you should multiply by 2.31:

$$(240 \text{ psig})(2.31 \text{ ft/psig}) = 554.4 \text{ ft of head}$$

3.

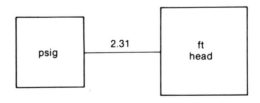

Moving from the larger box (ft) to the smaller box (psig), you should divide by 2.31:

$$\frac{32 \text{ ft}}{2.31 \text{ ft/psig}} = 13.85 \text{ psig}$$

4.

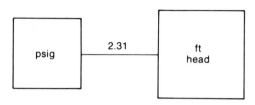

Converting pressure to feet, you are moving from the smaller box to the larger box. Therefore you should multiply by 2.31:

$$(105 \text{ psig})(2.31 \text{ ft/psig}) = 242.55 \text{ ft of head}$$

5.
$$\frac{\text{Force}}{\text{(lb)}} = \frac{\text{Pressure}}{\text{(psig)}} \times \frac{\text{Area}}{\text{(sq in.)}}$$
$$= (37 \text{ psig})(260 \text{ sq in.})$$
$$= 9620 \text{ lb total force}$$

6. To determine the force in pounds, you must know both the pressure in pounds per square inch gage and the area in square inches. The pressure is given in the problem, so only the area must be calculated. First calculate the area in square feet:

$$(10 \text{ ft})(12 \text{ ft}) = 120 \text{ sq ft}$$

Then convert to area in square inches:

$$(120 \text{ sq ft})(144 \text{ sq in./sq ft}) = 17280 \text{ sq in.}$$

Now determine the force on the surface:

$$\frac{\text{Force}}{\text{(lb)}} = \frac{\text{Pressure}}{\text{(psig)}} \times \frac{\text{Area}}{\text{(sq in.)}}$$
$$= 160 \text{ psig} \times 17280 \text{ sq in.}$$
$$= 2,764,800 \text{ lb total force}$$

Hydraulics 3

Piezometric Surface and
Hydraulic Grade Line

H3-1. Piezometric Surface

Many important hydraulic measurements are based on the difference in height between the FREE WATER SURFACE and some point in the water system. The PIEZOMETRIC SURFACE can be used to locate the free water surface in a container where it cannot be observed directly.

If you connect an open-ended tube (similar to a straw) to the side of a tank or pipeline, the water will rise in the tube. Such a tube, shown in Figure 16, is called a PIEZOMETER, and the level of the top of the water in the tube is called the piezometric surface. If the water-containing vessel is not under pressure (as is the case in Figure 16) the piezometric surface will be the same as the free water surface in the vessel, just as it would be if a soda straw (the piezometer) were left standing in a glass of water.

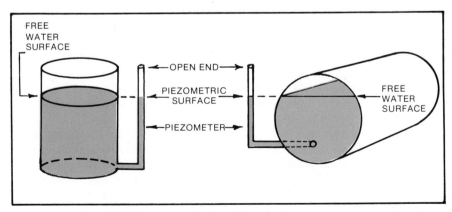

Figure 16. Piezometer and Piezometric Surface

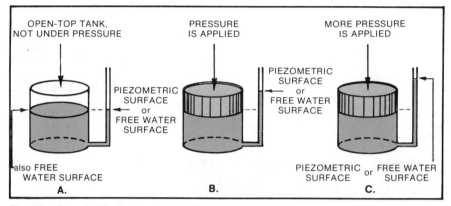

Fig. 17.   Piezometric Surface Varies With Pressure

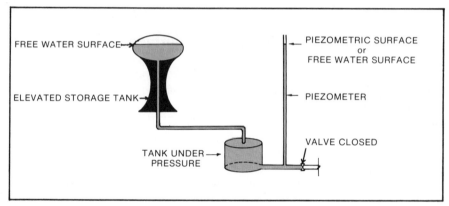

Figure 18.   Piezometric Surface Caused by Elevated Tank

If the tank and pipeline are under pressure, as they often are, the pressure will cause the piezometric surface to rise above the level of the water in the tank. The greater the pressure, the higher the piezometric surface (Figure 17).

Notice in Figure 17A that the free water surface shown by the piezometer is the same level as the water surface in the tank; but once a pressure is applied, as in figures 17B and 17C, the free water surface rises above the level of the tank water surface.

The applied pressure caused by a piston in figures 17B and 17C can also be caused by water standing in a connected tank at a higher elevation. For example, Figure 18 shows that the pressure caused by water in an elevated storage tank is transmitted down the standpipe, through the pipeline, into a closed, low-level tank, and into the piezometer. The pressure causes the water to rise in the piezometer to the height of the water surface in the storage tank. Notice that the relationship between the tank under pressure and the piezometric

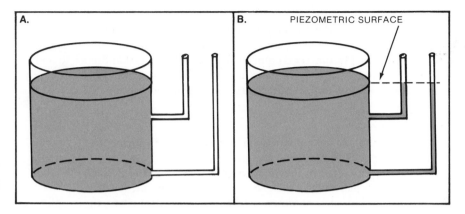

Figure 19. (Example 1.)

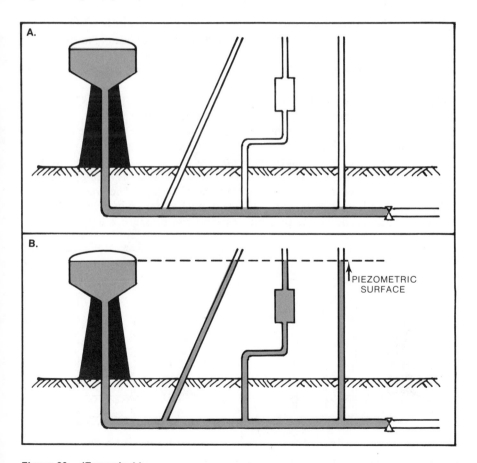

Figure 20. (Example 1.)

surface shown in Figure 18 is very similar to the relationship between the pressurized tank and the piezometric surface in Figure 17c. In both cases the piezometric surface is higher than the surface of the water in the pressurized tank to which the piezometer is connected. Note, however, that Figure 18 is also similar to Figure 17A, in that the piezometric surface in both figures is ultimately at the same level as the free water surface in the open-top tank.

Example 1

Locate the piezometric surface (free water surface) in Figures 19A and 20A.

The answers are shown in Figures 19B and 20B. In each case, the piezometric surface is the same as the water surface in the main body of water. This is true no matter where the piezometer is connected (Figure 19B) and no matter what slope or shape the piezometer takes (Figure 20B).

So far only the piezometric surface for a body of standing water (static water) has been considered. The piezometers have shown that the water always rises to the water level of the main body of water, *but only when the water is standing still.*

Changes in the piezometric surface occur when water is flowing. Figure 21 shows an elevated storage tank feeding a distribution system pipeline. When the valve is closed (Figure 21A), all the piezometric surfaces are the same height as the free water surface in storage. When the valve opens and water begins to flow (Figure 21B), the piezometric surfaces *drop.* The farther along the pipeline, the lower the piezometric surface, because some of the pressure is used up keeping the water moving over the rough interior surface of the pipe. The pressure that is lost (called head loss) is no longer available to push water up in the piezometer. As water continues down the pipeline, less and less pressure is left.

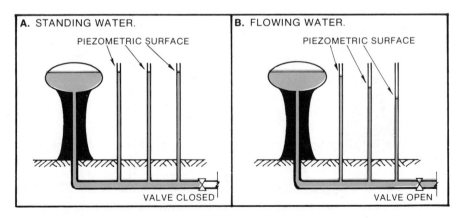

Figure 21.   Piezometric Surface Changes When Water Is Flowing

H3-2. Hydraulic Grade Line

The HYDRAULIC GRADE LINE (HGL) is the line which connects all the piezometric surfaces along the pipeline. It is important from an operating standpoint because it can be used to determine the pressure at any point in a water system.

To better understand HGL, you should know how the HGL is located and drawn. In this section two techniques are discussed:

- Locating HGL's from piezometric surface information
- Locating HGL's from pressure gage information

Locating HGL's From Piezometric Surface Information

First let's look at how to find the HGL of a STATIC WATER SYSTEM (water not moving). The pipeline in Figure 22 is fitted with four piezometers. With the valve closed, water rises in three of them, up to the free water surface. To find the HGL, draw a horizontal line from the free water surface through the piezometric surfaces, as shown.

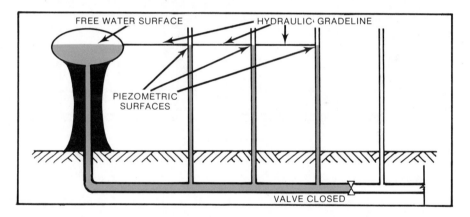

Figure 22. Hydraulic Gradeline (HGL) of Static Water System

This demonstrates two important facts about HGLs for *static* water systems:

- The static HGL is always horizontal.
- The static HGL is always at the same height as the free water surface.

Example 2
Locate and draw the HGL for the two static water conditions shown in Figures 23A and 23C.

First, the HGL will pass through the piezometric surfaces. Also, since the water is static, the following principles apply:

- The HGL will be horizontal
- The HGL will be at the free water surface

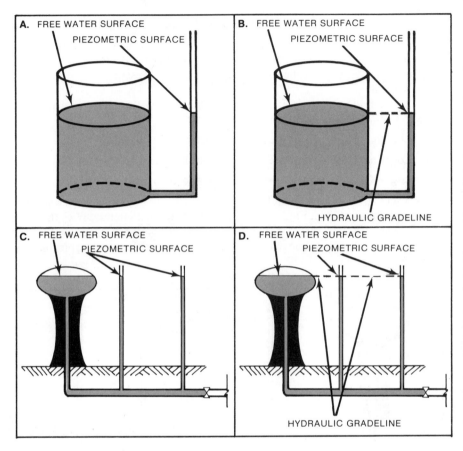

Figure 23.   (Example 2.)

Using this information, you can draw the HGL by connecting the piezometric surfaces to the free water surface of the reservoir with a horizontal line. The resulting HGLs are shown in Figures 23B and 23D.

Now consider how to find the HGL of a *dynamic water system* (water in motion). Figure 24 shows water moving from an elevated storage tank into the distribution system. As before, the HGL is drawn by connecting the free water surface and all the piezometric surfaces. The result can be one straight line, as shown in Figure 24, or it can be a series of connected straight lines at different angles, as shown in Figure 25.

The dashed line in Figure 24 shows the static HGL. The vertical distance between the static HGL and the dynamic HGL at any point along the pipeline is a measure of the amount of pressure the water has used flowing to that point.

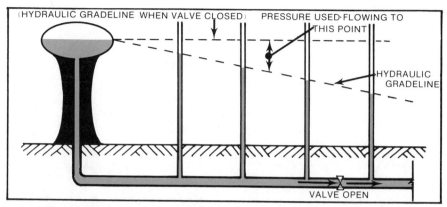

Figure 24. Hydraulic Gradeline of Water in Motion

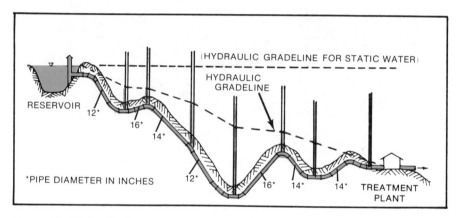

Figure 25. Hydraulic Gradeline of Water Transmission Line

Figure 25 illustrates how the HGL is established for a typical water transmission line. Notice in this case that the HGL is a series of straight lines at different slopes. There is head loss wherever the flow enters a smaller diameter pipe, as indicated by a downward change in HGL slope. Slope changes also show pressure loss depending on factors including flow rate, pipe material, and pipe age and roughness. The changing HGL indicates the combined effect of these factors.

As in Figure 24, the dashed line indicates where the HGL would be if the water were static. This would be the condition if, for example, flow was stopped by closing a valve just outside the treatment plant.

The concept of hydraulic grade line for dynamic systems also applies to artesian wells. When a well is drilled into an artesian aquifer, as shown by wells A and C in Figure 26, the water will rise in the well, with the well acting as a piezometer. The recharge area of the artesian aquifer acts like the elevated

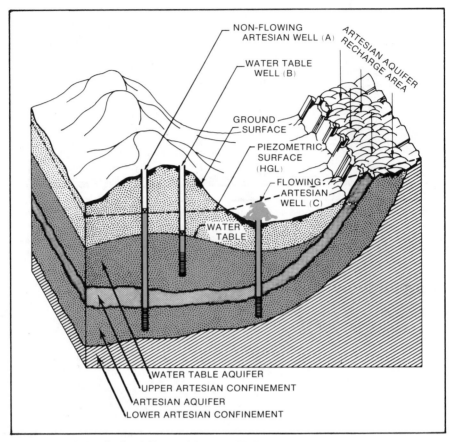

Figure 26. Hydraulic Gradeline and Artesian Wells

storage tank shown in Figure 24 and the rest of the aquifer and the artesian wells can be compared to the pipeline with the piezometers. The HGL slopes downward as the water moves through the aquifer.

Notice that one of the artesian wells (well C) is flowing. This happens because the wellhead (the opening of the well at the ground surface) is *below* the HGL and there is no pipe to contain the water rising to the piezometric surface. The same thing would happen in the system illustrated in Figure 25 if any one of the piezometers did not reach up to the HGL.

One situation you are likely to encounter in water system operations is shown in Figures 27 and 28: pumping from a lower reservoir to a higher reservoir. With the pump off (Figure 27), there are two separate hydraulic grade lines: the lower one representing the water level in Reservoir 1 and the higher one representing the water level in Reservoir 2.

When the pump is running (Figure 28), notice what happens to the HGL.

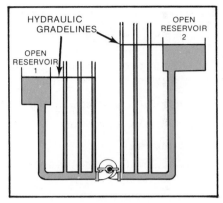

Figure 27. Pump Off

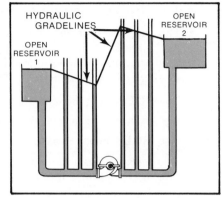

Figure 28. Pump On

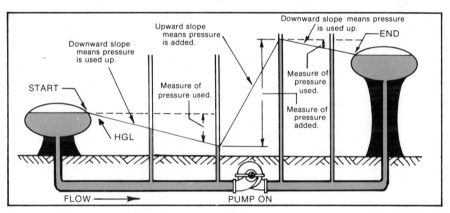

Figure 29. Five Basic Principles of HGLs in Dynamic Systems

Starting at Reservoir 1, the HGL slopes downward, indicating that the water is losing pressure as it flows from Reservoir 1 to the suction or upstream side of the pump. The HGL then slopes upward as the water moves through the pump, indicating that pressure has been *added* to the water. Finally, from the discharge side of the pump the HGL slopes downward again, since the water is losing pressure as it flows through the pipe from the pump to Reservoir 2. Notice that the HGL ends at the same elevation as the free water surface of Reservoir 2.

From these examples you can see that there are five important basic principles about HGLs for dynamic water systems:

- The HGL starts at the same elevation as the free water surface in the upstream reservoir.

- The HGL slopes *downward* in the direction of flow when pressure is used up.

- The HGL slopes upward as water gains pressure passing through a pump.

- The difference in height between any two points on a downward-sloping HGL shows the pressure used by the water between the two points. Similarly, the difference in height between any two points on an upward-sloping HGL shows the pressure added between those two points.

- The HGL ends at the same elevation as the downstream free water surface.

Figure 29 illustrates each of the five basic principles for locating an HGL.

Locating HGL's From Pressure Gage Information

There are many situations in water transmission, treatment, and distribution where the use of piezometers for pressure measurement or HGL location is totally impractical. In some cases, for example, piezometers hundreds of feet high would be required. Therefore in most practical applications pressure gage readings are used to locate the HGL. The following example illustrates the procedure:

Example 3

In Figure 30A, pressure gage readings taken along a transmission line are shown in pounds per square inch gage (psig). From this information, locate and draw the HGL.

To locate points to scale along the HGL so that a line may be drawn, you must first convert the pressure readings given from psig to pressures in feet. Use the following diagram to make the conversions:[11]

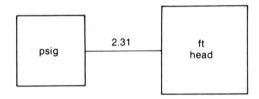

In each conversion from psig pressure to feet of head, you are moving from a smaller box to a larger box. Therefore you should multiply by 2.31. Converting the 65 psig reading at point 1 to feet of head

$$(65 \text{ psig})(2.31 \text{ ft/psig}) = 150.15 \text{ ft}$$

At point 2, with a gage reading of 39 psig:

$$(39 \text{ psig})(2.31 \text{ ft/psig}) = 90.09 \text{ ft}$$

Using the same method, the pressure in feet may be calculated for the remaining points:

---

[11]Hydraulics Section, Pressure and Force (Pressure).

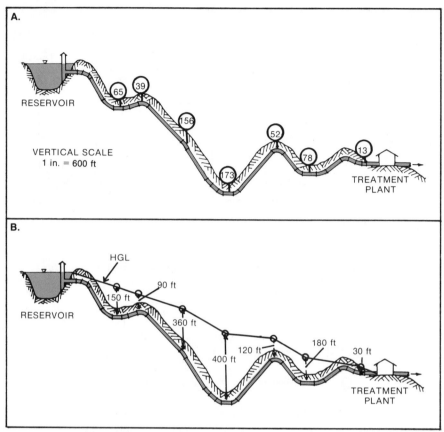

Figure 30.   (Example 3.)

Point 3, 156 psig; pressure $=$ 360.36 ft
Point 4, 173 psig; pressure $=$ 399.63 ft
Point 5, 52 psig; pressure $=$ 120.12 ft
Point 6, 78 psig; pressure $=$ 180.18 ft
Point 7, 13 psig; pressure $=$ 30.03 ft

Next, locate the points for pressure in feet *to scale* on the diagram, directly above the pressure gage locations. The HGL can now be drawn by connecting these points, as shown in Figure 30B.

Pressure gages are often the best way to locate the HGL in a pump system, as illustrated in the following example.

Example 4

Locate and draw the HGL for the pump system shown in Figure 31A. Pressure gage readings are in psig.

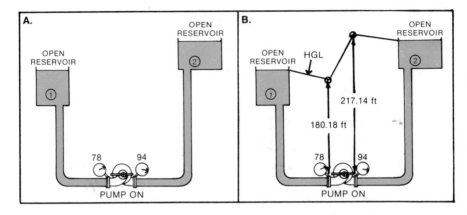

Figure 31.   (Example 4.)

As in the previous example, the gage pressures in psig are converted to pressure in feet:

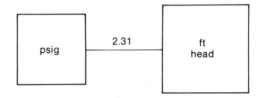

Converting from psig to feet of head, you are moving from a smaller box to a larger box. Therefore, multiplication by 2.31 is indicated. For a gage reading of 78 psig, the pressure in feet is

$$(78 \text{ psig})(2.31 \text{ ft/psig}) = 180.18 \text{ ft}$$

Converting the 94 psig reading to feet of head:

$$(94 \text{ psig})(2.31 \text{ ft/psig}) = 217.14 \text{ ft}$$

The two points are located (to scale) directly above the gage points, and the HGL is drawn, as shown in Figure 31B.

Review Questions

1. The piezometric surface is 87 ft above a point in the pipeline. What is the pressure in psig at that point in the pipeline?

2. The distance to water in a non-flowing artesian well is 59 ft. The driller's log indicates that the aquifer was first penetrated at a depth of 148 ft. How high has the water risen above the top of the artesian aquifer?

3. A cylindrical water storage tank 120 ft in diameter is equipped with a sight glass to monitor water depth in the tank. When full, the water in the tank is 48 ft deep. When the sight glass shows a piezometric surface of 33 ft, how many gallons of water are in the tank?

4. Explain the different levels of the piezometric surfaces in Figure 32A.

5. Define hydraulic grade line.

6. (a) Draw the HGL for the static water condition shown in Figure 33A.

   (b) What two basic concepts does Figure 33A illustrate regarding HGLs in static water systems?

7. Draw the HGL for the transmission line shown in Figure 34A.

8. Draw the HGL for the pump system shown in Figure 35A.

9. What does it mean when the HGL
   (a) slopes downward
   (b) slopes upward
   (c) is horizontal

10. If a pressure gage in a pipeline reads 175 psig, how high above the point is the HGL?

11. The pressure gage at ground level of a water tower reads 47 psig. The water in the tank is static (the tank is neither filling nor emptying). How high above the ground is the water surface in the tower?

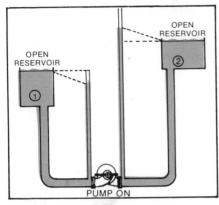

Figure 32A. (Review Question 4.)

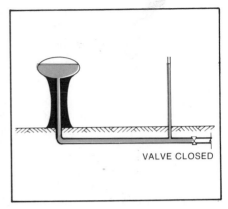

Figure 33A. (Review Question 6.)

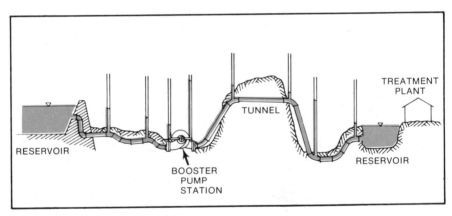

Figure 34A. (Review Question 7.)

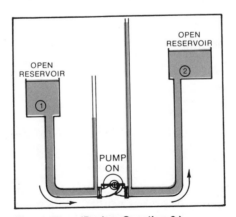

Figure 35A. (Review Question 8.)

Summary Answers

1. 37.67 psig

2. 89 ft

3. 2,790,279 gal

4. (See detailed answers.)

5. The hydraulic grade line is the line that connects all the piezometric surfaces.

6. (a) (See detailed answers.)
   (b)   1. The HGL is always horizontal.
         2. The HGL is always at the same height as the free water surface.

7. (See detailed answers.)

8. (See detailed answers)

9. (a) Pressure is used up
   (b) Pressure is added, such as by a pump.
   (c) No change in pressure—water is static, not moving.

10. 404.25 ft

11. 108.57 ft

Detailed Answers

1.

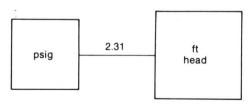

You are moving from the larger box to the smaller box so you should divide by 2.31:

$$\left(\frac{87}{2.31 \text{ ft}/\text{psig}}\right) = 37.67 \text{ ft}$$

2. First make a sketch of what information is given: (Figure 36)
   From this figure, then:

   The unknown distance $= 148$ ft $- 59$ ft
   $$= 89 \text{ ft}$$

3. First make a sketch of the problem: (Figure 37)

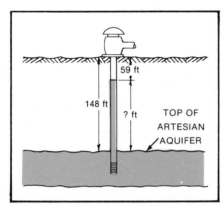

Figure 36.   (Review Question 2.)   **Figure 37.   (Review Question 3.)**

Using the water depth of 33 ft and the diameter of 120 ft, calculate the volume of water in the tank:[12]

$$\text{Volume} = (\text{Area})(\text{Third Dimension})$$
$$\text{Area of Circle} \quad \text{Depth}$$
$$= (0.785)(120 \text{ ft})(120 \text{ ft})(33 \text{ ft})$$
$$= 373,032 \text{ cu ft}$$

Then convert volume in cubic feet to volume in gallons:[13]

$$(373,032 \text{ cu ft})(7.48 \text{ gal/cu ft}) = 2,790,279 \text{ gal}$$

4. (See Figure 32B.)

5. Hydraulic grade line is the line that connects all the piezometric surfaces.

6.   (a) (See Figure 33B.)

     (b)
         (1) The static HGL is always horizontal.
         (2) The static HGL is always at the same height as the free water surface.

---

[12] Mathematics Section, Volume Measurement.

[13] Mathematics Section, Conversions (Cubic Feet to Gallons to Pounds).

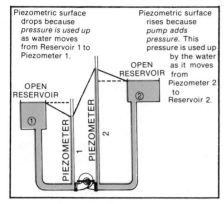

Piezometric surface drops because *pressure is used up* as water moves from Reservoir 1 to Piezometer 1.

Piezometric surface rises because *pump adds pressure*. This pressure is used up by the water as it moves from Piezometer 2 to Reservoir 2.

Figure 32B. (Review Question 4.)

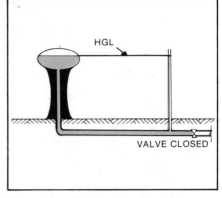

Figure 33B. (Review Question 6.)

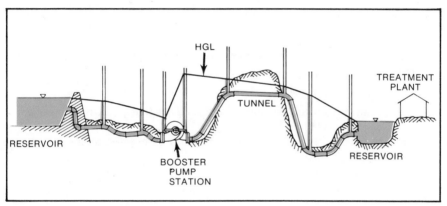

Figure 34B.   (Review Question 7.)

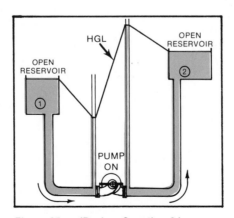

Figure 35B.   (Review Question 8.)

7. (See Figure 34B.)

8. (See Figure 35B.)

9.

    (a) Pressure is used up.
    (b) Pressure is added, such as by a pump.
    (c) No change in pressure—water is static, not moving.

10. The height of the HGL above the pipeline is the pressure at that point in feet of head. So to answer the question convert the pressure in psig to pressure in feet of head:

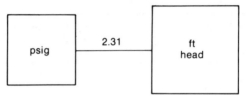

You are moving from the smaller box to the larger box, so you should multiply by 2.31:

$$(175 \ \text{psig})(2.31 \ \text{ft/psig}) = 404.25 \ \text{ft}$$

The distance from the pipeline to the HGL at that point is 404.25 ft.

11. To answer the question, convert the pressure in psig to pressure in feet of head

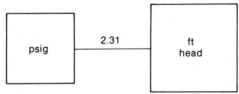

Moving from the smaller box to the larger box, you should multiply by 2.31:

$$(47 \ \text{psig})(2.31 \ \text{ft/psig}) = 108.57 \ \text{ft}$$

Hydraulics 4

Head

HEAD is one of the most important measurements in hydraulics. It is used to calculate the hydraulic forces acting in a pipeline, and to determine a pump's capacity to overcome or pump against these forces.

Head is a measurement of the energy possessed by the water at any particular location in the water system. In hydraulics, energy (and therefore head) is expressed in units of foot-pounds per pound, written

$$\text{Head} = \frac{\text{ft-lb}}{\text{lb}}$$

These somewhat cumbersome units of measure cancel out so that head can be expressed in feet[14]:

$$\text{Head} = \frac{\text{ft-lb}}{\text{lb}}$$

$$= \text{ft}$$

When head is expressed in feet, as it normally is, the measurement can always be considered to represent the height of water above some reference elevation. The height of water in feet can also be expressed as pressure in pounds per square inch gage (psig)[15] by dividing the feet-of-water height by 2.31.

H4-1. Types of Head

There are three types of head, as discussed in detail in the following sections:

- Pressure head

- Elevation head

- Velocity head

---

[14]Mathematics Section, Dimensional Analysis.

[15]Hydraulics Section, Pressure and Force.

Pressure Head

PRESSURE HEAD is a measurement of the amount of energy in water due to water pressure. As shown in Figure 38, it is the height above the pipe to which water will rise in a piezometer. It is easily measured by a pressure gage and the resulting pressure reading in psig may then be converted to pressure head in feet.

Pressure head describes the vertical distance from the point of measurement to the HGL.[16] So, if you were interested in locating the HGL for a particular pipeline, you would take gage pressure readings at several critical points along the pipeline, convert those pressure readings to feet of pressure head, and then follow the procedure in section 2.3 for plotting the HGL.

The normal range of pressure heads in water transmission and distribution systems can vary from as little as 50 ft (about 20 psig) to over 1000 ft (about 450 psig). At certain locations within the treatment plant, pressure heads can be very small, perhaps 2 ft (about 1 psig) or less.

Elevation Head

ELEVATION HEAD is a measurement of the amount of energy that water possesses because of its elevation. It is measured as the height in feet from some horizontal reference line or bench mark elevation (such as sea level) to the point of interest in the water system. For example, a reservoir located 500 ft above sea level is said to have an elevation head of 500 ft relative to sea level. The concept is illustrated in Figure 39. Elevation head is quite useful in design but has little day-to-day operating significance.

Velocity Head

VELOCITY HEAD is a measurement of the amount of energy in water due to its velocity, or motion. The greater the velocity, the greater the energy and, therefore, the greater the velocity head. Anything in motion has energy because of that motion.

You might think of the energy of motion in terms of the effort you would need to apply to stop the motion. For example, to stop a car traveling at 55 mph requires that the car's brakes do much more work than if the car were traveling at 5 mph, because the faster-moving car has more energy of motion.

Like all other heads, velocity head is expressed in feet. It is determined by multiplying velocity (in feet per second) by itself and then dividing by 64.4 feet per second squared, as follows:

$$\text{Velocity head} = \frac{V^2}{64.4 \text{ ft/sec}^2}$$

For example, if the velocity of the water was 30 ft/sec, the velocity head would be as follows:

---

[16] Hydraulics Section, Piezometric Surface and Hydraulic Gradeline.

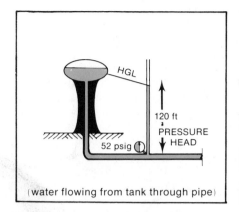

Figure 38.   Pressure Head

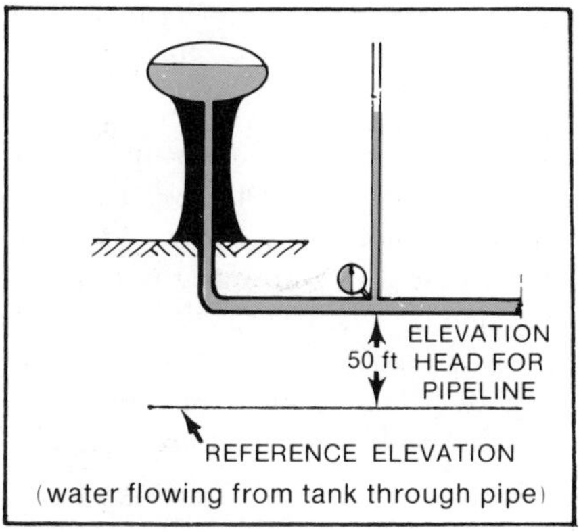

Figure 39.   Elevation Head.

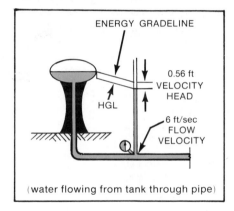

Pressure Head measures the amount
of energy in water due to water
pressure.

Velocity Head measures the amount of
energy in water due to its motion, or
velocity.

Elevation Head measures the amount
of energy in water due to elevation.

Figure 40.   Velocity Head.   **Definition of Heads**

$$\text{Velocity head} = \frac{(30)(30)}{64.4}$$

$$= 13.98 \text{ ft}$$

This means that if you poured water from a bucket held 13.98 ft in the air, then
the water would be moving 30 ft/sec when it hit the ground.

Velocity head is also a measurement of the vertical distance from the HGL to
the ENERGY GRADE LINE (EGL), as shown in Figure 40. Energy grade lines are an
advanced concept and will not be discussed further in this basic text.

The velocity, or speed, that water travels in a water system is usually at least 2
ft/sec and generally not more than 10 ft/sec. These speeds would produce
velocity heads of

$$\begin{array}{c}\text{Velocity head at} \\ \text{2 ft/sec velocity}\end{array} = \frac{(2)(2)}{64.4}$$

$$= 0.06 \text{ ft}$$

$$\frac{\text{Velocity head at}}{10 \text{ ft/sec velocity}} = \frac{(10)(10)}{64.4}$$

$$= 1.55 \text{ ft}$$

Because velocity heads are so small relative to pressure heads (which usually range from 50 to 1000 ft), they can usually be ignored in operations without causing a significant error. However, although velocity head can be ignored in most cases, it is important that you recognize that velocity head should be considered whenever it is greater than 1 or 2 percent of the pressure head—that is, whenever it is large enough to be a significant part of the total head. This will usually occur either in pipelines or pump systems when the pressure head is very small (less than 70 ft) and velocities are in the 7–10 ft/sec range.

H4-2. Calculating Heads

Of the three types of head discussed, pressure head is the most significant and useful from an operating standpoint. The following examples illustrate how pressure heads can be measured and used in operations.

Pressure heads can be measured whether water is standing still or moving. However, as discussed under the HGL explanation, dynamic pressure heads will be less than static heads, due to the pressures lost in moving the water.[17]

Example 1
Pressure readings were taken in the system in Figure 41A when the valve was open and again when it was closed:

	Valve Open	Valve Closed
Pressure at A	65 psig	76 psig
Pressure at B	54 psig	76 psig

Determine pressure heads in feet and show the hydraulic grade line for both conditions.

Valve closed:

The pressure heads at points A and B are the same, as shown by the gage readings:[18]

Converting from psig to feet of head, you are moving from the smaller box to the larger box; therefore you should multiply by 2.31:

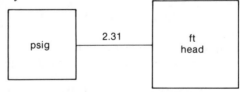

[17] Hydraulics Section, Piezometric Surface and Hydraulic Gradeline.

[18] Hydraulics Section, Pressure and Force.

Pressure head at A and B = (76 psig)(2.31 ft/psig)

= 175.56 ft

Valve open:

Calculate the pressure heads at points A and B using the same method shown for the valve-closed condition:

Pressure head at A = (65 psig)(2.31 ft/psig)

= 150.15 ft

Pressure head at B = (54 psig)(2.31 ft/psig)

= 124.74 ft

The two pressure heads and the HGL can be shown in Figures 41B and 41C.

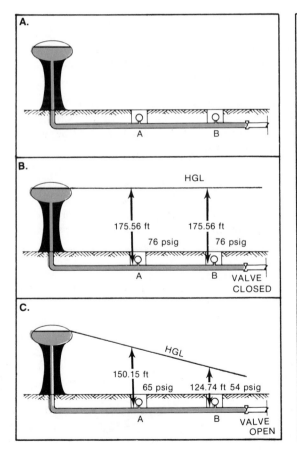

Figure 41.   (Example 1.)

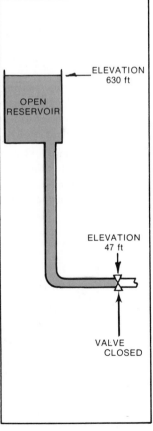

Figure 42.   (Example 2.)

The pressure heads of water standing still are higher than those of water moving. The difference represents the amount of energy used by the water in moving from the elevated storage tank to points A and B.

Often in water systems the static pressure head can be found from elevation information that was established by surveyors during construction.

Example 2

What is the pressure head at the closed valve in Figure 42? (Give the answer in feet.)

Since you know the elevation of both points, the pressure head in feet may be calculated:

$$630 \text{ ft elevation}$$
$$\underline{-47 \text{ ft elevation}}$$
$$583 \text{ ft difference in elevation}$$

Therefore, the pressure head at the valve is 583 ft. (Notice that you could not use this technique to determine the pressure head when the valve is open and the water moving.)

The final two examples of pressure head calculations concern the water transmission system shown in the next figure (Figure 43). In the examples you will see how important pressure head is, and how a pipeline could be seriously damaged if pressure head is not considered.

Example 3

The diagram shows a simplified water transmission line taking water from a reservoir to a treatment plant. With all valves open, pressure measurements (in psig) were made at each valve location. Also shown are selected elevations

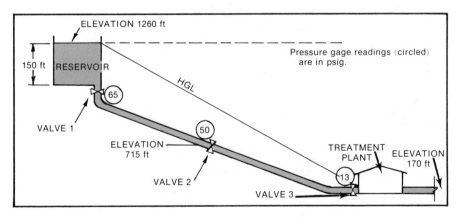

Figure 43.   (Examples 3 and 4.)

along the system. Based on this information, what is the pressure head at valve 3 when it is (a) open, and (b) closed?

(a) With the valve open, the pressure head at valve 3 is read directly from the gage:

Pressure head on valve 3 = 13 psig (valve open)

(b) When valve 3 is closed, however, the pressure head will increase dramatically. To find out exactly how much, calculate the height of water that the valve would be holding back. The valve is holding back all the water up to the reservoir water surface. The vertical distance can be calculated from the elevation information as follows:

$$
\begin{array}{ll}
1260 \text{ ft} & \text{Elevation of reservoir water surface} \\
-170 \text{ ft} & \text{Elevation of valve 3} \\
\hline
1090 \text{ ft} & \text{Difference in elevation}
\end{array}
$$

So the pressure head on the valve when closed is 1090 ft. Convert to pressure in psig:

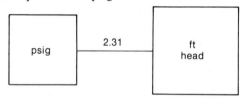

Moving from the larger box to the smaller box, you should divide by 2.31:

$$
\text{Pressure head in psig} = \frac{1090 \text{ ft}}{2.31 \text{ ft/psig}}
$$

$$
= 471.86 \text{ psig}
$$

This is more than 35 times the 13 psig of dynamic pressure head. If the pipeline and valve were designed for this high static pressure head, then no damage would occur when the valve was closed. However, if they were not designed to handle this static pressure head, either the valve or the pipeline just upstream might burst when the valve was closed.

Often it is too expensive to build a pipeline with the capability to handle such high static pressure heads. Instead, a series of isolating and vacuum relief valves are installed to prevent excessive static pressure heads such as would occur in the preceding example. (For description of these valves see *Introduction to Water Sources and Transmission,* Water Transmission Module, Valves.) The following example illustrates how such a system works.

Example 4

This problem utilizes the same figure as Example 3. In this case, however, when the system is shut down, the valves are closed in the sequence 1, 2, and then 3. After all three valves are closed in this sequence, what is the static pressure head in psig on (a) valve 1; (b) valve 2; and (c) valve 3?

(a) The static pressure head on valve 1 is equal to the height of water it supports, 150 ft. Therefore the pressure head in square inch gage pressure is

$$\text{Pressure in psig} = \frac{\text{Pressure in feet}}{2.31 \text{ ft/psig}}$$

$$= \frac{150 \text{ ft}}{2.31 \text{ ft/psig}}$$

$$= 64.93 \text{ psig}$$

pressure head on valve 1

(b) To determine the pressure head on valve 2, calculate the height of water it supports when it is closed. Since valve 1 is already closed, valve 2 only supports the water between valves 1 and 2.

The elevation of valve 1 is:

1260 ft	Elevation of reservoir water surface
−150 ft	Vertical distance to valve 1
1110 ft	Elevation of valve 1

Now determine the difference in height between the two valves:

1110 ft	Elevation of valve 1
−715 ft	Elevation of valve 2
395 ft	Water supported by valve 2

So the pressure head on valve 2 is 395 ft. Expressed in psig, the pressure head is:

$$\text{Pressure in psig} = \frac{\text{Pressure in feet}}{2.31 \text{ ft/psig}}$$

$$= \frac{395 \text{ ft}}{2.31 \text{ ft/psig}}$$

$$= 171 \text{ psig}$$

pressure head on valve 2

In an actual pipeline, the pressure head at valve 2 will be less than 171 psig because the vacuum relief valve (near valve 1) will allow water to drain from the line between points 1 and 2 after valve 1 is closed. If the closing of valves 1 and 2 is done manually by one operator, either walking or driving between the valves, then is it likely that the line from 1 to 2 will empty completely and that the pressure head at valve 2 after closing will be zero. However, if the valves are closed by remote control, the line will only have time to drain slightly between the closing of valves 1 and 2, and the pressure head at valve 2 will be very nearly 171 psig.

(c) Next, follow the same procedure for valve 3. First, find the height of water supported by valve 3. Since valve 2 is now closed, the height of the water supported by valve 3 is equal to the vertical distance between valves 2 and 3:

$$
\begin{array}{ll}
715 \text{ ft} & \text{Elevation of valve 2} \\
\underline{-170 \text{ ft}} & \text{Elevation of valve 3} \\
545 \text{ ft} & \text{Vertical distance between valves}
\end{array}
$$

Expressing this pressure in psig:

$$
\text{Pressure in psig} = \frac{\text{Pressure in feet}}{2.31 \text{ ft}/\text{psig}}
$$

$$
= \frac{545 \text{ ft}}{2.31 \text{ ft}/\text{psig}}
$$

$$
= 235.93 \text{ psig pressure head}
$$
$$
\text{on valve 3}
$$

Again the pressure head will be less than the calculated value depending on the elapsed time between the closing of valves 2 and 3.

Notice that 235.93 psig is only about one-half of the 471.86 psig pressure head on valve 3 which was calculated in Example 3. You can see that following the proper closing sequence of the valves can significantly reduce the pressure that valves and pipelines must withstand.

Review Questions

1. Express the following gage pressures as pressure head in feet.
   (a) 150 psig
   (b) 250 psig
   (c) 86 psig

2. What is the depth of water in the open tank in Figure 44 if a pressure gage at the bottom reads 9 psig?

3. Calculate the pressure heads in feet at points A and B, Figure 45, when the valve is open and closed, using these gage pressure readings:

	Valve Open	Valve Closed
Pressure at A	126 psig	174 psig
Pressure at B	108 psig	174 psig

4. Calculate the pressure head in pounds per square inch gage on valve 3, Figure 46, if it is closed
   (a) before valves 1 and 2
   (b) after valves 1 and 2

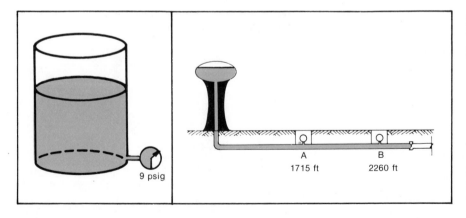

Figure 44. (Review Question 2.)   Figure 45.   (Review Question 3.)

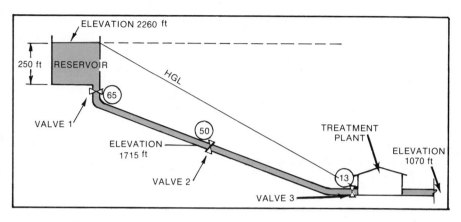

Figure 46.   (Review Question 4.)

Summary Answers

1. (a) 346.5 ft
    (b) 577.5 ft
    (c) 198.66 ft

2. 20.79 ft

3. For valve open:
    Pressure head at A = 291.06 ft
    Pressure head at B = 249.48 ft

    For valve closed:
    Pressure head at A = 401.94 ft
    Pressure head at B = 401.94 ft

4. (a) 515.15 psig
    (b) 279.22 psig

Detailed Answers

1. Pressure in feet = (Pressure in psig)(2.31 ft/psig)

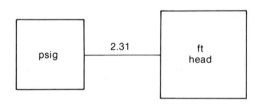

Converting from psig pressure to feet of head, you are moving from a smaller box to a larger box; therefore you should multiply by 2.31:

(a) Pressure in ft = (150 psig)(2.31 ft/psig)

= 346.5 ft

(b) Pressure in ft = (250 psig)(2.31 ft/psig)

= 577.5 ft

(c) Pressure in ft = (86 psig)(2.31 ft/psig)

= 198.66 psig

2. Pressure in feet = (Pressure in psig)(2.31 ft/psig)

To determine the depth of water in the tank, convert the pressure gage reading in psig pressure to feet of head:

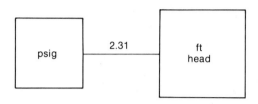

Converting from psig to feet of head, you are moving from a smaller box to a larger box; therefore you should multiply by 2.31:

$$= (9 \ \text{psig})(2.31 \ \text{ft/psig})$$
$$= 20.79 \ \text{ft}$$

Since pressure in feet is 20.79, the water depth in the tank is 20.79 ft.

3. With the valve open, pressure head expressed in feet is:

$$\text{Pressure in feet} = (\text{Pressure in psig})(2.31 \ \text{ft/psig})$$

$$\text{Pressure at A} = (126 \ \text{psig})(2.31 \ \text{ft/psig})$$

$$= 291.06 \ \text{ft}$$

$$\text{Pressure at B} = (108 \ \text{psig})(2.31 \ \text{ft/psig})$$

$$= 249.48 \ \text{ft}$$

With the valve closed, the pressure heads at A and B are the same and are equal to:

$$\text{Pressure head at A or B} = (\text{Pressure in psig})(2.31 \ \text{ft/psig})$$
$$= (174 \ \text{psig})(2.31 \ \text{ft/psig})$$
$$= 401.94 \ \text{ft}$$

4. (a) First calculate the static pressure head from the reservoir to valve 3:

2260 ft	Elevation of reservoir
−1070 ft	Elevation of valve 3
1190 ft	Static pressure head on valve 3

Now convert this to pressure head in feet to psig:

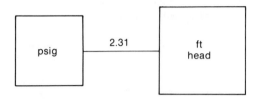

You are moving from a larger box to a smaller box, so divide by 2.31:

$$\text{Pressure head in psig} = \frac{\text{Pressure in feet}}{2.31 \ \text{ft/psig}}$$

$$= \frac{1190 \ \text{ft}}{2.31 \ \text{ft/psig}}$$

$$= 515.15 \text{ psig when valve 3 is closed } \textit{first}$$

(b) Now calculate the static pressure head from valve 2 to valve 3:

$$
\begin{array}{rl}
1715 \ \text{ft} & \text{Elevation of valve 2} \\
-1070 \ \text{ft} & \text{Elevation of valve 3} \\
\hline
645 \ \text{ft} & \text{Pressure head on valve 3}
\end{array}
$$

Convert this pressure head in feet to pressure head in psig:

$$\text{Pressure head} = \frac{\text{Pressure in feet}}{2.31 \ \text{ft/psig}}$$

$$= \frac{645 \ \text{ft}}{2.31 \ \text{ft/psig}}$$

$$= 279.22 \text{ psig when valve 3 closed } \textit{last.}$$

Hydraulics 5

Head Loss

HEAD LOSS is defined as the amount of energy used up by water moving from one point to another. The two most important categories of head loss are friction head loss and minor head loss.

H5-1. Friction Head Loss

The inside of pipe may feel quite smooth, but it is rough from a hydraulic standpoint. The friction caused by water moving over the rough surface causes an energy loss called FRICTION HEAD LOSS.

The amount of friction head loss occurring in a pipe varies, depending on:

- The roughness of the pipe, and
- The velocity of flow through the pipe

Pipe roughness varies widely depending on the type of pipe material and on the condition of the pipe. Plastic is one of the smoothest pipe materials, and steel is one of the roughest. A smooth interior surface results in lower friction loss; a rough interior surface causes high loss.

The condition of the pipe depends on several factors:

- Type of lining
- Age
- Degree of corrosion or scaling
- Slime growths
- Tuberculation
- Obstructions (such as mud, silt, sand, rocks, sticks)

Smooth linings reduce friction loss. The other factors listed increase friction loss, and cause friction loss to vary from point to point within the water system.

The velocity of the moving water also affects friction loss. Without exception, as the velocity increases, friction loss increases, even if the pipe roughness stays the same.

Some of the factors that cause friction loss, such as corrosion, scaling, slime growths, and obstructions, can be controlled by proper operation, maintenance, and treatment of the pipeline. However, the natural roughness of the pipe material or lining, and the natural roughening of that material or lining with age, are factors over which there is little control. Consequently, pipelines must be designed to operate with the additional head needed to overcome friction head loss.

Friction head loss is one of the simplest head losses to calculate. It can be determined using one of many readily available tables. Perhaps the most commonly used are the tables based on the Hazen-Williams formula, in which a smoothness coefficient C represents the smoothness of varying pipe materials of varying ages: the higher the C value, the smoother the pipe. For example the C value for cast iron pipe (CIP) may be as high as 140, for the very best new CIP, laid perfectly straight. A C of 130 is the average value for most new CIP. And as CIP ages, the C value drops, on the average, to $C = 120$ at five years, $C = 110$ at ten years, $C = 100$ at 17 years, $C = 90$ at 26 years, $C = 80$ at 37 years, and $C = 75$ or less for pipe 40 or more years old. Table 5 lists typical smoothness coefficients.

**Table 5.   Smoothness Coefficients
for
Various Pipe Materials**

Type of Pipe	C Value	
	New	10 yr old
Asbestos Cement Pipe	140	120-130
Cast Iron Pipe	130	100
Reinforced and Plain		
Concrete Pipe	140	120-130
Ductile Iron Pipe	130	100
Plastic Pipe	150	120-130
Steel Pipe	110	100

The following examples illustrate friction head loss calculations.

Example 1
A 12-in. diam water transmission line, 10,500 ft long, carries water from the supply reservoir to the treatment plant at a flow rate of 2.0 mgd. The steel transmission line has been in service for about ten years. Use Table 6 to determine the total friction head loss in the line.

To use Table 6 you need to know two numbers: the flow (2.0 mgd in this problem) and the C factor. According to the problem, the pipe is steel, about ten years old. Table 5 shows a smoothness coefficient for ten-year old steel pipe of $C = 100$.

Table 6. Friction Loss Factors for 12-in. Pipe

Discharge in		Veloc-ity	Veloc-ity Head	Loss of Head, ft per 1000 ft of length						
Gal per 24 hr	Cu ft per sec	ft per sec	ft	C=140	C=130	C=120	C=110	C=100	C=90	C=80
100,000	0.155	0.20	0.00	0.02	0.02	0.02	0.02	0.03	0.04	0.04
200,000	0.309	0.39	0.00	0.06	0.07	0.08	0.09	0.11	0.13	0.16
300,000	0.464	0.59	0.01	0.12	0.14	0.16	0.19	0.22	0.27	0.34
400,000	0.619	0.79	0.01	0.20	0.24	0.27	0.32	0.38	0.47	0.58
500,000	0.774	0.99	0.02	0.31	0.36	0.41	0.48	0.58	0.71	0.88
600,000	0.928	1.18	0.02	0.44	0.50	0.58	0.68	0.81	0.99	1.23
700,000	1.083	1.38	0.03	0.58	0.66	0.77	0.91	1.08	1.32	1.64
800,000	1.238	1.58	0.04	0.74	0.85	0.99	1.15	1.38	1.68	2.09
900,000	1.392	1.77	0.05	0.92	1.06	1.23	1.45	1.72	2.10	2.61
1,000,000	1.547	1.97	0.06	1.12	1.29	1.50	1.76	2.10	2.57	3.18
1,100,000	1.702	2.17	0.07	1.34	1.54	1.79	2.10	2.50	3.04	3.79
1,200,000	1.857	2.36	0.09	1.58	1.81	2.10	2.47	2.94	3.58	4.45
1,300,000	2.011	2.56	0.10	1.83	2.10	2.43	2.85	3.40	4.14	5.2
1,400,000	2.166	2.76	0.12	2.10	2.40	2.79	3.26	3.90	4.76	5.9
1,500,000	2.321	2.96	0.14	2.39	2.73	3.17	3.71	4.43	5.4	6.7
1,600,000	2.476	3.15	0.15	2.69	3.09	3.58	4.20	5.0	6.1	7.6
1,700,000	2.630	3.35	0.17	3.00	3.45	4.00	4.69	5.6	6.8	8.5
1,800,000	2.785	3.55	0.20	3.33	3.82	4.43	5.2	6.2	7.6	9.4
1,900,000	2.940	3.74	0.22	3.70	4.24	4.92	5.8	6.9	8.4	10.4
2,000,000	3.094	3.94	0.24	4.06	4.65	5.4	6.4	7.6	9.2	11.5
2,200,000	3.404	4.33	0.29	4.85	5.6	6.5	7.6	9.0	10.9	13.7
2,400,000	3.713	4.73	0.35	5.7	6.5	7.6	8.9	10.5	12.8	16.0
2,600,000	4.023	5.12	0.41	6.6	7.6	8.8	10.3	12.3	15.0	18.6
2,800,000	4.332	5.52	0.47	7.6	8.7	10.1	11.9	14.1	17.2	21.5
3,000,000	4.642	5.91	0.54	8.6	9.9	11.5	13.5	16.0	19.4	24.3
3,500,000	5.41	6.89	0.74	11.4	13.2	15.3	17.9	21.3	26.0	32.3
4,000,000	6.19	7.88	0.96	14.5	16.6	19.3	22.6	27.0	33.2	41
4,500,000	6.96	8.87	1.22	18.0	20.6	24.0	28.2	33.6	41.2	51
5,000,000	7.74	9.85	1.50	22.0	25.1	29.2	34.3	41.0	50.0	62
5,500,000	8.51	10.84	1.82	26.5	30.3	35.1	41.4	49.4	60	75
6,000,000	9.28	11.82	2.17	31.1	35.7	41.4	48.8	58	70	88
7,000,000	10.83	13.79	2.96	41.2	47.2	55	65	77	94	116
8,000,000	12.38	15.76	3.86	53	61	71	83	99	121	150
9,000,000	13.92	17.73	4.89	66	75	87	103	122	148	185
10,000,000	15.47	19.70	6.03	81	93	107	126	150	183	228

With this information, enter Table 6 in the first column at the flow rate 2,000,000 gpd. Then move across the table until you reach the column entitled $C = 100$. The number shown is 7.6.

This is the friction head loss, in feet, for every 1000-ft length of pipeline.

To determine the total friction head loss asked for in the

problem, you must determine how many 1000-ft lengths there are in 10,500 ft:

$$\frac{10,500 \text{ ft}}{1000 \text{ ft}} = 10.5 \text{ 1000-ft lengths of pipe}$$

Since there are 10.5 1000-ft lengths of steel pipe, the total friction head loss is

Total Friction Head Loss = (Loss per 1000 ft)(No. of 1000-ft increments)

= (7.6 ft)(10.5 increments)

= 79.8 ft head

The table provides other useful information not needed for the problem. The third column gives the velocity for the known flow rate (3.94 fps for 2.0 mgd) and the fourth column gives the velocity head (0.24 ft).

Let's see how the friction head loss would change if the flow rate were increased.

Example 2

What would the total friction head loss be in Example 1 if the flow rate were increased to 4.0 mgd? Compare the results with Example 1.

Using the same procedure as in Example 1, enter Table 6 at 4,000,000 gpd. Move across that line to the $C = 100$ column, then read the friction head loss per 1000-ft length of pipeline as 27.0 ft. Then calculate total friction head loss:

Total Friction Head Loss = (Loss per 1000 ft)(No. of 1000-ft increments)

= (27.0 ft)(10.5 increments)

= 283.5 ft

Now compare the results:

Flow Rate	Total Friction Loss
4,000,000 gpd	283.5 ft
2,000,000 gpd	79.8 ft

Although the flow rate was only doubled, the friction loss increased more than 3 1/2 times:

$$\frac{283.5}{79.8} = 3.55$$

The example illustrates how friction loss increases much faster than the increase in flow rate—an important fact to remember when you are considering installing higher capacity pumps in an existing pipeline. The higher capacity pump will need to overcome a friction head that may be many times greater than the increase in flow rate.

Table 7 shows another version of the Hazen-Williams Hydraulics Tables that you might encounter. The following example illustrates how the table is used:

Example 3

The raw water supply for a city is pumped 2.6 miles from a river intake to the treatment plant.

The first 0.6 mile is ten-year old cast iron pipe (CIP), 8 in. in diameter. The remaining 2 miles is 12-in. diameter CIP, also 10 years old. Use Table 7 to determine what total friction head loss can be expected if the pumping rate is 1000 gpm?

Table 7, unlike Table 6, is prepared for direct use with a C factor of 100 only. Since 10-year old CIP has a C of 100 (as indicated in Table 5), Table 7 can be readily used for this problem.

Enter the table at 1000 gpm, move across to find *loss in feet* for 8-in. pipe (2.97 ft), and then move further across to find *loss in feet* for 12-in. pipe (0.41 ft).

The table gives loss in feet per 100 ft of pipe length so the number of 100-ft increments of 8-in. and 12-in. pipe must be determined:

$$0.6 \text{ mile of 8-in. pipe} = (0.6 \text{ mile})(5280 \text{ ft/mile})$$
$$= 3168 \text{ ft}$$

$$\frac{3168 \text{ ft}}{100 \text{ ft}} = 31.68 \text{ 100-ft increments of 8-in. pipe}$$

$$2 \text{ mile of 12-in. pipe} = (2.0 \text{ miles})(5280 \text{ ft/mile})$$
$$= 10,560 \text{ ft}$$

Then

$$\frac{10,560}{100} = 105.6 \text{ 100-ft increments of 12-in. pipe}$$

Now determine the total friction loss in the 8-in. portion:

$$\text{Friction Head Loss} = (\text{Loss per 100 ft})(\text{No. of 100-ft increments})$$
$$= (2.97 \text{ ft})(31.68 \text{ increments})$$
$$= 94.09 \text{ ft head loss}$$

Make the same calculation for the 12-in. section:

$$\text{Friction Head Loss} = (\text{Loss per 100 ft})(\text{No. of 100-ft increments})$$
$$= (0.41 \text{ ft})(105.6 \text{ increments})$$
$$= 43.3 \text{ ft head loss}$$

Table 7. Friction Loss of Water in Feet per 100-ft Length of Pipe, Based on Williams & Hazen Formula Using C = 100.

	½-in. Pipe		¾-in. Pipe		1-in. Pipe		1¼-in. Pipe		1½-in. Pipe		2-in. Pipe		2½-in. Pipe		3-in. Pipe		4-in. Pipe		5-in. Pipe		6-in. Pipe		
US gal per min	Vel. ft per sec	Loss in ft	Vel. ft per sec	Loss in ft	Vel. ft per sec	Loss in ft	Vel. ft per sec	Loss in ft	Vel. ft per sec	Loss in ft	Vel. ft per sec	Loss in ft	Vel. ft per sec	Loss in ft	Vel. ft per sec	Loss in ft	Vel. ft per sec	Loss in ft	Vel. ft per sec	Loss in ft	Vel. ft per sec	Loss in ft	US gal per min
2	2.10	7.4	1.20	1.9																			2
4	4.21	27.0	2.41	7.0	1.49	2.14	.86	.57	.63	.26													4
6	6.31	57.0	3.61	14.7	2.23	4.55	1.29	1.20	.94	.56	.61	.20											6
8	8.42	98.0	4.81	25.0	2.98	7.8	1.72	2.03	1.26	.95	.82	.33	.52	.11									8
10	10.52	147.0	6.02	38.0	3.72	11.7	2.14	3.05	1.57	1.43	1.02	.50	.65	.17	.45	.07							10
12			7.22	53.0	4.46	16.4	2.57	4.3	1.89	2.01	1.23	.79	.78	.23	.54	.10							12
15			9.02	80.0	5.60	25.0	3.21	6.5	2.36	3.00	1.53	1.08	.98	.36	.68	.15							15
18			10.84	108.2	6.69	35.0	3.86	9.1	2.83	4.24	1.84	1.49	1.18	.50	.82	.21							18
20			12.03	136.0	7.44	42.0	4.29	11.1	3.15	5.20	2.04	1.82	1.31	.61	.91	.25	.51	.06					20
25					9.30	64.0	5.36	16.6	3.80	7.30	2.55	2.73	1.63	.92	1.13	.38	.64	.09					25
30					11.15	89.0	6.43	23.0	4.72	11.0	3.06	3.84	1.96	1.29	1.36	.54	.77	.13	.49	.04			30
35					13.02	119.0	7.51	31.2	5.51	14.7	3.57	5.10	2.29	1.72	1.59	.71	.89	.17	.57	.06			35
40					14.88	152.0	8.58	40.0	6.30	18.8	4.08	6.6	2.61	2.20	1.82	.91	1.02	.22	.65	.08			40
45							9.65	50.0	7.08	23.2	4.60	8.2	2.94	2.80	2.04	1.15	1.15	.28	.73	.09			45
50							10.72	60.0	7.87	28.4	5.11	9.9	3.27	3.32	2.27	1.38	1.28	.34	.82	.11	.57	.04	50
55							11.78	72.0	8.66	34.0	5.62	11.8	3.59	4.01	2.45	1.58	1.41	.41	.90	.14	.62	.05	55
60							12.87	85.0	9.44	39.6	6.13	13.9	3.92	4.65	2.72	1.92	1.53	.47	.98	.16	.68	.06	60
65							13.92	99.7	10.23	45.9	6.64	16.1	4.24	5.4	2.89	2.16	1.66	.53	1.06	.19	.74	.076	65
70							15.01	113.0	11.02	53.0	7.15	18.4	4.58	6.2	3.18	2.57	1.79	.63	1.14	.21	.79	.08	70
75							16.06	129.0	11.80	60.0	7.66	20.9	4.91	7.1	3.33	3.00	1.91	.73	1.22	.24	.85	.10	75
80							17.16	145.0	12.59	68.0	8.17	23.7	5.23	7.9	3.63	3.28	2.04	.81	1.31	.27	.91	.11	80
85							18.21	163.8	13.38	75.0	8.68	26.5	5.56	8.1	3.78	3.54	2.17	.91	1.39	.31	.96	.12	85
90							19.30	180.0	14.71	84.0	9.19	29.4	5.88	9.8	4.09	4.08	2.30	1.00	1.47	.34	1.02	.14	90
95									14.95	93.0	9.70	32.6	6.21	10.8	4.22	4.33	2.42	1.12	1.55	.38	1.08	.15	95
100									15.74	102.0	10.21	35.8	6.54	12.0	4.54	4.96	2.55	1.22	1.63	.41	1.13	.17	100
110									17.31	122.0	11.23	42.9	7.18	14.5	5.00	6.0	2.81	1.46	1.79	.49	1.25	.21	110
120	8″ Pipe								18.89	143.0	12.25	50.0	7.84	16.8	5.45	7.0	3.06	1.17	1.96	.58	1.36	.24	120
130									20.46	166.0	13.28	58.0	8.48	18.7	5.91	8.1	3.31	1.97	2.12	.67	1.47	.27	130
140	.90	.08							22.04	190.0	14.30	67.0	9.15	22.3	6.35	9.2	3.57	2.28	2.29	.76	1.59	.32	140
150	.96	.09									15.32	76.0	9.81	25.5	6.82	10.5	3.82	2.62	2.45	.88	1.70	.36	150
160	1.02	.10									16.34	86.0	10.46	29.0	7.26	11.8	4.08	2.91	2.61	.98	1.82	.40	160
170	1.08	.11									17.36	96.0	11.11	34.1	7.71	13.3	4.33	3.26	2.77	1.08	1.92	.45	170
180	1.15	.13									18.38	107.0	11.76	35.7	8.17	14.0	4.60	3.61	2.94	1.22	2.04	.50	180
190	1.21	.14	10″ Pipe								19.40	118.0	12.42	39.6	8.63	15.5	4.84	4.01	3.10	1.35	2.16	.55	190
200	1.28	.15									20.42	129.0	13.07	43.1	9.08	17.8	5.11	4.4	3.27	1.48	2.27	.62	200
220	1.40	.18	.90	.06							22.47	154.0	14.38	52.0	9.99	21.3	5.62	5.2	3.59	1.77	2.50	.73	220
240	1.53	.22	.98	.07							24.51	182.0	15.69	61.0	10.89	25.1	6.13	6.2	3.92	2.08	2.72	.87	240
260	1.66	.25	1.06	.08							26.55	211.0	16.99	70.0	11.80	29.1	6.64	7.2	4.25	2.41	2.95	1.00	260
280	1.79	.28	1.15	.09									18.30	81.0	12.71	33.4	7.15	8.2	4.58	2.77	3.18	1.14	280
300	1.91	.32	1.22	.11									19.61	92.0	13.62	38.0	7.66	9.3	4.90	3.14	3.40	1.32	300
320	2.05	.37	1.31	.12									20.92	103.0	14.52	42.8	8.17	10.5	5.23	3.54	3.64	1.47	320
340	2.18	.41	1.39	.14									22.22	116.0	15.43	47.9	8.68	11.7	5.54	3.97	3.84	1.62	340

Table 7. (Continued)

Velocity (V, ft/s) and friction head loss (hf) for the flows indicated. Pipe sizes labelled in the original: 12″, 14″, 16″, 20″, 24″ and 30″ Pipe.

GPM	2½″ V	2½″ hf	3″ V	3″ hf	4″ V	4″ hf	5″ V	5″ hf	6″ V	6″ hf	8″ V	8″ hf	10″ V	10″ hf	12″ V	12″ hf	14″ V	14″ hf	16″ V	16″ hf	20″ V	20″ hf	24″ V	24″ hf	30″ V	30″ hf
360	23.53	128.0	16.34	53.0	9.19	13.1	5.87	4.41	4.08	1.83	2.30	.45	1.47	.15												
380	24.84	142.0	17.25	59.0	9.69	14.0	6.19	4.86	4.31	2.00	2.43	.50	1.55	.17	1.08	.069										
400	26.14	156.0	18.16	65.0	10.21	16.0	6.54	5.4	4.55	2.20	2.60	.54	1.63	.19	1.14	.075										
450			20.40	78.0	11.49	19.8	7.35	6.7	5.11	2.74	2.92	.68	1.84	.23	1.28	.095										
500			22.70	98.0	12.77	24.0	8.17	8.1	5.68	2.90	3.19	.82	2.04	.28	1.42	.113	1.04	.06								
550			24.96	117.0	14.04	28.7	8.99	9.6	6.25	3.96	3.52	.97	2.24	.33	1.56	.135	1.15	.07								
600			27.23	137.0	15.32	33.7	9.80	11.3	6.81	4.65	3.84	1.14	2.45	.39	1.70	.159	1.25	.08								
650					16.59	39.0	10.62	13.2	7.38	5.40	4.16	1.34	2.65	.45	1.84	.19	1.37	.09								
700					17.87	44.9	11.44	15.1	7.95	6.21	4.46	1.54	2.86	.52	1.99	.22	1.46	.10								
750					19.15	51.0	12.26	17.2	8.50	7.12	4.80	1.74	3.06	.59	2.13	.24	1.58	.11								
800					20.42	57.0	13.07	19.4	9.08	7.96	5.10	1.90	3.26	.66	2.27	.27	1.67	.13								
850					21.70	64.0	13.89	21.7	9.65	8.95	5.48	2.20	3.47	.75	2.41	.31	1.79	.14	1.36	.08						
900					22.98	71.0	14.71	24.0	10.20	10.11	5.75	2.46	3.67	.83	2.56	.34	1.88	.16	1.44	.084						
950							15.52	26.7	10.77	11.20	6.06	2.87	3.88	.91	2.70	.38	2.00	.18	1.52	.095						
1000							16.34	29.2	11.34	12.04	6.38	2.97	4.08	1.03	2.84	.41	2.10	.19	1.60	.10	1.02	.04				
1100							17.97	34.9	12.48	14.55	7.03	3.52	4.49	1.19	3.13	.49	2.31	.23	1.76	.12	1.12	.04				
1200							19.61	40.9	13.61	17.10	7.66	4.17	4.90	1.40	3.41	.58	2.52	.27	1.92	.14	1.23	.05				
1300									14.72	18.4	8.30	4.85	5.31	1.62	3.69	.67	2.71	.32	2.08	.17	1.33	.06				
1400									15.90	22.60	8.95	5.50	5.71	1.87	3.98	.78	2.92	.36	2.24	.19	1.43	.064				
1500									17.02	25.60	9.58	6.24	6.12	2.13	4.26	.89	3.15	.41	2.39	.21	1.53	.07				
1600									18.10	26.9	10.21	7.00	6.53	2.39	4.55	.98	3.34	.47	2.56	.24	1.63	.08				
1800											11.50	8.78	7.35	2.95	5.11	1.21	3.75	.58	2.87	.30	1.84	.10	1.28	.04		
2000											12.78	10.71	8.16	3.59	5.68	1.49	4.17	.71	3.19	.37	2.04	.12	1.42	.05		
2200											14.05	12.78	8.98	4.24	6.25	1.81	4.59	.84	3.51	.44	2.25	.15	1.56	.06		
2400											15.32	14.2	9.80	5.04	6.81	2.08	5.00	.99	3.83	.52	2.45	.17	1.70	.07	1.09	.02
2600													10.61	5.81	7.38	2.43	5.47	1.17	4.15	.60	2.66	.20	1.84	.08	1.16	.027
2800													11.41	6.70	7.95	2.75	5.84	1.32	4.47	.68	2.86	.23	1.98	.09	1.27	.03
3000													12.24	7.62	8.52	3.15	6.01	1.49	4.79	.78	3.08	.27	2.13	.10	1.37	.037
3200													13.05	7.8	9.10	3.51	6.68	1.67	5.12	.88	3.27	.30	2.26	.12	1.46	.041
3500													14.30	10.08	9.95	4.16	7.30	1.97	5.59	1.04	3.59	.35	2.49	.14	1.56	.047
3800													15.51	13.4	10.80	4.90	7.98	2.36	6.07	1.20	3.88	.41	2.69	.17	1.73	.05
4200															11.92	5.88	8.76	2.77	6.70	1.44	4.29	.49	2.99	.20	1.91	.07
4500															12.78	6.90	9.45	3.22	7.18	1.64	4.60	.56	3.20	.22	2.04	.08
5000															14.20	8.40	10.50	3.92	8.01	2.03	5.13	.68	3.54	.27	2.26	.09
5500																	11.55	4.65	8.78	2.39	5.64	.82	3.90	.33	2.50	.11
6000																	12.60	5.50	9.58	2.79	6.13	.94	4.25	.38	2.73	.13
6500																	13.65	6.45	10.39	3.32	6.64	1.10	4.61	.45	2.96	.15
7000																	14.60	7.08	11.18	3.70	7.15	1.25	4.97	.52	3.18	.17
8000																			12.78	4.74	8.17	1.61	5.68	.66	3.64	.23
9000																			14.37	5.90	9.20	2.01	6.35	.81	4.08	.28
10000																			15.96	7.19	10.20	2.44	7.07	.98	4.54	.33
12000																					12.25	3.41	8.50	1.40	5.46	.48
14000																					14.30	4.54	9.95	1.87	6.37	.63
16000																							11.38	2.40	7.28	.81
18000																							12.76	2.97	8.18	1.02
20000																							14.20	3.60	9.10	1.23

Add the two friction losses to obtain the total friction head loss for the full 2.6 miles of pipeline:

$$\begin{array}{r} 94.09 \quad \text{ft} \\ +43.30 \quad \text{ft} \\ \hline 137.39 \quad \text{ft total head loss} \end{array}$$

Table 7 can also be used for C values other than 100 by first converting the actual flow rate to an *equivalent flow rate* for $C = 100$. The equation used to find the equivalent flow rate is:

$$\text{Equivalent Flow Rate} = \frac{(\text{Actual Flow Rate})(100)}{C \text{ value}}$$

The following problem shows how the equation is used.

Example 4

A 10-in. pipeline is 3000 ft long and has a smoothness coefficient of $C = 130$. Water is moving through the pipe at a flow rate of 3900 gpm. Use Table 7 to determine the friction head loss.

Since the C value is not 100, first use the equation given above to find an equivalent flow rate:

$$\text{Equivalent Flow Rate} = \frac{(\text{Actual Flow Rate})(100)}{C \text{ value}}$$

Fill in the known information:

$$\text{Equivalent Flow Rate} = \frac{(3900 \text{ gpm})(100)}{130}$$

And solve for the unknown value:

$$\text{Equivalent Flow Rate} = 3000 \quad \text{gpm}$$

Now use the Equivalent Flow Rate to find the friction head loss from Table 7. Enter the table at a flow rate of 3000 gpm and move across to the *10 in. pipe* column. The table shows a friction head loss of 7.62 ft of head per 100 ft of pipe. The pipeline in the problem is 3000 ft long, so calculate how many 100-ft lengths of pipe it contains:

$$\frac{3000 \text{ ft}}{100 \text{ ft}} = 30 \text{ lengths of 100-ft pipe}$$

Then calculate total friction head loss:

$$(30)(7.62) = 228.6 \text{ ft friction head loss}$$

Note that 228.6 ft is the same friction head loss that you could read directly from a table showing values for 10 in. pipe with $C = 130$. (No table covering those values is given in this book.)

H5-2. Minor Head Loss

MINOR HEAD LOSSES are energy losses caused by sudden changes in either the direction or the velocity of flow. Bends, ells, and 45-deg fittings are examples of fittings that cause sudden changes in direction; increasers, reducers, and valves are examples of fittings that cause sudden changes in velocity. Minor head loss also includes entrance head loss and exit head loss, where energy is used to force water into and out of pipes.

The word "minor" can be misleading. Minor head losses are indeed small for pipelines that include long reaches of straight pipe. However, in a treatment plant where there are short runs of pipe with many valves and fittings, the minor losses can be greater than the friction losses. As a rule of thumb, you can ignore minor losses only if the pipeline is longer than 500 times the pipe diameter. For instance, in a 12-in. diam pipeline minor losses should be considered if the pipe is less than 500 ft long:

$$(500)(1\text{-ft pipe diameter}) \; = \; 500 \text{ ft}$$

Using the same rule of thumb for 8-in. diameter pipe, minor losses should be considered for pipelines less than 335 ft long.

$$(500)(0.67\text{-ft pipe diameter}) \; = \; 335 \text{ ft}$$

One of the easiest ways to calculate minor losses is to use a table to find how long a section of straight pipe would be needed to cause the same head loss as the fitting, and then calculate the friction loss for that pipe length. Example 5 illustrates the procedure.

Example 5

A 6-in. diameter pipeline carries water from the filter backwash storage tank to the 24-in. diam pipe manifold feeding the filters. There is 100 ft of straight-run pipe, a standard elbow, and another 80 ft of straight-run pipe leading to the filter. Water flows through the pipe at a rate of

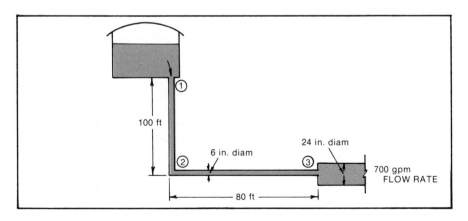

Figure 47.   (Example 5.)

700 gpm. What is the total head loss? (Use a $C = 100$.) Figure 47 illustrates the problem.

The largest part of the head loss will be caused by friction in the 180 ft of 6-in. pipe. But there are also three points where minor losses will occur:

(1) Entrance head loss from storage to 6-in. diameter pipe
(2) Standard elbow
(3) Sudden enlargement into the 24-in. diameter manifold (exit head loss)

Consider the minor losses first. In **Figure 48A** the fitting that best describes the situation at point 1 is called *ordinary entrance*. Draw a line on the table from the dot associated with *ordinary entrance* to the right hand scale entitled *nominal diameter* at the 6-in. mark. At the point where this line crosses the middle scale read the *length of straight pipe:*

$$\begin{array}{ccc} \text{Minor head loss} & & \text{Friction head loss} \\ \text{of} & = & \text{equivalent to} \\ \text{Ordinary Entrance} & & \text{9 ft of 6-in. diam pipe} \end{array}$$

This means that the head loss resulting from the water entering the 6-in. line is the same as the friction loss which would occur over 9 ft of 6-in. diam pipe.

Now go through the same steps for the standard elbow and the exit head loss. For the exit head loss use *Sudden Enlargement,* $d/D = 1/4$ since

$$\frac{6 \text{ in.}}{24 \text{ in.}} = \frac{1}{4}$$

$$\begin{array}{ccc} \text{Minor head loss} & & \text{Friction head loss} \\ \text{of} & = & \text{equivalent to} \\ \text{Standard elbow} & & \text{17 ft of 6-in. diam pipe} \end{array}$$

$$\begin{array}{ccc} \text{Minor head loss} & & \text{Friction head loss} \\ \text{of} & = & \text{equivalent to} \\ \text{Sudden Enlargement exit} & & \text{17 ft of 6-in. diam pipe} \end{array}$$

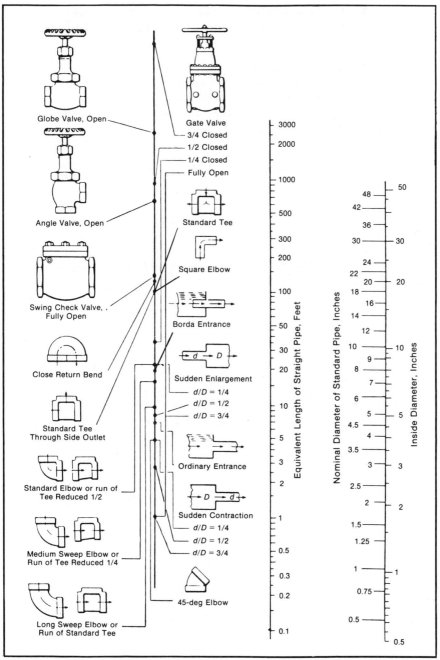

Figure 48A. Resistance of Valves and Fittings to Flow of Fluids

Add the minor losses expressed as equivalent pipe lengths to the actual pipe length before determining friction loss:

180 ft of 6-in. diam pipe	(actual pipe length)
9 ft of 6-in. diam pipe	(from *ordinary entrance* losses)
17 ft of 6-in. diam pipe	(from *standard elbow* losses)
17 ft of 6-in. diam pipe	(from *sudden enlargement* exit)
223 ft of 6-in. diam pipe	

From Table 7 find the head loss per 100 ft for 6-in. pipe at a flow of 700 gpm.

$$\text{Loss per 100 ft} = 6.21 \text{ ft head}$$

Now determine total head loss (friction + minor losses). First, calculate the number of 100-ft increments:

$$\frac{223 \text{ ft}}{100 \text{ ft}} = 2.23 \text{ increments}$$

Then calculate the total head loss:

$$\text{Total Head Loss} = (\text{Head loss per 100-ft increment})(\text{No. of Increments})$$
$$= (6.21 \text{ ft})(2.23 \text{ increments})$$
$$= 13.85 \text{ ft head loss}$$

Review Questions

1. (a) A 12-in. diam water transmission line carries water 7420 ft from a reservoir to a treatment plant. The pipeline is new reinforced concrete and the flow rate is 1.8 mgd. Use Tables 5 and 6 to determine the friction head loss in this line.

    (b) What would be the friction head loss if the flow rate increased to 4.5 mgd (2 1/2 times 1.8 mgd)?

2. A high elevation water tank continuously feeds a lower tank at a flow rate of 3500 gpm. The tanks are connected by 683 ft of 14-in. CIP, about 10 years old. The line contains the following fittings:

    an entrance (ordinary)
    a 45 degree elbow
    two gate valves, fully open
    two square elbows

    Use Figure 48A and Table 7 to determine the total head loss.

Summary Answers

1. (a) 24.71 ft
    (b) 133.56 ft

2. 17.63 ft

Detailed Answers

1. (a) Table 5 shows a C of 140 for new reinforced concrete pipe. Enter
         Table 6 at 1,800,000 gpd to find that the head loss for $C = 140$ is 3.33
         ft per 1000 ft of pipeline. Calculate the total friction head loss as:

    Total Friction Head Loss $=$ (Loss per 1000 ft)(No. of 1000-ft increments)

    $=$ (3.33 ft)(7.42 increments)

    $=$ 24.71 ft head loss

    (b) Enter Table 6 at 4,500,000 gpd to find that the head loss for $C = 140$
         is 18.0 ft per 1000 ft. Calculate the total friction head loss as:

    Total Friction Head Loss $=$ (Loss per 1000 ft)(No. of 1000-ft increments)

    $=$ (18 ft)(7.42 increments)

    $=$ 133.56 ft head loss

2. As shown in Figure 48B, the fittings cause minor head loss equivalent to the
    following lengths of straight pipe:

Ordinary entrance	20 ft of 14-in. diam pipe
45-deg elbow	18 ft of 14-in. diam pipe
Two open gate valves (8 ft ea)	16 ft of 14-in. diam pipe
Two square elbows (79 ft ea)	158 ft of 14-in. diam pipe
	212 ft of 14-in. diam pipe

    Add these equivalent pipe lengths to the actual length of pipe:

    Actual length of pipe $=$ 683 ft

    Equivalent pipe lengths $=$ 212 ft

    895 ft

    From Table 7, the head loss for a flow rate of 3500 gpm in a 14-in. pipe is
    1.97 ft/100 ft pipe. Since there are 8.95 increments of 100 ft, the total head
    loss is:

    Total Head Loss $=$ (Loss per 100 ft)(No. of 100-ft increments)

    $=$ (1.97 ft)(8.95 increments)

    $=$ 17.63 ft head loss

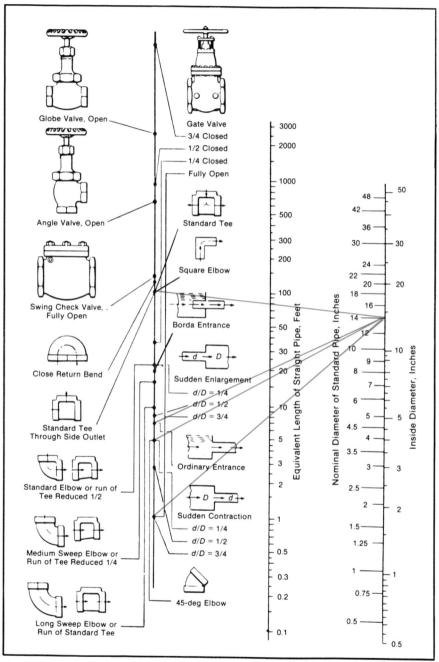

Globe Valve, Open

Angle Valve, Open

Swing Check Valve,
Fully Open

Close Return Bend

Standard Tee
Through Side Outlet

Standard Elbow or run of
Tee Reduced 1/2

Medium Sweep Elbow or
Run of Tee Reduced 1/4

Long Sweep Elbow or
Run of Standard Tee

Gate Valve
3/4 Closed
1/2 Closed
1/4 Closed
Fully Open

Standard Tee

Square Elbow

Borda Entrance

Sudden Enlargement
$d/D = 1/4$
$d/D = 1/2$
$d/D = 3/4$

Ordinary Entrance

Sudden Contraction
$d/D = 1/4$
$d/D = 1/2$
$d/D = 3/4$

45-deg Elbow

Equivalent Length of Straight Pipe, Feet

3000
2000

1000

500

300
200

100

50

30
20

10

5

3

2

1

0.5

0.3

0.2

0.1

Nominal Diameter of Standard Pipe, Inches

48
42
36
30
24
22
20
18
16
14
12
10
9
8
7
6
5
4.5
4
3.5
3
2.5
2
1.5
1.25

1

0.75

0.5

Inside Diameter, Inches

50

30

20

10

5

3

2

1

0.5

Figure 48B. Resistance of Valves and Fittings to Flow of Fluids

Hydraulics 6

Pumping Problems

H6-1. Pumping Rates

The rate of flow produced by a pump is expressed as the volume of water pumped during a given period of time.[19] The mathematical equation used in pumping rate problems can usually be determined from the verbal statement of the problem:

VERBAL: What is the pumping rate in "gallons *per* minute"?

$$\text{MATH: pumping rate} = \frac{\text{gallons}}{\text{minutes}}$$

VERBAL: What is the pumping rate in "gallons *per* hour"?

$$\text{MATH: pumping rate} = \frac{\text{gallons}}{\text{hours}}$$

The number of gallons pumped during a period can be determined either with a flow meter or by measuring the number of gallons pumped into or out of a tank.

Example 1

The totalizer of the meter on the discharge side of your pump reads in hundreds of gallons. If the totalizer shows a reading of 108 at 1:00 PM and 312 at 1:30 PM, what is the pumping rate expressed in gallons per minute?

The problem asks for pumping rate in *gallons per minute* (gpm), so the mathematical setup is

$$\text{Pumping Rate} = \frac{\text{gallons}}{\text{minutes}}$$

---

[19]Hydraulics Section, Flow Rate Problems.

To solve the problem, fill in the blanks (number of gallons and number of minutes) in the equation. The total gallons pumped is determined from the totalizer readings:

$$\begin{array}{r} 31{,}200 \ \ \text{gal} \\ -10{,}800 \, . \, \text{gal} \\ \hline 20{,}400 \ \ \text{gal} \end{array}$$

The volume was pumped between 1:00 PM and 1:30 PM, for a total of 30 min. From this information calculate the gallons-per-minute pumping rate:

$$\text{Pumping Rate} = \frac{20{,}400 \ \ \text{gal}}{30 \ \ \text{min}}$$

$$= 680 \ \ \text{gpm} \ \ \text{pumping} \ \ \text{rate}$$

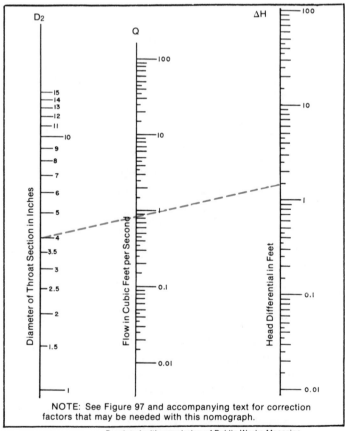

NOTE: See Figure 97 and accompanying text for correction factors that may be needed with this nomograph.

Figure 49.   Flow Rate Nomograph for Venturi Meter

Instead of using totalizer readings to calculate the average pumping rate for a period of a few minutes or hours (as in Example 1), you can read the *instantaneous* pumping rate or flow rate—the flow rate or pumping at *one particular moment*[20]—directly from many flow meters. Other flow meters require that you perform calculations to determine the instantaneous flow rate, as illustrated in the following example.

Example 2

The venturi meter on the discharge side of your pump has a throat diameter of 4 in. The head differential between the high-pressure tap and the low-pressure tap is 1.5 ft. Use the nomograph shown in Figure 49 to determine the pumping rate in gallons per minute.

To determine the gallons-per-minute pumping rate, first determine the cubic-feet-per-second (cfs) pumping rate. Then convert to gallons per minute.

Draw a line on the nomograph from 4 in. on the *diameter of throat section in inches* scale to 1.5 ft on the *head differential in feet* scale. The point where the line crosses the middle scale indicates the approximate flow of 0.85 cfs. Now convert the cubic-feet-per-second pumping rate to gallons-per-minute pumping rate:[21]

$$(0.85 \text{ cfs})(7.48 \text{ gal/cu ft})(60 \text{ sec/min}) = 381.48 \text{ gpm pumping rate}$$

When there is no meter on the discharge side of the pump, the pumping rate can be determined by measuring the number of gallons pumped into or out of a tank during a given time period. Let's look at three examples of determining total gallons pumped, illustrated by figures 50A, 50B, and 50C.

In the first example (Figure 50A), the pump is discharging into an empty tank with the outlet valve to the tank closed. The total gallons pumped during the given time is the number of gallons in the tank at the end of the pumping test.

Since it is not always possible or practical to pump into an empty tank, the pumping test is sometimes conducted by pumping into a tank that already contains water (Figure 50B). The outlet valve is closed and the total gallons (cross-hatched area) pumped during the given time is determined from the *rise in water level*.

In Figure 50C the pump is located on the discharge or outlet side of the tank. To conduct a pumping test, the inlet valve of the tank is shut off. When the pump is turned on, the total gallons (cross-hatched area) pumped during the given time is determined from the *fall in water level*.

Example 3

During a 15-min pumping test, 15,820 gal is pumped into

---

[20]Hydraulics Section, Flow Rate Problems.

[21]Mathematics Section, Conversions (Flow Conversions).

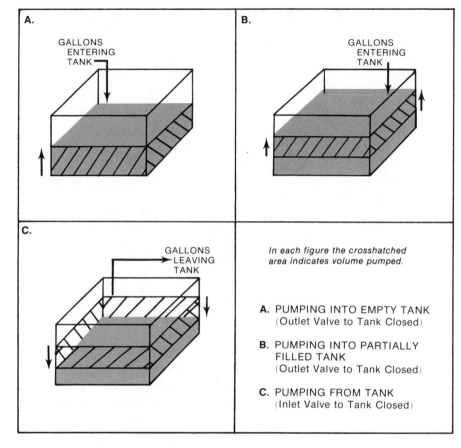

Figure 50. Determining Pumping Rate Using Tank

an empty rectangular tank (Figure 51). What is the pumping rate in gallons per minute?

Since the problem asks for the pumping rate in gallons per minute, the mathematical setup is

$$\text{Pumping Rate} = \frac{\text{gallons}}{\text{minutes}}$$

Fill in the information given and perform the calculations to complete the problem:

$$\text{Pumping Rate} = \frac{15{,}820 \text{ gal}}{15 \text{ min}}$$

$$= 1055 \text{ gpm pumping rate}$$

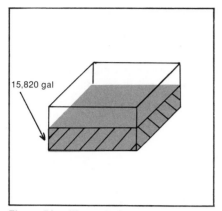

Figure 51.   (Example 3.)

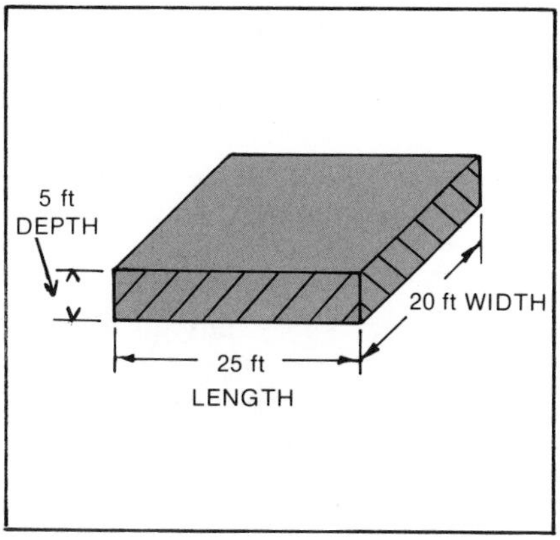

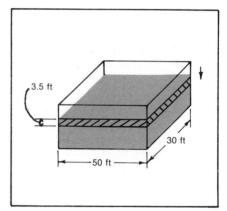

Figure 52.   (Example 4.)

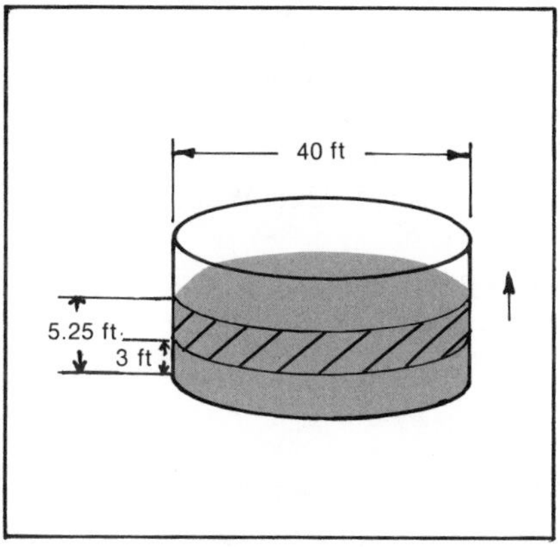

Figure 53.   (Example 5.)

Figure 54.   (Example 6.)

Example 4

An empty rectangular tank 25 ft long and 20 ft wide (Figure 52) can hold water to a depth of 5 ft. If this tank is filled by a pump in 55 min, what is the pumping rate in gallons per minute?

In this example, the entire tank was filled during the 55-min pumping test. Therefore the total gallons pumped is equal to the capacity of the tank in gallons:[22]

$$\text{Volume of Tank} = (\text{Area of Rectangle})(\text{Depth})$$
$$= (25 \ \text{ft})(20 \ \text{ft})(5 \ \text{ft})$$
$$= 2500 \ \text{cu ft}$$

---

[22]Mathematics Section, Volume Measurements.

Convert cubic-feet volume to gallons:[23]

$$(2500 \ \text{cu ft})(7.48 \ \text{gal/cu ft}) = 18,700 \ \text{gal}$$

Then use the total volume pumped and the time period of the pumping test to calculate the gallons-per-minute pumping rate:

$$\text{Pumping Rate} = \frac{\text{gallons}}{\text{minutes}}$$

$$= \frac{18,700 \ \text{gal}}{55 \ \text{min}}$$

$$= 340 \ \text{gpm pumping rate}$$

Example 5

A tank 40 ft in diameter is filled with water to a depth of 3 ft (Figure 53). To conduct a pumping test, the outlet valve to the tank is closed and the pump is allowed to discharge into the tank. After 1 hr 15 min the water level is 5.25 ft. What is the pumping rate in gallons per minute?

In this problem, the total gallons pumped is represented by the cross-hatched area on the diagram (tank diameter = 40 ft and water depth = 2.25 ft). Calculate the volume in cubic feet:

$$\text{Volume Pumped} = (\text{Area of Circle})(\text{Depth})$$

$$= (0.785)(40 \ \text{ft})(40 \ \text{ft})(2.25 \ \text{ft})$$

$$= 2826 \ \text{cu ft}$$

And convert the cubic-feet volume to gallons:

$$(2826 \ \text{cu ft})(7.48 \ \text{gal/cu ft}) = 21,138 \ \text{gal}$$

The pumping test was conducted over a period of 1 hr 15 min. The problem asks for the pumping rate in gallons per minute, so convert to minutes:

$$\begin{array}{r} 60 \ \text{min} \\ +15 \ \text{min} \\ \hline 75 \ \text{min} \end{array}$$

Using the volume pumped (in gallons) and the time (in minutes) it took to pump the volume, calculate the pumping rate in gallons per minute:

---

[23] Mathematics Section, Conversions.

$$\text{Pumping Rate} = \frac{\text{gallons}}{\text{minutes}}$$

$$= \frac{21,138 \text{ gal}}{75 \text{ min}}$$

$$= 281.84 \text{ gpm}$$

pumping rate

Example 6

A 1-hr pumping test is run on a pump located on the outlet side of a tank. The inlet valve is closed and the pump is started. At the end of the test the water level in the tank has dropped 3.5 ft. If the tank (Figure 54) is rectangular, that is, 50 ft by 30 ft, with a water depth before pumping of 6 ft, what is the gallons-per-minute pumping rate?

The total gallons pumped is represented by the cross-hatched area on the diagram. (Notice that information pertaining to the water depth before pumping is not needed in solving this problem. The drop in the water level is the essential depth information.) Calculate the total volume pumped:

$$\text{Volume Pumped} = (\text{Area of Rectangle})(\text{Depth})$$

$$= (50 \text{ ft})(30 \text{ ft})(3.5 \text{ ft})$$

$$= 5250 \text{ cu ft}$$

Convert cubic-feet volume to gallons:

$$(5250 \text{ cu ft})(7.48 \text{ gal/cu ft}) = 39,270 \text{ gal}$$

Now determine the pumping rate:

$$\text{Pumping Rate} = \frac{\text{gallons}}{\text{minutes}}$$

$$= \frac{39,270 \text{ gal}}{60 \text{ min}}$$

$$= 654.5 \text{ gpm pumping rate}$$

H6-2. Pump Heads

Pump head measurements are used to determine the amount of energy a pump can or must impart to the water. These heads, measured in feet, are a special application of the pressure and velocity heads covered in Section H4-1. Specific terms are used to describe heads measured at various points and under different conditions of the pump system.

Suction and discharge. In pump systems the words *suction* and *discharge* identify the inlet and outlet sides of the pump. As shown in Figure 55, the *suction side* of the pump is the *inlet* or *low pressure* side. The *discharge side* is the *outlet*

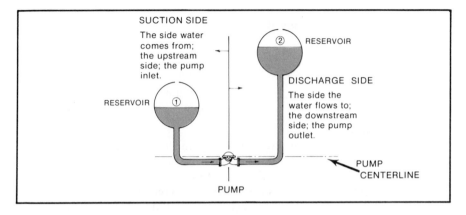

Figure 55. Suction and Discharge Sides of a Pump

or *high pressure* side. Any time a pump term includes the words *suction* or *discharge* you can recognize immediately which side of the pump system is being discussed.

The PUMP CENTERLINE (¢) is represented by a horizontal line drawn through the center of the pump. This line is important because it is the reference line from which pump head measurements are made.

The terms *suction* and *discharge* help establish the location of the particular pump head measurement. The terms *static* and *dynamic,* however, describe the condition of the system when the measurement is taken. Static heads are measured when the pump is off, and therefore the water is not moving. Dynamic heads are measured with the pump running and water flowing through the system.

Static Heads. Figure 56 illustrates the two basic pumping configurations found in water systems. The labeled vertical distances show the four types of static pump head:

- Static suction head
- Static suction lift
- Static discharge head
- Total static head

In a system where the reservoir feeding the pump is higher than the pump (Figure 56A), the difference in elevation (height) between the pump centerline and the free water surface of the reservoir feeding the pump is termed STATIC SUCTION HEAD. But in a system where the reservoir feeding the pump is lower than the pump (Figure 56B), the difference in elevation between the centerline and the free water surface of the reservoir feeding the pump is termed STATIC SUCTION LIFT. Notice that a single system will have either a static head measurement or a static suction lift measurement, but not both.

STATIC DISCHARGE HEAD is defined as the difference in height between the pump

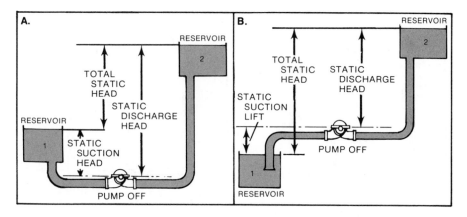

Figure 56.   Static Heads

centerline and the level of the discharge free water surface (Figure 56).

TOTAL STATIC HEAD is the total height that the pump must lift the water when moving it from *reservoir 1* to *reservoir 2* (Figure 56). It is defined precisely as the vertical distance from the suction free water surface to the discharge free water surface. In Figure 56A total static head is found by subtracting static suction head from static discharge head, whereas in Figure 56B it is the sum of the static suction lift and static discharge head.

Static heads can be calculated from measurements of reservoir and pump elevations, or from pressure gage readings taken when the pump is not running.

Example 7

In Figure 57A locate, label, and calculate (in feet)

(a) Static suction head

(b) Static discharge head

(c) Total static head

First locate and label the three types of static head, as shown in Figure 57B. Then, using the elevation information given for the problem, calculate each head as follows:

$$\text{Static Suction Head} = 742 \text{ ft} - 722 \text{ ft}$$
$$= 20 \text{ ft}$$
$$\text{Static Discharge Head} = 927 \text{ ft} - 722 \text{ ft}$$
$$= 205 \text{ ft}$$

Total static head can be found in two ways:

$$\text{Total Static Head} = \text{Elevation of Reservoir 1}$$
$$- \text{Elevation of Reservoir 2}$$
$$= 927 \text{ ft} - 742 \text{ ft}$$
$$= 185 \text{ ft}$$

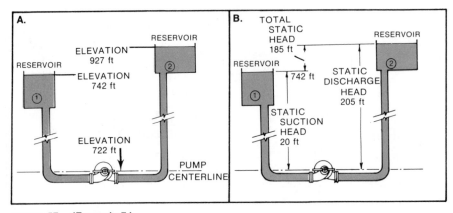

Figure 57.   (Example 7.)

or:

$$\begin{aligned}
\text{Total Static Head} &= \text{Static Discharge Head} \\
&\quad - \text{Static Suction Head} \\
&= 205 \ \text{ft} - 20 \ \text{ft} \\
&= 185 \ \text{ft}
\end{aligned}$$

Example 8

In Figure 58A, locate, label, and calculate (in feet)
(a) Static suction lift
(b) Static discharge head
(c) Total static head

First, locate and label the three heads as shown in Figure 58B. Then using the elevation information given, calculate the heads, in feet, as follows:

$$\begin{aligned}
\text{Static suction lift} &= 971 \ \text{ft} - 962 \ \text{ft} \\
&= 9 \ \text{ft} \\
\text{Static discharge head} &= 1578 \ \text{ft} - 971 \ \text{ft} \\
&= 607 \ \text{ft} \\
\text{Total static head} &= 607 \ \text{ft} + 9 \ \text{ft} \\
&= 616 \ \text{ft}
\end{aligned}$$

or:

$$\begin{aligned}
\text{Total static head} &= 1578 \ \text{ft} - 962 \ \text{ft} \\
&= 616 \ \text{ft}
\end{aligned}$$

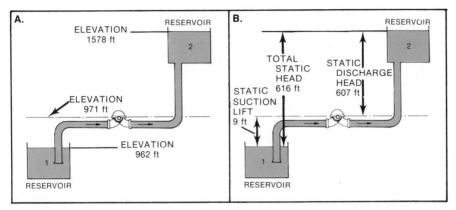

Figure 58.   (Example 8.)

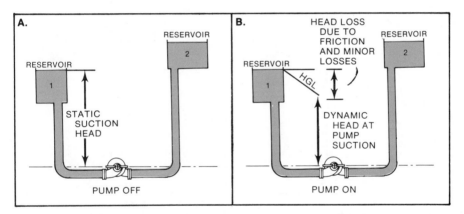

Figure 59.   Effect of Friction and Minor Head Losses on Suction Head

If static heads were the only heads involved in pump operations, then any force exerted by the pump greater than the static head would cause water to be pumped. However, this doesn't happen. As soon as the pump is turned on, the pressures and energies within the system change. Measurements made while the pump is running, called DYNAMIC HEADS, are used to describe the total energy that the pump must develop for pumping to take place.

Dynamic Heads. When water flows through a pipe, the water rubs against the walls of the pipe, and energy is lost to friction. In addition, there is a certain amount of resistance to flow as the water passes through valves, fittings, inlets, and outlets of a piping system. These additional energy losses that the pump must overcome are called FRICTION HEAD LOSSES and MINOR HEAD LOSSES.[24]

---

[24]Hydraulics Section, Head Loss.

Let's look at the effect of these additional energy losses on each side of the pump. Figure 59 shows a comparison of the head that exists on the suction side of the pump before and after the pump is turned on. Notice that loss of head is caused by friction losses when the pump is turned on.

Part of the static suction head (which could otherwise aid the pump in pumping the water up to reservoir 2) is lost because of friction and minor losses as the water moves from reservoir 1 to the pump.

Similarly, friction and minor head losses develop on the discharge side of the pump as the water moves from the discharge side of the pump to reservoir 2. This creates an additional load or head against which the pump must operate (Figure 60).

If the pump does not add enough energy to overcome the friction and minor head losses that occur on both the suction and discharge sides of the pump, the pump will not be able to move the water.

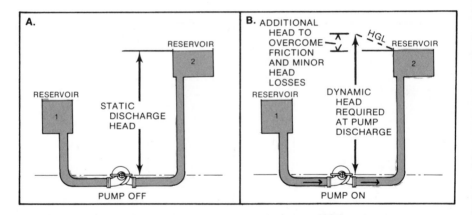

Figure 60.   Affect of Friction and Minor Head Losses on Discharge Head

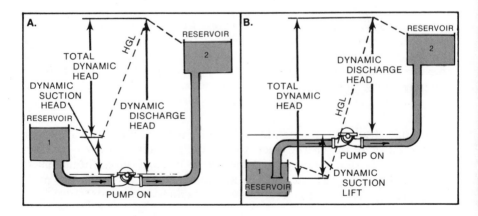

Figure 61.   Dynamic Heads

Now that you have a basic understanding of the reasons for a difference between static and dynamic conditions, consider the four measurements of dynamic head, shown in Figure 61.

- Dynamic suction head
- Dynamic suction lift
- Dynamic discharge head
- Total dynamic head

In a system where the HGL on the suction side is higher than the pump (Figure 61A), the DYNAMIC SUCTION HEAD is measured from the pump centerline at the *suction* of the pump to the point on the HGL directly *above* it.

In a system where the HGL on the suction side is lower than the pump (Figure 61B), the DYNAMIC SUCTION LIFT is measured from the pump centerline at the *suction* of the pump to the point on the HGL directly *below* it.

A single system will have a dynamic suction head or a dynamic suction lift, but not both. The location of the HGL, not of the reservoir, determines which condition exists. As shown in Figure 62, it is possible to have a dynamic suction lift condition even though the reservoir feeding the pump is above the pump centerline, providing the HGL on the suction side is below the pump centerline when the pump is running.

The DYNAMIC DISCHARGE HEAD is measured from the pump centerline at the *discharge* of the pump to the point on the HGL directly *above* it.

The TOTAL DYNAMIC HEAD is the difference in height between the HGL on the discharge side of the pump and the HGL on the suction side of the pump. This head is a measure of the total energy that the pump must impart to the water to move it from reservoir 1 to reservoir 2.

Dynamic pump heads can be calculated in either of two ways:

- By adding friction and minor losses to static head. (Static heads can be determined using existing elevation information such as shown in

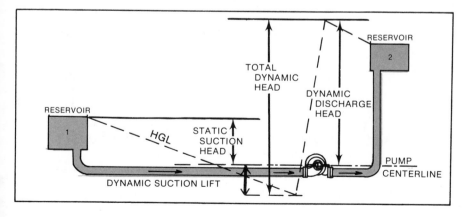

Figure 62. Reservoir Above Pump Can Still Result in Dynamic Suction Lift

examples 7 and 8 above; and friction and minor losses can be found by using tables such as those shown in the head loss section.)

- By direct measurement using pressure gages. (Pressure gages cannot measure the velocity head component. As discussed in Section H4-3, though, velocity head can often be ignored without significant error.)

Examples 9 and 10 illustrate how dynamic pump heads can be measured using pressure gage information. As noted before, this method of determining heads can also be used to determine static heads by simply turning the pump off before taking the gage readings.

Example 9

Using the pressure gage information taken when the pump was running, determine the dynamic heads. Gage readings are shown on Figure 63A in pounds per square inch gage (psig).

Find the dynamic suction and dynamic discharge heads as shown by converting the pressure gage reading to pressure in feet.[25]

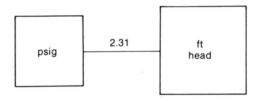

Moving from a smaller box to a larger box, you should multiply by 2.31.

Pressure in ft $=$ (pressure in psig)(2.31 ft/psig)

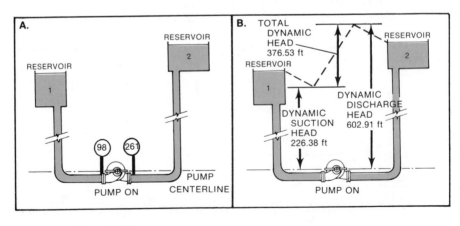

Figure 63.   (Example 9.)

[25]Hydraulics Section, Pressure and Force.

Dynamic Suction Head = (98 psig)(2.31 ft/psig)

$\qquad\qquad\qquad\quad$ = 226.38 ft

Dynamic Discharge Head = (261 psig)(2.31 ft/psig)

$\qquad\qquad\qquad\qquad\quad$ = 602.91 ft

The total dynamic head in this problem is the *difference* between the head on the discharge side of the pump and the head on the suction side of the pump (Figure 63B).

Total Dynamic Head = 602.91 ft − 226.38 ft

$\qquad\qquad\qquad\quad$ = 376.53 ft

Example 10

Using the two gage pressure readings given in Figure 64A, calculate the dynamic heads. Gage readings are in psig and were taken when the pump was operating.

To calculate the dynamic heads shown, first convert the pressure gage readings to feet of head.

Dynamic Suction Lift = (6 psig)(2.31 ft/psig)

$\qquad\qquad\qquad\quad$ = 13.86 ft

The minus sign associated with the 6 psig reading in the diagram is just a reminder that the dynamic suction lift is a vertical distance *below* the pump centerline.

Dynamic Discharge Head = (48 psig)(2.31 ft/psig)

$\qquad\qquad\qquad\qquad\quad$ = 110.88 ft

In a situation involving a suction lift, the total dynamic head against which the pump must operate is the *sum* of the

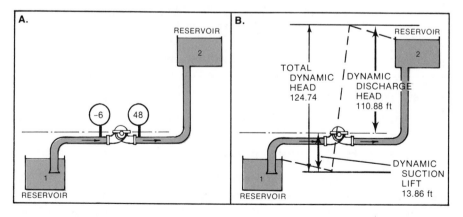

Figure 64.   (Example 10.)

heads on the suction and discharge sides of the pump (Figure 64B).

$$\text{Total Dynamic Head} = 110.88 \text{ ft} + 13.86 \text{ ft}$$
$$= 124.74 \text{ ft}$$

The following example combines most of the pump head terms.

Example 11

Using Figure 65A, draw and label the static and dynamic heads. Also label the two distances that represent the combined effects of friction head loss and minor head loss.

First, draw the pump centerline. Next, extend the two free water surfaces, shown by the dashed lines. Then draw the vertical lines representing the three static and three dynamic heads. Finally, find the friction and minor head losses by measuring the difference between the dynamic suction head and the static suction head (this is the friction and minor loss on the suction side), by measuring the difference between the static discharge head and the dynamic discharge head (this is the friction and minor loss on the discharge side).

The completed diagram is shown in Figure 65B.

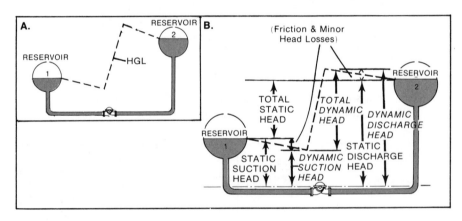

Figure 65.   (Example 11.)

H6-3.   Horsepower and Efficiency

Calculations of pump horsepower and efficiency are made in conjunction with many water transmission, treatment, and distribution operations. The selection of a pump or combination of pumps with an adequate pumping capacity will depend upon the flow rate required and the effective height through which the

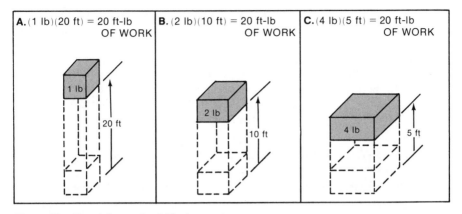

A. (1 lb)(20 ft) = 20 ft-lb OF WORK   **B.** (2 lb)(10 ft) = 20 ft-lb OF WORK   **C.** (4 lb)(5 ft) = 20 ft-lb OF WORK

Figure 66.   Equal Amounts of Work

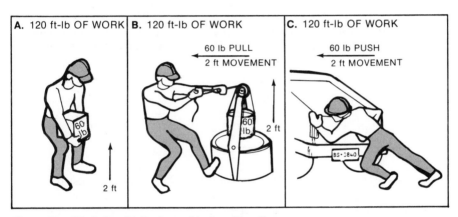

A. 120 ft-lb OF WORK   **B.** 120 ft-lb OF WORK   **C.** 120 ft-lb OF WORK

Figure 67.   Work Can Be Performed in Any Direction

flow must be pumped. EFFECTIVE HEIGHT is defined as the total feet of head against which the pump must work.[26]

Horsepower

To understand horsepower, you must first understand the technical meaning of the term WORK. Work is defined as the operation of a force over a specific distance; for example, lifting a *one-pound* object *one foot*. Thus, the *amount of work* done is measured in foot-pounds (ft-lb):

$$(feet)(pounds) = foot\text{-}pounds$$

Because it always requires the same amount of work to lift a 1-lb object 1 ft straight up, that amount of work is used as a standard measure; 1 ft-lb. However, work performed on an object of any weight can be measured in foot-pounds (Figure 66). Work can be performed in any direction (Figure 67). And, as shown

---

[26]Hydraulics Section, Head.

in Figure 67c, work doesn't need to involve lifting—an engine pushing a car along a level highway and a pump pushing water through a level pipeline are both examples of work.

The rate of doing work, that is, the measure of how much work is done in a given time, is called POWER. To make power calculations, therefore, the time required to perform the work must be known. The basic unit for power measurement is foot-pounds per minute (ft-lb/min—note that this is work performed *per* time). One equation for calculating power in foot-pounds per minute is

$$\left(\begin{array}{c}\text{Head} \\ \text{in} \\ \text{feet}\end{array}\right)\left(\begin{array}{c}\text{Flow rate} \\ \text{in} \\ \text{pounds per minute}\end{array}\right) = \left(\begin{array}{c}\text{Power} \\ \text{in} \\ \text{foot-pounds per minute}\end{array}\right)$$

You will often work with measurements of power expressed in horsepower (hp), which is related to foot-pounds per minute by the conversion equation

$$1 \text{ hp} = 33{,}000 \text{ ft-lb/min}$$

When doing horsepower problems using the basic power equation, you must first convert the measurements of head, flow rate, and power into the proper units:

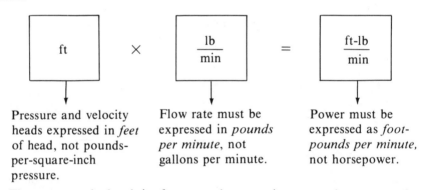

| ft | × | $\dfrac{\text{lb}}{\text{min}}$ | = | $\dfrac{\text{ft-lb}}{\text{min}}$ |

Pressure and velocity heads expressed in *feet* of head, not pounds-per-square-inch pressure.

Flow rate must be expressed in *pounds per minute*, not gallons per minute.

Power must be expressed as *foot-pounds per minute*, not horsepower.

The power calculated in foot-pounds per minute can be converted to horsepower by dividing by 33,000. (The conversion equation is 1 hp = 33,000 ft-lb/min.) This is called WATER HORSEPOWER (WHP), since it is the amount of horsepower required to lift the water. Another equation that can be used to calculate water horsepower is

$$\text{whp} = \frac{(\text{gpm Flow Rate})(\text{Total Head in ft})}{3960}$$

However, only the equation first shown is used in this section since it is more flexible and more closely related to the concept of power.

Example 12

A pump must pump 2000 gpm against a total head of 20 ft. What horsepower (water horsepower) will be required to do the work?

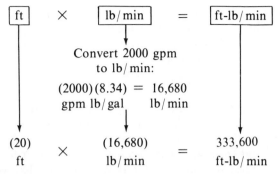

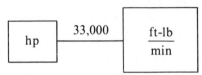

Now convert the foot-pounds per minute to horsepower. Since 1 hp = 33,000 ft-lb/min; in asking, "How many horsepower does this represent?" you are asking, "How many 33,000's are there in 333,600?" Mathematically, this is written as:

$$\frac{333,600 \ \text{ft-lb/min}}{33,000 \ \text{ft-lb/min/hp}} = 10.11 \ \text{hp}$$

As an alternative, you can use the box diagram method[27] to convert foot-pounds per minute to horsepower.

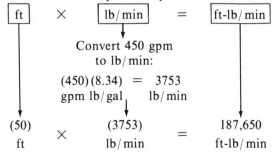

You are moving from a larger box to a smaller box. Therefore, division by 33,000 is indicated:

$$\frac{333,600 \ \text{ft-lb/min}}{33,000 \ \text{ft-lb/min/hp}} = 10.11 \ \text{hp}$$

Example 13
A flow of 450 gpm must be pumped against a head of 50 ft. What is the water horsepower required?

---

[27] Mathematics Section, Conversions.

Convert 187,650 ft-lb/min to horsepower (1 hp = 33,000 ft-lb/min) by asking, "How many 33,000's are there in 187,650?" Mathematically, this is written:

$$\frac{187,650}{33,000} = 5.69 \text{ hp}$$

Examples 12 and 13 above illustrated the calculation of horsepower when *water* is being pumped. And in almost all horsepower calculations pertaining to water supply, treatment, and distribution, water will be the liquid being pumped. However, you should know what happens to the horsepower calculation when a liquid is being pumped with a specific gravity[28] different than water.

Example 14

Given the same flow rate and feet of head as described in Example 13, what horsepower would be required if gasoline (specific gravity = 0.75) were being pumped rather than water?

For this problem use the same equation to calculate horsepower. The principal difference between this problem and Examples 12 and 13 is the conversion of gallons per minute to pounds per minute. Because gasoline has a different specific gravity than water, one gallon does not weigh 8.34 lb. Instead, one gallon of gasoline weighs

$$(8.34 \text{ lb/gal})(0.75) = 6.26 \text{ lb/gal}$$

Therefore, in making the conversion from gallons per minute to pounds per minute, use 6.26 lb/gal rather than 8.34 lb/gal. The calculation of required horsepower is as follows:

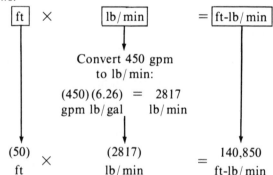

Then convert foot-pounds per minute to horsepower.

$$\frac{140,850 \text{ ft-lb/min}}{33,000 \text{ ft-lb/min/hp}} = 4.27 \text{ hp}$$

---

[28] Hydraulics Section, Density and Specific Gravity.

Because gasoline is not as dense as water, the horsepower required to pump gasoline is less than the horsepower required to pump water.

Since most of the liquids being pumped in water supply and treatment have a specific gravity of 1 or very nearly 1 (0.98 to 1.2), in the remaining example problems we will assume the specific gravity to be 1.

Example 15

Suppose a pump is pumping against a total head of 46.2 ft. If 800 gpm are to be pumped, what is the water horsepower requirement?

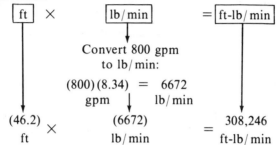

$$\boxed{\text{ft}} \times \boxed{\text{lb/min}} = \boxed{\text{ft-lb/min}}$$

Convert 800 gpm
to lb/min:

$$(800)(8.34) = \underset{\text{lb/min}}{6672}$$
$$\text{gpm}$$

$$\underset{\text{ft}}{(46.2)} \times \underset{\text{lb/min}}{(6672)} = \underset{\text{ft-lb/min}}{308,246}$$

Then convert foot-pounds per minute to horsepower.

$$\frac{308,246 \text{ ft-lb/min}}{33,000 \text{ ft-lb/min/hp}} = 9.34 \text{ hp}$$

In all four example problems, the unknown value was horsepower (and therefore, indirectly, foot-pounds per minute. However, either of the other two factors in the equation (feet or pounds per minute) might be the unknown value in a particular problem. The basic approach to the problem will remain unchanged regardless of which of the factors is unknown. The following two examples illustrate this concept.

Example 16

A pump is putting out 5 whp and delivering a flow of 430 gpm. What is the total feet of head against which the pump is operating?

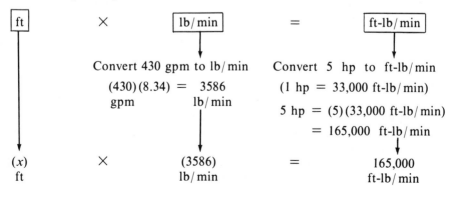

$$\boxed{\text{ft}} \times \boxed{\text{lb/min}} = \boxed{\text{ft-lb/min}}$$

Convert 430 gpm to lb/min   Convert 5 hp to ft-lb/min

$$(430)(8.34) = \underset{\text{lb/min}}{3586} \qquad (1 \text{ hp} = 33,000 \text{ ft-lb/min})$$
$$\text{gpm}$$

$$5 \text{ hp} = (5)(33,000 \text{ ft-lb/min})$$
$$= 165,000 \text{ ft-lb/min}$$

$$\underset{\text{ft}}{(x)} \times \underset{\text{lb/min}}{(3586)} = \underset{\text{ft-lb/min}}{165,000}$$

Now solve for the unknown value in the equation:

$$(x \text{ ft})(3586) = 165,000$$

$$x \text{ ft} = \frac{165,000}{3586}$$

$$x \text{ ft} = 46.01 \text{ ft total head}$$

Example 17

What is the maximum pumping rate (in gallons per minute) of a pump that is producing 15 whp against a head of 65 ft?

First calculate the maximum flow rate in pounds per minute:

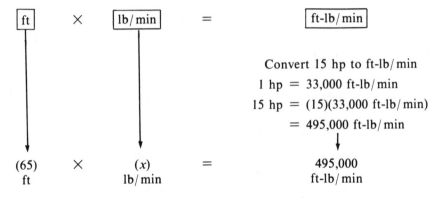

Solve for the unknown value:

$$(65)(x \text{ lb/min}) = 495,000$$

$$x \text{ lb/min} = \frac{495,000}{65}$$

$$x \text{ lb/min} = 7615 \text{ lb/min}$$

Now express this maximum pumping rate in gallons per minute:

$$\frac{7615 \text{ lb/min}}{8.34 \text{ lb/gal}} = 913.70 \text{ gpm}$$

Efficiency

In the preceding examples, you learned how to calculate water horsepower, the amount of power that must be applied directly to water to move it at a given rate against a given head. To apply the power to the water requires a pump, which in turn must be driven by a motor, which is powered by electric current.

Neither the pump nor the motor will ever be 100 percent efficient (Figure 68). This means that not all of the power supplied by the motor to the pump (called

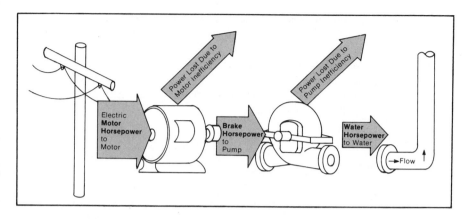

Figure 68.   Power Loss Due to Motor and Pump Inefficiency

brake horsepower) will be used to lift the water (*water horsepower*)—some of the power is used to overcome friction within the pump. Similarly, not all of the power of the electric current driving the motor (called *motor horsepower*) will be used to drive the pump—some of the current is used to overcome friction within the motor, and some current is lost in the conversion of electrical energy to mechanical power.

Depending on size and type, pumps are usually 50-85 percent efficient, and motors are usually 80-95 percent efficient. The efficiency of a particular motor or pump is given in the manufacturer's information accompanying the unit.

In some installations you will only know the combined efficiency of the pump and motor. This is called the WIRE-TO-WATER EFFICIENCY and is obtained by multiplying the motor and pump efficiencies together. For example, if a motor is 82 percent efficient and a pump is 67 percent efficient, the overall, or wire-to-water, efficiency is 55 percent ($0.82 \times 0.67 = 0.55$).

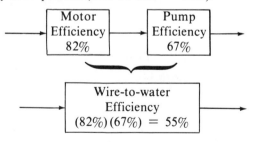

In practical horsepower calculations, you must take into account pump and motor efficiencies so that you can determine what size pump and motor and how much electric current will be necessary to move the water at the desired flow rate.

To show how horsepower and efficiency calculations are used, consider the system used for Example 15, in which the pump was pumping 800 gpm against a total head of 46.2 ft. The water horsepower required to perform this work was calculated to 9.34 whp. Suppose that the pump was only 85 percent efficient and the motor 95 percent efficient. To account for these inefficiencies, more than 9.34 whp would have to be supplied to the pump, and even more horsepower would have to be supplied to the motor.

Techniques for calculating the required horsepower are discussed in the next series of examples. To use the techniques, however, you must know that power input to the motor is usually expressed in terms of electrical power (watts or kilowatts) rather than horsepower.

$$\left.\begin{array}{c} \text{mhp} \\ \text{or watts} \\ \text{or kilowatts} \end{array}\right\} \longrightarrow \boxed{\text{MOTOR}}$$

The relationship between horsepower and watts or kilowatts is given below. With these equations, you can convert from one term to the other.

> 1 Horsepower = 746 watts power
>
> 1 Horsepower = 0.746 kilowatts power

Example 18

If a pump is to deliver 460 gpm of water against a total head of 95 ft, and the pump has an efficiency of 75 percent, what horsepower must be supplied to the pump?

In this problem, brake horsepower is to be calculated. First calculate the water horsepower, based on the work to be accomplished. Then determine the brake horsepower required.

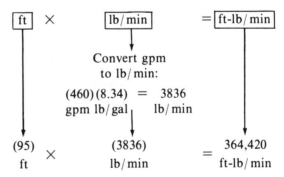

Convert the foot-pounds per minute to water horsepower:

$$\frac{364{,}420 \text{ ft-lb/min}}{33{,}000 \text{ ft-lb/min/hp}} = 11.04 \text{ whp}$$

This calculation shows that 11.04 whp (water horsepower) is required.

Now determine the amount of horsepower that must be supplied to the pump (brake horsepower):

? bhp →| PUMP |→ 11.04 whp

(75% Efficient)

As stated in the problem, the pump is 75 percent efficient. Therefore the brake horsepower will have to be *more than* *11.04 hp* in order to accomplish the 11.04 hp work. Since the usable brake horsepower (75 percent) must equal 11.04 hp, the mathematical equation is[29]

$$(75\%)(bhp) = 11.04 \text{ whp}$$

Restating the percent as a decimal number:

$$(0.75)(x \text{ bhp}) = 11.04 \text{ whp}$$

And then solving for the unknown value:

$$(0.75)(x \text{ bhp}) = 11.04$$

$$x \text{ bhp} = \frac{11.04}{0.75}$$

$$x \text{ bhp} = 14.72 \text{ bhp}$$

Therefore, the 75 percent efficient pump must be supplied with 14.72 hp in order to accomplish the 11.04 hp of work:

14.72 bhp →| PUMP |→ 11.04 whp

(75% Efficient)

Example 19

The motor nameplate indicated that the output of a certain motor is 15 hp. How much horsepower must be supplied to the motor if the motor is 90 percent efficient?

In this problem, the brake horsepower is known and the motor horsepower must be calculated:

? mhp →| MOTOR |— 15 bhp →| PUMP |

(90% Efficient)

Since only 90 percent of the horsepower supplied to the motor is usable, you want that 90 percent to equal 15 hp. Mathematically, this is stated as:

$$(90\%)(mhp) = 15 \text{ bhp}$$

---

[29]Mathematics Section, Percent.

Express the percent as a decimal number:

$$(0.90)(x \text{ mhp}) = 15 \text{ bhp}$$

And then solve for the unknown value:

$$(0.90)(x \text{ mhp}) = 15$$

$$x \text{ mhp} = \frac{15}{0.90}$$

$$x \text{ mhp} = 16.67 \text{ mhp}$$

Example 20

You have calculated that a certain pumping job will require 6 whp. If the pump is 80 percent efficient and the motor is 92 percent efficient, what motor horsepower will be required?

Often, when you are given both the motor and pump efficiencies, the information is combined and treated as one overall efficiency, as follows:

0.92	×	0.80	=	0.74
Motor Efficiency		Pump Efficiency		Overall Efficiency

This problem, then, becomes very similar to the two previous examples:

? mhp → | MOTOR AND PUMP | → 6 whp

(74% Efficient)

Since 74 percent of the motor horsepower must equal 6 hp, the mathematical setup is

$$(74\%)(\text{mhp}) = 6 \text{ whp}$$

Restate the percent as a decimal number:

$$(0.74)(x \text{ mhp}) = 6 \text{ whp}$$

And then solve for the unknown value:

$$(0.74)(x \text{ mhp}) = 6$$

$$x \text{ mhp} = \frac{6}{0.74}$$

$$x \text{ mhp} = 8.11 \text{ hp}$$

In the preceding problems the pump and motor efficiencies were used to calculate the required brake and motor horsepower. Occasionally, however, the motor, brake, and water horsepowers will be known and the *efficiencies* unknown.

A calculation of motor or pump efficiency is essentially a percent calculation. The general equation for percent calculations is [30]

$$\text{Percent} = \frac{\text{Part}}{\text{Whole}} \times 100$$

In calculations of efficiency, the "whole" is the total horsepower supplied to the unit, and the "part" is the horsepower output of the unit:

$$\text{Percent Efficiency} = \frac{\text{hp output}}{\text{hp supplied}} \times 100$$

From this general equation, the specific equations to be used for motor, pump, and overall efficiency calculations are

$$\text{Percent Motor Efficiency} = \frac{\text{bhp}}{\text{mhp}} \times 100$$

$$\text{Percent Pump Efficiency} = \frac{\text{whp}}{\text{bhp}} \times 100$$

$$\text{Percent Overall Efficiency} = \frac{\text{whp}}{\text{mhp}} \times 100$$

Let's look at three examples of efficiency calculations.

Example 21

Based on the gallons per minute to be pumped and the total head the pump must pump against, the water horsepower requirement was calculated to be 8.5 whp. If the motor supplies the pump with 11 hp, what must be the efficiency of the pump?

A diagram will help sort out the information given in the problem:

$$\boxed{\text{MOTOR}} \xrightarrow{\text{11 bhp}} \boxed{\text{PUMP}} \xrightarrow{\text{8.5 whp}}$$

The diagram indicates the horsepower going into the pump and the horsepower coming out of the pump. From this information, calculate the pump efficiency:

$$\text{Percent Pump Efficiency} = \frac{\text{hp output}}{\text{hp supplied}} \times 100$$

$$= \frac{8.5 \text{ whp}}{11 \text{ bhp}} \times 100$$

$$= 0.77 \times 100$$

$$= 77\% \text{ Pump Efficiency}$$

---

[30]Mathematics Section, Percent.

Example 22

What is the wire-to-water efficiency if electric power equivalent to 20 hp is supplied to the motor and 13.5 hp of work is accomplished by the pump?

In this problem, overall efficiency must be determined. The approach to solving the problem is similar to the previous example. Diagram the information given in the problem:

$$20\ \text{mhp} \longrightarrow \boxed{\begin{array}{c} \text{MOTOR} \\ \text{AND} \\ \text{PUMP} \end{array}} \ 13.5\ \text{whp} \longrightarrow$$

Calculate the percent overall efficiency:

$$\text{Percent Overall Efficiency} = \frac{\text{hp output}}{\text{hp supplied}} \times 100$$

$$= \frac{13.5\ \text{whp}}{20\ \text{mhp}} \times 100$$

$$= 0.68 \times 100$$

$$= 68\% \text{ wire-to-water efficiency}$$

Example 23

Suppose that 11 kilowatts (kW) power is supplied to the motor. If the brake horsepower is 13 bhp, what is the efficiency of the motor?

First, diagram the information given in the problem:

$$11\ \text{kW} \longrightarrow \boxed{\text{MOTOR}} \ 13\ \text{bhp} \longrightarrow \boxed{\text{PUMP}}$$

To calculate the percent efficiency of the motor, the kilowatts power is converted to horsepower.

Using the equation 1 hp = 0.746 kW, the box diagram method can be used to perform the conversion:

$$1\ \text{hp} = 0.746\ \text{kW}$$

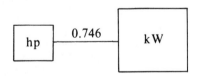

In converting from kilowatts to horsepower, you are moving from a larger box to a smaller box. Therefore *division* by 0.746 is indicated:

$$\frac{11\ \text{kW}}{0.746\ \text{kW/hp}} = 14.75\ \text{hp}$$

Now calculate the percent efficiency of the motor:

$$\text{Percent Efficiency} = \frac{\text{hp output}}{\text{hp supplied}} \times 100$$

$$= \frac{13 \text{ bhp}}{14.75 \text{ mhp}} \times 100$$

$$= 0.88 \times 100$$

$$= 88\% \text{ Efficiency}$$

Pumping Costs

Power is generally sold as kilowatt-hours (kWhr). If a motor draws 1 kW of power and runs for 1 hr, then the electric company will charge for 1 kWhr. Therefore, to calculate pumping costs, you will need to know the power requirements (power demand) of the motor, and the length of time the motor runs. In most situations, you will know, or be able to determine, the total head and flow rate against which the pump is working, so the water horsepower can be calculated. Then, using the wire-to-water efficiency of the motor and pump, the motor horsepower and the kilowatts of power demand can be determined. (Power demand can also be calculated from a meter reading, but that calculation is not covered in this text.) The following three examples illustrate how pumping costs can be calculated.

Example 24

The motor horsepower required for a particular pumping job is 25 hp. If your power cost is $0.03/kWhr, what is the cost of operating the motor for one hour?

First, convert the motor horsepower demand to kilowatt power demand:

$$1 \text{ hp} = 0.746 \text{ kW}$$

Therefore,

$$25 \text{ hp} = (25)(0.746 \text{ kW})$$

$$= 18.65 \text{ kW power demand}$$

Since the kilowatt demand is 18.65 kW, in one hour 18.65 kWhr is required. Therefore, the cost of operating the pump for an hour is

$$(18.65 \text{ kWhr})(\$0.03/\text{kWhr}) = \$0.56$$

Example 25

You have calculated that the minimum motor horsepower requirement for a particular pumping problem is 15 mhp. If the cost of power is $0.025/kWhr, what is the power cost in operating the pump for 12 hr?

To determine power costs, as in the previous example,

first express the motor horsepower requirement or demand in terms of kilowatts demand:

$$1 \text{ hp} = 0.746 \text{ kW}$$

$$15 \text{ hp} = (15)(0.746 \text{ kW})$$

$$= 11.19 \text{ kW power demand}$$

In 12 hr, the kWhr required to operate the pump is

$$(11.19 \text{ kW})(12 \text{ hr}) = 134.28 \text{ kWhr}$$

Thus, the power cost associated with operating the pump for 12 hr is

$$(134.28 \text{ kWhr})(\$0.025/\text{kWhr}) = \$3.36$$

Example 26

A pump is discharging 1200 gpm against a head of 65 ft. The wire-to-water efficiency is 68 percent. If the cost of power is $0.022/kWhr, what is the cost of the power consumed during a week in which the pump runs 78 hr?

To determine the power cost, first calculate the water horsepower required and then the motor horsepower demand of the motor:

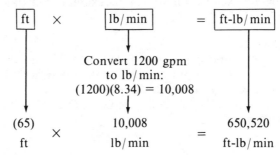

Express foot-pounds per minute as horsepower:

$$\frac{650,520 \text{ ft-lb/min}}{33,000 \text{ ft-lb/min/hp}} = 19.71 \text{ whp}$$

Now the water horsepower requirement has been determined. And from the statement of the problem, you know wire-to-water (overall) efficiency. A diagram illustrates what information is now known:

```
? mhp →    ┌─────────┐ ┌──────┐   19.71 whp
           │  MOTOR  ├─┤ PUMP ├──────────→
           └─────────┘ └──────┘
```

68% Efficiency

Since only 68 percent of the horsepower supplied to the motor is usable as water horsepower, 68 percent mhp must equal 19.71 hp. Mathematically, this is stated as:

$$(68\%)(x \text{ mph}) = 19.71 \text{ whp}$$

Changing the percent to a decimal number:

$$(0.68)(x \text{ mhp}) = 19.71 \text{ whp}$$

Solving for the unknown value:

$$(0.68)(x \text{ mhp}) = 19.71$$

$$x \text{ mhp} = \frac{19.71}{0.68}$$

$$x \text{ mhp} = 28.99 \text{ mhp}$$

If this was merely a horsepower problem, the problem would end right here, as in similar problems of the previous section. However, this is where a cost problem begins.

To calculate power costs, the motor horsepower demand will have to be expressed in terms of kilowatts power demand:

$$1 \text{ hp} = 0.746 \text{ kW}$$

Therefore,

$$28.99 \text{ hp} = (28.99)(0.746 \text{ kW})$$

$$= 21.63 \text{ kW power demand}$$

For 78 hr, the kWhr required to operate the pump is

$$(21.63 \text{ kW})(78 \text{ hr}) = 1687 \text{ kWhr}$$

And the power cost associated with operating the pump for that week is

$$(1687 \text{ kWhr})(\$0.022/\text{kWhr}) = \$37.11$$

H6-4.   Reading Pump Curves

There is a great deal of information written about pumps, pump performance, and pump curves. The following section covers just those basic topics needed to understand and use pump curves.

Description of Pump Curves

A pump curve is a graph showing the *characteristics* of a particular pump. For this reason these graphs are commonly called "pump characteristic curves." The characteristics commonly shown on a pump curve are

- Capacity (flow rate)
- Total head
- Power (brake horsepower)
- Efficiency

If a pump is designed to be driven by a variable speed motor, then "speed" is also a characteristic shown on the graph. For the purposes of this discussion we will assume that the pump is driven by a constant speed motor, so the graphs used will have only four curves.

The four pump characteristics (capacity, head, power, and efficiency) are related to each other. This is extremely important, for it is the interrelationship that enables the four pump curves to be plotted on the same graph.

Experience has shown that the *capacity* (flow rate) of a pump changes as the *head* against which the pump is working changes. Pump capacity also changes as the *power* supplied to the pump changes. Finally, pump capacity changes as *efficiency* changes. Consequently head, power, and efficiency can all be graphed "as a function of" pump capacity. This simply means that *capacity* is shown along the horizontal (bottom) scale of the graph, and *head, power,* and *efficiency* (any one or a combination of them) are shown along the vertical (side scales) of the graph. The following material discusses each of these relationships individually.

The *H-Q* Curve. The *H-Q* curve shows the relationship between total head *H*

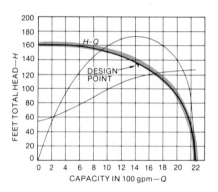

Figure 69. *H-Q* **Curve**

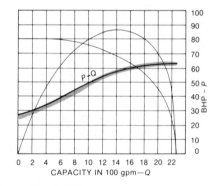

Figure 70. *P-Q* **Curve**

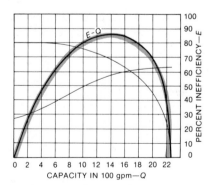

Figure 71. *E-Q* **Curve**

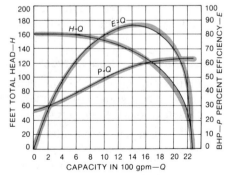

Figure 72A. **Complete Pump Curve**

against which the pump must operate and pump capacity Q. A typical H-Q curve is shown in Figure 69. The curve indicates what flow rate the pump will produce at any given total head.

For example, if the total head is 100 ft, the pump will produce a flow of about 1910 gpm. The H-Q curve also indicates that the capacity of the pump *decreases* as the total head *increases*. (Generally speaking, when the force against which the pump must work increases, the flow rate decreases.) The way total head controls the capacity is a characteristic of a particular pump.

The H-Q curve also identifies two very important operational facts. Normally on the curve you will find a mark like this: ⌐. The mark defines the DESIGN POINT, the head and capacity at which the pump is intended to operate for best efficiency in a particular installation.

The P-Q Curve. The P-Q curve (Figure 70) shows the relationship between power P and capacity Q. As usual, pump capacity is measured as gallons per minute, and power is measured as brake horsepower. For example, if you pump at a rate of 1480 gpm, then the power used to drive the pump is about 58 bhp. This is valuable information. As explained in the preceding section on horsepower calculations, if you know the brake horsepower and the motor efficiency, then you can determine the motor horsepower (the power required to drive the motor).

Knowledge of what power the pump requires is valuable for checking the adequacy of an existing pump and motor system. The knowledge is also important when you find yourself scrambling to rig a temporary system from spare or surplus parts.

The E-Q Curve. The E-Q curve shows the relationship between pump efficiency E and capacity Q, as shown in Figure 71. (Sometimes the Greek letter eta—η—is used to stand for efficiency, and the efficiency-capacity curve is designated the η-Q curve.) In sizing a pump system, the engineer attempts to select a pump that will produce the desired flow rate at or near peak pump efficiency.

The more efficient your pump is, the less costly it is to operate. Stated another way, the more efficient your pump is, the more water you can pump for each dollar's worth of power. Knowledge of pump efficiency allows you to compute the cost of pumping water. Therefore, from an operational point of view, pump efficiency is a "must know" item.

Reading the Curve

Figure 72A is the complete pump curve. By using the pump curve *any three* of the four pump characteristics (capacity, head, efficiency, and horsepower) can be determined when given information about the fourth characteristic. (Normally the total head or the capacity is the known characteristic.) The following examples illustrate how to read pump curves.[31]

---

[31]Mathematics Section, Graphs and Tables.

Example 27

Using the pump curve given in Figure 72A, determine the pump head, power, and efficiency when operating at a flow rate of 1600 gpm.

The key to reading pump curves is to find a certain vertical line and to note where the line intersects all three pump characteristic curves.

In this problem, the known characteristic is *capacity* (1600 gpm). To determine head, power, and efficiency, you should first draw a vertical line upward from the number 16 on the horizontal scale (Figure 72B).

Note that the vertical line intersects each of the three curves, as indicated by the three circled dots. Following the line upward, you first reach the intersection with the *P-Q* (power) curve. To determine what the power is at this point, move *horizontally to the right* toward the *BHP* scale; you should hit the scale at about 60 bhp.

Now return to the vertical line (1600 gpm) and move to the intersection with the *H-Q* (head) curve. To determine the head at this point, move *horizontally to the left*, toward the *feet head* scale; you should hit the scale at about 128 ft.

Finally, look at the point where the vertical line (1600 gpm) intersects the *E-Q* (efficiency) curve. To determine the

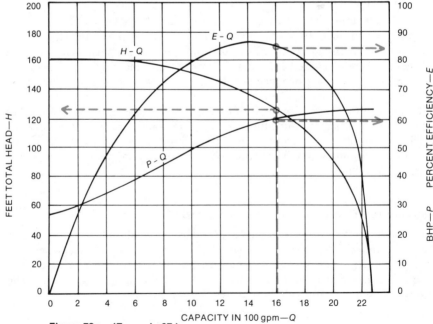

Figure 72B.   (Example 27.)

efficiency at this point, move *horizontally to the right* toward the efficiency scale. The reading is about 85 percent.

From the known pump capacity of 1600 gpm, then, the following characteristics are determined:

$$\text{Brake Horsepower } P = 60 \text{ bhp}$$

$$\text{Pump Head } H = 128 \text{ ft}$$

$$\text{Pump Efficiency } E = 85\%$$

Example 28

The total head against which a pump must operate is 35 ft. Using the pump curve shown in Figure 73, determine pump capacity, power, and efficiency at this head.

As mentioned in the previous example, the key in pump curve problems is to find a certain vertical line and to note where it intersects all three characteristic curves. Since the given characteristic is head (35 ft), locate *35 ft* on the *feet total head* scale and draw a horizontal line to intersect with the head curve (*H-Q*). Draw a vertical line through the point where the horizontal line intersects the curve.

The vertical line enables you to determine the three unknown characteristics.

First, the line falls halfway between 2.5 and 3 on the

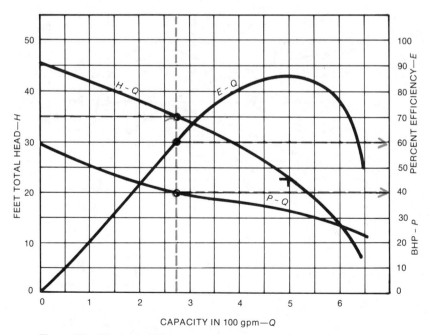

Figure 73.   (Example 28.)

capacity scale. Therefore the reading on the scale is 2.75, which indicates a capacity of 2750 gpm (the scale is given in 1000 gpm).

Following the vertical line upward, you will first come to the intersection on the *P-Q* curve. As before, to determine the power at this point move *horizontally to the right* toward the *bhp* scale. About 40 bhp is indicated.

The next intersection point up on the vertical line is the *E-Q* curve. To determine pump efficiency at this point, move *horizontally to the right* toward the efficiency scale. The reading is about 60 percent.

The last intersection with the vertical line is at the *H-Q* curve. You already know that characteristic to be 35 ft.

Given a head of 35 ft, then, the following characteristics were determined:

$$\text{Pump Capacity} = 2750 \text{ gpm}$$

$$\text{Brake Horsepower} = 40 \text{ bhp}$$

$$\text{Pump Efficiency} = 60\%$$

Example 29

Your pump is producing 5500 gpm, your motor is 85 percent efficient. Use the pump curve in Figure 74 to

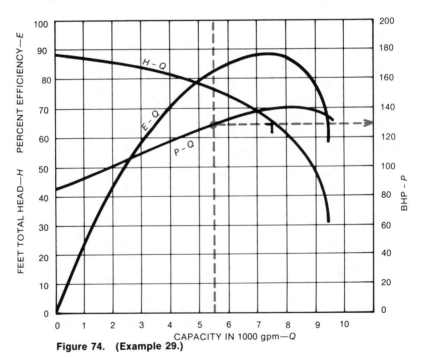

Figure 74. (Example 29.)

determine brake horsepower, and then calculate the motor horsepower.

First draw the vertical line at the point corresponding with 5500 gpm. Where this line intersects the *P-Q* curve, brake horsepower can be determined.

On this graph, brake horsepower is given on the right scale. Therefore the brake horsepower that corresponds with a flow rate of 5500 gpm is about 129 bhp.

So far, the brake horsepower and motor efficiency are known:

$$\xrightarrow{\text{? mhp}} \boxed{\text{MOTOR}} \xrightarrow{\text{129 bhp}} \boxed{\text{PUMP}}$$

(85% Efficient)

Use this information to calculate the motor horsepower:

$$(85\%)(x \text{ mhp}) = 129 \text{ bhp}$$

Express the percent as a decimal number:

$$(0.85)(x \text{ mhp}) = 129 \text{ bhp}$$

And then solve for the unknown value:

$$x \text{ mhp} = \frac{129}{0.85}$$

$$x = 151.76 \text{ mhp}$$

Example 30

A pump operates against a total head of 32 ft. The motor is 78 percent efficient. Power costs \$0.03/kWhr.

Use the pump curve in Figure 75 to determine brake horsepower required under the given conditions. Then calculate the cost of operating the pump for 100 hr.

To locate the brake horsepower that corresponds with a total head of 32 ft, first draw a horizontal line from *32* on the *feet total head* scale until the line intersects the *H-Q* curve. Draw a vertical line through the intersection.

The vertical line intersects the *P-Q* curve at a point corresponding to about 75 bhp. Using this figure and the motor efficiency given in the problem, calculate the motor horsepower required.

$$(78\%)(x \text{ mhp}) = 75 \text{ bhp}$$

Express the percent as a decimal number:

$$(0.78)(x \text{ mhp}) = 75 \text{ bhp}$$

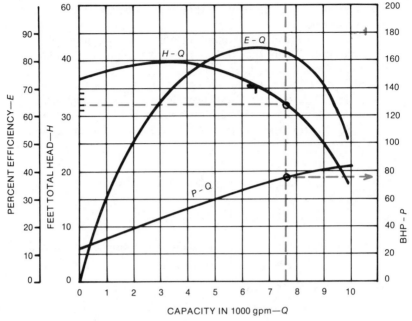

Figure 75. (Example 30.)

And then solve for the unknown value:

$$x \text{ mhp} = \frac{75}{0.78}$$

$$x \text{ mhp} = 96.15 \text{ mhp}$$

Once the motor horsepower requirement has been determined, power demand can be expressed in terms of kilowatts:

$$1 \text{ hp} = 0.746 \text{ kW}$$

$$96.15 \text{ hp} = (96.15)(0.746 \text{ kW})$$

$$= 71.73 \text{ kW demand}$$

In 100 hr then, the kilowatt hours used are

$$(71.73 \text{ kW})(100 \text{ hr}) = 7173 \text{ kWhr}$$

At a cost of $0.03/kWhr, the cost of operating the pump for one day is

$$(7173 \text{ kWhr})($0.03/kWhr) = $215.19$$

Review Questions

1. The totalizer of the meter on the discharge side of your pump reads in hundreds of gallons. At 3:10 PM the totalizer reads 272; at 4:40 PM it reads 635. Use this information to determine the pumping rate in gallons per minute.

2. During a 30-min pumping test, 3680 gal are pumped into a tank which has a diameter of 10 ft. The water level before the pumping test was 3 ft. What is the pumping rate expressed in gallons per minute?

3. A 50-ft diameter tank has water to a depth of 6 ft. The inlet valve is closed and a 2-hr pumping test is begun. If the water level in the tank at the end of the test is 2.3 ft, what is the pumping rate in gallons per minute?

4. A tank is 50 ft long and 25 ft wide with water to a depth of 4.5 ft. The outlet valve to the tank is closed and a pump test is conducted. If the water level in the tank rises 18 in. during the 1-hr test, what is the pumping rate in gallons per minute?

5. List the four types of static pump head and locate three of them on Figure 76A.

6. There are four terms used to describe dynamic pumping heads. What are they and where would three of them appear in Figure 77A?

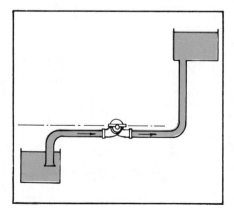

Figure 76A. (Review Question 5.)

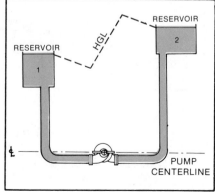

Figure 77A. (Review Question 6.)

7. A pump is moving water from Reservoir 1 to Reservoir 2 as shown in Figure 78A. Locate, label, and calculate in feet

   (a) Static suction head
   (b) Static discharge head
   (c) Total static head

8. Using the pressure gage information in Figure 79A, determine the dynamic heads. Gage readings are in pounds per square inch (gage) and were taken when the pump was operating.

9. Using the gage pressure readings shown in Figure 80A, calculate the dynamic heads. Gage readings are in pounds per square inch (gage) and were taken while the pump was operating.

10. Use the information given in Figure 81A to calculate:

   Static suction head
   Static discharge head
   Total static head
   Dynamic suction head
   Dynamic discharge head
   Total dynamic head
   Combined friction and minor head losses, suction side
   Combined friction and minor head losses, discharge side

   The pressure gage readings were taken when the pump was on.

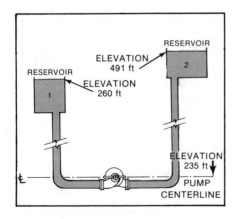

Figure 78A. (Review Question 7.)

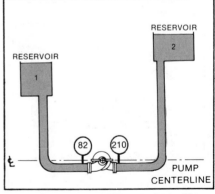

Figure 79A. (Review Question 8.)

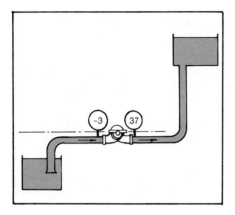

Figure 80A. (Review Question 9.)

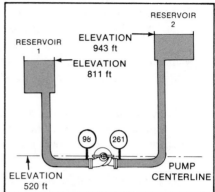

Figure 82A. Figure 82A. (Review Question 11.)

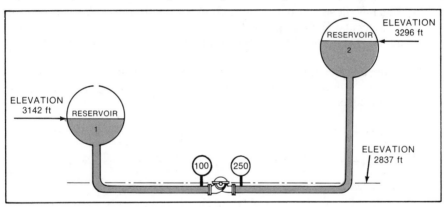

Figure 81A. (Review Question 10.)

11. Using the information in Figure 82A, calculate the static and dynamic heads, and the combined friction and minor head losses, in feet. Then locate and label each head. Gage pressure readings were taken with the pump on.

12. A flow of 1840 gpm is to be pumped against a total head of 40 ft. What is the water horsepower required?

13. Suppose you wish to pump a flow of 380 gpm against a total head of 35 ft. If the specific gravity of the liquid to be pumped is 1.2, what is the water horsepower required?

14. A pump is putting out 10 whp and delivering a flow of 810 gpm. What is the total head (in feet) the pump is working against? Assume the pump and motor are 100 percent efficient.

15. What is the maximum pumping rate (in gallons per minute) of a pump producing 8 whp working against a total head of 43 ft?

16. If a pump which has an efficiency of 80 percent is discharging 590 gpm against a total head of 77 ft, what is the brake horsepower?

17. A power of 19 hp (whp) is required for a particular pumping application. If the pump is 80 percent efficient and the motor is 90 percent efficient, what is the motor horsepower required?

18. Given a motor output of 24.8 hp and a water horsepower of 20 hp, what is the pump efficiency?

19. You have determined that the minimum motor horsepower requirement of a particular pumping situation is 12 hp. If the cost of power is $0.031/kWhr, what is the hourly cost of power used by the motor?

20. Given a brake horsepower of 18.5, a motor efficiency of 88 percent and a cost of power of $0.015/kWhr, determine the daily power cost for operating a pump.

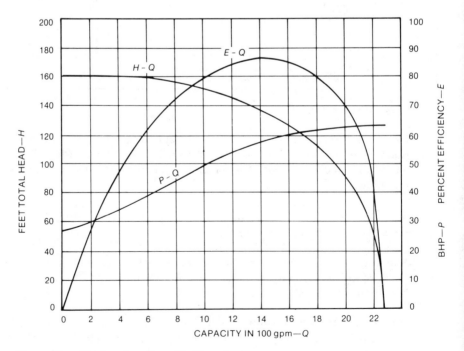

Figure 83A.   (Review Questions 22, 23, and 24.)

21. A particular pump is discharging 900 gpm against a total head of 82 ft. The wire-to-water efficiency is 83 percent and the power rate is $0.02/kWhr. What is the power cost for 14 hr?

22. Using the pump curve in Figure 83A, if the pump is producing 1800 gpm, find the corresponding total head, power, and efficiency.

23. For a total head of 140 ft and using the pump curve (Figure 83A), find the capacity, power, and efficiency of the pump.

24. Power costs $0.03/kWhr, total head is 132 ft, and motor efficiency is 85 percent. How much does it cost to operate the pump represented in the pump curve (Figure 83A) for 20 hr?

Summary Answers

1. 403.33 gpm pumping rate

2. 122.67 gpm pumping rate

3. 452.6 gpm pumping rate

4. 233.75 gpm pumping rate

5. • Static suction head
   • Static suction lift
   • Static discharge head
   • Total static head
   (Diagram with labeled heads given in Detailed Answers.)

6. • Dynamic suction head
   • Dynamic suction lift
   • Dynamic discharge head
   • Total dynamic head
   (Diagram with labeled heads given in Detailed Answers.)

7. Static suction head = 25 ft
   Static discharge head = 256 ft
   Total static head = 231 ft
   (Diagram with labeled heads given in Detailed Answers.)

8. Dynamic suction head = 189.42 ft
   Dynamic discharge head = 485.1 ft
   Total dynamic head = 295.68 ft

9. Dynamic suction lift = 6.93 ft
   Dynamic discharge head = 85.47 ft
   Total dynamic head = 92.4 ft

10. Static suction head = 305 ft
    Static discharge head = 459 ft
    Total static head = 154 ft
    Dynamic suction head = 231 ft
    Dynamic discharge head = 577.5 ft
    Total dynamic head = 346.5 ft
    Friction and minor head losses, suction = 74 ft
       (Includes entrance loss)
    Friction and minor head losses, discharge = 118.5 ft
       (Includes exit loss)

11. Static suction head = 291 ft
    Static discharge head = 423 ft
    Total static head = 132 ft
    Dynamic suction head = 226.38 ft
    Dynamic discharge head = 602.91 ft
    Total dynamic head = 376.53 ft
    Friction and minor head losses, suction = 64.62 ft
      (Includes entrance loss)
    Friction and minor head losses, discharge = 179.91 ft
      (Includes exit loss)
    (Diagram with labeled heads given in Detailed Answers.)

12. 18.6 whp

13. 4.03 whp

14. 48.85 ft

15. 736.21 gpm

16. 14.35 bhp

17. 26.39 mhp

18. 81% efficiency

19. $0.28 in one hour

20. $5.64

21. $4.69

22. Total head = 110 ft
        Power = 62 bhp
    Efficiency = 80%

23.   Capacity = 1300 gpm
         Power = 54 bhp
    Efficiency = 84%

24. $30.48

Detailed Answers

1. The total gallons pumped during the pumping test are calculated from the totalizer readings:

$$\begin{array}{r} 63,500 \text{ gallons} \\ - \ 27,200 \text{ gallons} \\ \hline 36,300 \text{ gallons} \end{array}$$

The pump test was run from 3:10 PM to 4:40 PM (1 1/2 hr). So total time of pumping was 90 min. With this information the gallons-per-minute pumping rate can be calculated:

$$\text{Pumping Rate} = \frac{36,300 \text{ gal}}{90 \text{ min}}$$

$$= 403.33 \text{ gpm pumping rate}$$

2. In this problem, both the total gallons pumped and the total minutes are given directly. No other information is necessary in calculating the gpm pumping rate. (The diameter and original water level information are not needed to solve this problem.)

$$\text{Pumping Rate} = \frac{3680 \text{ gal}}{30 \text{ min}}$$

$$= 122.67 \text{ gpm pumping rate}$$

3. The water level in the tank before the pumping test was 6 ft. After the pumping test, the water level was at 2.3 ft. Therefore, the *drop* in the water level (cross-hatched area in Figure 84) was 6 ft $-$ 2.3 ft $=$ 3.7 ft. With this

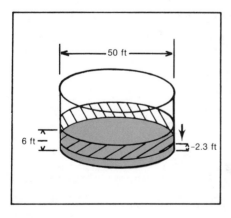

Figure 84.   (Review Question 3.)

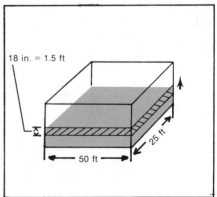

Figure 85.   (Review Question 4.)

depth and the tank diameter, the total cubic feet of water pumped can be calculated:

$$\text{Volume Pumped} = (\text{Area of Circle})(\text{Depth})$$
$$= (0.785)(50 \text{ ft})(50 \text{ ft})(3.7 \text{ ft})$$
$$= 7261 \text{ cu ft}$$

Converting cubic-feet volume to gallons:

$$(7261 \text{ cu ft})(7.48 \text{ gal/cu ft}) = 54{,}312 \text{ gal}$$

The total pumping time was 2 hr (120 min). With both total gallons pumped and total minutes, the gallons-per-minute pumping rate can be calculated:

$$\text{Pumping Rate} = \frac{54{,}312 \text{ gal}}{120 \text{ min}}$$
$$= 452.6 \text{ gpm pumping rate}$$

4. In this problem, to calculate the total gallons pumped during the pumping test, it is only necessary to know the dimensions of the tank and the water level *rise* during the test (See Figure 85). The original depth of the water is of no importance. Therefore the total gallons pumped is:

$$\text{Volume Pumped} = (\text{Area of Rectangle})(\text{Depth})$$
$$= (50 \text{ ft})(25 \text{ ft})(1.5 \text{ ft})$$
$$= 1875 \text{ cu ft}$$

And converting cubic-foot volume to gallons

$$(1875 \text{ cu ft})(7.48 \text{ gal/cu ft}) = 14{,}025 \text{ gal}$$

Given the total gallons and total time, the pumping rate can now be calculated:

$$\text{Pumping Rate} = \frac{14{,}025 \text{ gal}}{60 \text{ min}}$$
$$= 233.75 \text{ gpm pumping rate}$$

5. From Figure 76B:
   - Static suction head
   - Static suction lift
   - Static discharge head
   - Total static head

6. From Figure 77B
   - Dynamic Suction Head
   - Dynamic Suction Lift
   - Dynamic Discharge Head
   - Total Dynamic Head

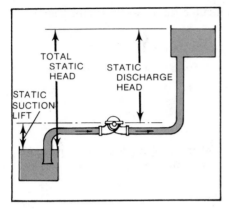

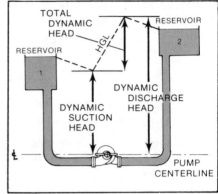

Figure 76B.   (Review Question 5.)   **Figure 77B.   (Review Question 6.)**

7. First, locate and label the three types of static head (Figure 78B). Then, using the elevation information given for the problem, calculate the feet of head in each case, as follows.

$$\text{Static Suction Head} = 260 \text{ ft} - 235 \text{ ft}$$
$$= 25 \text{ ft}$$
$$\text{Static Discharge Head} = 491 \text{ ft} - 235 \text{ ft}$$
$$= 256 \text{ ft}$$

The total static head can be calculated using either of two methods:

$$\text{Total Static Head} = \text{Elevation of Reservoir} - \text{Elevation of Reservoir 1}$$
$$= 491 \text{ ft} - 260 \text{ ft}$$
$$= 231 \text{ ft}$$

or:

$$\text{Total Static Head} = \text{Static Discharge Head} - \text{Static Suction Head}$$
$$= 256 \text{ ft} - 25 \text{ ft}$$
$$= 231 \text{ ft}$$

8. The dynamic suction and dynamic discharge heads as shown in Figure 79B can be found by converting the pressure gage readings to pressure in feet:

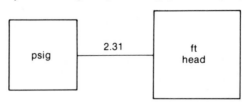

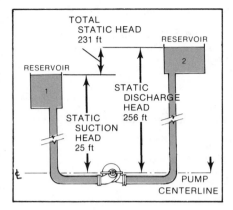

Figure 78B. (Review Question 7.)

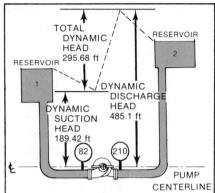

Figure 79B. (Review Question 8.)

You are moving from the smaller to the larger box, so multiply by 2.31:

$$\text{Pressure in ft} = (\text{pressure in psig})(2.31 \text{ ft}/\text{psig})$$

Therefore,

$$\text{Dynamic Suction Head} = (82 \text{ psig})(2.31 \text{ ft}/\text{psig})$$
$$= 189.42 \text{ ft}$$
$$\text{Dynamic Discharge Head} = (210 \text{ psig})(2.31 \text{ ft}/\text{psig})$$
$$= 485.1 \text{ ft}$$

The total dynamic head in this problem is the difference between the head on the discharge side of the pump and the head on the suction side of the pump:

$$\text{Total Dynamic Head} = 485.1 \text{ ft} - 189.42 \text{ ft}$$
$$= 295.68 \text{ ft}$$

9. To calculate the dynamic heads shown in Figure 80B, first convert pressure gage readings to pressure in feet:

$$\text{Pressure in ft} = (\text{pressure in psig})(2.31 \text{ ft}/\text{psig})$$

Therefore,

$$\text{Dynamic Suction Lift} = (3 \text{ psig})(2.31 \text{ ft}/\text{psig})$$
$$= 6.93 \text{ ft}$$

(The minus sign associated with 3 psig in the figure merely indicates that the feet of head is a vertical distance *below* the pump centerline.)

$$\text{Dynamic Discharge Head} = (37 \text{ psig})(2.31 \text{ ft}/\text{psig})$$
$$= 85.47 \text{ ft}$$

The total dynamic head against which the pump must operate is the sum of the vertical distances of the suction lift and discharge head:

$$\begin{aligned}\text{Total Dynamic Head} &= \text{Dynamic Suction Lift} \\ &\quad + \text{Dynamic Discharge Head} \\ &= 6.93 \text{ ft} + 85.47 \text{ ft} \\ &= 92.4 \text{ ft}\end{aligned}$$

10. First find the static suction and static discharge heads:

$$\begin{aligned}\text{Static Suction Head} &= 3142 \text{ ft} - 2837 \text{ ft} \\ &= 305 \text{ ft} \\ \text{Static Discharge Head} &= 3296 \text{ ft} - 2837 \text{ ft} \\ &= 459 \text{ ft}\end{aligned}$$

Then,

$$\begin{aligned}\text{Total Static Head} &= \text{Static Discharge Head} \\ &\quad - \text{Static Suction Head} \\ &= 459 \text{ ft} - 305 \text{ ft} \\ &= 154 \text{ ft}\end{aligned}$$

Next, convert the gage pressure readings to head in feet, as follows:

$$\begin{aligned}\text{Dynamic Suction Head} &= (\text{Pressure in psig})(2.31 \text{ ft/psig}) \\ &= (100 \text{ psig})(2.31 \text{ ft/psig}) \\ &= 231 \text{ ft} \\ \text{Dynamic Discharge Head} &= (\text{Pressure in psig})(2.31 \text{ ft/psig}) \\ &= (250 \text{ psig})(2.31 \text{ ft/psig}) \\ &= 577.5 \text{ ft}\end{aligned}$$

Then,

$$\begin{aligned}\text{Total Dynamic Head} &= \text{Dynamic Discharge Head} \\ &\quad - \text{Dynamic Suction Head} \\ &= 577.5 \text{ ft} - 231 \text{ ft} \\ &= 346.5 \text{ ft}\end{aligned}$$

Next, calculate the suction and discharge friction and minor head losses:

$$\begin{aligned}\frac{\text{Friction and minor}}{\text{losses—suction side}} &= 305 \text{ ft} - 231 \text{ ft} \\ &= 74 \text{ ft} \\ \frac{\text{Friction and minor}}{\text{losses—discharge side}} &= 577.5 \text{ ft} - 459 \text{ ft} \\ &= 118.5 \text{ ft}\end{aligned}$$

Figure 81B summarizes the answers.

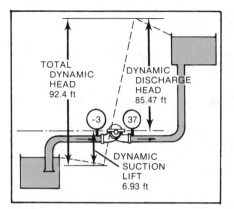

Figure 80B. (Review Question 9.)

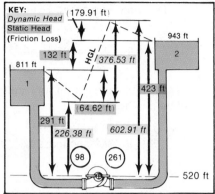

Figure 82B. (Review Question 11.)

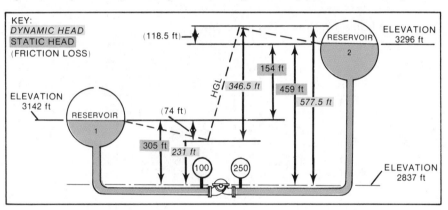

Figure 81B. (Review Question 10.)

11. First, calculate the static heads using the elevation information given:

$$\text{Static Suction Head} = 811 \text{ ft} - 520 \text{ ft}$$
$$= 291 \text{ ft}$$
$$\text{Static Discharge Head} = 943 \text{ ft} - 520 \text{ ft}$$
$$= 423 \text{ ft}$$
$$\text{Total Static Head} = 943 \text{ ft} - 811 \text{ ft}$$
$$= 132 \text{ ft}$$

Next, based on the pressure gage readings, calculate the dynamic heads:

$$\text{Dynamic Suction Head} = (98 \text{ psig})(2.31 \text{ ft/psig})$$
$$= 226.38 \text{ ft}$$

$$\text{Dynamic Discharge Head} = (261 \text{ psig})(2.31 \text{ ft/psig})$$
$$= 602.91 \text{ ft}$$
$$\text{Total Dynamic Head} = 602.91 \text{ ft} - 226.38 \text{ ft}$$
$$= 376.53 \text{ ft}$$

Then, find the friction and minor head losses by subtracting static and dynamic heads:

$$\begin{array}{c}\text{Friction and minor}\\ \text{losses—suction side}\end{array} = 291 \text{ ft} - 226.38 \text{ ft}$$

$$= 64.62 \text{ ft}$$

$$\begin{array}{c}\text{Friction and minor}\\ \text{losses—discharge side}\end{array} = 602.91 \text{ ft} - 423 \text{ ft}$$

$$= 179.91 \text{ ft}$$

These heads are shown in Figure 82B.

12.

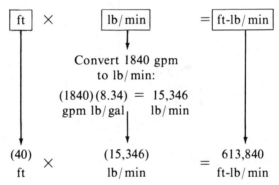

Converting foot-pounds per minute to water horsepower:

$$\frac{613{,}840 \text{ ft-lb/min}}{33{,}000 \text{ ft-lb/min/hp}} = 18.6 \text{ whp}$$

13. The impact of specific gravity on the horsepower calculation is in the conversion of gallons-per-minute flow to pounds-per-minute flow. Since water weighs 8.34 lb/gal, normally the gallons-per-minute flow is multiplied by 8.34. However, when a liquid with a specific gravity different than 1 is being converted from gallons per minute to pounds per minute, the factor must be adjusted. In this case then, instead of 8.34 lb/gal, the factor is adjusted:

$$(8.34 \text{ lb/gal})(1.2) = 10 \text{ lb/gal}$$

Now continue with the basic horsepower calculation:

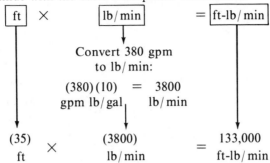

And convert foot-pounds per minute to water horsepower:

$$\frac{133,000 \text{ ft-lb/min}}{33,000 \text{ ft-lb/min/hp}} = 4.03 \text{ whp}$$

14.

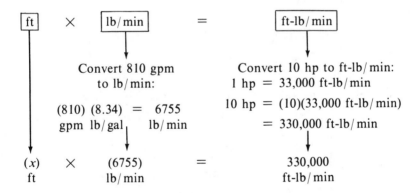

And now solve for the unknown value:

$$(x)(6755) = 330,000$$
$$x = \frac{330,000}{6755}$$
$$x = 48.85 \text{ ft}$$

15. First calculate the pounds per minute flow rate, then convert the flow rate to gallons per minute:

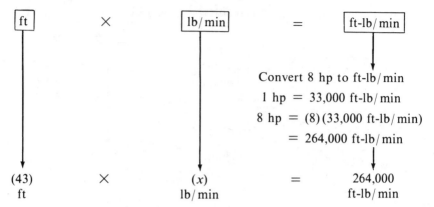

And now solve for the unknown value:

$$(43)(x \text{ lb/min}) = 264,000$$

$$x \text{ lb/min} = \frac{264,000}{43}$$

$$x \text{ lb/min} = 6140$$

Finally, convert the pounds-per-minute flow rate to gallons per minute as requested in the problem:

$$\frac{6140 \text{ lb/min}}{8.34 \text{ lb/gal}} = 736.21 \text{ gpm}$$

16. First calculate the water horsepower, then calculate the brake horsepower:

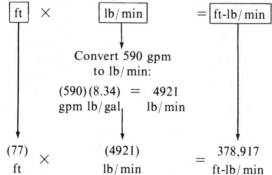

Convert foot-pounds per minute to water horsepower:

$$\frac{378,917 \text{ ft-lb/min}}{33,000 \text{ ft-lb/min/hp}} = 11.48 \text{ whp}$$

Knowing the water horsepower, and given the pump efficiency, the information can be diagrammed as follows:

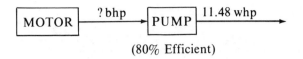

(80% Efficient)

Only 80 percent of the brake horsepower will be usable horsepower. Therefore 80 percent of the brake horsepower must equal at least 11.48 hp. Mathematically, this is written:

$$(80\%)(x \text{ bhp}) = 11.48 \text{ whp}$$

Express the percent as a decimal number:

$$(0.80)(x \text{ bhp}) = 11.48 \text{ whp}$$

And then solve for the unknown value:

$$x \text{ bhp} = \frac{11.48}{0.80}$$

$$= 14.35 \text{ bhp}$$

17. First express the efficiencies of the pump and motor as a *combined* efficiency:

$$\begin{array}{ccccc} 0.90 & \times & 0.80 & = & 0.72 \\ \text{Motor} & & \text{Pump} & & \text{Combined} \\ \text{Efficiency} & & \text{Efficiency} & & \text{Efficiency} \end{array}$$

$$= 72\% \text{ Combined Efficiency}$$

The information known can now be diagrammed:

? mhp ⟶ | MOTOR |—| PUMP | 19 whp ⟶

(72% Efficient)

The problem is now very similar to Problem 16. To calculate the motor horsepower, then:

$$(72\%)(x \text{ mhp}) = 19 \text{ whp}$$

Express the percent as a decimal:

$$(0.72)(x \text{ mhp}) = 19 \text{ whp}$$

And then solve for the unknown value:

$$x \text{ mhp} = \frac{19}{0.72}$$

$$x \text{ mhp} = 26.39 \text{ mhp}$$

18. Diagram the information given:

$$\boxed{\text{MOTOR}} \xrightarrow{\text{24.8 hp}} \boxed{\text{PUMP}} \xrightarrow{\text{20 hp}}$$

$$\text{Percent Efficiency} = \frac{\text{hp output}}{\text{hp supplied}} \times 100$$

$$= \frac{20 \text{ hp}}{24.8 \text{ hp}} \times 100$$

$$= 0.81 \times 100$$

$$= 81\% \text{ Pump Efficiency}$$

19. To determine pumping cost, the motor horsepower requirement must be converted to kilowatts power demand.

$$1 \text{ hp} = 0.746 \text{ kW}$$

$$12 \text{ hp} = (12)(0.746 \text{ kW})$$

$$= 8.95 \text{ kW}$$

In one hour then, the power consumed is 8.95 kWhr. And at a cost of $0.031/kWhr, the hourly cost of power used by the motor is

$$(8.95 \text{ kWhr})(\$0.031/\text{kWhr}) = \$0.28$$

20. First calculate the motor horsepower. Then convert the motor horsepower requirement to kilowatts power and determine the cost. Diagram the given information:

$$\xrightarrow{x \text{ mhp}} \boxed{\text{MOTOR}} \xrightarrow{\text{18.5 bhp}} \boxed{\text{PUMP}}$$
$$(88\% \text{ Efficient})$$

88 percent of the mhp must equal 18.5:

$$(0.88)(x \text{ mhp}) = 18.5$$

And solve for the unknown value:

$$x \text{ mhp} = \frac{18.5}{0.88}$$

$$x \text{ mhp} = 21.02 \text{ hp}$$

Now convert the motor horsepower required to kilowatts demand:

$$1 \text{ hp} = 0.746 \text{ kW}$$

$$21.02 \text{ hp} = (21.02)(0.746 \text{ kW})$$

$$= 15.68 \text{ kW}$$

At this demand, in one day the kilowatt hours of power consumed is

$$(15.68 \ kW)(24 \ hr) = 376.32 \ kWhr$$

The cost for the power is $0.015/kWhr; therefore the total daily cost for power consumed by this pump is

$$(376.32 \ kWhr)(\$0.015/kWhr) = \$5.64$$

21. In this problem calculate the water horsepower first. Then calculate the motor horsepower, kilowatts power demand, and cost.

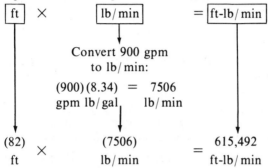

Convert foot-pounds per minute to water horsepower:

$$\frac{615,492 \ \text{ft-lb}/\text{min}}{33,000} = 18.65 \ \text{whp}$$

Now diagram the information known at this point in the problem:

```
  ? mhp                                18.65 whp
─────────▶│ MOTOR │─│ PUMP │──────────▶
          (83% Efficient)
```

From this information calculate the motor horsepower; 83 percent of the motor horsepower should equal at least 18.65 horsepower:

$$(0.83)(x \ \text{mhp}) = 18.65 \ \text{whp}$$

$$x \ \text{mhp} = \frac{18.65}{0.83}$$

$$x \ \text{mhp} = 22.47 \ \text{mhp}$$

To calculate pumping cost, the motor horsepower requirement will first have to be expressed as kilowatts demand:

$$1 \ \text{hp} = 0.746 \ \text{kW}$$

Therefore:

$$22.47 \ \text{hp} = (22.47)(0.746 \ \text{kW})$$

$$= 16.76 \ \text{kW demand}$$

In 14 hr then, the kilowatt hours consumed by the motor is:
$$(16.76 \ \text{kW})(14 \ \text{hr}) = 234.64 \ \text{kWhr}$$
With a power cost of $0.02/kWhr, the cost for 14 hr is:
$$(234.64 \ \text{kWhr})(\$0.02/\text{kWhr}) = \$4.69$$

22. From Figure 83B:

Total head = 110 ft
   Power = 62 bhp
Efficiency = 80%

23. From Figure 83C:

Capacity = 1300 gpm
   Power = 54 bhp
Efficiency = 84%

24. To determine pumping costs, first determine the motor horsepower
requirement, then convert motor horsepower to kilowatts power demand.

With the pump curve (Figure 83D), there are two methods that can be used
to determine motor horsepower for this problem. The first method is the
most direct.

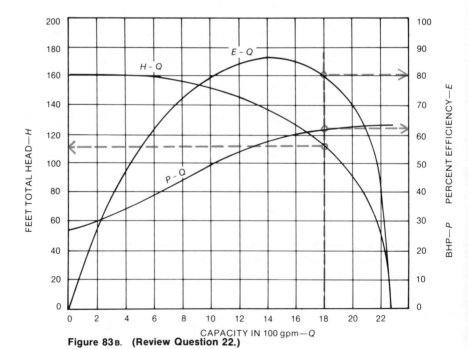

Figure 83B. (Review Question 22.)

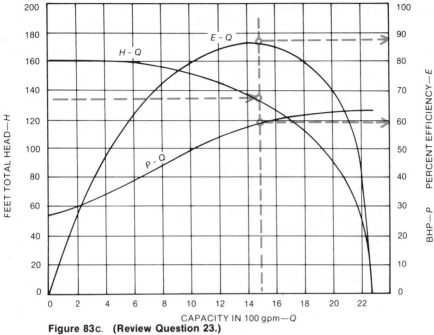

Figure 83c.   (Review Question 23.)

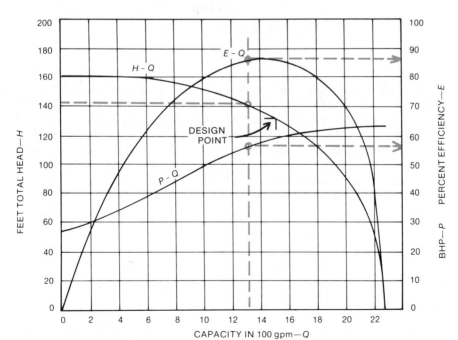

Figure 83d.   (Review Question 24.)

Method 1: (Using brake horsepower information)
From the pump curve, determine the brake horsepower. First draw a horizontal line from *132* on the *Feet Total Head* scale until the line intersects the *H-Q* curve. This locates the position of the vertical line. The point at which the vertical line intersects the *P-Q* curve indicates the brake horsepower in this problem: about 59 bhp.
The information known thus far in the problem can be diagrammed:

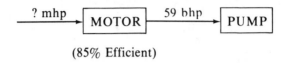

(85% Efficient)

From this information calculate the motor horsepower:
$$(85\%)(x \text{ mhp}) = 59 \text{ bhp}$$

Express the percent as a decimal:
$$(0.85)(x \text{ mhp}) = 59 \text{ bhp}$$

Then solve for the unknown value:
$$x \text{ mhp} = \frac{59 \text{ bhp}}{0.85}$$
$$x \text{ mhp} = 69.41 \text{ mhp}$$

Method 2: (Using pump efficiency and capacity, or flow rate, information)
From the pump curve shown above, the pump efficiency and capacity (flow rate) can be determined. First, draw a horizontal line from *132* on the *Feet Total Head* scale to the *H-Q* curve, and then draw a vertical line through the point of intersection. The points at which the vertical line intersects the *E-Q* curve and the *capacity* scale indicate the pump efficiency and capacity (86% and about 1490 gpm).
The information known thus far in the problem is

$$\text{Head} = 132 \text{ ft}$$
$$\text{Flow Rate} = 1490 \text{ gpm}$$
$$\text{Motor Efficiency} = 85\%$$
$$\text{Pump Efficiency} = 86\%$$

Use this information to calculate first the water horsepower, then the motor horsepower:

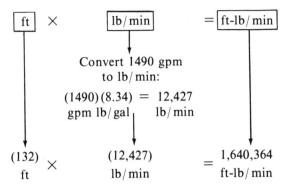

$$\frac{(132)}{ft} \times \frac{(12,427)}{lb/min} = \frac{1,640,364}{ft\text{-}lb/min}$$

And convert foot-pounds per minute to water horsepower:

$$\frac{1,640,364 \text{ ft-lb/min}}{33,000 \text{ ft-lb/min/hp}} = 49.71 \text{ whp}$$

Diagram the information known at this point in the problem:

? mhp ⟶ | MOTOR |—| PUMP | ⟶ 49.71 whp

$$(0.85) \quad (0.86) \ = \ 0.73$$
$$= 73\% \text{ Overall Efficiency}$$

Calculate the motor horsepower requirement:

$$(0.73)(x \text{ mhp}) = 49.71 \text{ whp}$$
$$x \text{ mhp} = \frac{49.71}{0.73}$$
$$x \text{ mhp} = 68.1 \text{ mhp}$$

The slight difference in answers between Method 1 (69.41 mhp) and Method 2 (68.1 mhp) is due to rounding and to the limited accuracy of readings taken from the graph. The following calculations use 68.1 mhp.

Once the motor horsepower requirement has been determined using either Method 1 or Method 2, convert motor horsepower to kilowatts power demand:

$$1 \text{ hp} = 0.746 \text{ kW}$$
$$68.1 \text{ hp} = (68.1)(0.746 \text{ kW})$$
$$= 50.8 \text{ kW power demand}$$

In 20 hr the power consumed is:

$$(50.8 \text{ kW})(20 \text{ hr}) = 1016 \text{ kWhr}$$

And the power cost is therefore:

$$(1016 \text{ kWhr})(\$0.03/\text{kWhr}) = \$30.48$$

Hydraulics 7

Flow Rate Problems

The measurement of water flow rate through a treatment plant is essential for efficient operation. Flow rate information is used in determining chemical dosages, assessing the cost of treatment, evaluating the plant efficiency, determining the amounts of water used by various commercial and industrial sources, and planning for future expansion of the treatment facility.

Two types of flow rates are important in water treatment operations: (1) INSTANTANEOUS FLOW RATES (flow rates at a particular moment) and (2) AVERAGE FLOW RATES (the average of the instantaneous flow rates over a given period of time, such as a day). Instantaneous flow rates are calculated from the cross sectional area and velocity of water in a channel or pipe. Such calculations are discussed in the following section. Average flow rates are calculated from records of instantaneous flow rates or from records of time and total volume, as discussed in the Mathematics Section of this book.[32]

Flow rates in and through a treatment plant are measured using various metering devices, such as weirs or Parshall flumes for open-channel flow, and venturi or orifice meters for closed conduit flow. The second part of this section illustrates how graphs and tables are used in determining flow rates through these devices.

H7-1. Instantaneous Flow Rate Calculations

The flow rate of water through a channel or pipe at a particular moment depends on the cross-sectional area[33] and the velocity of the water moving through it. This is stated mathematically as follows:

---

[32]Mathematics Section, Average Daily Flow.
[33]Mathematics Section, Area Measurements.

$$
\begin{array}{ccc}
Q & A & V \\
\text{flow rate} & = \text{area} & \times \text{velocity}
\end{array}
$$

Figure 86A illustrates the $Q = AV$ equation as it pertains to flow in an open channel. Since flow rate and velocity must be expressed for the same unit of time, the flow rate in the open channel is expressed as

$$
\begin{array}{cccc}
Q & & A & V \\
\text{flow rate} & = (\text{width}) & (\text{depth}) & (\text{velocity}) \\
\text{cu ft/time} & \text{ft} & \text{ft} & \text{ft/time}
\end{array}
$$

In using the $Q = AV$ equation, the time units given for the velocity must match the time units for cubic-feet flow rate. For example, if the velocity is expressed as feet per second (fps), then the resulting flow rate must be expressed as cubic feet per second (cfs). If the velocity is expressed as feet per minute (fpm), then the resulting flow rate must be expressed as cubic feet per minute (cfm). And if the velocity is expressed as feet per day (fps), then the resulting flow rate must be expressed as cubic feet per day (cfd). Figures 86B—86D illustrate this concept.

Using the $Q = AV$ equation, if you know the cross sectional area of the water in the channel (the width and depth of the rectangle), and you know the velocity, then you will be able to calculate the flow rate. This approach to flow rate calculations can be used if the problem involves an open channel flow with a rectangular cross section, as shown previously, or if the problem involves a pipe flow with a circular cross section, as shown in Figure 86E. Only the calculation of the cross sectional area will differ.

If a circular pipe is flowing full (as it will be in most situations in water supply and treatment), the resulting flow rate is expressed as

$$
\begin{array}{cccc}
Q & & A & V \\
(\text{flow rate}) & = (0.785) & (\text{diameter}^2) & (\text{velocity}) \\
\text{cu ft/time} & & \text{ft}^2 & \text{ft/time}
\end{array}
$$

As in the case of the rectangular channel, if the velocity is expressed as feet per second, then the resulting flow rate must be expressed as cubic feet per second. If the velocity is expressed as feet per minute, then the resulting flow rate must be expressed as cubic feet per minute. If the velocity is expressed as feet per day, then the resulting flow rate must be expressed as cubic feet per day.

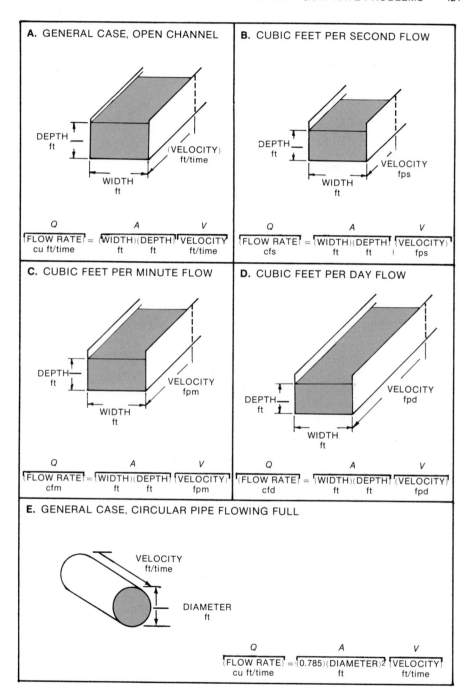

Figure 86.   Use of the Q = AV Equation

Example 1

A channel is 3 ft wide with water flowing to a depth of 2 ft. The velocity in the channel is found to be 1.8 fps. What is the cubic-feet-per-second flow rate in the channel?

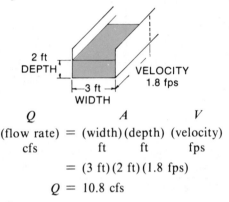

$$
\begin{array}{ccc}
Q & A & V \\
(\text{flow rate}) = (\text{width})\,(\text{depth}) & (\text{velocity}) \\
\text{cfs} & \text{ft} \quad \text{ft} & \text{fps}
\end{array}
$$

$$= (3 \text{ ft})(2 \text{ ft})(1.8 \text{ fps})$$

$$Q = 10.8 \text{ cfs}$$

Example 2

A 15-in. diameter pipe is flowing full. What is the gallons-per-minute flow rate in the pipe if the velocity is 110 fpm?

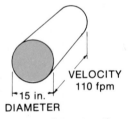

To use the $Q = AV$ equation, the diameter and velocity terms should be expressed using feet. Therefore, convert the 15-in. diameter to feet:[34]

$$\frac{15 \text{ in.}}{12 \text{ in.}/\text{ft}} = 1.25 \text{ ft}$$

Now use the $Q = AV$ equation to calculate the flow rate. Since the velocity is expressed in *feet per minute,* first calculate the cubic-feet-per-minute flow rate, then convert to gallons per minute:

$$
\begin{array}{ccc}
Q & A & V \\
(\text{flow rate}) = (0.785)\,(\text{diameter}^2) & (\text{velocity}) \\
\text{cfm} & \text{ft}^2 & \text{fpm}
\end{array}
$$

$$= (0.785)(1.25 \text{ ft})(1.25 \text{ ft})(110 \text{ fpm})$$

$$Q = 134.92 \text{ cfm}$$

---

[34]Mathematics Section, Conversions (Linear Measurement).

Converting cubic feet per minute to gallons per minute:[35]

(134.92 cfm)(7.48 gal/cu ft) = 1009 gpm

Example 3

What is the million-gallons-per-day flow rate through a channel that is 3 ft wide with the water flowing to a depth of 16 in. at a velocity of 2 fps?

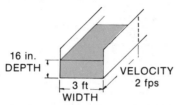

Before beginning the $Q = AV$ calculation, convert the 16-in. depth to feet:

$$\frac{16 \text{ in.}}{12 \text{ in./ft}} = 1.33 \text{ ft}$$

Now use the $Q = AV$ equation to calculate flow rate through the channel. Since the velocity is given in feet per second, first calculate the cubic-feet-per-second flow rate, then convert to gallons per day:

$$\begin{array}{ccc} Q & A & V \\ \text{(flow rate)} = & \text{(width)(depth)} & \text{(velocity)} \\ \text{cfs} & \text{ft} \quad \text{ft} & \text{fps} \end{array}$$

$$= (3 \text{ ft}) (1.33 \text{ ft})(2 \text{ fps})$$

$$Q = 7.98 \text{ cfs}$$

Convert cubic feet per second to gallons per day:

(7.98 cfs)(60 sec/min)(7.48 gal/cu ft)(1440 min/day) = 5,157,251 gpd

The "millions comma" in a gallons-per-day flow rate always indicates the million-gallons-per-day flow rate. In this example,

5,157,251 gpd = 5.16 mgd flow rate

↑

millions comma

In examples 1 through 3 above, Q was the unknown value in the $Q = AV$ formula. However, in some problems V or A might be the unknown value. Such problems would be set up in the same basic manner as the previous examples, then the equations would be solved for the unknown value.[36]

---

[35]Mathematics Section, Conversions (Flow Conversions).
[36]Mathematics Section, Solving for the Unknown Value.

Example 4
A 12-in. diameter pipe flowing full is carrying 560 gpm. What is the velocity of the water (in feet per minute) through the pipe?

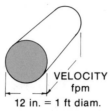

VELOCITY
fpm

12 in. = 1 ft diam.

You are asked to determine the velocity V given the flow rate Q and diameter. To use the $Q = AV$ formula to calculate velocity in a *pipe*, you should use the following mathematical setup:

$$
\underset{\substack{\text{(flow rate)} \\ \text{cfm}}}{Q} = (0.785)\underset{\substack{(\text{diameter}^2) \\ \text{ft}^2}}{A} \underset{\substack{(\text{velocity}) \\ \text{fpm}}}{V}
$$

Since you want to know velocity in feet per minute, the flow rate must also be expressed per minute (cubic feet per minute, as shown above). The information given in the problem expresses the flow rate as *gallons per minute*. Therefore, the flow rate must be converted to cubic feet per minute before beginning the $Q = AV$ calculation:

$$
\frac{560 \text{ gpm}}{7.48 \text{ gal/cu ft}} = 74.87 \text{ cfm}
$$

Now determine the velocity in feet per minute using the $Q = AV$ equation:

$$
\underset{74.87 \text{ cfm}}{Q} = (0.785)(1 \text{ ft})(1 \text{ ft})\underset{(x \text{ fpm})}{}
$$

And then solve for the unknown value:

$$
\frac{74.87}{(0.785)(1)(1)} = x \text{ fpm}
$$

$$
95.38 \text{ fpm} = x \text{ fpm}
$$

Example 5
A channel is 3 ft wide. If the flow in the channel is 7.5 mgd and the velocity of the flow is 185 fpm, what is the depth (in feet) of water in the channel?

The velocity is given in *feet per minute*. The corresponding flow rate is therefore *cubic feet per minute*. So before continuing with the $Q = AV$ calculation, convert the

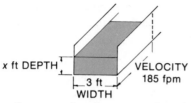

x ft DEPTH   3 ft   WIDTH   VELOCITY 185 fpm

million-gallons-per-day flow rate given in the problem to gallons-per-day flow rate and then to cubic-feet-per-minute flow rate:

$$7.5 \text{ mgd} = 7,500,000 \text{ gpd}$$

millions comma

Converting gallons-per-day flow rate to cubic feet per minute:

$$\frac{7,500,000 \text{ gpd}}{(1440 \text{ min/day})(7.48 \text{ gal/cu ft})} = 696.3 \text{ cfm}$$

Now that the flow rate is expressed in the correct units, use the $Q = AV$ equation to determine the depth of water in the channel:

$$\begin{array}{ccc} Q & A & V \\ (\text{flow rate}) = & (\text{width})(\text{depth}) & (\text{velocity}) \\ \text{cfm} & \text{ft} \quad \text{ft} & \text{fpm} \end{array}$$

$$696.3 \text{ cfm} = (3 \text{ ft})(x \text{ ft})(185 \text{ fpm})$$

Then solve for the unknown value:

$$\frac{696.3}{(3)(185)} = x \text{ ft}$$

$$1.25 \text{ ft} = x \text{ ft}$$

In estimating flow in an open channel you will usually be able to determine the approximate cross-sectional area A without much difficulty. In addition, you will often have to estimate the velocity V of the flow in the channel so that flow rate Q may be determined.

Flow velocity may be estimated by placing a float (a stick or a cork) in the water and recording the time it takes to travel a measured distance. Usually the flow on the surface is slightly greater or less than the average flow velocity, so that the calculated velocity will be fairly inaccurate.

Example 6

A float is placed in a channel. It takes 2.5 min to travel 300 ft. What is the flow velocity in feet per minute in the channel? (Assume the float is traveling at the average velocity of the water.)

This problem asks for the velocity in feet per minute. The statement of the problem helps to indicate the mathematical setup:

Verbal: Velocity = "feet per minute"

$$Math:\ \text{Velocity} = \frac{\text{feet}}{\text{minute}}$$

Fill in the information given:

$$\text{Velocity} = \frac{300 \text{ ft}}{2.5 \text{ min}}$$

$$= 120 \text{ fpm}$$

Example 7

A cork is placed in a channel and travels 30 ft in 20 sec. What is the velocity of flow in feet per second?

In this problem, the velocity is requested in feet per second:

Verbal: Velocity = "feet per second"

$$Math:\ \text{Velocity} = \frac{\text{feet}}{\text{second}}$$

Fill in the given information:

$$= \frac{30 \text{ ft}}{20 \text{ sec}}$$

$$= 1.5 \text{ fps}$$

Example 8

A channel is 4 ft wide with water flowing to a depth of 2.3 ft. If a float placed in the channel takes 3 min to travel a distance of 500 ft, what is the cubic-feet-per-minute flow rate in the channel?

Before using the $Q = AV$ equation to determine the flow rate in the channel, calculate the estimated velocity:

$$\text{Velocity} = \frac{\text{feet}}{\text{min}}$$

$$= \frac{500 \text{ ft}}{3 \text{ min}}$$

$$= 166.67 \text{ fpm}$$

Now use this information, along with channel width and depth, to determine the cubic-feet-per-minute flow rate in the channel:

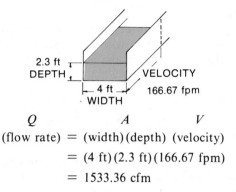

Q A V

(flow rate) = (width)(depth) (velocity)

= (4 ft)(2.3 ft)(166.67 fpm)

= 1533.36 cfm

Once you have learned how to solve $Q = AV$ problems you can solve more complex problems dealing with flow rate. The RULE OF CONTINUITY states that the flow Q that enters a system must also be the flow that leaves the system. Mathematically, this can be stated as follows:

$$Q_1 = Q_2$$

or

$$A_1 V_1 = A_2 V_2$$

The following examples illustrate the rule of continuity.

Example 9

The flow entering the leg of a tee connection is 9 cfs, as shown in the diagram below. If the flow through one branch of the tee is 5 cfs, what is the flow through the other branch?

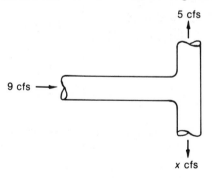

According to the rule of continuity, the flow rate Q going into the tee connection will be the same as the flow rate Q coming out of the tee connection. Mathematically, this is:

$$9 \text{ cfs} = 5 \text{ cfs} + x \text{ cfs}$$

Therefore, x cfs = 4 cfs.

Example 10

If the velocity in the 10-in. diameter section of pipe shown in the following diagram is 3.5 fps, what is the feet-per-second velocity in the 8-in. diameter section?

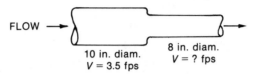

FLOW →

10 in. diam.
V = 3.5 fps

8 in. diam.
V = ? fps

According to the rule of continuity, the flow rate Q in the 10-in. diameter section must equal the flow rate Q in the 8-in. section:

$$Q_1 = Q_2 \text{ and } A_1 V_1 = A_2 V_2$$

To calculate the velocity V_2 in the 8-in. pipe, fill in the information given in the problem and solve for the unknown value. But first, as in the previous examples, express the pipe diameters in feet:

$$\frac{10 \text{ in.}}{12 \text{ in.}/\text{ft}} = 0.83 \text{ ft} \qquad \frac{8 \text{ in.}}{12 \text{ in.}/\text{ft}} = 0.67 \text{ ft}$$

Now fill in the information given in the problem:

$$\underbrace{(0.785)(0.83 \text{ ft})(0.83 \text{ ft})}_{A_1} \underbrace{(3.5 \text{ fps})}_{V_1} = \underbrace{(0.785)(0.67 \text{ ft})(0.67 \text{ ft})}_{A_2} \underbrace{(x \text{ fps})}_{V_2}$$

Solve for the unknown value:

$$\frac{(0.785)(0.83)(0.83)(3.5)}{(0.785)(0.67)(0.67)} = x \text{ fps}$$

$$\frac{1.89}{0.35} = x \text{ fps}$$

$$5.4 \text{ fps} = x \text{ fps}$$

Example 11

Given the following diagram and information, determine the velocity in feet per second at points A, B, and C.

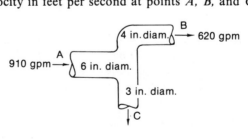

B
4 in. diam. → 620 gpm

A
910 gpm → 6 in. diam.

3 in. diam.

C

To determine the velocity at points *A*, *B*, and *C*, you will have to know the other two values *Q* and *A* at each point. First determine the flow rate at *C*. The flow rate going into the system is 910 gpm. Therefore, the flow rate coming out of the system equals 910 gpm:

$$910 \text{ gpm} = 620 \text{ gpm} + x \text{ gpm}$$
$$\text{(Point } A) \quad \text{(Point } B) \quad \text{(Point } C)$$

The flow at point *C* is:

$$910 \text{ gpm}$$
$$-620 \text{ gpm}$$
$$\overline{290 \text{ gpm}}$$

You now know the flow rate *Q* and the diameter (needed to calculate *A*) at each of the three points. Therefore, the $Q = AV$ equation can be used to solve for the velocity at each point.

Since the velocity in this problem must be expressed as feet per second, each of the flow rates should be converted to the corresponding units (cubic feet per second). In addition, the pipe diameters given in inches should be converted to feet.[37]

gpm to cfs
conversions

$$\frac{910 \text{ gpm}}{(7.48 \text{ gal/cu ft})(60 \text{ sec/min})} = 2.03 \text{ cfs}$$

$$\frac{620 \text{ gpm}}{(7.48 \text{ gal/cu ft})(60 \text{ sec/min})} = 1.38 \text{ cfs}$$

$$\frac{290 \text{ gpm}}{(7.48 \text{ gal/cu ft})(60 \text{ sec/min})} = 0.65 \text{ cfs}$$

in. to ft
conversions

$$\frac{6 \text{ in.}}{12 \text{ in./ft}} = 0.5 \text{ ft diameter}$$

$$\frac{4 \text{ in.}}{12 \text{ in./ft}} = 0.33 \text{ ft diameter}$$

$$\frac{3 \text{ in.}}{12 \text{ in./ft}} = 0.25 \text{ ft diameter}$$

---

[37]Mathematics Section, Conversions.

Based on these calculations, the revised diagram is as follows:

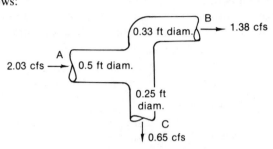

Now calculate the velocity at each of the three points by filling in the given information for each point:

At point A:
$$Q_A = A_A V_A$$
$$2.03 \text{ cfs} = (0.785)(0.5 \text{ ft})(0.5 \text{ ft})(V_A \text{ fps})$$
$$\frac{2.03}{(0.785)(0.5)(0.5)} = V_A \text{ fps}$$

$$10.34 \text{ fps} = V_A \text{ fps}$$

At point B:
$$Q_B = A_B V_B$$
$$1.38 \text{ cfs} = (0.785)(0.33)(0.33)(V_B \text{ fps})$$
$$\frac{1.38}{(0.785)(0.33)(0.33)} = V_B \text{ fps}$$

$$16.14 \text{ fps} = V_B \text{ fps}$$

At point C:
$$Q_C = A_C V_C$$
$$0.65 \text{ cfs} = (0.785)(0.25 \text{ ft})(0.25 \text{ ft})(V_C \text{ fps})$$
$$\frac{0.65}{(0.785)(0.25)(0.25)} = V_C \text{ fps}$$

$$13.25 \text{ fps} = V_C \text{ fps}$$

H7-2   Flow Measuring Devices

Most flow rate measuring devices can be equipped with instrumentation that will give a continuous recording of flow rates entering or leaving the treatment plant. To calibrate the instrumentation, and in some situations where there is no automatic continuous monitoring, you must be able to calculate the flow rates through the measuring devices. Although the most accurate interpretation of the readings from the devices usually involves calculations beyond the scope of this handbook, graphs and tables are available for day-to-day use in determining flow rates. The following discussion provides examples and illustrates the use of a few such graphs and tables.[38]

Weirs

The two most commonly used weirs are the **V**-notch and rectangular weirs, illustrated in figures 87A and 87B. To read a flow rate graph or table pertaining to a weir, you must know two measurements: (1) the height *H* of the water above the weir crest; and (2) the *angle* of the weir (**V**-notch weir) *or* the *length* of the crest (rectangular weir).

Example 12

Given the discharge curve for a 60-deg **V**-notch weir shown in Figure 88, what is the flow rate (in gallons per day) when the height of the water above the weir crest is 3 in.?

Since the height or head above the weir crest is 3 in., enter the *head* scale at *3*. This head indicates a flow of 30,000 gpd.

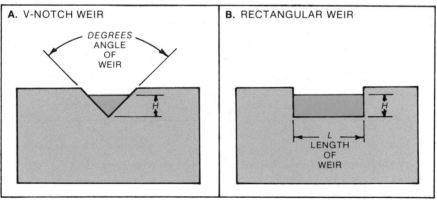

Adapted from Public Works Magazine, *1968*

Figure 87.   Types of Weirs

---

[38]Mathematics Section, Graphs and Tables.

Figure 88.   Discharge Curve, 60-deg V-notch Weir

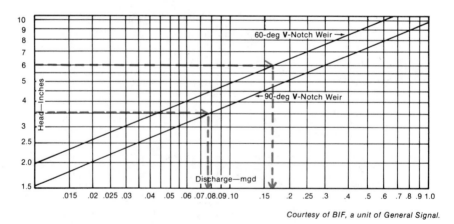

Figure 89.   Discharge Curves, 60- and 90-deg V-notch Weirs

Example 13

Given the discharge curves shown in Figure 89 for 60-deg and 90-deg V-notch weirs, (a) what is the flow rate (in million gallons per day) when the height (head of the water above the 60-deg V-notch weir crest) is 6 in.? (b) What is the flow rate (in gallons per day) when the height of water above the 90-deg **V**-notch weir crest is 3.5 in.?

First, notice that the part of the graph for 60-deg **V**-notch weirs presents information similar to the graph used in Example 12. The principal difference between the two graphs is that the graph in Example 12 uses arithmetic scales whereas the graph in this problem uses logarithmic scales.

(a) Enter the *head* scale at 6 in.; this head indicates a million-gallons-per-day flow rate somewhere between 0.15 and 0.2 mgd for the 60-deg **V**-notch weir.

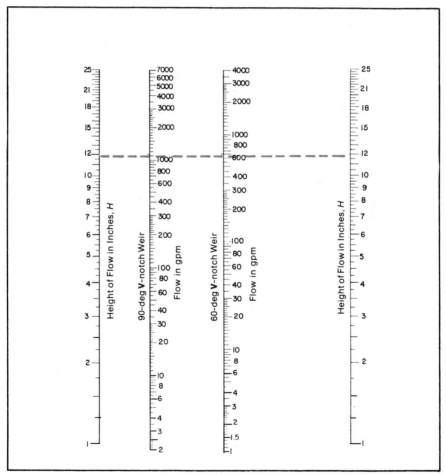

Reprinted with permission of Public Works Magazine, *September, October, and November 1968 Issues, Copyright* ©1968.

Figure 90.   Flow Rate Nomograph for 60- and 90-deg V-notch Weirs.

(b) The flow rate that corresponds with a head of 3.5 in. is 0.078 mgd for the 90-deg **V**-notch weir. Converting this to gallons per day:

$$0.078 \text{ mgd} = 78,000 \text{ gpd}$$

Example 14

A nomograph for 60-deg and 90-deg **V**-notch weirs is given in Figure 90. Using this nomograph, determine (a) the flow rate in gallons per minute if the height of water above the 60-deg **V**-notch weir crest is 12 in.; (b) the gallons-per-day flow rate over a 90-deg **V**-notch weir when the height of the water is 12 in. over the crest.

(a) As in the previous example, the scales used on this graph are logarithmic. This information is important since it determines how interpolation should be performed when the indicated flow falls between two known values.

First, draw a horizontal line from *12 in.* on the *height* scale on the left to *12 in.* on the *height* scale on the right. Then on the *60-deg **V**-notch* scale, read the flow rate indicated by 12 in. head: the flow rate falls between 600 and 700 gpm at approximately 650 gpm.

(b) On the *90-deg **V**-notch* scale, the indicated flow rate is between 1000 and 2000 gpm. More precisely, it falls between 1100 and 1200 gpm at a reading of about 1150 gpm. Converting gallons-per-minute rate to gallons per day:

$$(1150 \text{ gpm})(1440 \text{ min/day}) = 1,656,000 \text{ gpd}$$

Example 15

Table 8 pertains to the discharge of 45-deg **V**-notch weirs. Use the table to determine (a) flow rate in cubic feet per second when the head above the crest is 0.75 ft; (b) the gallons-per-day flow rate when the head is 1.5 ft.

(a) In the table, part of the head (0.7) is given on the vertical scale, and the remainder (0.05) is given on the horizontal scale (0.7 + 0.05 = 0.75). The cubic-feet-per-second flow rate indicated by a head of 0.75 ft is 0.504 cfs.

(b) A head of 1.5 ft is read as 1.5 on the vertical scale and 0.00 on the horizontal scale (1.5 + 0.00 = 1.50). The million-gallons-per-day flow rate indicated by this head is 1.84 mgd. This is equal to a flow rate of 1,840,000 gpd. (The *mgd* column was read in this problem since it is easier to convert to gallons per day from million gallons per day than from cubic feet per second.)

Table 8. Discharge of 45-deg V-Notch Weirs

Head Ft.	.00 CFS	.00 MGD	.01 CFS	.01 MGD	.02 CFS	.02 MGD	.03 CFS	.03 MGD	.04 CFS	.04 MGD	.05 CFS	.05 MGD	.06 CFS	.06 MGD	.07 CFS	.07 MGD	.08 CFS	.08 MGD	.09 CFS	.09 MGD
0.1	.003	.002	.004	.003	.005	.003	.006	.004	.008	.005	.009	.006	.011	.007	.012	.008	.014	.009	.016	.011
0.2	.019	.012	.021	.014	.024	.015	.026	.017	.029	.019	.032	.021	.036	.023	.039	.025	.043	.028	.047	.030
0.3	.051	.033	.055	.036	.060	.039	.065	.042	.070	.045	.075	.048	.081	.052	.086	.056	.092	.060	.098	.064
0.4	.105	.068	.111	.072	.118	.077	.126	.081	.133	.086	.141	.091	.149	.096	.157	.101	.165	.107	.174	.112
0.5	.183	.118	.192	.124	.202	.130	.212	.137	.222	.143	.232	.150	.243	.157	.254	.164	.265	.171	.277	.179
0.6	.289	.187	.301	.194	.313	.203	.326	.211	.339	.219	.353	.228	.366	.237	.380	.246	.395	.255	.410	.265
0.7	.425	.274	.440	.284	.455	.294	.471	.305	.488	.315	.504	.326	.521	.337	.539	.348	.556	.360	.574	.371
0.8	.593	.383	.611	.395	.630	.407	.650	.420	.670	.433	.690	.446	.710	.459	.731	.472	.752	.486	.774	.500
0.9	.796	.514	.818	.529	.841	.543	.864	.558	.887	.573	.911	.589	.935	.604	.960	.620	.985	.636	1.01	.653
1.0	1.04	.669	1.06	.686	1.09	.703	1.11	.721	1.14	.738	1.17	.756	1.20	.774	1.23	.793	1.25	.811	1.28	.830
1.1	1.31	.849	1.34	.869	1.37	.888	1.41	.908	1.44	.929	1.47	.949	1.50	.970	1.53	.991	1.57	1.01	1.60	1.03
1.2	1.63	1.06	1.67	1.08	1.70	1.10	1.74	1.12	1.77	1.15	1.81	1.17	1.84	1.19	1.88	1.22	1.92	1.24	1.96	1.26
1.3	1.99	1.29	2.03	1.31	2.07	1.34	2.11	1.36	2.15	1.39	2.19	1.42	2.23	1.44	2.27	1.47	2.32	1.50	2.36	1.52
1.4	2.40	1.55	2.44	1.58	2.49	1.61	2.53	1.64	2.58	1.66	2.62	1.69	2.67	1.72	2.71	1.75	2.76	1.78	2.81	1.81
1.5	2.85	1.84	2.90	1.87	2.95	1.91	3.00	1.94	3.05	1.97	3.10	2.00	3.15	2.03	3.20	2.07	3.25	2.10	3.30	2.13
1.6	3.35	2.17	3.41	2.20	3.46	2.23	3.51	2.27	3.57	2.30	3.62	2.34	3.68	2.38	3.73	2.41	3.79	2.45	3.84	2.48
1.7	3.90	2.52	3.96	2.56	4.02	2.60	4.08	2.63	4.13	2.67	4.19	2.71	4.25	2.75	4.32	2.79	4.38	2.83	4.44	2.87
1.8	4.50	2.91	4.56	2.95	4.63	2.99	4.69	3.03	4.75	3.07	4.82	3.11	4.89	3.16	4.95	3.20	5.02	3.24	5.08	3.29
1.9	5.15	3.33	5.22	3.37	5.29	3.42	5.36	3.46	5.43	3.51	5.50	3.55	5.57	3.60	5.64	3.64	5.71	3.69	5.78	3.74
2.0	5.86	3.79	5.93	3.83	6.00	3.88	6.08	3.93	6.15	3.98	6.23	4.03	6.31	4.08	6.38	4.13	6.46	4.18	6.54	4.23

Reprinted with permission of Leupold & Stevens, Inc., P.O. Box 688, Beaverton, Oregon 97005, from *Stevens Water Resources Data Book.*

Formula: $CFS = 1.035 H^{5/2}$; $MGD = CFS \times 0.646$

Example 16

A discharge curve for a 15-in. contracted rectangular weir is given in Figure 91. Using this curve, determine the gallons-per-day flow rate if the head above the weir crest is 2.75 in.

Since the scales for the discharge curve are arithmetic, it is much easier to determine the indicated flow rate when the arrow falls between two known values. In this case, a head of 3.75 ft indicates a flow rate between 250,000 and 300,000 gpd, about 285,000 gpd.

Example 17

A nomograph for rectangular weirs (contracted and suppressed) is given in Figure 92. Using this nomograph, determine (a) the flow in gallons per minute over a suppressed rectangular weir if the length of the weir is 3 ft and the height of the water over the weir is 4 in.; (b) the flow in gallons per minute over a contracted rectangular weir, for the same weir length and head as in (a).

To use the nomograph, you must know the difference between a contracted rectangular weir (one *with* end contractions) and a suppressed rectangular weir (one

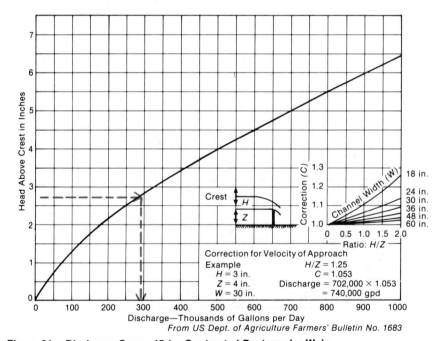

From US Dept. of Agriculture Farmers' Bulletin No. 1683

Figure 91.   Discharge Curve, 15-in. Contracted Rectangular Weir

without end contractions). As shown in Figure 93, the contracted rectangular weir comes in somewhat from the side of the channel before the crest cutout begins. On a suppressed rectangular weir, however, the crest cutout stretches from one side of the channel to the other.

(a) To determine the flow over the suppressed weir, draw a line on Figure 92 from $L = 3$ ft on the left-hand scale, through $H = 4$ in. (right side of the middle scale). A flow rate of 850 gpm is indicated where the line crosses the right-hand scale. This is the flow over the suppressed rectangular weir.

(b) To determine the flow rate over a contracted rectangular weir using the nomograph, first determine the flow rate over a suppressed weir given the weir length and

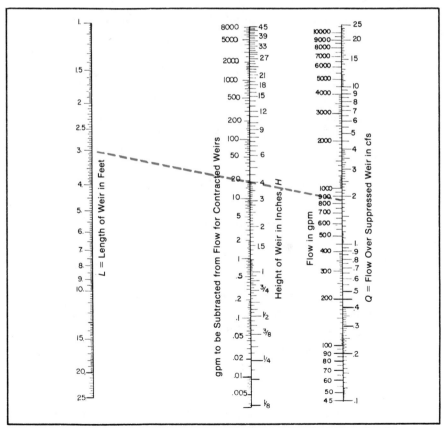

Figure 92.   Flow Rate Nomograph for Rectangular Weirs

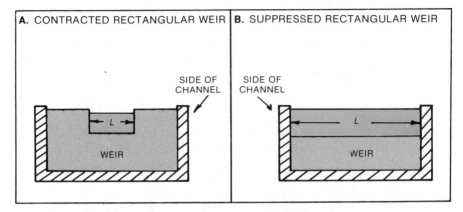

Figure 93. Two Types of Rectangular Weirs

head, as in part (a). Then subtract the flow indicated on the middle scale.

In this example, the flow rate over a 3-ft long suppressed weir with a head of 4 in. is 850 gpm. To determine the flow rate over a contracted weir 3 ft long with a head of 4 in., a correction factor must be subtracted from the 850 gpm. As indicated by the middle scale, the correction factor is 20 gpm.

$$\begin{array}{l} 850 \text{ gpm—suppressed rectangular weir} \\ \underline{-\ \ 20 \text{ gpm}} \\ \ 830 \text{ gpm—contracted rectangular weir} \end{array}$$

Example 18

Use Table 9 to determine the flow rate (in million gallons per day) over a contracted rectangular weir if the length of the weir crest is 3 ft and the head is 0.58 ft.

Enter the table under *Head ft* at 0.58; move right until you come under the *3* heading for *Length of Weir Crest in Feet*. The indicated flow rate is 2.739 mgd.

A few of the charts and graphs available for determining flows in Parshall flumes, venturi meters, and orifice meters are shown in the remaining examples of this section. Most such graphs and tables are very similar to the ones discussed in previous examples.

Table 9.   Flow Through Contracted Rectangular Weirs

LENGTH OF WEIR CREST IN FEET

Ft.	1 CFS	1 MGD	1½ CFS	1½ MGD	2 CFS	2 MGD	3 CFS	3 MGD	4 CFS	4 MGD	5 CFS	5 MGD
.36	.667	.431	1.026	.663	1.386	.895	2.105	1.360	2.824	1.824	3.543	2.289
.37	.695	.448	1.070	.690	1.445	.932	2.195	1.416	2.945	1.900	3.695	2.384
.38	.721	.465	1.111	.717	1.501	.969	2.281	1.473	3.061	1.976	3.841	2.480
.39	.748	.483	1.153	.745	1.559	1.006	2.370	1.530	3.181	2.054	3.992	2.577
.40	.775	.500	1.196	.772	1.617	1.044	2.459	1.588	3.301	2.132	4.143	2.676
.41	.802	.518	1.239	.800	1.676	1.083	2.550	1.647	3.424	2.216	4.298	2.776
.42	.830	.536	1.283	.829	1.736	1.121	2.642	1.707	3.548	2.292	4.454	2.877
.43	.858	.554	1.327	.857	1.797	1.160	2.736	1.767	3.675	2.373	4.614	2.979
.44	.886	.572	1.372	.886	1.858	1.200	2.830	1.827	3.802	2.455	4.774	3.082
.45	.915	.591	1.417	.915	1.920	1.240	2.925	1.889	3.930	2.538	4.935	3.187
.46	.943	.609	1.462	.945	1.982	1.280	3.021	1.951	4.060	2.621	5.099	3.292
.47	.972	.628	1.508	.974	2.045	1.320	3.118	2.013	4.191	2.706	5.264	3.399
.48	1.001	.646	1.554	1.004	2.108	1.361	3.215	2.076	4.322	2.791	5.429	3.506
.49	1.030	.665	1.601	1.034	2.172	1.403	3.314	2.140	4.456	2.878	5.598	3.615
.50	1.059	.684	1.647	1.064	2.236	1.444	3.413	2.204	4.590	2.965	5.767	3.725
.51	1.089	.703	1.695	1.095	2.302	1.486	3.515	2.269	4.728	3.052	5.941	3.835
.52	1.119	.722	1.743	1.126	2.368	1.529	3.617	2.335	4.866	3.141	6.115	3.947
.53	1.149	.742	1.791	1.156	2.434	1.571	3.719	2.401	5.004	3.230	6.289	4.060
.54	1.178	.761	1.838	1.188	2.499	1.614	3.820	2.467	5.141	3.321	6.462	4.174
.55	1.209	.781	1.888	1.219	2.567	1.658	3.925	2.534	5.283	3.411	6.641	4.288
.56	1.240	.800	1.938	1.251	2.636	1.701	4.032	2.602	5.428	3.503	6.824	4.404
.57	1.270	.820	1.986	1.282	2.703	1.745	4.136	2.670	5.569	3.595	7.002	4.520
.58	1.300	.840	2.035	1.314	2.771	1.790	4.242	2.739	5.713	3.689	7.184	4.638
.59	1.331	.859	2.085	1.347	2.840	1.830	4.349	2.808	5.858	3.783	7.367	4.757
.60	1.362	.879	2.136	1.380	2.910	1.879	4.458	2.879	6.006	3.877	7.554	4.876
.61	1.393	.899	2.186	1.412	2.980	1.924	4.567	2.948	6.154	3.972	7.741	4.997
.62	1.424	.920	2.237	1.444	3.050	1.970	4.676	3.019	6.302	4.068	7.928	5.118
.63	1.455	.940	2.287	1.477	3.120	2.015	4.785	3.090	6.450	4.165	8.115	5.240
.64	1.487	.960	2.339	1.510	3.192	2.061	4.897	3.162	6.602	4.262	8.307	5.363
.65	1.518	.980	2.390	1.544	3.263	2.107	5.008	3.234	6.753	4.360	8.498	5.487
.66	1.550	1.001	2.443	1.577	3.336	2.153	5.122	3.306	6.908	4.459	8.694	5.612
.67	1.581	1.021	2.494	1.611	3.407	2.200	5.233	3.379	7.059	4.558	8.885	5.738
.68	1.613	1.042	2.546	1.644	3.480	2.247	5.347	3.453	7.214	4.658	9.081	5.864
.69	1.646	1.062	2.600	1.680	3.555	2.295	5.464	3.527	7.373	4.759	9.282	5.991
.70	1.677	1.083	2.652	1.713	3.627	2.342	5.577	3.601	7.527	4.860	9.477	6.120
.71	1.709	1.104	2.705	1.747	3.701	2.390	5.693	3.676	7.685	4.962	9.677	6.249
.72	1.741	1.124	2.758	1.781	3.775	2.438	5.809	3.751	7.843	5.065	9.877	6.379
.73	1.774	1.145	2.812	1.816	3.851	2.486	5.928	3.827	8.005	5.168	10.08	6.510

Parshall Flumes

The Parshall flume is another type of flow metering device designed to measure flows in open channels. Top and side views of the flume are shown in Figure 94.

To use most nomographs pertaining to flow in a Parshall flume, you must know the width of the throat section of the flume (W), the upstream depth of water (depth H_a in tube of stilling well), and the depth of water at the foot of the throat section (depth H_b in tube of stilling well).

At normal flow rates, the nomograph may be used directly to determine the flow through the Parshall flume. At higher flows, however, the throat begins to become submerged (as shown in the side view of Figure 94). When this occurs, a correction to the nomograph reading is required. For a 1-in. throat width flume, this correction factor is not required until the percent of submergence, ($H_b/H_a \times 100$), is greater than 50 percent. For a 1-ft throat width flume, the correction is

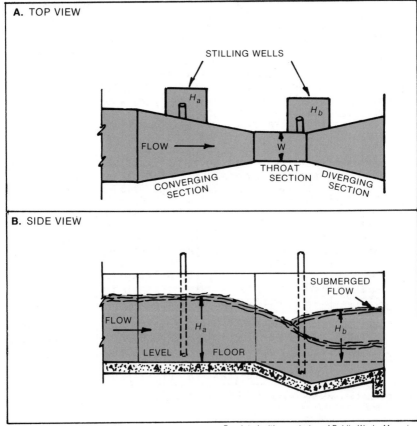

Figure 94.   Parshall Flume

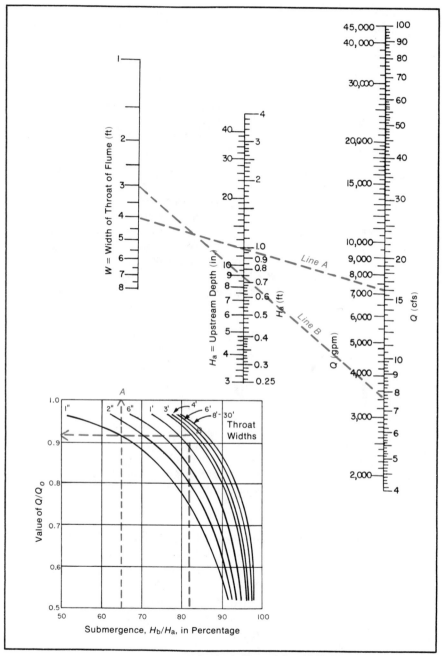

Figure 95.   Parshall Flume Flow Rate Nomograph and Corrections Graph

not required until the percent of submergence exceeds 70 percent. The following two examples use the Parshall flume nomograph and corrections graph shown in Figure 95.

Example 19

Suppose that your treatment plant has a 4-ft throat width Parshall flume. If the upstream depth of the water (depth H_a) is 1 ft and the depth of water in the throat section (depth H_b) is 0.65 ft, what is the gallons-per-minute flow rate through the flume?

First, determine the approximate flow rate using the flume throat width and the upstream (H_a) depth reading. Then use the correction graph if required.

Line A is drawn on the nomograph from 4 ft on the *throat width* (left) scale through 1 ft on the H_a *depth* scale (right side of the middle scale); the flow rate indicated where the line crosses the right scale is about 7200 gpm.

To determine whether a correction to this flow rate is required, first calculate the percent submergence of the throat:[39]

$$\text{Percent Submergence} = \frac{H_b}{H_a} \times 100$$

$$= \frac{0.65 \text{ ft}}{1 \text{ ft}} \times 100$$

$$= 0.65 \times 100$$

$$= 65\% \text{ submergence}$$

The correction line corresponding to a 4-ft throat width does not even begin until the percent submergence reaches about 78 percent. Therefore, in this problem no correction to the 7200-gpm flow rate is required.

Example 20

A flume has a throat width of 3 ft. If the upstream depth of the water (depth H_a) is 9 in., and the depth of water in the throat section (depth H_b) is 7.5 in., what is the gallons-per-day flow rate through the flume?

As for Example 19, first determine the approximate flow rate in the flume using the flume throat width and the upstream (H_a) depth reading. Then use the correction graph if needed.

Line B is drawn on the nomograph from 3 ft on the *throat*

---

[39] Mathematics Section, Percent.

width (left) scale through 9 in. on the H_a *depth* scale (left side of the middle scale); the flow rate indicated where the line crosses the right scale is 3400 gpm. Convert this flow rate to gallons per day:

$$(3400 \text{ gpm})(1440 \text{ min/day}) = 4,896,000 \text{ gpd}$$

To determine whether a correction to this flow rate is needed, first calculate the percent submergence of the throat:

$$\text{Percent Submergence} = \frac{H_b}{H_a} \times 100$$

$$= \frac{7.5 \text{ in.}}{9 \text{ in.}} \times 100$$

$$= 0.83 \times 100$$

$$= 83\% \text{ submergence}$$

Looking at the correction line corresponding to a 3-ft throat width, you can see that a correction is needed once the submergence exceeds about 77 percent. Since the submergence calculated is 83 percent, a correction is required. Draw a vertical line up from the indicated 83 percent until it crosses the 3-ft correction line (Point B on the correction graph). Then, move directly to the left until you reach the Q/Q_0 scale. Calculate the corrected flow through the flume by multiplying the flow read from the nomograph by the Q/Q_0 value.

In this case, the Q/Q_0 value is about 0.92. Multiply this times the flow rate obtained from the nomograph:

$$(4,896,000 \text{ gpd})(0.92) = 4,504,320 \text{ gpd corrected flow}$$

Venturi Meters

Venturi meters are designed to measure flow in closed conduit or pressure pipeline systems. As shown in Figure 96, flow measurement in this type of meter is based on differences in pressure between the upstream section (D_1) and throat section (D_2) of the meter. One type of nomograph available for use with venturi meters is shown in Figure 97. The graph applies to meters for which the diameter at D_1 divided by the diameter at D_2 is equal to 0.5. It may be used for meters having D_1/D_2 values as high as 0.75, but the results will be about 1 percent high. For high flow rates, the nomograph may be used with no correction factors necessary. For low flow rates, however, the correction factors graph shown in Figure 97 must be used to find a C' value. The value of Q obtained from the nomograph is then multiplied by the C' value to obtain the final answer.

Example 21

Using the nomograph in Figure 97, what is the flow rate in cubic feet per second in a venturi meter with a 1-in. throat, if the difference in head between D_1 and D_2 is 4 ft?

First, determine the cubic-feet-per-second flow rate using the nomograph, then use a correction factor if needed.

Line A is drawn from 1 in. on the *throat diameter* (left) scale to 4 ft on the *head differential* (right) scale. The flow rate indicated where the line crosses the middle scale is about 0.09 cfs.

To determine if a flow correction is required, first locate 0.09 on the flow (bottom) scale of the correction graph. Then move upward until you cross the *1-in. throat diameter* line (Point A). The flow rate corresponds with a C' value of 1.0. Therefore, the flow obtained from the nomograph requires no correction, since:

Corrected Flow $=$ (Q from nomograph)(C' from correction graph)

$\qquad\qquad\quad = (0.09 \text{ cfs})(1.0)$

$\qquad\qquad\quad = 0.09 \text{ cfs}$

The corrected flow is the same as the flow obtained from the nomograph.

Examine the correction graph closely, and notice that any flow *at or above* 0.06 cfs on the *1-in. throat diameter* scale would need no correction, since the C' values at these points equal 1.0. Any flow less than 0.06 cfs would require correction.

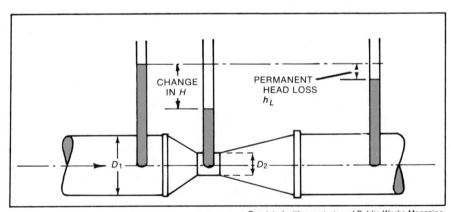

Figure 96.   Venturi Meter

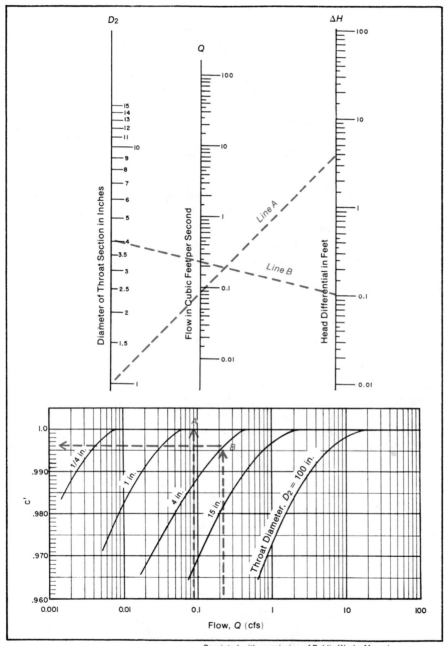

Figure 97.   Venturi Meter Flow Rate Nomograph and Corrections Graph

Example 22

The venturi meter at your treatment plant has a throat diameter of 4 in. If the pressure differential between D_1 and D_2 is 0.1 ft, what is the flow rate in cubic feet per second?

As in the previous example, first determine the cubic-feet-per-second flow rate using the nomograph. Then use a correction factor if necessary.

On the nomograph in Figure 97, line *B* is drawn from 4 in. on the *throat diameter* (left) scale to 0.1 ft on the *head differential* (right) scale; the flow rate indicated on the middle scale is about 0.22 cfs.

Now, to determine if a correction to this flow rate is required, locate 0.22 cfs on the *flow* scale of the correction graph, and move upward until you intersect the *4-in. throat diameter* line at *Point B*. The flow corresponds to a *C'* value of 0.996. Therefore, the corrected flow rate is:

Corrected Flow $= (Q$ from nomograph$)(C'$ from correction graph$)$

$= (0.22 \text{ cfs})(0.996)$

$= 0.219 \text{ cfs}$

The corrected flow is essentially the same as the flow indicated on the nomograph (0.219 cfs compared with 0.22 cfs).

Orifice Meters

Orifice meters are another type of head differential meter designed for use in closed conduits. As shown in Figure 98, there is a pressure tap on each side of the orifice plate, and flow rate calculations are based on the difference in pressure between the upstream or high pressure tap (Point 1), and the downstream or low

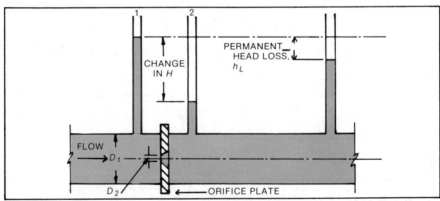

Figure 98.   Orifice Meter

pressure tap (Point 2). The nomograph shown in Figure 99 can be used to estimate the flow in orifice meters.

Example 23

If the head differential between pressure taps 1 and 2 is 10.2 ft, the diameter of the pipe is 6 in., and the diameter of the orifice is 3 in., what is the flow rate in the pipeline in cubic feet per second?

First calculate the ratio of the pipe diameter to the orifice diameter and locate the result on the right side of the middle scale of the nomograph in Figure 99.

$$\frac{D_1}{D_2} = \frac{6 \text{ in.}}{3 \text{ in.}}$$

$$= 2.0$$

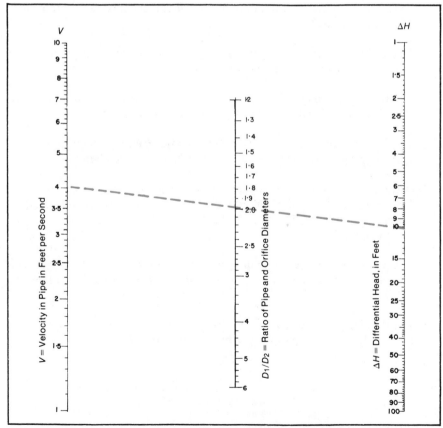

Figure 99. Flow Rate Nomograph for Orifice Meter

Now draw a line from 10.2 on the right-hand scale, through 2.0 on the right side of the middle scale, and across to the velocity scale. The indicated velocity is 4 fps.

The diameter of the pipeline is known (so that area can be calculated), and the velocity of the flow is known, so use the $Q = AV$ equation to determine the flow rate in the pipe:

$$Q = AV$$
$$= (0.785)(0.5 \text{ ft})(0.5 \text{ ft})(4 \text{ fps})$$
$$= 0.79 \text{ cfs}$$

Review Questions

1. Water in a channel is flowing at the rate of 9.2 fps. If the channel is 2.5 ft wide and has water to a depth of 14 in., what is the flow rate in cubic feet per second in the channel?

2. Determine the flow rate in gallons per minute of water passing through an 8-in. diameter pipe if the velocity of flow is 6 fps.

3. The flow rate through a 10-in. diameter pipe is 2.8 mgd. The pipe is flowing full. What is the velocity in feet per minute in the line?

4. The flow rate in a 3-ft wide, 16-in. deep channel measures 5400 gpm. What is the average water velocity in feet per second in the channel?

5. A float has been placed in a channel to estimate the velocity of flow. The float moved a distance of 160 ft in 2 min. The channel is 2 ft wide and has water to a depth of 1.5 ft. What is the flow rate in million gallons per day in the channel?

6. Assuming the flow rate remains constant at 3 cfs, how is the velocity of flow affected in a change from a 6-in. diameter pipe to a 4-in. diameter pipe?

7. For the diagram below, determine the velocity in feet per second at point A and the flow rate in gallons per minute at points B and C.

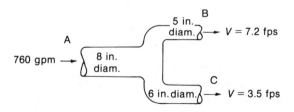

8. Use the discharge curve for a 30-deg **V**-notch weir shown in Figure 100A to determine the flow rate in gallons per day for a head of 3.3 in.

Figure 100A.   Discharge Curve for 30-deg V-notch Weir (Review Question 8)

9. Use the discharge curves for 60-deg and 90-deg **V**-notch weirs shown in Figure 101A to determine (a) the flow rate in million gallons per day when the head above a 60-deg **V**-notch weir crest is 3.5 in.; (b) the flow rate in gallons per minute when the height of the water above a 90-deg **V**-notch weir crest is 2.4 in.

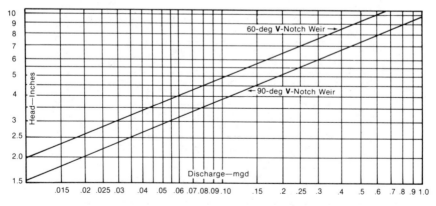

Figure 101A.   Discharge Curves for 60- and 90-deg V-notch Weirs (Review Question 9)

Table 10. Discharge of 22½-deg V-Notch Weirs

Head	.00		.01		.02		.03		.04		.05		.06		.07		.08		.09	
Ft.	CFS	MGD	CFS	MGD	CFS	MGD	CFS	MGD	CFS	MGD	CFS	MGD	CFS	MGD	CFS	MGD	CFS	MGD	CFS	MGD
0.1	.002	.001	.002	.001	.002	.002	.003	.002	.004	.002	.004	.003	.005	.003	.006	.004	.007	.004	.008	.005
0.2	.009	.006	.010	.006	.011	.007	.013	.008	.014	.009	.016	.010	.017	.011	.019	.012	.021	.013	.023	.015
0.3	.025	.016	.027	.017	.029	.019	.031	.020	.034	.022	.036	.023	.039	.025	.041	.027	.044	.029	.047	.031
0.4	.050	.033	.054	.035	.057	.037	.060	.039	.064	.041	.068	.044	.071	.046	.075	.049	.079	.051	.084	.054
0.5	.088	.057	.092	.060	.097	.063	.102	.066	.107	.069	.112	.072	.117	.075	.122	.079	.127	.082	.133	.086
0.6	.139	.090	.145	.093	.151	.097	.157	.101	.163	.105	.169	.109	.176	.114	.183	.118	.190	.123	.197	.127
0.7	.204	.132	.211	.137	.219	.141	.226	.146	.234	.151	.242	.157	.250	.162	.259	.167	.267	.173	.276	.178
0.8	.285	.184	.294	.190	.303	.196	.312	.202	.322	.208	.331	.214	.341	.220	.351	.227	.361	.233	.372	.240
0.9	.382	.247	.393	.254	.404	.261	.415	.268	.426	.275	.437	.283	.449	.290	.461	.298	.473	.306	.485	.313
1.0	.497	.321	.510	.329	.523	.338	.535	.346	.549	.355	.562	.363	.575	.372	.589	.381	.603	.390	.617	.399
1.1	.631	.408	.646	.417	.660	.427	.675	.436	.690	.446	.705	.456	.721	.466	.736	.476	.752	.486	.768	.496
1.2	.784	.507	.801	.518	.818	.528	.834	.539	.851	.550	.869	.561	.886	.573	.904	.584	.922	.596	.940	.607
1.3	.958	.619	.977	.631	.996	.643	1.01	.656	1.03	.668	1.05	.681	1.07	.693	1.09	.706	1.11	.719	1.13	.732
1.4	1.15	.745	1.17	.759	1.19	.772	1.22	.786	1.24	.800	1.26	.814	1.28	.828	1.30	.842	1.32	.856	1.35	.871
1.5	1.37	.886	1.39	.901	1.42	.915	1.44	.931	1.46	.946	1.49	.961	1.51	.977	1.54	.993	1.56	1.01	1.58	1.02
1.6	1.61	1.04	1.64	1.06	1.66	1.07	1.69	1.09	1.71	1.11	1.74	1.12	1.77	1.14	1.79	1.16	1.82	1.18	1.85	1.19
1.7	1.87	1.21	1.90	1.23	1.93	1.25	1.96	1.26	1.99	1.28	2.01	1.30	2.04	1.32	2.07	1.34	2.10	1.36	2.13	1.38
1.8	2.16	1.40	2.19	1.42	2.22	1.44	2.25	1.46	2.28	1.48	2.31	1.50	2.35	1.52	2.38	1.54	2.41	1.56	2.44	1.58
1.9	2.47	1.60	2.51	1.62	2.54	1.64	2.57	1.66	2.61	1.68	2.64	1.71	2.67	1.73	2.71	1.75	2.74	1.77	2.78	1.79
2.0	2.81	1.81	2.85	1.84	2.88	1.86	2.92	1.89	2.96	1.91	2.99	1.93	3.03	1.96	3.07	1.98	3.10	2.01	3.14	2.03

Reprinted with permission of Leupold & Stevens, Inc., P.O. Box 688, Beaverton, Oregon 97005, from Stevens Water Resources Data Book.

Formula: $CFS = 0.497\ H^{5/2}$ $MGD = CFS \times 0.646$

10. Table 10 pertains to discharge of 22 1/2 deg **V**-notch weirs. Use the table to determine the flow rate in gallons per day when the head above the crest is 1.27 ft.

11. Use the nomograph for rectangular weirs given in Figure 102A to determine (a) the flow rate in gallons per minute over a suppressed rectangular weir if the length of the weir is 1.6 ft and the height of the water over the weir is 8 in. and (b) the flow rate in gallons per minute over a contracted rectangular weir, given the same weir length and head.

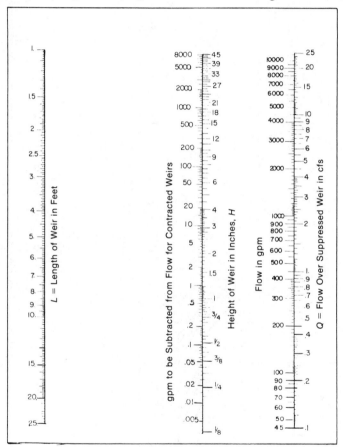

Reprinted with permission of Public Works Magazine,
September, October, and November 1968 Issues, Copyright ©1968.

Figure 102A.   Nomograph for Rectangular Weirs (Review Question 11)

12. A treatment plant has a 3 ft throat width Parshall flume. Use the nomograph and graph in Figure 103A to determine the flow rate in gallons per minute through the flume if the upstream depth of the water (depth at H_a) is 1.2 ft and the depth of the water in the throat section (depth H_b) is 1.1 ft.

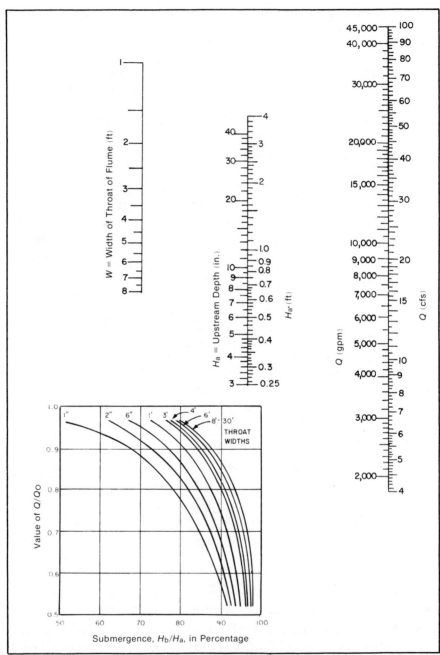

Figure 103A.   Nomograph and Corrections for Parshall Flume (Review Question 12)

13. Use the nomograph in Figure 104A to determine the flow rate in cubic feet per second in an 8-in. pipeline when the head differential between pressure taps 1 and 2 is 8 ft and the diameter of the meter orifice is 3 in. (D_1 = pipe diameter and D_2 = orifice diameter).

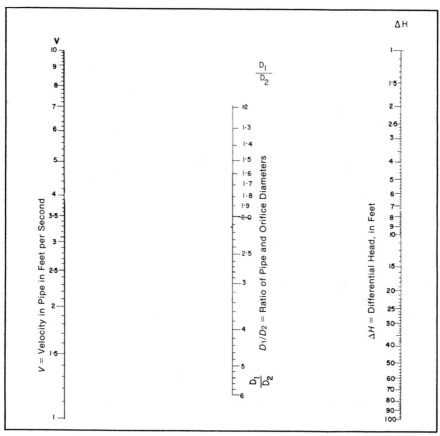

Figure 104A.   Nomograph for Orifice Meter (Review Question 13)

Summary Answers

1. 26.91 cfs

2. 946.97 gpm

3. 480.69 fpm

4. 3.02 fps

5. 2.59 mgd

6. The velocity more than doubles (from 15.29 fps to 35.09 fps) as the flow travels from the 6-in. diameter line to the 4-in. diameter line.

7. At point A: 4.8 fps; at point B: 448.8 gpm; at point C: 309.67 gpm

8. 17,500 gpd

9. (a) 0.042 mgd, (b) 20.83 gpm

10. 584,000 gpd

11. (a) 1300 gpm, (b) 1190 gpm

12. 5402 gpm

13. 0.67 cfs

Detailed Answers

1. Before beginning the $Q = AV$ calculation, convert the depth in inches to depth in feet:

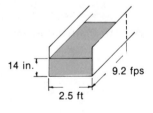

$$\frac{14 \text{ in.}}{12 \text{ in.}/\text{ft}} = 1.17 \text{ ft}$$

Now use the $Q = AV$ equation to calculate the cubic-feet-per-second flow rate:

$$Q = \quad A \qquad V$$
$$= (2.5 \text{ ft})(1.17 \text{ ft}) \ (9.2 \text{ fps})$$
$$= 26.91 \text{ cfs}$$

2. Since the velocity is expressed in feet per second, first calculate cubic-feet-per-second flow rate, then convert to gallons per minute:

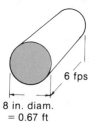

8 in. diam.
= 0.67 ft

$$Q = \qquad A \qquad\qquad V$$
$$= (0.785)(0.67 \text{ ft})(0.67 \text{ ft}) \ (6 \text{ fps})$$
$$= 2.11 \text{ cfs}$$

Convert to gallons per minute:

$$(2.11 \text{ cfs})(7.48 \text{ gal/cu ft})(60 \text{ sec/min}) = 946.97 \text{ gpm}$$

3. Because the velocity is requested in feet per minute, first convert the 2.8 mgd flow rate to cubic feet per minute:

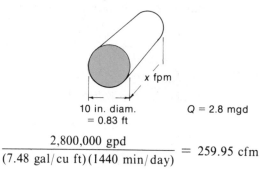

10 in. diam. $Q = 2.8 \text{ mgd}$
= 0.83 ft

$$\frac{2{,}800{,}000 \text{ gpd}}{(7.48 \text{ gal/cu ft})(1440 \text{ min/day})} = 259.95 \text{ cfm}$$

Now use the $Q = AV$ equation to determine the feet-per-minute velocity. Begin by filling in the given information:

$$Q \qquad = \qquad\qquad A \qquad\qquad V$$
$$259.95 \text{ cfm} = (0.785)(0.83 \text{ ft})(0.83 \text{ ft}) \ (x \text{ fpm})$$

Then solve for the unknown value:

$$\frac{259.95}{(0.785)(0.83)(0.83)} = x \text{ fpm}$$
$$480.69 \text{ fpm} = x \text{ fpm}$$

4. Since the velocity of the water must be expressed as feet per second, first convert the gallons-per-minute flow rate to cubic feet per second:

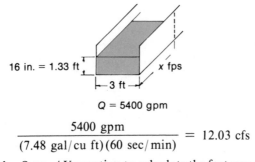

16 in. = 1.33 ft

x fps

3 ft

$Q = 5400$ gpm

$$\frac{5400 \text{ gpm}}{(7.48 \text{ gal/cu ft})(60 \text{ sec/min})} = 12.03 \text{ cfs}$$

Now use the $Q = AV$ equation to calculate the feet-per-second velocity in the channel:

$$Q = A \qquad V$$
$$12.03 \text{ cfs} = (3 \text{ ft})(1.33 \text{ ft}) \ (x \text{ fps})$$

Solve for the unknown value:

$$\frac{12.03}{(3)(1.3)} = x \text{ fps}$$

$$3.02 \text{ fps} = x \text{ fps}$$

5. First simplify velocity information given in the problem:

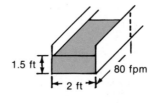

1.5 ft

2 ft

80 fpm

$$\text{Velocity} = \frac{\text{feet}}{\text{minute}}$$

$$= \frac{160 \text{ ft}}{2 \text{ min}}$$

$$= 80 \text{ fpm}$$

Since velocity is given in feet per minute, first calculate the cubic-feet-per-minute flow rate, then convert to million gallons per day.

$$Q = A \qquad V$$
$$Q \text{ cfm} = (2 \text{ ft})(1.5 \text{ ft}) \ (80 \text{ fpm})$$
$$= 240 \text{ cfm}$$

Convert cubic-feet-per-minute flow rate to gallons per day and then to million gallons per day:

$$(240 \text{ cfm})(7.48 \text{ gal/cu ft})(1440 \text{ min/day}) = 2{,}585{,}088 \text{ gpd}$$
$$= 2.59 \text{ mgd}$$

6. When a figure is not given with the statement of the problem, it is usually a good idea to draw a figure based on the description in the problem. In this problem, the velocity in a 6-in. diameter pipe is compared to that in a 4-in. diameter pipe:

6 in. diam.   4 in. diam.
= 0.5 ft diam.   = 0.33 ft diam.

Use the $Q = AV$ equation to solve first for V_1, then for V_2:

$$Q_1 = A_1 V_1$$
$$3 \text{ cfs} = (0.785)(0.5 \text{ ft})(0.5 \text{ ft})(x \text{ fps})$$
$$\frac{3}{(0.785)(0.5)(0.5)} = x \text{ fps}$$
$$15.29 \text{ fps} = x \text{ fps}$$

$$Q_2 = A_2 V_2$$
$$3 \text{ cfs} = (0.785)(0.33 \text{ ft})(0.33 \text{ ft})(x \text{ fps})$$
$$\frac{3}{(0.785)(0.33)(0.33)} = x \text{ fps}$$
$$35.09 \text{ fps} = x \text{ fps}$$

The velocity increases from 15.29 fps to 35.09 fps, more than double, in the change from the 6-in. diameter pipe to the 4-in. diameter pipe.

7. To solve this problem, use the $Q = AV$ equation to find the velocity at point A and flow rates at points B and C.

At Point A

Since the velocity must be expressed in feet per second, first convert the gallons per minute to cubic feet per second:

$$\frac{760 \text{ gpm}}{(7.48 \text{ gal/cu ft})(60 \text{ sec/min})} = 1.69 \text{ cfs}$$

And now perform $Q = AV$ calculation:

$$Q_A = A_A V_A$$
$$1.69 \text{ cfs} = (0.785)(0.67 \text{ ft})(0.67 \text{ ft})(x \text{ fps})$$
$$\frac{1.69}{(0.785)(0.67)(0.67)} = x \text{ fps}$$
$$4.80 \text{ fps} = x \text{ fps}$$

At Point B

Although the problem asked for gallons-per-minute flow rate, first calculate the flow rate in cubic feet per second. Then convert cubic-feet-per-second flow rate to gallons-per-minute flow rate.

$$Q_B = A_B V_B$$
$$Q_B \text{ cfs} = (0.785)(0.42 \text{ ft})(0.42 \text{ ft})(7.2 \text{ fps})$$
$$= 1.00 \text{ cfs}$$

Convert this flow rate to gallons per minute:

$$(1.00 \text{ cfs})(7.48 \text{ gal/cu ft})(60 \text{ sec/min}) = 448.8 \text{ gpm}$$

At Point C

To solve for flow rate at point C, follow the same basic steps as for point B:

$$Q_C = A_C V_C$$
$$Q_C \text{ cfs} = (0.785)(0.5 \text{ ft})(0.5 \text{ ft})(3.5 \text{ fps})$$
$$= 0.69 \text{ cfs}$$

Convert to gallons per minute:

$$(0.69 \text{ cfs})(7.48 \text{ gal/cu ft})(60 \text{ sec/min}) = 309.67 \text{ gpm}$$

At Point C—Alternate solution

Rather than making the full calculation at point C, you could have used the rule of continuity, which states:

$$Q_A = Q_B + Q_C$$

Therefore, the flow rate at Q_C would be:

$$\begin{array}{r} 760.0 \text{ gpm} \\ -448.8 \text{ gpm} \\ \hline 311.2 \text{ gpm} \end{array}$$

In the first calculation for flow at point *C*, the answer was only 309.67 gpm. This answer is less than the alternate solution (311.2 gpm) as a result of rounding in the calculation.

8. Since the head above the weir crest is 3.3 in., enter the *head in inches* scale at *3.3*. As marked on the graph in Figure 100B, this head indicates a flow of about 17,500 gpd

9. (a) (See Figure 101B.) Enter the *head–inches* scale at *3.5*, and move to the right until you intersect the *60-deg V-notch weir* line. Read the indicated flow directly below this point as between 0.04 and 0.05 mgd; the indicated flow at this head is about 0.042 mgd.

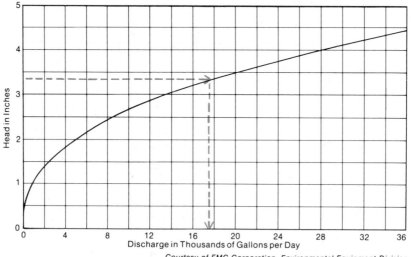

Courtesy of FMC Corporation, Environmental Equipment Division.

Figure 100B.   Curve for 30-deg V-notch Weir (Review Question 8.)

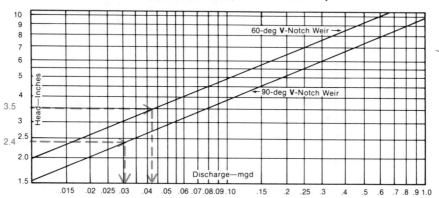

Figure 101B.   Curves for 60- and 90-deg V-notch Weirs (Review Question 8)

(b) As shown by the dotted arrows on the graphs, the flow that is indicated by a head of 2.4 in. is 0.03 mgd for a 90-deg **V**-notch weir. Convert this flow rate to gallons per day, then to gallons per minute:

$$0.03 \text{ mgd} = 30,000 \text{ gpd}$$

$$\frac{30,000 \text{ gpd}}{1440 \text{ min/day}} = 20.83 \text{ gpm}$$

10. The head is 1.27. In the table, part of the head (1.2) is given on the vertical scale, and the remainder (0.07) is given on the horizontal scale. Locate *1.2* in the far left column of the table, then move from there across the table until you reach the columns under *.07*. Read the flow in the column titled *MGD* as 0.584 mgd. Convert this to gallons per day:

$$0.584 \text{ mgd} = 584,000 \text{ gpd}$$

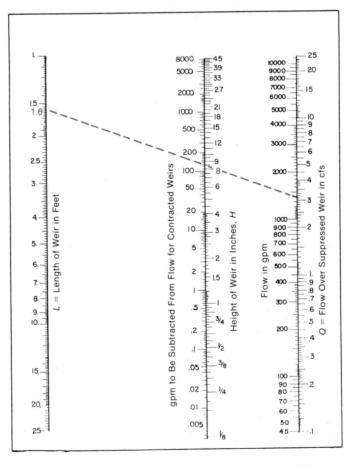

Figure 102B. Nomograph for Rectangular Weirs (Review Question 11)

11. (a) (See Figure 102B.) To determine the flow over a suppressed rectangular weir, draw a line from $L = 1.6$ ft on the left-hand scale through $H = 8$ in. on the right side of middle scale; the flow rate is indicated on the left side of the right-hand scale as approximately 1300 gpm.

(b) To determine the flow over a contracted rectangular weir of the same length and with the same head, read the value where the drawn line crosses the left side of the middle scale; subtract this value from the flow rate found in part (a):

1300 gpm (suppressed rectangular weir flow rate)
− 110 gpm (from middle scale left side)
1190 gpm (contracted rectangular weir flow rate)

12. (See Figure 103B.) First determine the approximate flow rate in the flume with the nomograph. Then use a correction factor if required.

Draw a line on the nomograph from 3 ft on the *throat width* (left) scale through 1.2 ft on the H_a *depth* scale (right side of middle scale); the indicated flow is about 7300 gpm.

To determine whether a correction to this flow rate is needed, calculate the percent submergence of the flume throat:

$$\text{Percent Submergence} = \frac{H_b}{H_a} \times 100$$

$$= \frac{1.1 \text{ ft}}{1.2 \text{ ft}} \times 100$$

$$= 0.92 \times 100$$

$$= 92\% \text{ submergence}$$

To locate the required correction factor, if any, draw a vertical line up from 92 percent on the *submergence* scale of the correction graph until it crosses the 3-ft correction line (*Point A* on the correction graph). Move directly to the left to the Q/Q_0 scale; the indicated correction factor is about 0.74. Then multiply the correction factor by the nomograph flow rate:

$$(7300 \text{ gpm})(0.74) = 5402 \text{ gpm corrected flow}$$

13.   To calculate the cubic-feet-per-second flow rate through the pipe, the nomograph is first used to calculate the velocity of flow through the pipe (see Figure 104B.) Then using the velocity and pipe diameter information, the flow rate is calculated using the $Q = AV$ equation. First calculate the ratio of the pipe diameter to the orifice diameter and locate the result on the right side of the middle scale:

$$\frac{D_1}{D_2} = \frac{8 \text{ in.}}{3 \text{ in.}}$$

$$= 2.67$$

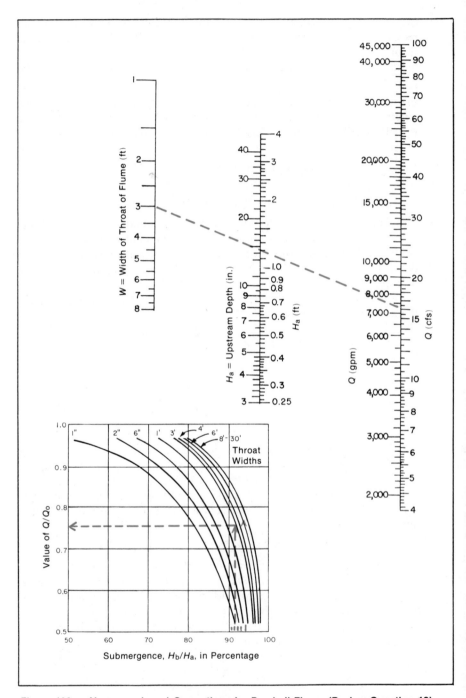

Figure 103B. Nomograph and Corrections for Parshall Flume (Review Question 12)

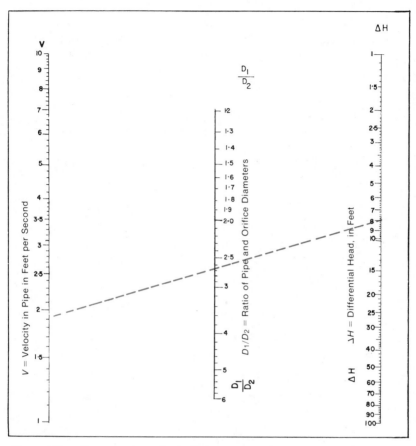

Figure 104B. Nomograph for Orifice Meter (Review Question 13)

Now draw a line, from 8 on the right-hand scale through 2.67 on the right side of the middle scale and across to the velocity (left) scale. The velocity indicated is about 1.9 fps.

The diameter of the pipeline (8 in. = 0.67 ft) and the velocity of the flow are known, so use the $Q = AV$ equation to determine the flow rate in the pipe:

$$Q = AV$$
$$= (0.785)(0.67 \text{ ft})(0.67 \text{ ft})(1.9 \text{ fps})$$
$$= 0.67 \text{ cfs}$$

Hydraulics 8

Thrust Control

Water under pressure and water in motion can exert tremendous forces inside a pipeline. One of these forces, THRUST, pushes against fittings, valves, and hydrants, causing couplings to leak or to pull apart entirely. Although it is not possible to eliminate thrust, it is possible and absolutely necessary to control it.

Thrust can be caused by any factor that produces a pressure or force on a fitting, but it is primarily caused by water pressure and WATER HAMMER. As shown in Figure 105, it most always acts perpendicular (at 90 deg) to the inside surface it pushes against. Note in the figure how thrust acts against the outer curve of the fitting, tending to push the fitting away from both sections of pipeline. You can picture the force as being caused by water as it "rounds the bend." If the flow direction were reversed, the direction of thrust would be unchanged. Uncontrolled, the thrust can cause movement in the fitting or pipeline which will lead to leakage or complete separation at coupling A or B, or

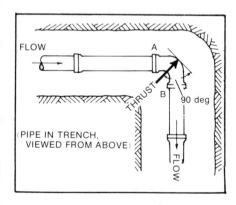

Figure 105. Direction of Thrust

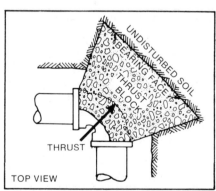

Reprinted with permission of Johns-Manville Sales Corporation, Denver, CO, from Design Manual: Johns-Manville Transite Transmission Pipe.

Figure 106. Thrust Block, 90-deg Bend

at some other nearby coupling upstream or downstream of the fitting.

There are two devices commonly used to control thrust: (1) thrust blocks and (2) thrust anchors. A THRUST BLOCK is a mass of concrete, cast in place between the fitting being restrained and the undisturbed soil at the side or bottom of the pipe trench. An example is diagrammed in Figure 106. This type of thrust control can be used either to control thrust forces that act horizontally or to control thrusts that act downward.

A THRUST ANCHOR is a massive block of concrete, often a cube, cast in place below the fitting to be anchored. As shown in Figure 107, imbedded steel shackle rods anchor the fitting to the concrete block, effectively resisting upward thrusts.

The size and shape of a thrust control device depend primarily on five factors:

- Type of fitting,
- Diameter of fitting,
- Water pressure,
- Water hammer, and
- Soil type.

If you are given information on these five factors, then you can solve many thrust control sizing problems by reading and interpreting tables, such as the ones shown in Tables 11 through 15. Be cautious in applying the information given in this section in the field. Although the following procedures will work quite adequately in many situations, they should not be used to substitute for professional engineering design, particularly in situations involving large diameter pipelines (greater than 12 in.), high velocity situations (greater than 10 ft/sec), or soils where soil type or stability may be questionable. (One of the more common causes of thrust block failure is the installation of thrust blocks in unstable soil, or in locations too close to trenches for other pipelines.)

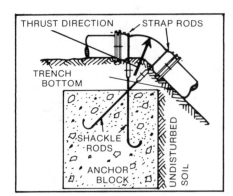

Figure 107.   Thrust Anchor, 45-deg Bend

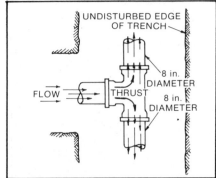

Reprinted with permission of Johns-Manville Sales Corporation, Denver, CO, from Design Manual: Johns-Manville Transite Transmission Pipe.

Figure 108.   (Example 1.)

H8-1. Thrust Block Calculations

To determine the proper size of thrust block for a particular fitting, you will first have to determine the total thrust in pounds against the fitting. You can then determine the area of the bearing face based on the type of soil and how much pressure it can withstand. The following two examples illustrate how to size thrust blocks using this method.

Example 1

Thrust is developed at the tee fitting shown in Figure 108. The pipeline has an 8 in. diameter and will operate at a maximum pressure of 200 psig.

The undisturbed soil type in the trench is alluvial soil. Determine (a) thrust on the tee, in pounds, and (b) area and dimensions of the necessary thrust block bearing face.

(a) First determine the thrust on the 8-in. tee using Table 11. Move across the row beginning with 8-in. diameter and read the answer from the last column (tees and dead ends) as 5000 lb. Notice that Table 11 gives thrust values *per 100 psig of pressure*. The maximum thrust against the fitting depends on the maximum pressure in the pipeline. The maximum water pressure that is expected to occur is the sum of the maximum operating pressure plus an allowance for pressure resulting from water hammer. In this problem, the

**Table 11. Fitting Thrust in Pounds
at 100 psig Water Pressure**

Diameter, inches	Type of Fitting				
	11¼-deg Bend	22½-deg Bend	45-deg Bend	90-deg Bend	Tees and Dead Ends
3	140	280	540	1,000	710
4	250	490	960	1,800	1,300
6	550	1,100	2,200	4,000	2,800
8	990	2,000	3,800	7,100	5,000
10	1,500	3,100	6,000	11,100	7,900
12	2,200	4,400	8,700	16,000	11,300
14	3,000	6,000	11,800	21,800	15,400
16	3,900	7,800	15,400	28,400	20,100
18	5,000	9,900	19,500	36,000	25,400
20	6,200	12,300	24,000	44,400	31,400
24	7,500	14,800	29,100	53,800	38,000
30	13,900	27,600	54,100	100,000	70,700
36	20,000	40,000	77,900	144,000	102,000
42	27,200	54,100	106,000	196,000	139,000
48	35,500	70,600	138,000	256,000	181,000
54	44,900	89,400	175,000	324,000	229,000
60	55,400	110,000	216,000	400,000	283,000

Table 12. Allowances for Water Hammer

Pipeline Diameter inches	Water Hammer Pressure Allowance psig
3–10	120
12–14	110
16–18	100
20	90
24	85
30	80
36	75
42–60	70

maximum operating pressure is stated as 200 psig. From Table 12, determine that the water hammer allowance for an 8-in. pipeline is 120 psig. Adding these two pressures, you know that the total water pressure which may occur is

Maximum operating pressure	200 psig
Water hammer allowance	+120 psig
Maximum total water pressure	320 psig

Using this water pressure information, the maximum expected thrust against the fitting can be calculated. As determined from Table 11, the thrust against the fitting is 5000 lb for every 100 psig of pressure. Since $320/100 = 3.2$, there are 3.2 units of 100 in the number 320. Therefore, the maximum thrust is

(5000 lb per 100 psig)(3.2 units of 100 psig) = 16,000 lb maximum thrust

(b) To find the area and dimensions of the thrust block bearing face, first find out how much bearing pressure the soil can support (the allowable soil bearing pressure) from Table 13 for alluvial soil:

Allowable soil bearing pressure = 1000 lb/sq ft

Table 13. Allowable Bearing Pressure for Soil Types

Soil Type	Bearing Pressure lb/sq ft
Peat or muck	0
Alluvial soil	1,000
Soft clay	2,000
Sand	4,000
Sand and Gravel	6,000
Sand and Gravel with Clay	8,000
Shale	12,000
Rock	20,000

This means that each square foot of soil should have no more than 1000 lb of force against it. For a maximum thrust of 16,000 lb, the area needed for the bearing face is

$$\frac{16,000 \text{ lb}}{1000 \text{ lb/sq ft}} = 16 \text{ sq ft}$$

It is desirable to have square bearing faces. The dimensions for such faces can be found by taking the square root of the area (square root tables or calculators with the square root function are available for finding the square root)

$$\sqrt{16 \text{ sq ft}} = 4 \text{ ft on a side}$$

The correctly sized thrust block is shown in Figure 109.

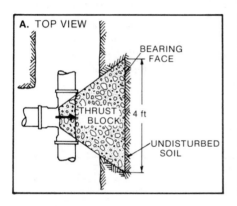

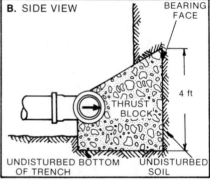

Reprinted with permission of Johns-Manville Sales Corporation, Denver, CO, from Design Manual: Johns-Manville Transite Transmission Pipe.

Figure 109.   (Example 1.)

Often pipelines are pressure-tested after construction is complete. The test pressure is always greater than the pipeline would experience in normal day-to-day operation. Since such tests normally last for two hours or so, the fittings need to be designed to withstand the test pressures. Therefore, in such cases you should use the test pressure instead of the operating pressure in determining the size of thrust blocks. The following example will illustrate the procedure.

Example 2

The 10-in. pipeline shown in Figure 110 is designed to operate at 75 psig. After construction is completed the pipeline will be tested at twice the design operating pressure. The soil type is sand. Determine the size of the thrust block required.

Follow the same procedure as in Example 1. First find the thrust against the 10-in. diameter 90-deg bend using Table 11. The thrust listed is 11,100 lb for each 100 psig. To determine the maximum thrust against the fitting, you will have to know the maximum water pressure that could occur in the pipeline:

Maximum test pressure (2×75)	150 psig
Water hammer allowance (from Table 12)	+120 psig
Maximum total water pressure	270 psig

Using this pressure information, the total thrust against the fitting can be calculated. As determined from Table 11, the thrust against the fitting is 11,100 lb for every 100 psig of pressure. Since $270/100 = 2.7$, there are 2.7 units of 100 in the number 270. Therefore, the maximum thrust is

$$(11{,}100 \text{ lb}/100 \text{ psig})(2.7 \text{ units of } 100 \text{ psig}) = 29{,}970 \text{ lb maximum thrust}$$

From Table 13, determine that the allowable soil bearing pressure for sand is 4000 lb/sq ft. This means that each square foot of soil should have no more than 4000 lb of force against it. For a total thrust of 29,970 lb, the area of the thrust block bearing face should be

$$\frac{29{,}970 \text{ lb}}{4000 \text{ lb/sq ft}} = 7.49 \text{ sq ft}$$

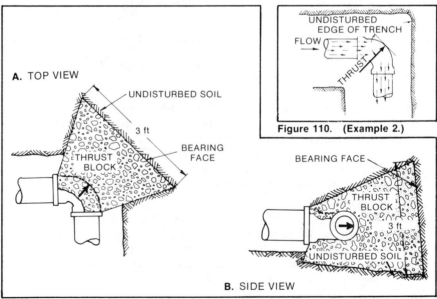

A. TOP VIEW

UNDISTURBED SOIL

3 ft

THRUST BLOCK

BEARING FACE

UNDISTURBED EDGE OF TRENCH

FLOW

THRUST

Figure 110.   (Example 2.)

BEARING FACE

THRUST BLOCK

3 ft

UNDISTURBED SOIL

B. SIDE VIEW

Figure 111.   (Example 2.)

Since the bearing face should be square, the dimensions can be found by taking the square root of the area:

$$\sqrt{7.49 \text{ sq ft}} = 2.74 \text{ ft}$$

Rounding up to the nearest foot, you find that the correctly-sized thrust block is 3 ft on a side (Figure 111).

H8-2.   Thrust Anchor Calculations

Sometimes standard engineering plans and specifications are prepared for a variety of routinely-used devices such as thrust anchors. Tables like the one

Table 14.   Thrust Anchoring for 11¼-, 22½-,
30-, and 45-deg Vertical Bends—
Only Applicable to "Type X"
Trench Conditions

Pipe Size Nom. Diameter—in.	Test Pressure psi	Vertical Bend deg	Volume of Concrete Blocking cu ft	Side of Cube ft	Diameter of Shackle Rods in.*	Depth of Rods in Concrete ft
4	300	11¼	8	2	¾	1.5
		22½	11	2.2	¾	2.0
		30	17	2.6	¾	2.0
		45	30	3.1	¾	2.0
6	300	11¼	11	2.2	¾	2.0
		22½	25	2.9	¾	2.0
		30	41	3.5	¾	2.0
		45	68	4.1	¾	2.0
8	300	11¼	16	2.5	¾	2.0
		22½	47	3.6	¾	2.0
		30	70	4.1	¾	2.5
		45	123	5.0	¾	2.0
12	250	11¼	32	3.2	¾	2.0
		22½	88	4.5	⅞	3.0
		30	132	5.1	⅞	3.0
		45	232	6.1	¾	2.5
16	225	11¼	70	4.1	⅞	3.0
		22½	184	5.7	1⅛	4.0
		30	275	6.5	1¼	4.0
		45	478	7.8	1⅛	4.0
20	200	11¼	91	4.5	⅞	3.0
		22½	225	6.1	1¼	4.0
		30	330	6.9	1⅜	4.5
		45	560	8.2	1¼	4.0
24	200	11¼	128	5.0	1	3.5
		22½	320	6.8	1⅜	4.5
		30	480	7.9	1⅝	5.5
		45	820	9.4	1⅜	4.5

*45 deg: 4 rods; all others: 2 rods.

shown in Table 14 allow the non-engineer to size thrust anchors for routine situations with a minimum of effort. Note that such standard tables are prepared for a particular set of trench conditions, and can only be used when those conditions exist. (The set of trench conditions applicable for Table 14 information have arbitrarily been denoted as "Type X" conditions.) Other standard tables would have to be used for differing trench conditions. There is no single tabular procedure that applies to all thrust anchor problems, and there is no set of tables that will replace a competent design engineer. The following two examples illustrate how to size thrust anchors using standard tables.

Example 3

Thrust acts on a 22-1/2 deg vertical bend, as shown in Figure 112. Assuming Type X trench conditions, use Table 14 to determine how the fitting can be properly anchored to resist upward thrust. The fitting is 8 in. in diameter and the pipeline will be tested after construction at a pressure of 300 psig.

Using Table 14, thrust anchor sizing is a simple task. Enter the table at *8* inches in the *Pipe Size Nom. Diameter —in.* column; then read the sizing needs from the last four columns, opposite *22½ deg*, as follows:

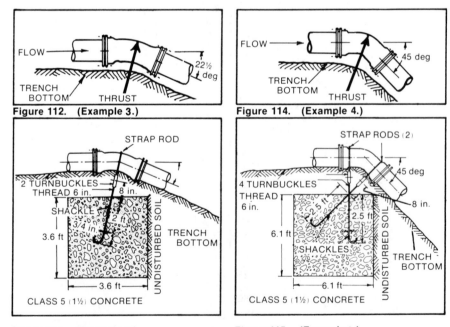

Figure 112.   (Example 3.)

Figure 114.   (Example 4.)

Figure 113.   (Example 3.)

Figure 115.   (Example 4.)

Volume of concrete block . . . . 47 cu ft
Dimension of cubic concrete block . . . . 3.6 ft on a side
Diameter of shackle rods (2 required) . . . . . 3/4 in.
Imbedded depth of rods in concrete . . . . . . 2 ft

This information is now used to draw the thrust anchor shown in Figure 113. Notice that two shackles are required (one is directly behind the one shown). One shackle is connected to each threaded end of the strap rod by a turnbuckle.

Example 4

Thrust acts on a 45-deg vertical bend as shown in Figure 114. Based on thrust anchor data shown in Table 14, make a drawing of the correct thrust anchor for Type X trench conditions and label the dimensions. The fitting is 12 in. in diameter and will be tested at a pressure of 250 psig.

Enter Table 14 at a diameter of 12 in. and read the sizing information from the last four columns as follows:

Volume of concrete block . . . . 232 cu ft
Dimension of cubic concrete block . . . . 6.1 ft on a side
Diameter of shackle rods (4 required) . . . . . 3/4 in.
Imbedded depth of rods in concrete . . . . . . 2.5 ft

Now make a drawing of the information as shown in Figure 115. Notice that two of the four required shackles are hidden directly behind the two shown.

Review Questions

1. What is thrust and why is it a concern in pipelines?

2. What is the purpose of thrust control?

3. The size and shape of a thrust block is based primarily on five factors. List them.

4. You are installing a horizontal replacement section of 6-in. PVC with push-on joints. The section includes one tee and several 90-deg and 11-1/4 deg bends. Upon completion the line will be tested at 300 psig. The soil type throughout the replacement area is sand and gravel. Find the dimensions of the thrust control device required at each type fitting.

Summary Answers

1. (See detailed answers.)

2. (See detailed answers.)

3. (See detailed answers.)

4. For tee: 1.4 ft
        or about 1 1/2 ft on a side
   For 90-deg bends: 1.67 ft
        or about 1 2/3 to 2 ft on a side
   For 11-1/4 deg bends: 0.62 ft
        or about 2/3 to 1 ft on a side

Detailed Answers

1. A force exerted on fittings, valves and hydrants within a pipeline, caused primarily by water pressure and water hammer. Uncontrolled, thrust can cause couplings to leak or separate completely.

2. To restrain the movement of fittings, valves, and hydrants in order to prevent leakage or uncoupling.

3. Type of fitting, Diameter of fitting, Water pressure, Water hammer, and Soil type.

4. First, using Table 11, find the thrust against each type of 6-in. diameter fitting:

Fitting	Thrust
Tee	2800 lb per 100 psig
90 deg bend	4000 lb per 100 psig
11 1/4 deg bend	550 lb per 100 psig

Then to determine the maximum thrust acting on each fitting you will have to know the maximum water pressure which could occur in the pipeline. Find water hammer allowance from Table 12, then add:

Maximum test pressure	300 psig
Water hammer allowance	+120 psig
Maximum total water pressure	420 psig

Using this pressure information, the maximum thrust against the fitting can now be calculated. Since $420/100 = 4.2$, there are 4.2 units of 100 psig. The thrust at each fitting is expressed per 100 psig, so the maximum thrusts are as follows:

Maximum thrust at tee $= (2800 \text{ lb per } 100 \text{ psig}) (4.2 \text{ units of } 100 \text{ psig})$

$\qquad\qquad\qquad\qquad = (2800)(4.2)$

$\qquad\qquad\qquad\qquad = 11,760 \text{ lb}$

Maximum thrust at
90-deg bends $= (4000 \text{ lb per } 100 \text{ psig}) (4.2 \text{ units of } 100 \text{ psig})$

$\qquad\qquad\qquad\qquad = (4000)(4.2)$

$\qquad\qquad\qquad\qquad = 16,800 \text{ lb}$

Maximum thrust at
11-1/4 deg bends $= (550 \text{ lb per } 100 \text{ psig}) (4.2 \text{ units of } 100 \text{ psig})$

$\qquad\qquad\qquad\qquad = (550)(4.2)$

$\qquad\qquad\qquad\qquad = 2310 \text{ lb}$

To find the area and dimensions of the thrust block bearing face, first find the allowable soil bearing pressure from Table 13 for sand and gravel:

Allowable soil bearing pressure $= 6000 \text{ lb per sq ft}$

This means that each square foot of soil should have no more than 6000 lb of force against it. For the maximum expected thrusts at the tee, 90-deg bends, and 11-1/4 deg bends, the areas needed for the bearing faces are:

$$\text{For the tee} = \frac{11,760 \text{ lb}}{6000 \text{ lb/sq ft}}$$

$$= 1.96 \text{ sq ft}$$

$$\text{For the 90-deg bends} = \frac{16,800 \text{ lb}}{6000 \text{ lb/sq ft}}$$

$$= 2.8 \text{ sq ft}$$

$$\text{For the 11-1/4 deg bends} = \frac{2310 \text{ lb}}{6000 \text{ lb/sq ft}}$$

$$= 0.39 \text{ sq ft}$$

The dimensions of the desired square bearing faces are found by taking the square root of the areas:

$$\text{Face dimension for tee} = \sqrt{1.96 \text{ sq ft}}$$

$$= 1.4 \text{ ft,}$$
$$\text{or about } 1\ 1/2 \text{ ft on a side}$$

$$\text{Face dimension for 90-deg bends} = \sqrt{2.8 \text{ sq ft}}$$

$$= 1.67 \text{ ft,}$$
$$\text{or about } 1\ 2/3 \text{ to } 2 \text{ ft on a side}$$

$$\text{Face dimension}$$
$$\text{for 11-1/4 deg bends} = \sqrt{0.39 \text{ sq ft}}$$

$$= 0.62 \text{ ft,}$$
$$\text{or about } 2/3 \text{ to } 1 \text{ ft on a side}$$

Basic Science
Concepts and
Applications

Chemistry

Chemistry 1

The Structure of Matter

If you could take a sample of an ELEMENT and divide it into smaller and smaller pieces, you would eventually come down to a tiny particle that, if subdivided any more, would no longer show the characteristics of the original element. The smallest particle that still retains the characteristics of the element is called an ATOM, from the Greek word *atomos* meaning "uncut" or "indivisible."

Although an atom is the smallest particle that still retains the characteristics of the element it is taken from, and is so small that it can't be seen with today's most powerful microscopes, the atom itself can be broken down into even smaller particles called *subatomic* particles. The different number and arrangement of subatomic particles distinguish the atoms of one element from those of another, and give each element specific qualities. A general understanding of the structure of atoms is basic to an understanding of chemical reactions and equations.

C1-1. Atomic Structure

A great many subatomic particles have been identified, many of which exist for only a fraction of a second. In the study of chemistry and chemical reactions, however, the structure of the atom is adequately explained on the basis of three fundamental particles: the PROTON, the NEUTRON, and the ELECTRON.

As shown in Figure 1, the center of the atom, called the NUCLEUS (plural: NUCLEI), is made up of positively charged particles called protons and uncharged particles called neutrons. Negatively charged particles called electrons occupy the space surrounding the nucleus and make up most of the volume of the atom. The electrons are said to occupy the space around the nucleus in "shells."

The basic defining characteristic of the atoms of any one element is the number of protons in the nucleus. An atom of carbon, for example, always has six protons in the nucleus; and any atom with exactly six protons in the nucleus

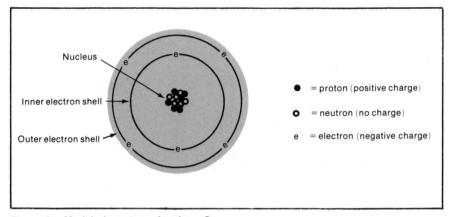

Figure 1.   Model of an atom of carbon, C

must be a carbon atom. Boron atoms have five protons in the nucleus, whereas nitrogen atoms have seven. The number of protons in the nucleus of an atom is called the ATOMIC NUMBER. Therefore, the atomic number of carbon is six; boron, five; and nitrogen, seven.

The nucleus is extremely small in comparison to the total size of the atom. If the atom were the size of a football stadium, the nucleus would be no larger than a small insect flying in the middle. Nonetheless, because neutrons and protons (which have nearly identical weights) are much heavier than electrons, the nucleus contains most of the mass (weight) of the atom. The ATOMIC WEIGHT of an atom is defined as the sum of the number of protons and the number of neutrons in the nucleus. In the nucleus of a carbon atom there are six protons and six neutrons; therefore, the atomic weight of the atom is 12.

Atomic weights are not "weights" in the usual sense of the word. They do not indicate the number of pounds or grams an atom weighs, but merely how the weight of one atom *compares* with the weight of another. For example, the atomic weight of hydrogen is 1 (the hydrogen nucleus has one proton and no neutrons) and the atomic weight of carbon is 12. Therefore, an atom of carbon weighs twelve times as much as an atom of hydrogen.

Isotopes

All atoms of a given element have the same number of protons in the nucleus, but the number of neutrons may vary. Atoms of the same element, but containing varying numbers of neutrons in the nucleus, are called ISOTOPES of that element. The atoms of each element with the numbers of neutrons found most commonly are called the PRINCIPLE ISOTOPES of the element.

The atomic weight of an element is generally given in tables as a whole number with decimals, the result of averaging together the atomic weights of the principle (most common) isotopes. The average also takes into account how often each isotope occurs. The element chlorine, for example, has a listed atomic weight of 35.45—an average of the principal isotopes of chlorine. (About 75

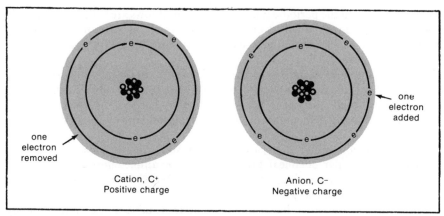

Figure 2.   Models of ionized carbon atoms

percent of all chlorine atoms have atomic weight 35, and about 25 percent have atomic weight 37.)

Ions

As shown in Figure 1, there are the same number of electrons (negative charges) surrounding the carbon atom as there are protons (positive charges) in the nucleus. The atom is said to be electrically stable.

Now consider the effect on the atom's charge if an electron is removed from the outer electron shell, or if an electron is added to the outer shell. (See Figure 2.) In either case, the charges on the atom are no longer balanced. With one electron removed from the outer shell, there are six protons (positive charges) counterbalanced by only five electrons (negative charges), resulting in a net charge on the atom of *plus 1*. On the other hand, with one electron added to the outer shell, there are six protons (positive charges) counterbalanced by seven electrons (negative charges), resulting in a net charge on the atom of *minus 1*.

When the charges on the atom are *not* balanced, the atom is no longer stable (that is, it has a plus or minus charge). In this unstable condition, the atom is called an ION. When the net charge on an atom is positive (more protons than electrons), the ion is called CATION. When the net charge on an atom is negative (more electrons than protons), the ion is called an ANION.

Of the three fundamental particles—protons, neutrons, and electrons—the electron is the most important particle in understanding basic chemistry. During chemical reactions the nucleus of an atom remains unchanged; only the electrons of atoms interact, and only those in the outermost shell.

C1-2.   The Periodic Table

The elements can be arranged according to the number of electron shells they have, and according to similarities of chemical properties. When so arranged, a table is formed called the PERIODIC TABLE (see Appendix B for the complete periodic table). In the periodic table, the horizontal rows are called PERIODS, and the vertical columns are called GROUPS. Elements of the same period have the same number of electron shells; elements in the same group have similar chemical properties.

As an example of members of the same period, hydrogen and helium are members of the first horizontal row (period) in the periodic table. Each has only one electron shell. Lithium, berylium, boron, carbon, nitrogen, oxygen, fluorine, and neon are members of the second period, and all have two electron shells.

The vertical columns (groups) are important since the elements within groups tend to have similar chemical properties. For example, notice that the two elements chlorine and iodine, both of which can be used for disinfecting water, are in the same chemical group.

Although the individual boxes of a large periodic table may contain as many as nine or more kinds of information about each element (including element name, symbol, electron structure, atomic number and weight, oxidation states, boiling points, melting points, and density), the four basic kinds of information included in almost all periodic tables are: (1) atomic number, (2) element symbol, (3) element name, and (4) atomic weight. Consider the carbon atom model again, this time as it appears in the periodic table (see Figure 3).

The top number in the box is the atomic number—the number of protons in the nucleus of an atom of that element. (It also indicates the number of electrons in a stable atom of the element, since the number of protons is the same as the number of electrons in a stable atom.) The letter in the box is the standard abbreviation, or symbol, that has been assigned to the element. Often, two letters are used because a single letter has already been assigned to a specific element. For example, *H* stands for hydrogen; *Hg* for mercury. The number at the bottom of the box is the atomic weight of the element. As discussed before, this number represents the average weight of the common isotopes of the element.

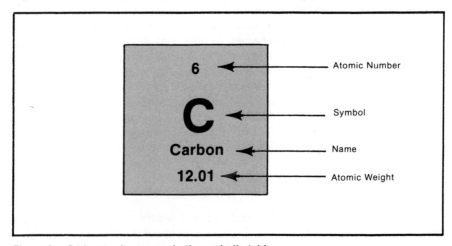

Figure 3.   Carbon as it appears in the periodic table

Review Questions

1. Briefly define the terms nucleus, proton, neutron, and electron as they refer to the structure of an atom.

2. What is the charge of each of the following atomic particles: proton? neutron? electron?

3. What is an isotope?

4. What is an ion?

5. What is the difference between a cation and an anion?

6. In addition to the name of the element, the periodic table has at least four basic types of information, as shown below. Label each arrow.

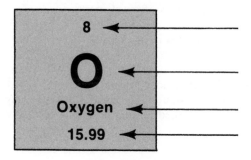

Summary Answers

(See Detailed Answers.)

Detailed Answers

1. The nucleus is the center of the atom and is made of protons and neutrons. Electrons occupy the space around the nucleus and make up the greater part of the volume of the atom. Electrons are said to occupy the space around the nucleus in shells.

2. Proton: positive charge.
   Neutron: no charge (neutral).
   Electron: negative charge.

3. Atoms that contain different numbers of neutrons, but the same number of protons, in their nuclei are called isotopes of an element.

4. An atom that has lost or gained electrons in its outer shell so that it is no longer stable (neutral) is called an ion.

5. When the charge on an atom is positive (more protons than electrons), the atom is called a cation. When the charge on an atom is negative (more electrons than protons), the atom is called an anion.

6.

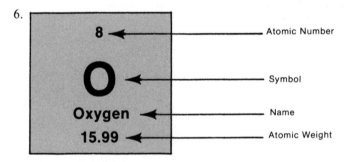

Chemistry 2

The Classification of Matter

Matter is anything that occupies space and has weight (mass). Matter includes subatomic particles—protons and electrons—as well as the atoms that such particles form. Matter also includes everything formed by atoms—nearly everything you encounter in the world.

Matter exists in three forms: (1) solids, (2) liquids, and (3) gases. Solids, liquids, and gases may exist in pure form, may combine chemically with other elements to form compounds, or may be mixed together without chemically combining to form mixtures.

C2-1. Pure Elements

As discussed previously, elements are matter built up from subatomic particles, with properties determined by their nucleus (protons and neutrons) and by their electron shells. All of the atoms of an element have the same number of protons in their nuclei. Elements do not break down into simpler elements. There are over 100 elements known; 92 occur naturally, and others have been produced in the laboratory. Elements important in water chemistry are listed in Table 1.

A few elements exist in pure form. Carbon is an example—a diamond is pure carbon in a particular arrangement. Oxygen in the air is another example of an element in its pure form. However, most elements are unstable and are usually found combined with other elements in the form of compounds.

C2-2. Compounds

COMPOUNDS are two or more elements that are "stuck" (bonded) together by a chemical reaction (explained in the next module). A compound can only be broken down into its original elements by reversing the chemical reaction. The

Table 1. Elements Important in Water Treatment

Element	Symbol	Element	Symbol	Element	Symbol
Aluminum	Al	Chromium	Cr	Oxygen	O
Arsenic*	As	Fluorine†	F	Phosphorus	P
Barium*	Ba	Hydrogen	H	Potassium	K
Boron	B	Iodine	I	Radium*	Ra
Bromine	Br	Iron	Fe	Selenium*	Se
Cadmium*	Cd	Lead*	Pb	Silicon	Si
Calcium	Ca	Magnesium	Mg	Silver*	Ag
Carbon	C	Manganese	Mn	Sodium	Na
Chlorine	Cl	Mercury*	Hg	Strontium*	Sr
Copper	Cu	Nitrogen	N	Sulfur	S

*This element must be monitored according to the requirements of the Safe Drinking Water Act.

†Fluoride, an anion of the element flourine, must be monitored according to the requirements of the Safe Drinking Water Act.

weight of the atoms of any one element in a compound is always a definite fraction (or proportion of the weight of the entire compound). For example, in any given weight of water, $2/18$ of the weight is atoms of hydrogen.

When atoms of two or more elements are bonded together to form a compound, the resulting particle is called a MOLECULE. A molecule may be only two atoms of one or more elements bonded together; or it may be dozens of atoms bonded together, and may consist of several elements. For example, when two atoms of hydrogen and one of oxygen combine, a molecule of water is formed. When one atom of carbon and two of oxygen combine, a molecule of carbon dioxide is formed. When two atoms of oxygen combine, a molecule of oxygen is formed.

Other examples of compounds are

- Salt (sodium and chlorine),
- Sulphuric acid (hydrogen, sulphur, and oxygen),
- Ammonia (nitrogen and hydrogen),
- Rust (iron and oxygen),
- Lime (calcium, oxygen, and hydrogen), and
- Sand (silicon and oxygen).

The number of compounds that can be formed by chemical reaction between elements is enormous. Well over two million compounds have been identified by chemists, and the number is still increasing.

C2-3.   Mixtures

When two or more elements, compounds, or both, are mixed together and no chemical reaction (bonding between individual particles) occurs, then the result is a MIXTURE. No new compounds are formed, and the elements or compounds may be mixed in any proportion. Any mixture can be separated into its original elements or compounds by "physical" means, such as filtering, settling, or distillation. For example, a mixture of salt water can be separated into its compounds, salt and water, by the process of distillation—heating the mixture causes the water to evaporate, leaving the salt behind.

Other examples of mixtures are

- Air (mostly oxygen, carbon dioxide, and nitrogen),
- Glass (sand, various metals, and borax),
- Steel (primarily iron and carbon), and
- Concrete (lime, sand, and water).

It is helpful to think of the differences between elements, compounds, and mixtures by considering Figure 4.

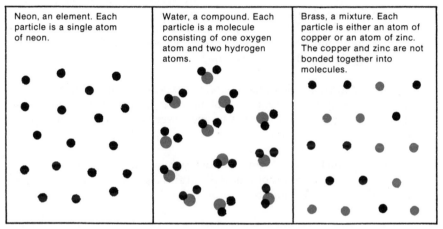

Figure 4.   Models of an element, a compound, and a mixture

Review Questions

1. Give two examples of elements that exist in pure form.

2. Define (a) Compound, (b) Mixture.

3. What is a molecule? Give one example.

4. Give two examples of compounds and two examples of mixtures.

Summary Answers

(See Detailed Answers)

Detailed Answers

1. Oxygen and carbon.

2. (a) A compound is a stable combination of two or more elements in definite proportion by weight. The elements are bonded (stuck together) by chemical reaction.
   (b) A mixture is a blend of two or more elements or compounds, in no definite proportion by weight. The mixed elements or compounds are not bonded by chemical reaction.

3. A molecule is the bonded union of two or more elements. Molecules may be composed of only two atoms, or may be composed of hundreds of atoms. A water molecule consists of one oxygen atom and two hydrogen atoms.

4. Compounds include salt, sand, water, carbon dioxide, ammonia, rust, lime, sand, and thousands of others. Mixtures include concrete, air, glass, steel, salt water, and thousands of others.

Chemistry 3

Valence, Chemical Formulas, and Chemical Equations

As explained in the previous section, atoms of elements can combine to form molecules of compounds. Experience has shown that only certain combinations of atoms will react (bond) together. For example, two atoms of iron will bond to three atoms of oxygen to form a molecule of ferric oxide (rust), but atoms of iron will not bond to atoms of magnesium. Iron and magnesium can be blended into a mixture—an alloy—but not combined into a compound.

It has also been found through experiment that the number of atoms of each element in a molecule is very definite. A water molecule is formed of two hydrogen atoms and one oxygen atom—no other combination of hydrogen and oxygen makes water, and other compounds of hydrogen and oxygen can be formed only under special circumstances. Similarly, exactly one atom of hydrogen is required to combine with one atom of chlorine to form a molecule of hydrogen chloride.

The following paragraphs contain a brief discussion of why only certain molecules occur, and of how chemists describe the characteristics of an atom that determine which chemical combinations it can enter into.

C3-1. Valence

In nature, atoms of elements tend to form molecules whenever the molecule is more chemically stable than the individual element. The number of electrons on the very outside of each atom (in the outermost shell) is the most important factor in determining which atoms will combine with which other atoms to form greater stability. The electrons in the outermost shell are called the VALENCE ELECTRONS.

Based on experience, chemists have assigned to every element in the periodic table one or more numbers, indicating the ability of the element to react with other elements. The numbers, which depend on the number of valence electrons, are called the VALENCES of the element.

In the formation of chemical compounds from the elements, the valence electrons are transferred from the outer shell of one atom to the outer shell of another atom, or they are shared among the outer shells of the combining atoms. When electrons are transferred the process is called ionic bonding. If electrons are shared it is called covalent bonding. This rearrangement of electrons produces CHEMICAL BONDS. The actual number of electrons that an atom gains, loses, or shares in bonding with one or more other atoms is the valence of the atom. For example, if an atom gives away one electron in a reaction, then it has a valence of +1. Similarly, if an atom must gain one electron to complete a reaction, then it has a valence of −1.

The following example of ionic bonding (transfer of electrons) shows how valence works. A diagram of the sodium and chlorine atoms and how they react to form sodium chloride (NaCl) is shown in Figure 5. Sodium has an atomic number of 11, indicating that it has 11 protons in a nucleus surrounded by 11 electrons. As illustrated, there is only one electron in the outermost shell, or ring. Chlorine, with an atomic number of 17, has 17 protons in the nucleus, surrounded by 17 electrons; 7 of the chlorine electrons are in the outermost shell. As the diagram indicates, to form a molecule of sodium chloride, the sodium atom transfers one electron to the chlorine atom. The molecule has greater chemical stability than the separate elements, so when sodium and chlorine are mixed together (under conditions that make the electron transfer possible), sodium chloride molecules will form. *CAUTION: Mixing pure sodium with chlorine will cause a violent explosion.*

Covalent bonding is a process similar to ionic bonding, but the electrons are shared rather than transferred, as illustrated for hydrogen chloride in Figure 6. Since electrons are not lost or gained, the valence of an atom involved in a covalent bond is expressed without a + or − sign. In the example, the valence of

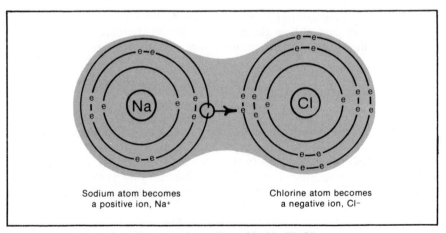

Sodium atom becomes
a positive ion, Na+

Chlorine atom becomes
a negative ion, Cl−

Figure 5.   Ionic bonding, illustrated by sodium chloride (NaCl)

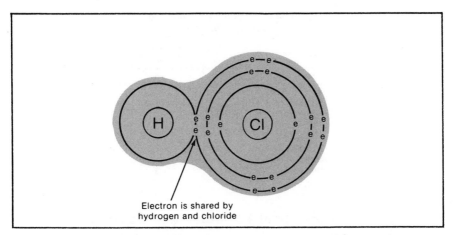

Figure 6.   Covalent bonding, illustrated by hydrogen chloride (HCl)

both hydrogen and chlorine is 1, since the bond is formed by sharing a single electron.

Many elements (iron and copper, for example) have more than one valence. The number of electrons involved in a reaction—the valence number—depends on several factors, such as the conditions under which the reaction occurs and the other elements involved. Iron has a valence of +2 when it forms $FeSO_4$, ferrous sulfate; but it has a valence of +3 when it forms Fe_2O_3, common red rust.

Some groups of elements bond together and act like single atoms or ions in forming compounds. Such groups of elements are called RADICALS. For example, the sulfate ion (SO_4^{-2}) and the nitrate ion (NO_3^{-1}) are important radicals in drinking water quality. The valences of some common elements and radicals are listed in Tables 2 and 3.

Table 2.   Oxidation Numbers of Various Elements

Element	Common Valences	Element	Common Valences
Aluminum (Al)	+3	Lead (Pb)	+2, +4
Arsenic (As)	+3, +5	Magnesium (Mg)	+2
Barium (Ba)	+2	Manganese (Mn)	+2, +4
Boron (B)	+3	Mercury (Hg)	+1, +2
Bromine (Br)	−1	Nitrogen (N)	+3, −3, +5
Cadmium (Cd)	+2	Oxygen (O)	−2
Calcium (Ca)	+2	Phosphorus (P)	−3
Carbon (C)	+4, −4	Potassium (K)	+1
Chlorine (Cl)	−1	Radium (Ra)	+2
Copper (Cu)	+1, +2	Selenium (Se)	−2, +4
Chromium (Cr)	+3	Silicon (Si)	+4
Fluorine (F)	−1	Silver (Ag)	+1
Hydrogen (H)	+1	Sodium (Na)	+1
Iodine (I)	−1	Strontium (Sr)	+2
Iron (Fe)	+2, +3	Sulfur (S)	−2, +4, +6

**Table 3. Oxidation Numbers of
Common Radicals**

Radical	Common Valences
Ammonium (NH_4)	+1
Bicarbonate (HCO_3)	−1
Hydroxide (OH)	−1
Nitrate (NO_3)	−1
Nitrite (NO_2)	−1
Carbonate (CO_3)	−2
Sulfate (SO_4)	−2
Sulfite (SO_3)	−2
Phosphate (PO_4)	−3

C3-2. Chemical Formulas and Equations

A group of chemically bonded atoms forms a particle called a molecule. The simplest molecules contain only one type of atom, such as when two atoms of oxygen combine (O_2) or when two atoms of chlorine combine (Cl_2). Molecules of compounds are made up of the atoms of at least two different elements; for example, one oxygen atom and two hydrogen atoms form a molecule of the compound water (H_2O). "H_2O" is called the FORMULA of water. The formula is a shorthand way of writing *what* elements are present in a molecule of a compound, and *how many* atoms of each element are present in each molecule.

Reading Chemical Formulas

The following are examples of chemical formulas and what they indicate.

Example 1
The chemical formula for calcium carbonate is

$$CaCO_3$$

According to the formula, what is the chemical makeup of the compound?

First, the letter symbols given in the formula indicate the three elements that make up the calcium carbonate compound:

$$Ca = calcium$$
$$C = carbon$$
$$O = oxygen$$

Second, the subscripts (numbers) in the formula indicate how many atoms of each element are present in a single molecule of the compound. There is no number just to the right of the Ca or C symbols; this indicates that only one atom of each is present in the molecule. The subscript *3* to the right of the 0 symbolizing oxygen indicates that there are three oxygen atoms in each molecule.

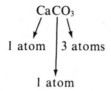

$$CaCO_3$$

1 atom   3 atoms

1 atom

Example 2

The chemical formula for sulfuric acid is H_2SO_4. On the basis of this formula, what can be said about the chemical makeup of the compound?

The chemical symbols in the formula indicate that sulfuric acid contains three elements:

$$H = hydrogen$$
$$S = sulfur$$
$$O = oxygen$$

The formula also indicates the number of atoms of each element present:

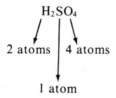

$$H_2SO_4$$

2 atoms   4 atoms

1 atom

In some chemical formulas you will see a subscript just outside a parentheses, as shown in the two following examples:

$$Ca(HCO_3)_2 \qquad Al_2(SO_4)_3$$

Notice that in both examples *radicals* are inside the parentheses. As explained previously, radicals are a group of atoms bonded together into a unit and acting as a single atom (ion). The parentheses and subscripts are used to indicate how many of these units are involved in the reaction. In the case of $Ca(HCO_3)_2$, one calcium atom reacts with two bicarbonate units (ions):

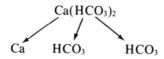

$$Ca(HCO_3)_2$$

Ca HCO_3 HCO_3

Therefore, the number of atoms of each element in a molecule of the compound are as follows:

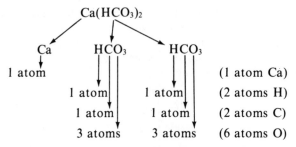

Notice that there are 6 atoms of oxygen, because there are 3 atoms of oxygen present in each bicarbonate radical (HCO_3), and there are 2 bicarbonate radicals present in each molecule of $Ca(HCO_3)_2$. Multiplying the subscripts, $3 \times 2 = 6$.

In the case of $Al_2(SO_4)_3$, two aluminum atoms react with three sulfate ions:

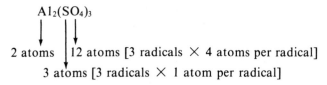

Therefore, the number of atoms of each element in a molecule of the compound is

$Al_2(SO_4)_3$

2 atoms | 12 atoms [3 radicals $\times$ 4 atoms per radical]

3 atoms [3 radicals $\times$ 1 atom per radical]

Example 3

The chemical formula for calcium hydroxide (lime) is $Ca(OH)_2$. Determine the number of atoms of each element in a molecule of the compound.

The formula indicates that one atom of calcium reacts with two hydroxyl ions:

$Ca(OH)_2$

Ca   OH   OH

The number of atoms of each element is

$Ca(OH)_2$

1 atom | 2 atoms

2 atoms

Determining Percent by Weight of Elements in a Compound

If 100 lb of sodium chloride (NaCl) were separated into the elements that make up the compound, there would be 39.3 lb of pure sodium (Na) and 60.7 lb of pure chlorine (Cl). We say that sodium chloride is 39.3 *percent* sodium *by weight* and that it is 60.7 *percent* chlorine *by weight.*[1] The PERCENT BY WEIGHT of each element in a compound can be calculated using the compound's chemical formula and atomic weights from the periodic table.

The first step in calculating percent by weight of an element in a compound is to determine the MOLECULAR WEIGHT (sometimes called FORMULA WEIGHT) of the compound. The molecular weight of a compound is defined as the sum of the atomic weights of all the atoms in the compound.

For example, to determine the molecular weight of sodium chloride, first count how many atoms of each element a single molecule contains:

$$\text{Na} \qquad \text{Cl}$$

$$1 \text{ atom} \qquad 1 \text{ atom}$$

Next, find the atomic weight of each atom, using the periodic table:

$$\text{Atomic weight of Na} = 22.99$$

$$\text{Atomic weight of Cl} = 35.45$$

Finally, multiply each atomic weight by the number of atoms of that element in the molecule, and total the weights:

	No. of atoms	*Atomic weight*	*Total weight*
Sodium (Na)	1	$\times$ 22.99 =	22.99
Chlorine (Cl)	1	$\times$ 35.45 =	35.45
Molecular weight of NaCl =			58.44

Once the molecular weight of a compound is determined, the percent by weight of each element in the compound can be found with the following formula:

$$\text{Percent element by weight} = \frac{\text{Weight of element in compound}}{\text{Molecular weight of compound}} \times 100$$

Using the formula, first calculate the percent by weight of sodium in the compound:

$$\text{Percent Na by weight} = \frac{\text{Weight of Na in compound}}{\text{Molecular weight of compound}} \times 100$$

$$= \frac{22.99}{58.44} \times 100$$

$$= 0.393 \times 100$$

$$= 39.3\% \text{ sodium by weight}$$

---

[1]Mathematics Section, Percent.

Then, calculate percent by weight of chlorine in the compound:

$$\text{Percent Cl by weight} = \frac{\text{Weight of Cl in compound}}{\text{Molecular weight of compound}} \times 100$$

$$= \frac{35.45}{58.44} \times 100$$

$$= 0.607 \times 100$$

$$= 60.7\% \text{ chlorine by weight}$$

To check the calculations, add the percents—the total should be 100:

$$\begin{array}{r} 39.3\% \text{ Na} \\ + \ 60.7\% \text{ Cl} \\ \hline 100.0\% \text{ NaCl} \end{array}$$

Example 4

The formula for calcium carbonate is $CaCO_3$. What percent by weight of the compound is calcium, what percent by weight is carbon, and what percent by weight is oxygen?

To calculate the percent by weight of each of the elements, first find the atomic weight of each element in the compound, then calculate the molecular weight of $CaCO_3$.

	No. of atoms		Atomic weight		Total weight
Calcium (Ca)	1	×	40.08	=	40.08
Carbon (C)	1	×	12.01	=	12.01
Oxygen (O)	3	×	16.00	=	48.00
Molecular weight of $CaCO_3$				=	100.09

Now calculate the percent by weight for each element:

$$\text{Percent Ca by weight} = \frac{\text{Weight of Ca in compound}}{\text{Molecular weight of compound}} \times 100$$

$$= \frac{40.08}{100.09} \times 100$$

$$= 0.40 \times 100$$

$$= 40\% \text{ Ca by weight}$$

$$\text{Percent C by weight} = \frac{\text{Weight of C in compound}}{\text{Molecular weight of compound}} \times 100$$

$$= \frac{12.01}{100.09} \times 100$$

$$= 0.12 \times 100$$

$$= 12\% \text{ C by weight}$$

$$\text{Percent O} \atop \text{by weight} = \frac{\text{Weight of O in compound}}{\text{Molecular weight of compound}} \times 100$$

$$= \frac{48.00}{100.09} \times 100$$

$$= 0.48 \times 100$$

$$= 48\% \text{ O by weight}$$

Check the calculations by making sure the sum of the percents is 100:

$$\begin{array}{r} 40\% \text{ Ca} \\ 12\% \text{ C} \\ + \ 48\% \text{ O} \\ \hline 100\% \text{ CaCO}_3 \end{array}$$

Example 5

The formula for calcium bicarbonate is $Ca(HCO_3)_2$. What percent of the weight of this compound is calcium? What are the percents of hydrogen, carbon, and oxygen?

To calculate the percent of the weight represented by each element, first determine the molecular weight of the compound:

	No. of atoms		Atomic weight		Total weight
Calcium (Ca)	1	×	40.08	=	40.08
Hydrogen (H)	2	×	1.01	=	2.02
Carbon (C)	2	×	12.01	=	24.02
Oxygen (O)	6	×	16.00	=	96.00

Molecular weight of $Ca(HCO_3)_2$ = 162.12

Now calculate the percent weight represented by each of the elements:

$$\text{Percent Ca} \atop \text{by weight} = \frac{\text{Weight of Ca in compound}}{\text{Molecular weight of compound}} \times 100$$

$$= \frac{40.08}{162.12} \times 100$$

$$= 0.247 \times 100$$

$$= 24.7\% \text{ Ca by weight}$$

$$\text{Percent H by weight} = \frac{\text{Weight of H in compound}}{\text{Molecular weight of compound}} \times 100$$

$$= \frac{2.02}{162.12} \times 100$$

$$= 0.012 \times 100$$

$$= 1.2\% \text{ H by weight}$$

$$\text{Percent C by weight} = \frac{\text{Weight of C in compound}}{\text{Molecular weight of compound}} \times 100$$

$$= \frac{24.02}{162.12} \times 100$$

$$= 0.148 \times 100$$

$$= 14.8\% \text{ C by weight}$$

$$\text{Percent O by weight} = \frac{\text{Weight of O in compound}}{\text{Molecular weight of compound}} \times 100$$

$$= \frac{96.00}{162.12} \times 100$$

$$= 0.592 \times 100$$

$$= 59.2\% \text{ O by weight}$$

Checking these calculations shows the sum of the percents equal to 99.9% instead of 100%. This slight difference is due to rounding of the answers.

Once you have calculated the percent composition by weight, if you know how many pounds of the chemical you have, then you can calculate the actual weight of any element present. The following examples illustrate the procedure.

Example 6
If you have 50 lb of $Ca(HCO_3)_2$, then how many pounds of each element in the compound do you have? (Use the percent information determined in Example 5: Ca = 24.7%; H = 1.2%; C = 14.8%; and O = 59.3%.)

The percents given indicate what percent of the total weight each element makes up. For example, 24.7% of the total weight of the compound is calcium. This means that 24.7% of the 50 lb of $Ca(HCO_3)_2$ is calcium:

$$[0.247][50 \text{ lb } Ca(HCO_3)_2] = 12.4 \text{ lb calcium}$$

Similar calculations may be made for the other elements:

$$[0.012][50 \text{ lb Ca(HCO}_3)_2] = 0.6 \text{ lb hydrogen}$$

$$[0.148][50 \text{ lb Ca(HCO}_3)_2] = 7.4 \text{ lb carbon}$$

$$[0.593][50 \text{ lb Ca(HCO}_3)_2] = 29.7 \text{ lb oxygen}$$

The total of these numbers should be 50 lb. In this case the total is 50.1 lb due to rounding.

Example 7

Suppose you have 75 lb of sodium carbonate (Na_2CO_3). How many pounds of sodium, carbon, and oxygen are in this much sodium carbonate?

To determine the weight of each element, you will first have to determine *what percent* of the total weight is represented by each element. First find the molecular weight:

	No. of atoms	Atomic weight	Total weight
Sodium (Na)	2 $\times$	22.99 =	45.98
Carbon (C)	1 $\times$	12.01 =	12.01
Oxygen (O)	3 $\times$	16.00 =	48.00

Molecular weight of Na_2CO_3 = 105.99

Next calculate the percent composition by weight for each element:

$$\frac{\text{Percent Na}}{\text{by weight}} = \frac{\text{Weight of Na in compound}}{\text{Molecular weight of compound}} \times 100$$

$$= \frac{45.98}{105.99} \times 100$$

$$= 0.434 \times 100$$

$$= 43.4\% \text{ Na by weight}$$

$$\frac{\text{Percent C}}{\text{by weight}} = \frac{\text{Weight of C in compound}}{\text{Molecular weight of compound}} \times 100$$

$$= \frac{12.01}{105.99} \times 100$$

$$= 0.113 \times 100$$

$$= 11.3\% \text{ C by weight}$$

$$\begin{aligned} \text{Percent O} \atop \text{by weight} &= \frac{\text{Weight of O in compound}}{\text{Molecular weight of Na}} \times 100 \\ &= \frac{48.00}{105.99} \times 100 \\ &= 0.453 \times 100 \\ &= 45.3\% \text{ O by weight} \end{aligned}$$

Now that the percent of the weight represented by each element has been calculated, determine how much of the 75 lb is made up by each element:

43.4% Sodium: $(0.434)(75 \text{ lb Na}_2\text{CO}_3) = 32.6$ lb sodium

11.3% Carbon: $(0.113)(75 \text{ lb Na}_2\text{CO}_3) = 8.5$ lb carbon

45.3% Oxygen: $(0.453)(75 \text{ lb Na}_2\text{CO}_3) = 34.0$ lb oxygen

Check the calculation by adding together the weights of the individual elements—the sum is 75.1 lb, differing from the 75 lb total because of rounding.

Chemical Equations

A CHEMICAL EQUATION is a shorthand way, using chemical formulas, of writing the reaction that takes place when certain chemicals are brought together. As shown in the following example, the left side of the equation indicates the REACTANTS, or chemicals that will be brought together; the arrow indicates which direction the reaction occurs; and the right side of the equation indicates the PRODUCTS, or results, of the chemical reaction.

calcium bicarbonate	*plus*	calcium hydroxide	*react to form*	calcium carbonate	*plus* water
$Ca(HCO_3)_2$ + $Ca(OH)_2$			$\longrightarrow$	$2CaCO_3$ + $2H_2O$	
REACTANTS				PRODUCTS	

The 2 in front of $CaCO_3$ is called a *coefficient*. A coefficient indicates the relative number of molecules of the compound which are involved in the chemical reaction. If no coefficient is shown, then only one molecule of the compound is involved. For example, in the equation above, one molecule of calcium bicarbonate reacts with one molecule of calcium hydroxide to form two molecules of calcium carbonate and two molecules of water. Without the coefficients, the equation could be written

$$Ca(HCO_3)_2 + Ca(OH)_2 \longrightarrow CaCO_3 + CaCO_3 + H_2O + H_2O$$

If you count the atoms of calcium (Ca) on the left side of the equation, and then count the ones on the right side, you will find that the numbers are the same. In fact, for each element in the equation, as many atoms are shown on the left side as on the right. An equation for which this is true is said to be BALANCED. A balanced equation accurately represents what really happens in a chemical reaction: because matter is neither created nor destroyed, the number of atoms of each element going into the reaction must be the same as the number coming out. Coefficients allow balanced equations to be written compactly.

Coefficients and subscripts can be used to calculate the molecular weight of each term in an equation, as illustrated in the following example.

Example 8

Calculate the molecular weights for each of the four terms in the following equation:

$$Ca(HCO_3)_2 + Ca(OH)_2 \longrightarrow 2CaCO_3 + 2H_2O$$

First, calculate the molecular weight of $Ca(HCO_3)_2$:

	No. of atoms		Atomic weight		Total weight
Calcium (Ca)	1	×	40.08	=	40.08
Hydrogen (H)	2	×	1.01	=	2.02
Carbon (C)	2	×	12.01	=	24.02
Oxygen (O)	6	×	12.00	=	96.00

Molecular weight of $Ca(HCO_3)_2$ = 162.12

The molecular weight for $Ca(OH)_2$ is

	No. of atoms		Atomic weight		Total weight
Calcium (Ca)	1	×	40.08	=	40.08
Oxygen (O)	2	×	16.00	=	32.00
Hydrogen (H)	2	×	1.01	=	2.02

Molecular weight for $Ca(OH)_2$ = 74.10

The 2 in front of the next term of the equation $(2CaCO_3)$ indicates that two molecules of $CaCO_3$ are involved in the reaction. First find the weight of *one molecule*, then find the weight of *two molecules*:

	No. of atoms		Atomic weight		Total weight
Calcium (Ca)	1	×	40.08	=	40.08
Carbon (C)	1	×	12.01	=	12.01
Oxygen (O)	3	×	16.00	=	48.00

Weight of one molecule $CaCO_3$ = 100.09

$$\text{Weight of two molecules } CaCO_3 = (2)(100.09)$$
$$= 200.18$$

The coefficient in front of the fourth term in the equation ($2H_2O$) also indicates that two molecules are involved in the reaction. As in the last calculation, first determine the weight of one molecule of H_2O, then the weight of two molecules:

	No. of atoms	Atomic weight	Total weight
Hydrogen (H)	2	× 1.01 =	2.02
Oxygen (O)	1	× 16.00 =	16.00

$$\text{Weight of one molecule } H_2O = 18.02$$
$$\text{Weight of two molecules } H_2O = (2)(18.02)$$
$$= 36.04$$

In summary, the weights that correspond to each term of the equation are

$$Ca(HCO_3)_2 + Ca(OH)_2 \longrightarrow 2CaCO_3 + 2H_2O$$
$$162.12 \qquad 74.10 \qquad 200.18 \quad 36.04$$

Notice that the total weight on the left side of the equation (236.22) is equal to the total weight on the right side of the equation (236.22).

The practical importance of the weights of each term of the equation is that the chemicals shown in the equation will always react in the proportions indicated by their weights.

For example, from the calculation above you know that $Ca(HCO_3)_2$ reacts with $Ca(OH)_2$ in the ratio 162.12:74.10. This means that, given 162.12 lb of $Ca(OH)_2$, you must add 74.10 lb of $Ca(HCO_3)_2$ for a complete reaction. Given twice the amount of $Ca(HCO_3)_2$ (that is, 324.24 lb), you must add twice the amount of $Ca(OH)_2$ (equal to 148.20 lb) to achieve complete reaction. The next two examples illustrate more complicated calculations using the same principle.

Example 9

If 25 g of $Ca(OH)_2$ were added to some $Ca(HCO_3)_2$, how many grams of $Ca(HCO_3)_2$ would react with the $Ca(OH)_2$? The molecular weight of $Ca(HCO_3)_2$ is 162.12; of $Ca(OH)_2$, 74.10.

The molecular weights and the chemical equation indicate the weight ratio in which the two compounds will react:

$$Ca(HCO_3)_2 + Ca(OH)_2$$
$$162.12 \qquad 74.10$$

The equation for the reaction is

$$Ca(HCO_3)_2 + Ca(OH)$$

Use this information to set up a proportion[2] in order to determine how many grams of $Ca(HCO_3)_2$ will react with the $Ca(OH)_2$:

Known ratio	*Desired ratio*
$\dfrac{74.10 \text{ grams } Ca(OH)_2}{162.12 \text{ grams } Ca(HCO_3)_2}$ =	$\dfrac{25 \text{ grams } Ca(OH)_2}{x \text{ grams } Ca\ (HCO_3)_2}$

Next solve for the unknown value[3]

$$\frac{74.10}{162.12} = \frac{25}{x}$$

$$\frac{(x)(74.10)}{162.12} = 25$$

$$x = \frac{(25)(162.12)}{74.10}$$

$$x = 54.7 \text{ g } Ca(HCO_3)_2$$

Given the molecular weights and the chemical equation indicating the ratio by which the two chemicals would combine, we were able to calculate that 54.7 g of $Ca(HCO_3)_2$ would react with 25 g of $Ca(OH)_2$.

Example 10

The equation of the reaction between calcium carbonate ($CaCO_3$) and carbonic acid (H_2CO_3) is shown. If 10 lb of H_2CO_3 are to be used in the reaction, how many pounds of $CaCO_3$ will react with the H_2CO_3?

$$CaCO_3 + H_2CO_3 \longrightarrow Ca(HCO_3)_2$$

To determine how many pounds of $CaCO_3$ will react with the H_2CO_3, first determine the weight *ratios* in the reaction:

CaCO₃:	*No. of atoms*		*Atomic weight*		*Total weight*
Calcium (Ca)	1	×	40.08	=	40.08
Carbon (C)	1	×	12.01	=	12.01
Oxygen (O)	3	×	16.00	=	48.00

Molecular weight of $CaCO_3$ = 100.09

---

[2] Mathematics Section, Ratios and Proportions.

[3] Mathematics Section, Solving for the Unknown Value.

H_2CO_3:

	No. of atoms		Atomic weight		Total weight
Hydrogen (H)	2	$\times$	1.01	=	2.02
Carbon (C)	1	$\times$	12.01	=	12.01
Oxygen (O)	3	$\times$	16.00	=	48.00

Molecular weight of H_2CO_3 =   62.03

The reacting weight ratios of $CaCO_3$ and H_2CO_3 are

$$CaCO_3 + H_2CO_3 \longrightarrow Ca(HCO_3)_2$$
$$100.09 \quad\quad 62.03$$

Now set up a proportion to solve for the amount of $CaCO_3$ which will react with 10 lb of H_2CO_3:

Known Ratio		Desired Ratio
$\dfrac{100.09 \text{ lbs } CaCO_3}{62.03 \text{ lbs } H_2CO_3}$	=	$\dfrac{x \text{ lbs } CaCO_3}{10 \text{ lbs } H_2CO_3}$

And solve for the unknown value:

$$\frac{100.09}{62.03} = \frac{x}{10}$$

$$\frac{(10)(100.09)}{62.03} = x$$

$$16.1 \text{ lb } CaCO_3 = x$$

A list of compounds and chemical equations common to water treatment is given in Appendix C.

Definition of Mole

You may sometimes find chemical reactions described in terms of MOLES of a substance reacting. The measurement "mole" (an abbreviation for "gram-mole") is closely related to molecular weight. The molecular weight of water, for example, is 18.02—and one mole of water is defined to be 18.02 grams of water. The general definition of a mole is as follows:

A mole of a substance is a number of grams of that substance, where the number equals the substance's molecular weight.

In Example 8 you saw that the following equation and molecular weights were correct:

$$Ca(HCO_3)_2 + Ca(OH)_2 \longrightarrow 2CaCO_3 + 2H_2O$$
$$162.12 \quad\quad 74.10 \quad\quad 2(100.09) \quad 2(18.02)$$

If 162.12 grams of $Ca(HCO_3)_2$ were used in the reaction, then the ratio equations given in Example 9 would show that the weights of each of the substances in the reaction were

$$Ca(HCO_3)_2 \;+\; Ca(OH)_2 \;\longrightarrow\; 2CaCO_3 \;+\; 2H_2O$$

| 162.12 g | 74.10 g | (2)(100.09) g | (2)(18.02) g |

Because of the way a mole is defined, this could also be written in the more compact form:

$$Ca(HCO_3)_2 \;+\; Ca(OH)_2 \;\longrightarrow\; 2CaCO_3 \;+\; 2H_2O$$

| 1 mole | 1 mole | 2 moles | 2 moles |

Reading this information, a chemist could state that, "one mole of $Ca(HCO_3)_2$ is needed to react with one mole of $Ca(OH)_2$, and the reaction yields two moles of $CaCO_3$ and two moles of water."

When measuring chemicals in moles, always remember that *the weight of a mole of a substance depends on what the substance is.* One mole of water weighs 18.02 g—but one mole of calcium carbonate weighs 100.09 g.

Example 11

A lab procedure calls for 3.0 moles of sodium bicarbonate ($NaHCO_3$) and 0.10 mole of potassium chromate (K_2CrO_4). How many grams of each compound are required?

To find grams required of $NaHCO_3$, first determine the weight of 1 mole of the compound:

	No. of atoms		Atomic weight		Total weight
Sodium (Na)	1	×	22.99	=	22.99
Hydrogen (H)	1	×	1.01	=	1.01
Carbon (C)	1	×	12.01	=	12.01
Oxygen (O)	3	×	16.00	=	48.00
Molecular weight of $NaHCO_3$ =					84.01

Therefore, 1 mole of $NaHCO_3$ weighs 84.01 g.

Next, multiply the weight of 1 mole by the number of moles required:

3 moles of $NaHCO_3$ are required.

3 moles $NaHCO_3$ weighs (3)(84.01 g) = 252.03 g.

To find grams required of K_2CrO_4, first determine the weight of 1 mole of the compound:

	No. of atoms	Atomic weight		Total weight
Potassium (K)	2	× 39.10	=	78.20
Chromium (Cr)	1	× 52.00	=	52.00
Oxygen (O)	4	× 16.00	=	64.00

Molecular weight of K_2CrO_4 = 194.20

Therefore, 1 mole of K_2CrO_4 weighs 194.20 g.

Next, multiply the weight of 1 mole by the number of moles required:

0.10 mole of K_2CrO_4 required.

0.10 mole K_2CrO_4 weighs $(0.10)(194.20 \text{ g}) = 19.42 \text{ g}$

Review Questions

1. What is a radical?

2. For the following chemical formulas, list the elements present and indicate how many atoms of each element are present in one molecule of each compound:
   (a) HOCl   (d) $Ca(OCl)_2$
   (b) NH_3 (e) $Al_2(OH)_6$
   (c) $CaCO_3$ (f) $Al_2(SO_4)_3$

3. The formula for calcium hydroxide is $Ca(OH)_2$. Calcium represents what percent of the weight of the compound? What percent is represented by calcium and hydrogen? (Use the periodic table in the appendix to find atomic weights.)

4. The formula for ammonia is NH_3. What percent (by weight) of the compound is represented by nitrogen? What percent by hydrogen? (Use the periodic table in the appendix to find atomic weights.)

5. The percent composition by weight for each of the elements present in $CaSO_4$ is Ca = 29.4%; S = 23.6%; 0 = 47%. If you had 150 lb of $CaSO_4$, how many pounds of each element would you have?

6. How many pounds of calcium, carbon, and oxygen are there in 60 lb of pure calcium carbonate? (Use the periodic table in the appendix to find atomic weights.)

7. What is the molecular weight for each of the four terms in the following equation. (Use the periodic table in the appendix to find atomic weights.)

$$4NH_3 + 3O_2 \qquad 2N_2 + 6H_2O$$

8. The equation for the reaction between calcium hypochlorite $(Ca(OCl)_2)$ and sodium carbonate (Na_2CO_3) is

$$Ca(OCl)_2 + Na_2CO_3 \longrightarrow 2NaOCl + CaCO_3$$

If 20 lb of $Ca(OCl)_2$ are to be used in the reaction, how many pounds of Na_2CO_3 will be used? (Molecular weights: $Ca(OCl)_2$ = 142.98; Na_2CO_3 = 105.99; NaOCL = 74.44; $CaCO_3$ = 100.09).

9. Find the weight in grams of each of the following:
   (a) 0.015 moles sodium arsenite, $NaAsO_2$
   (b) 2.5 moles zinc sulfate, $ZnSO_4$
   (c) 0.001 mole potassium chloroplatinate, K_2PtCl_6

Summary Answers

1. A group of atoms which bond together and behave like a single ion.

2. (a) $HOCl$: hydrogen = 1 atom; oxygen = 1 atom; chlorine = 1 atom
   (b) NH_3: nitrogen = 1 atom; hydrogen = 3 atoms
   (c) $CaCO_3$: calcium = 1 atom; carbon = 1 atom; oxygen = 3 atoms
   (d) $Ca(OCl)_2$: calcium = 1 atom; oxygen = 2 atoms; chlorine = 2 atoms
   (e) $Al_2(OH)_6$: aluminum = 2 atoms; oxygen = 6 atoms; hydrogen = 6 atoms
   (f) $Al_2(SO_4)_3$: aluminum = 2 atoms; sulfur = 3 atoms; oxygen = 12 atoms

3. 54.1% calcium; 43.2% oxygen; 2.7% hydrogen

4. 82.2% nitrogen; 17.8% hydrogen

5. 44.1 lb calcium; 35.4 lbs sulfur; 70.5 lb oxygen

6. 24 lb calcium; 7.2 lb carbon; 28.8 lb oxygen

7. $4NH_3 + 3O_2 \longrightarrow 2N_2 + 6H_2O$
   68.16   96.00   56.04   108.12

8. 14.8 lb Na_2CO_3

9. (a) 1.95 g
   (b) 403.58 g
   (c) 0.49 g

Detailed Answers

1. A group of atoms which bond together and behave as a single ion.

2. $HOCl$: The elements in the compound are

$$H = \text{hydrogen}$$
$$O = \text{oxygen}$$
$$Cl = \text{chlorine}$$

And the number of atoms of each element which make up the compound are

HOCl

1 atom | 1 atom

1 atom

NH_3: The elements in the compound are

N = nitrogen

H = hydrogen

The number of atoms of each element in the compound are

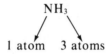

1 atom   3 atoms

$CaCO_3$: The elements in the compound are

Ca = calcium

C = carbon

O = oxygen

The number of atoms of each element in the compound are

CaCO₃

1 atom | 3 atoms

1 atom

$Ca(OCl)_2$: The elements in the compound are

Ca = calcium

O = oxygen

Cl = chlorine

The formula for this chemical compound involves a radical (hypochlorite). Another way of expressing the formula is

$$Ca(OCl)_2 = Ca\ OCl\ OCl$$

Therefore, the number of atoms of each element in the compound is

Ca(OCl)₂

1 atom | 2 atoms

2 atoms

$Al_2(OH)_6$: The elements in the compound are

$$Al = aluminum$$
$$O = oxygen$$
$$H = hydrogen$$

This compound involves a radical (hydroxyl). Another way of expressing the formula is

$$Al_2(OH)_6 = Al_2 \ OH \ OH \ OH \ OH \ OH \ OH$$

Therefore, the number of atoms of each element in the compound is

$$Al_2(OH)_6$$

2 atoms | 6 atoms

6 atoms

$Al_2(SO_4)_3$: The elements in the compound are

$$Al = aluminum$$
$$S = sulfur$$
$$O = oxygen$$

This compound also involves a radical (sulfate). Another way of expressing the formula is

$$Al_2(SO_4)_3 = Al_2 \ SO_4 \ SO_4 \ SO_4$$

And therefore, the number of atoms of each element in the compound in

$$Al_2(SO_4)_3$$

2 atoms | 12 atoms

3 atoms

3. To calculate the percent by weight of each element, first determine the atomic weight of each element and the molecular weight of $Ca(OH)_2$:

	No. of atoms	Atomic weight	Total weight
Calcium (Ca)	1	$\times$ 40.08 $=$	40.08
Oxygen (O)	2	$\times$ 16.00 $=$	32.00
Hydrogen (H)	2	$\times$ 1.01 $=$	2.02

Molecular weight of $Ca(OH)_2 = $ 74.10

Now calculate the percent of the weight of the compound represented by each element:

$$\text{Percent Ca by weight} = \frac{\text{Weight of Ca in compound}}{\text{Molecular weight of compound}} \times 100$$

$$= \frac{40.08}{74.10} \times 100$$

$$= 0.541 \times 100$$

$$= 54.1\% \text{ Ca by weight}$$

$$\text{Percent O by weight} = \frac{\text{Weight of O in compound}}{\text{Molecular weight of compound}} \times 100$$

$$= \frac{32.00}{74.10} \times 100$$

$$= 0.432 \times 100$$

$$= 43.2\% \text{ O by weight}$$

$$\text{Percent H by weight} = \frac{\text{Weight of H in compound}}{\text{Molecular weight of compound}} \times 100$$

$$= \frac{2.02}{74.10} \times 100$$

$$= 0.027 \times 100$$

$$= 2.7\% \text{ H by weight}$$

4. To calculate the percent by weight of each element in the compound, first determine the atomic weight of each element and the molecular weight of NH_3:

	No. of atoms		Atomic weight		Total weight
Nitrogen (N)	1	×	14.01	=	14.01
Hydrogen (H)	3	×	1.01	=	3.03

Molecular weight of NH_3 =   17.04

Now calculate the percent of the weight of the compound represented by each element:

$$\text{Percent N by weight} = \frac{\text{Weight of N in compound}}{\text{Molecular weight of compound}} \times 100$$

$$= \frac{14.01}{17.04} \times 100$$

$$= 0.822 \times 100$$

$$= 82.2\% \text{ N by weight}$$

$$\begin{aligned}
\text{Percent H} \atop \text{by weight} &= \frac{\text{Weight of H in compound}}{\text{Molecular weight of compound}} \times 100 \\
&= \frac{3.03}{17.04} \times 100 \\
&= 0.178 \times 100 \\
&= 17.8\% \text{ H by weight}
\end{aligned}$$

5. The percents given in the problem indicate what percent of the weight is made up by each element. Therefore the actual weight of each element may be calculated as follows:

   29.4% Ca: $(0.294)(150 \text{ lb } CaSO_3) = 44.1 \text{ lb calcium}$

   23.6% S: $(0.236)(150 \text{ lb } CaSO_3) = 35.4 \text{ lb sulfur}$

   47% O: $(0.47)(150 \text{ lb } CaSO_3) = 70.5 \text{ lb oxygen}$

6. To determine the *number of pounds* of each element present, you must first determine *what percent* of the total weight is represented by each element:

	No. of atoms		Atomic weight		Total weight
Calcium (Ca)	1	$\times$	40.08	=	40.08
Carbon (C)	1	$\times$	12.01	=	12.01
Oxygen (O)	3	$\times$	16.00	=	48.00

Molecular weight of $CaCO_3$ = 100.09

The percent composition by weight for each element is

$$\begin{aligned}
\text{Percent Ca} \atop \text{by weight} &= \frac{\text{Weight of Ca in compound}}{\text{Molecular weight of compound}} \times 100 \\
&= \frac{40.08}{100.09} \times 100 \\
&= 0.40 \times 100 \\
&= 40\% \text{ Ca by weight}
\end{aligned}$$

$$\begin{aligned}
\text{Percent C} \atop \text{by weight} &= \frac{\text{Weight of C in compound}}{\text{Molecular weight of compound}} \times 100 \\
&= \frac{12.01}{100.09} \times 100 \\
&= 0.12 \times 100 \\
&= 12\% \text{ C by weight}
\end{aligned}$$

$$\begin{aligned}
\frac{\text{Percent O}}{\text{by weight}} &= \frac{\text{Weight of O in compound}}{\text{Molecular weight of compound}} \times 100 \\
&= \frac{48.00}{100.09} \times 100 \\
&= 0.48 \times 100 \\
&= 48\% \text{ O by weight}
\end{aligned}$$

Now that the percent of the weight represented by each element has been determined, determine the actual number of pounds of each element in the 60 lb of $CaCO_3$:

40% Calcium: $(0.40)(60 \text{ lb } CaCO_3) = 24$ lb calcium

12% Carbon: $(0.12)(60 \text{ lb } CaCO_3) = 7.2$ lb carbon

48% Oxygen: $(0.48)(60 \text{ lb } CaCO_3) = 28.8$ lb oxygen

7. $4NH_3$: The coefficient of 4 in front of the molecule indicates that 4 molecules of NH_3 are involved in the reaction. First determine the weight of one molecule, then calculate the weight of four molecules:

	No. of atoms		Atomic weight		Total weight
Nitrogen (N)	1	×	14.01	=	14.01
Hydrogen (H)	3	×	1.01	=	3.03
Weight of 1 molecule NH_3				=	17.04
Weight of 4 molecules NH_3				=	(4)(17.04)
				=	68.16

$3O_2$: The coefficient 3 in front of the molecule indicates that 3 molecules of O_2 are involved in the reaction. First, calculate the weight of one molecule, then the weight of 3 molecules:

	No. of atoms		Atomic weight		Total weight
Oxygen (O)	2	×	16.00	=	32.00
Weight of 1 molecule of O_2				=	32.00
Weight of 3 molecules of O_2				=	(3)(32.00)
				=	96.00

$2N_2$: First calculate the weight of one molecule of N_2, then the weight of two molecules:

	No. of atoms	Atomic weight		Total weight
Nitrogen (N)	2	$\times$ 14.01	=	28.02

Weight of one molecule of N_2 = 28.02
Weight of two molecules of N_2 = (2)(28.02)
= 56.04

$6H_2O$: First calculate the weight of one molecule of H_2O, then the weight of six molecules:

	No. of atoms	Atomic weight		Total weight
Hydrogen (H)	2	$\times$ 1.01	=	2.02
Oxygen (O)	1	$\times$ 16.00	=	16.00

Weight of one molecule of H_2O = 18.02
Weight of six molecules of H_2O = (6)(18.02)
= 108.12

In summary, the weights that correspond to each term of the equation are as follows:

$$4NH_3 + 3O_2 \longrightarrow 2N_2 + 6H_2O$$
$$68.16 \quad 96.00 \qquad 56.04 \quad 108.12$$

8. In this problem you are concerned with the only reactants in the equation

$$\underline{Ca(OCl)_2 + Na_2CO_3} \longrightarrow 2NaOCl + CaCO_3$$
REACTANTS

The molecular weights of these reactants indicate the weight *ratio* in which the two compounds will react:

$$Ca(OCl)_2 + Na_2CO_3$$
$$142.98 \qquad 105.99$$

Using this information, set up a proportion to determine the amount of $NaCO_3$ which will react with the 20 lbs of $Ca(OCl)_2$:

$$\frac{142.98 \text{ lb } Ca(OCl)_2}{105.99 \text{ lb } Na_2CO_3} = \frac{20 \text{ lb } Ca(OCl)_2}{x \text{ lb } Na_2CO_3}$$

Then solve for the unknown value:

$$\frac{142.98}{105.99} = \frac{20}{x}$$

$$\frac{(x)(142.98)}{105.99} = 20$$

$$x = \frac{(20)(105.99)}{142.98}$$

$$= 14.8 \text{ lb } Na_2CO_3$$

9. (a) *0.015 mole of NaAsO₂:*
First, determine the weight of 1 mole of the compound:

	No. of atoms		Atomic weight		Total weight
Sodium (Na)	1	×	22.99	=	22.99
Arsenic (As)	1	×	74.92	=	74.92
Oxygen (O)	2	×	16.00	=	32.00

Molecular weight of $NaAsO_2$ = 129.91
Therefore, 1 mole of $NaAsO_2$ weighs 129.91 g.
Next, multiply the weight of 1 mole by the number of moles required:
0.015 mole of $NaAsO_2$ is required.
0.015 mole $NaAsO_2$ weighs (0.015)(129.91 g) = 1.95 g.

(b) *2.5 moles of ZnSO₄:*
First, determine the weight of 1 mole of the compound:

	No. of atoms		Atomic weight		Total weight
Zinc (Zn)	1	×	65.37	=	65.37
Sulpher (S)	1	×	32.06	=	32.06
Oxygen (O)	4	×	16.00	=	64.00

Molecular weight of $ZnSO_4$ = 161.43
Therefore, 1 mole of $ZnSO_4$ weighs 161.43 g.
Next, multiply the weight of 1 mole by the number of moles required.
2.5 moles are required.
2.5 moles $ZnSO_4$ weighs (2.5)(161.43 g) = 403.58 g.

(c) *0.001 mole of K_2PtCl_6:*

First, determine the weight of 1 mole of the compound:

	No. of atoms	Atomic weight	Total weight
Potassium (K)	2	$\times$ 39.10 =	78.20
Platinum (Pt)	1	$\times$ 195.09 =	195.09
Chlorine (Cl)	6	$\times$ 35.45 =	212.70

Molecular weight of K_2PtCl_6 = 485.99

Therefore, 1 mole of K_2PtCl_6 weighs 485.99 g.

Next, multiply the weight of 1 mole by the number of moles required:

0.001 mole of K_2PtCl_6 is required.

0.001 mole K_2PtCl_6 weighs 0.48599 g = 0.49 g (rounded)

Chemistry 4

Solutions

A SOLUTION consists of two parts: a SOLVENT and a SOLUTE, which are completely and evenly mixed (they form a HOMOGENOUS mixture[4]). The solute part of the solution is dissolved in the solvent (Figure 7).

In a true solution, the solute will remain dissolved and will not settle out. Salt water is a true solution; salt is the solute and water is the solvent. On the other hand, sand mixed into water does not form a solution—the sand will settle out when the water is left undisturbed.

In water treatment, the most common solvent is water. Before it is dissolved, the solute may be solid (such as dry alum), liquid (such as sulfuric acid), or gaseous (such as chlorine).

The CONCENTRATION of a solution is a measure of the amount of solute dissolved in a given amount of solvent. A concentrated (strong) solution is a solution in which a relatively great amount of solute is dissolved in the solvent. A dilute (or weak) solution is one in which a relatively small amount of solute is dissolved in the solvent.

There are many ways of expressing the concentration of a solution, including

- Grains per gallon,
- Milligrams per litre,
- Percent strength,
- Molarity, and
- Normality.

C4-1. Grains per Gallon and Milligrams per Litre

The measurements GRAINS PER GALLON (GPG) and MILLIGRAMS PER LITRE (MG/L) each express the *weight* of solute dissolved in a given *volume* of solution. The

---

[4]Chemistry Section, The Classification of Matter (Mixtures).

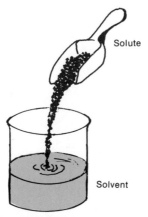

Figure 7.   Solutions are composed of a solute and a solvent

mathematics needed to deal with grains per gallon[5] and milligrams per litre[6] are covered in other sections of this book, and the measurements will not be discussed further here.

C4-2.   Percent Strength

The percent strength of a solution can be expressed as percent-by-weight or percent-by-volume. The percent-by-weight calculation is used more often in water treatment.

Conversions between milligrams per litre and percent are discussed in another section of this book.[7]

Percent Strength by Weight

The equation used to calculate percent by weight is as follows:

$$\text{Percent strength (by weight)} = \frac{\text{Weight of solute}}{\text{Weight of solution}} \times 100$$

NOTE:
$$\text{Weight of solution} = \text{Weight of solute} + \text{Weight of solvent}$$

Use of the equation is illustrated in the following examples.

Example 1
If 25 lb of chemical is added to 400 lb of water, what is the percent strength of the solution by weight?

$$\text{Percent strength (by weight)} = \frac{\text{Weight of solute}}{\text{Weight of solution}} \times 100$$

---

[5]Mathematics Section, Conversions (Milligrams-per-Litre, Grains-per-Gallon, and Parts per Million).

[6]Chemistry Section, Chemical Dosage Problems.

[7]Chemistry Section, Chemical Dosage Problems (Milligrams-per-Litre to Percent Conversions).

The weight of the chemical (the solute) is given in the problem as 25 lb. The weight of the solution, however, is the combined weight of the solute and the solvent:

$$\begin{array}{c}\text{Weight} \\ \text{of solution}\end{array} = \begin{array}{c}\text{Weight} \\ \text{of solute}\end{array} + \begin{array}{c}\text{Weight} \\ \text{of solvent}\end{array}$$

$$= \quad 25 \text{ lb} \quad + \quad 400 \text{ lb}$$

$$= 425 \text{ lb solution}$$

Using this information, calculate the percent concentration:

$$\begin{array}{c}\text{Percent strength} \\ \text{(by weight)}\end{array} = \frac{\text{Weight of solute}}{\text{Weight of solution}} \times 100$$

$$= \frac{25 \text{ lb chemical}}{425 \text{ lb solution}} \times 100$$

$$= 0.059 \times 100$$

$$= 5.9\% \text{ strength solution}$$

Example 2

If 40 lb of chemical is added to 120 gal of water, what is the percent strength of the solution by weight?

First, calculate the weight of the solution. The weight of the solution is equal to the weight of the solute plus the weight of solvent. To calculate this, first convert 120 gal of water to pounds of water:[8]

$$(120 \text{ gal})(8.34 \text{ lbs/gal} = 1001 \text{ lb water}$$

Then calculate the weight of solution:

$$\begin{array}{c}\text{Weight} \\ \text{of solution}\end{array} = \begin{array}{c}\text{Weight} \\ \text{of solute}\end{array} + \begin{array}{c}\text{Weight} \\ \text{of solvent}\end{array}$$

$$= \quad 40 \text{ lb} \quad + \quad 100 \text{ lb}$$

$$= 1041 \text{ lb}$$

Now calculate the percent strength of the solution:

$$\begin{array}{c}\text{Percent strength} \\ \text{(by weight)}\end{array} = \frac{\text{Weight of solute}}{\text{Weight of solution}} \times 100$$

$$= \frac{40 \text{ lb chemical}}{1041 \text{ lb solution}} \times 100$$

$$= 0.038 \times 100$$

$$= 3.8\% \text{ strength solution}$$

---

[8]Mathematics Section, Conversions (Cubic Feet to Gallons to Pounds).

Example 3

In 80 gal of a solution there are 15 lb of chemical. What is the percent strength of the solution? Assume the solution has the same density as water—8.34 lb/gal.[9]

First express the 80 gal of solution as pounds:

$$(80 \text{ gal})(8.34 \text{ lb/gal}) = 667.2 \text{ lb solution}$$

Now calculate the percent strength. Notice that in this problem the gallons measure the amount of *solution,* not merely the amount of water.

$$\begin{aligned}
\text{Percent strength} \atop \text{(by weight)} &= \frac{\text{Weight of solute}}{\text{Weight of solution}} \times 100 \\
&= \frac{15 \text{ lb chemical}}{667.2 \text{ lb solution}} \times 100 \\
&= 0.022 \times 100 \\
&= 2.2\% \text{ strength solution}
\end{aligned}$$

In the previous three examples, percent strength was the unknown quantity. However, the same equation can be used to determine the pounds of chemical required, provided the percent strength and weight (or percent strength, volume, and density) of the solution are known. The next two examples illustrate the procedure.

Example 4

You need to prepare 25 gal of a 2.5 percent strength solution. Assume the solution will have the same density as water: 8.34 lb/gal. How many pounds of chemical will you need to dissolve in the water?

Use the same equation of percent strength as in the previous problems. First, write into the equation the information given in the problem:

$$\begin{aligned}
\text{Percent strength} \atop \text{(by weight)} &= \frac{\text{Weight of solute}}{\text{Weight of solution}} \times 100 \\
2.5\% &= \frac{x \text{ lb chemical}}{(25)(8.34) \text{ lb solution}} \times 100
\end{aligned}$$

Then solve for the unknown value:[10]

$$2.5 = \frac{x}{(25)(8.34)} \times 100$$

---

[9]Hydraulics Section, Density and Specific Gravity.
[10]Mathematics Section, Solving for the Unknown Value.

This equation may be restated as

$$\frac{2.5}{1} = \frac{(x)(100)}{(25)(8.34)}$$

Then,

$$\frac{(25)(8.34)(2.5)}{1} = (x)(100)$$

$$\frac{(25)(8.34)(2.5)}{100} = x$$

$$5.21 \text{ lb} = x$$
chemical

Example 5

You wish to prepare 50 gal of a 4 percent strength solution. How much water and chemical should be mixed together? Assume the solution will have the same density as water: 8.34 lb/gal.

First calculate the pounds of chemical by writing the given information into the equation:

$$\frac{\text{Percent strength}}{\text{(by weight)}} = \frac{\text{Weight of solute}}{\text{Weight of solution}} \times 100$$

$$4\% = \frac{x \text{ lb chemical}}{(50)(8.34) \text{ lb solution}} \times 100$$

Then solve for the unknown value:

$$4 = \frac{x}{(50)(8.34)} \times 100$$

This can be restated as

$$\frac{4}{1} = \frac{(x)(100)}{(50)(8.34)}$$

$$\frac{(50)(8.34)(4)}{1} = (x)(100)$$

$$\frac{(50)(8.34)(4)}{100} = x$$

$$16.68 \text{ lb} = x$$
chemical

To calculate the pounds of water, first calculate the total pounds of solution:

$$(50 \text{ gal solution})(8.34 \text{ lb/gal}) = 417 \text{ lb solution}$$

The 417 lb of solution contains both water and chemical. Since the pounds of chemical are known, the pounds of water can be determined:

$$417 \text{ lb} = \text{? lb} + 16.68 \text{ lb}$$
$$\text{solution} \quad \text{water} \quad \text{chemical}$$

Therefore,

$$\begin{array}{r} 417.00 \text{ lb solution} \\ -16.68 \text{ lb chemical} \\ \hline 400.32 \text{ lb water} \end{array}$$

Dilution Calculations

Sometimes a particular strength solution will be made up by diluting a strong solution with a weak solution of the same chemical. The new solution will have a concentration somewhere between the weak and the strong solutions.

Although there are several methods available for determining what amounts of the weak and strong solutions are needed, perhaps the easiest is the "Rectangle Method" (sometimes called the "Dilution Rule"), shown in the following diagram:

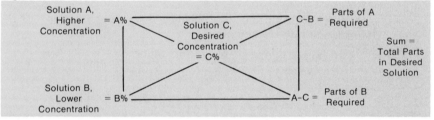

The following three examples illustrate how the rectangle method is used.

Example 6

What volumes of a 3 percent solution and an 8 percent solution must be mixed to make 400 gal of a 5 percent solution?

Use the rectangle method to solve the problem. First write the given concentrations into the proper places in the rectangle:

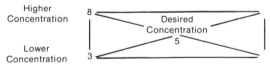

Now complete the rectangle by subtraction:

The circled numbers on the right side of the rectangle indicate the volume *ratios* of the solutions to be mixed. The sum of the circled numbers are a total of five parts to be added (2 parts + 3 parts = 5 parts total).

Two parts out of the five parts (2/5) of the new solution should be made up of the 8 percent solution. And three parts out of the five parts (3/5) of the new solution should be made up of the 3 percent solution.

Using the ratios, the *number of gallons* of each solution to be mixed can be determined:

$$\left(\frac{2}{5}\right)(400 \text{ gal}) = 160 \text{ gal of } 8\% \text{ solution}$$

$$\left(\frac{3}{5}\right)(400 \text{ gal}) = 240 \text{ gal of } 3\% \text{ solution}$$

Mixing these amounts will result in 400 gal of a 5 percent solution.

Example 7

What weights of a 7 percent solution and a 4 percent solution should be added together to obtain 300 lb of a 5 percent solution?

Use the rectangle method. First write in the given concentrations:

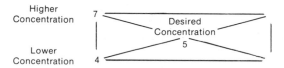

Now complete the right side of the rectangle:

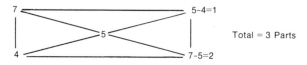

The circled numbers on the right side of the rectangle indicate the weight *ratios* by which the two solutions are to be mixed. There are a total of three parts to the new solution. One part out of three (1/3) comes from the 7 percent solution, and two parts out of three (2/3) come from the 4 percent solution.

The *number of pounds* of each of the two solutions which must be used in making the desired 5 percent solution are calculated as follows:

$$\left(\frac{1}{3}\right)(300 \text{ lb}) = 100 \text{ lb of 7 percent solution}$$

$$\left(\frac{2}{3}\right)(300 \text{ lb}) = 200 \text{ lb of 4 percent solution}$$

When these amounts are mixed, the result will be 300 lb of a 5 percent solution.

Example 8

How many gallons of water and a 6 percent solution should be used to make 80 gal of a 4 percent solution?

The rectangle method can also be used to solve this problem. Water is considered a 0 percent solution; therefore, the problem can be restated as, "How many gallons of a 0 percent solution and a 6 percent solution must be mixed to obtain 80 gal of a 4 percent solution?"

First write the given information into the rectangle:

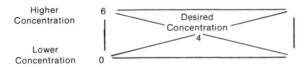

Now, complete the right side of the rectangle:

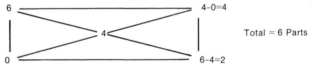

As indicated on the right side of the rectangle, four parts out of six (4/6) is the *ratio* by which the 6 percent solution must be added; and two parts out of six (2/6) is the ratio by which the water must be added.

The *number of gallons* of each of the two liquids which must be used in making the 4 percent solution are calculated as follows:

$$\left(\frac{4}{6}\right)(80 \text{ gal}) = 53.33 \text{ gal of the } 6\% \text{ solution}$$

$$\left(\frac{2}{6}\right)(80 \text{ gal}) = 26.67 \text{ gal of water}$$

When these amounts are mixed, the resulting solution will be 80 gal of a 4 percent solution.

Although the percent strength method of expressing solution concentration is sufficiently accurate in many applications of water treatment, a more precise measurement of solution concentration is required in the laboratory.

C4-3.   Moles and Molarity

A more accurate way of expressing the concentration of a solution than percent strength is MOLARITY (usually designated by the letter M). Molarity is defined as the number of MOLES[11] of solute per litre of solution. A mole (abbreviation for GRAM-MOLE) is the quantity of a compound that has a weight in grams equal to the compound's molecular weight. For example, the molecular weight of $CaCO_3$ is 100.09. If you had 100.09 g of $CaCO_3$, then you would have one mole of $CaCO_3$. If you had 200.18 g of $CaCO_3$, then you would have two moles of $CaCO_3$.

To determine the number of moles you have of a given compound, compare the total weight of the compound in grams with the compound's molecular weight:

$$\frac{\text{Number}}{\text{of Moles}} = \frac{\text{Total weight}}{\text{Molecular weight}}$$

Example 9

If 150 g of sodium hydroxide ($NaOH$) is mixed into water to make a solution, how many moles of solute have been used? (Use the periodic table given in Appendix B for atomic weights.)

First determine the molecular weight of $NaOH$:

	No. of atoms		Atomic weight		Total weight
Sodium (Na)	1	×	22.99	=	22.99
Oxygen (O)	1	×	16.00	=	16.00
Hydrogen (H)	1	×	1.01	=	1.01

Molecular weight of $NaOH$ = 40.00

Next calculate the number of moles of $NaOH$ used in making up the solution:

$$\frac{\text{Number}}{\text{of Moles}} = \frac{\text{Total weight}}{\text{Molecular weight}}$$

$$= \frac{150 \text{ g}}{40.00 \text{ g}}$$

$$= 3.75 \text{ moles of } NaOH$$

Example 10

If 45 g of sulfuric acid (H_2SO_4) were used in making up a solution, how many moles of H_2SO_4 were used?

To determine the number of moles used, first calculate the molecular weight of H_2SO_4:

---

[11]Chemistry Section, Valence, Chemical Formulas, and Chemical Equations.

	No. of atoms		Atomic weight		Total weight
Hydrogen (H)	2	×	1.01	=	2.02
Sulfur (S)	1	×	32.06	=	32.06
Oxygen (O)	4	×	16.00	=	64.00

$$\text{Molecular weight of } H_2SO_4 = 98.08$$

Now determine the number of moles:

$$\frac{\text{Number}}{\text{of Moles}} = \frac{\text{Total weight}}{\text{Molecular weight}}$$

$$= \frac{45 \text{ g}}{98.08 \text{ g}}$$

$$= 0.46 \text{ moles of } H_2SO_4$$

Once the number of moles of solute has been determined, the molarity of a solution may be calculated using the following equation:

$$\text{Molarity} = \frac{\text{Moles of solute}}{\text{Litres of solution}}$$

The following two examples illustrate the use of the equation.

Example 11

If 0.4 moles of NaOH is dissolved in 2 L of solution, what is the molarity of the solution?

In this problem, the number of moles in the solution has already been calculated. Therefore, the molarity of the solution is

$$\text{Molarity} = \frac{\text{Moles of solute}}{\text{Litres of solution}}$$

$$= \frac{0.4 \text{ moles}}{2 \text{ L solution}}$$

$$= 0.2 \text{ molarity}$$

$$= 0.2 \; M \text{ solution}$$

Example 12

In 800 mL of solution, 200 g of NaCl is dissolved. What is the molarity of the solution? (Use the periodic table in Appendix B for atomic weights.)

First determine the number of moles of NaCl in the solution:

	No. of atoms		Atomic weight		Total weight
Sodium (Na)	1	×	22.99	=	22.99
Chlorine (Cl)	1	×	35.45	=	35.45

$$\text{Molecular weight of NaCl} = 58.44$$

The number of moles used in the solution is therefore

$$\frac{\text{Number}}{\text{of Moles}} = \frac{\text{Total weight}}{\text{Molecular weight}}$$

$$= \frac{200}{58.44}$$

$$= 3.42 \text{ moles}$$

The chemical is dissolved in 800 mL of solution. Millilitres must be expressed as litres for the calculation of molarity:[12]

$$800 \text{ mL} = 0.8 \text{ L}$$

Now calculate the molarity of the solution:

$$\text{Molarity} = \frac{\text{Moles of solute}}{\text{Litres of solution}}$$

$$= \frac{3.4 \text{ moles}}{0.8 \text{ L solution}}$$

$$= 4.25 \text{ molarity}$$

$$= 4.25 \text{ } M \text{ solution}$$

Example 13

You need to mix 2.25 L of a 1.5 M solution of H_2SO_4. (a) How many moles of H_2SO_4 must be dissolved in the 2.25 L? (b) How many grams of H_2SO_4 is this? The molecular weight of H_2SO_4 is 98.08.

(a) Use the same molarity equation as in the previous examples. Write in the given information:

$$\text{Molarity} = \frac{\text{Moles of solute}}{\text{Litres of solution}}$$

$$1.5 \text{ } M = \frac{x \text{ moles } H_2SO_4}{2.25 \text{ litres solution}}$$

$$(2.25)(1.5) = x$$

$$3.38 \text{ moles } H_2SO_4 = x$$

---

[12]Mathematics Section, Conversions (Metric System Conversions).

(b) Calculating how many grams of H_2SO_4 there are in 3.4 moles. Note that 1 mole H_2SO_4 weighs 98.08 g. Therefore, 3.38 moles of H_2SO_4 weighs

$$(3.38)(98.08 \text{ g}) = 331.51 \text{ g } H_2SO_4$$

C4-4.   Equivalent Weights and Normality

Another method of expressing the concentration of a solution is NORMALITY. Normality depends in part on the valence of an element or compound. An element or compound may have more than one valence, and it is not always clear which valence (and therefore what concentration) a given normality represents. Because of this problem, normality is being replaced by molarity as the expression of concentration used for chemicals in the lab.

In the lab you will often have detailed, step-by-step instructions for preparing a solution of a needed normality. Nonetheless, it is useful to have a basic idea of what the measurement means. To understand normality, you must first understand equivalent weights.

Equivalent Weights

The EQUIVALENT WEIGHT of an element or compound is *the weight of that element or compound which, in a given chemical reaction, has the same combining capacity as 8 g of oxygen or as 1 g of hydrogen.* The equivalent weight may vary with the reaction being considered. However, one equivalent weight of a reactant will always react with one equivalent weight of the other reactant in a given reaction.

Although you are not expected to know how to determine the equivalent weights of various reactants at this level, it will help if you remember one characteristic: the equivalent weight of a reactant either will be *equal* to the reactant's molecular weight, or it will be a *simple fraction* of the molecular weight. For example, if the molecular weight of a compound is 60.00 g, then the equivalent weight of the compound in a reaction will be 60.00 g or a simple fraction (usually 1/2, 1/3, 1/4, 1/5, or 1/6) of 60.00 g.

Normality

NORMALITY is defined as the *number of equivalent weights of solute per litre of solution.* Therefore, to determine the normality of a solution, you must first determine how many equivalent weights of solute are contained in the total weight of dissolved solute. Use the following equation:

$$\frac{\text{Number of}}{\text{Equivalent weights}} = \frac{\text{Total weight}}{\text{Equivalent weight}}$$

Example 14

If 90 g of sodium hydroxide (NaOH) were used in making up a solution, how many equivalent weights were used? Use 40.00 g as the equivalent weight for NaOH.

$$\frac{\text{Number of}}{\text{Equivalent weights}} = \frac{\text{Total weight}}{\text{Equivalent weight}}$$

$$= \frac{90 \text{ g}}{40 \text{ g}}$$

$$= 2.25 \text{ equivalent weights}$$

Example 15

If 75 g of sulfuric acid (H_2SO_4) were used in making up a solution, how many equivalent weights of H_2SO_4 were used? Use 49.04 as the equivalent weight for H_2SO_4.

$$\frac{\text{Number of}}{\text{Equivalent weights}} = \frac{\text{Total weight}}{\text{Equivalent weight}}$$

$$= \frac{75 \text{ g}}{49.04 \text{ g}}$$

$$= 1.53 \text{ equivalent weights}$$

When you have determined the number of equivalent weights of dissolved solute, you can determine the normality of the solution using the following equation:

$$\text{Normality} = \frac{\text{Number of equivalent weights of solute}}{\text{Litres of solution}}$$

Example 16

If 2.1 equivalents of NaOH were used in making up 1.75 L of solution, what is the normality of the solution?

$$\text{Normality} = \frac{\text{Number of equivalent weights of solute}}{\text{Liters of solution}}$$

$$= \frac{2.1 \text{ equivalents}}{1.75 \text{ L solution}}$$

$$= 1.2 \text{ normality}$$

$$= 1.2 \ N \text{ solution}$$

Example 17

If 0.5 equivalents of Na_2CO_3 were used in making up 750 mL of solution, what is the normality of the solution?

$$\text{Normality} = \frac{\text{Number of equivalent weights of solute}}{\text{Litres of solution}}$$

$$= \frac{0.5 \text{ equivalents}}{0.750 \text{ L solution}}$$

$$= 0.67 \text{ normality}$$

$$= 0.67 \ N \text{ solution}$$

C4-5. Hardness

Sometimes in water chemistry it is not important to know exactly which impurities a sample of water contains but merely what combined effect those impurities have. A common example of this is hardness.

Hardness is a measurement of the effects that water impurities such as magnesium and calcium have on corrosion, scaling, and soap. Water from one source might contain a great deal of magnesium and little else; or it might contain a great deal of calcium and little else; or it might contain some magnesium, some calcium, and traces of some similarly behaving elements, such as strontium—in each case it would be a "hard water;" a water with noticeable effects associated with hardness. Furthermore, the laboratory test used to determine hardness would show the same result, whether the hardness was the result of calcium, magnesium, or a combination of elements.

Calcium carbonate ($CaCO_3$) is one of the more common causes of hardness, and the total hardness of water is usually expressed "in terms of $CaCO_3$." For example, a lab report might read:

total hardness: 180 mg/ L as $CaCO_3$

This means that the lab has not determined exactly what chemicals are causing the water's hardness but that their combined effect is the same as if the water contained exactly 180 mg/ L of $CaCO_3$ and no other chemicals. By expressing the hardness of every sample in terms of how much calcium carbonate it might contain, the hardness of any two samples can be compared more easily.

In fact, measuring hardness in terms of $CaCO_3$ is convenient that even when the lab determines exactly what chemicals are in a sample, the result may still be expressed as $CaCO_3$. Thus, a lab report could read:

Calcium:	125 mg/ L as $CaCO_3$
Total hardness:	150 mg/ L as $CaCO_3$
Bicarbonate alkalinity:	122 mg/ L as $CaCO_3$
Total alkalinity:	130 mg/ L as $CaCO_3$
Carbon dioxide:	23 mg/ L as $CaCO_3$

In softening and other water treatment operations, you may need to convert hardness expressed in terms of one chemical to hardness expressed in terms of another, usually $CaCO_3$. In a problem where hardness is expressed as the effect of 100 mg/ L of chemical A and you want it expressed as the effect of (an unknown weight) / L of chemical B, you are interested in what weight of chemical B has the same chemical combining power as 100 mg of chemical A. The question can be expressed in equivalent weights: you need as many equivalent weights of B as there are equivalent weights of A in 100 mg of A. The following examples illustrate the calculations needed.

Example 18

A water sample contains 136 mg/ L Ca as Ca. What is the concentration expressed as $CaCO_3$? For the equivalent weight of Ca, use 20.00 g; for the equivalent weight of $CaCO_3$, use 50.00 g.

First find the number of equivalent weights of Ca in 136 mg:

$$\begin{array}{l} \text{Number of} \\ \text{equivalent weights} \\ \text{of Ca} \end{array} = \frac{\text{Total weight of Ca}}{\text{Equivalent weight of Ca}}$$

$$= \frac{136 \text{ mg}}{20.00 \text{ g}}$$

$$= \frac{0.136 \text{ g}}{20.00 \text{ g}}$$

$$= 0.0068 \text{ equivalent weights of Ca}$$

Because the sample behaves as if it has 0.0068 equivalent weights of Ca per litre, it will behave as if it has 0.0068 equivalent weights of $CaCO_3$ per litre. So, calculate the total weight of 0.0068 equivalent weights of $CaCO_3$:

$$\begin{array}{l} \text{Total weight} \\ \text{of } CaCO_3 \end{array} = \left(\begin{array}{l} \text{Number of} \\ \text{equivalent weights} \\ \text{of } CaCO_3 \end{array} \right) (\text{Equivalent weight of } CaCO_3)$$

$$= (0.0068)(50.00 \text{ g})$$

$$= 0.34 \text{ g}$$

$$= 340 \text{ mg}$$

Therefore, the hardness measured as 136 mg/L as Ca can also be expressed as a hardness of 340 mg/L as $CaCO_3$.

The calculations just performed demonstrate the meaning of "equivalent weights." In practice, a simpler calculation is commonly used, combining the two previous equations into a single equation that gives the same result. The simpler calculation is performed as follows: set up the ratio[13] of the equivalent weight of the desired ("new") measure to the equivalent weight of the given ("old") measure; then multiply by the "old" concentration:

$$\left(\frac{\begin{array}{c} \text{Equivalent weight of} \\ \text{new measure} \end{array}}{\begin{array}{c} \text{Equivalent weight of} \\ \text{old measure} \end{array}} \right) (\text{old concentration}) = (\text{new concentration})$$

In Example 18, filling in the values gives

$$\left(\frac{\begin{array}{c} \text{Equivalent weight} \\ \text{of } CaCO_3 \end{array}}{\begin{array}{c} \text{Equivalent weight} \\ \text{of Ca} \end{array}} \right) (\text{concentration as Ca}) = (\text{concentration as } CaCO_3)$$

---

[13]Mathematics Section, Ratios and Proportions.

$$\left(\frac{50.00 \text{ g}}{20.00 \text{ g}}\right)(136 \text{ mg/L as Ca}) = x \text{ mg/L as CaCO}_3)$$

$$= 340 \text{ mg/L as CaCO}_3$$

Example 19
A lab report shows:

$$\text{Magnesium:} \quad 17 \text{ mg/L}$$
$$\text{Carbon dioxide:} \quad 3 \text{ mg/L}$$

Express the concentrations as $CaCO_3$. Use 12.16 g for the equivalent weight of magnesium (Mg); use 22.00 g for the equivalent weight of carbon dioxide (CO_2); and use 50.00 g for the equivalent weight of $CaCO_3$.

First note that the lab report does not state that the concentrations are expressed "as" anything. This means that the concentrations are simply measurements of the chemicals listed—there actually is 17 mg of magnesium in every litre of the water. You could say that the magnesium concentration is expressed "as magnesium," and that the carbon dioxide concentration is expressed "as carbon dioxide."

To convert the measurements to $CaCO_3$, use the equation

$$\left(\frac{\text{Equivalent weight of new measure}}{\text{Equivalent weight of old measure}}\right)(\text{old concentration}) = (\text{new concentration})$$

Fill in the values and solve for the magnesium concentration:

$$\left(\frac{\text{Equivalent weight of CaCO}_3}{\text{Equivalent weight of Mg}}\right)(\text{concentration as Mg}) = (\text{concentration as CaCO}_3)$$

$$\left(\frac{50.00 \text{ g}}{12.16 \text{ g}}\right)(17 \text{ mg/L as Mg}) = (x \text{ mg/L as CaCO}_3)$$

$$= 69.9 \text{ mg/L as CaCO}_3$$

Fill in the values and solve for the carbon dioxide concentration:

$$\left(\frac{\text{Equivalent weight of CaCO}_3}{\text{Equivalent weight of CO}_2}\right)(\text{concentration as CO}_2) = (\text{concentration as CaCO}_3)$$

$$\left(\frac{50.00 \text{ g}}{22.00 \text{ g}}\right) (3 \text{ mg/L as } CO_2) = (x \text{ mg/L as } CaCO_3)$$

$$= 6.82 \text{ mg/L as } CaCO_3$$

C4-6. Nitrogen Compounds

Nitrogen is one of the more common elements associated with the biochemical processes of life. Chemists sometimes find it convenient to use a special measurement for concentrations of nitrogen compounds. For example, when a chemist measuring nitrate (NO_3^-) writes that a sample contains "8 mg/L NO_3^- as N," the phrase "as N" means that each litre of the sample contains 8 mg of nitrogen atoms; not that it contains 8 mg of NO_3^- ions. Because each nitrogen atom is bonded to three oxygen atoms, the weight of nitrate (NO_3^-) in each litre of solution is equal to the 8 mg of nitrogen atoms *plus* the weight of all the oxygen atoms to which the nitrogen is bonded. In fact, the weight of nitrate in each litre is 35.4 mg; therefore, the nitrate concentration can also be written, "35.4 mg/L NO_3^- as NO_3^-."

To convert between measurements of nitrogen compounds "as N" and measurements of the compounds "as [the compound]," set up a ratio between the molecular weight of the "new" measurement and the molecular weight of the "old" measurement:

$$\left(\frac{\text{Molecular weight of new measurement}}{\text{Molecular weight of old measurement}}\right) (\text{old concentration}) = (\text{new concentration})$$

When one of the measurements is "as N," the atomic weight of nitrogen—14.01—is used for that measurement's molecular weight. The following example illustrates the use of the equation.

Example 20

Under the Interim Primary Regulations of the Safe Drinking Water Act, the MCL for nitrate (NO_3^-) is 10 mg/L as N. A lab reports that a water sample contains 42 mg/L NO_3^- as NO_3^-. Does the concentration of nitrate exceed the MCL?

First find the molecular weight of nitrate:

	No. of atoms		Atomic weight		Total weight
Nitrogen	1	×	14.01	=	14.01
Oxygen	3	×	16.00	=	48.00
			Molecular weight of NO_3^- =		62.01

Next set up the ratio, fill in the values, and solve:

$$\left(\frac{\text{Molecular weight of new measurement}}{\text{Molecular weight of old measurement}}\right)(\text{old concentration}) = (\text{new concentration})$$

$$\left(\frac{\text{Molecular weight of N}}{\text{Molecular weight of } NO_3^-}\right)\text{Concentration as } NO_3^-) = (\text{concentration as N})$$

$$\left(\frac{14.01}{62.01}\right)(42 \text{ mg/L as } NO_3^-) = (x \text{ mg/L as N})$$

$$= 9.48 \text{ mg/L as N}$$

Therefore, the concentration does not exceed the MCL.

When expressing a nitrogen compound "as N," there are several ways the expression may be written. The following all mean the same:

10 mg/L NO_3^- as N

10 mg/L nitrate as nitrogen

10 mg/L NO_3^-–N

10 mg/L nitrate–nitrogen

C4-7. Standard Solutions

A STANDARD SOLUTION is any solution with an accurately known concentration. Although there are many uses of standard solutions, they are often used to determine the concentration of substances in other solutions. Standard solutions are generally made up in one of three ways:

- Weight per unit volume
- Dilution
- Reaction

Weight Per Unit Volume

When a standard solution is made up by weight per unit volume, a pure chemical is accurately weighed and then dissolved in some solvent. By adding more solvent, the amount of solution is increased to a given volume. The concentration of the standard is then determined in terms of molarity or normality, as discussed previously.

Dilution

When a given volume of an existing standard solution is diluted with a measured amount of solvent, the concentration of the resulting (more dilute) solution can be determined using the following equation:

$$\left(\begin{matrix}\text{Normality of} \\ \text{Solution 1}\end{matrix}\right)\left(\begin{matrix}\text{Volume of} \\ \text{Solution 1}\end{matrix}\right) = \left(\begin{matrix}\text{Normality of} \\ \text{Solution 2}\end{matrix}\right)\left(\begin{matrix}\text{Volume of} \\ \text{Solution 2}\end{matrix}\right)$$

This equation can be abbreviated as

$$(N_1)(V_1) = (N_2)(V_2)$$

When using this equation it is important to remember that both volumes—V_1 and V_2—must be expressed in the same units. That is, both must be in litres (L) or both must be in millilitres (mL).

Example 21

You have a standard 1.4 N solution of H_2SO_4. How much water must be added to 100 mL of the standard solution to produce a 1.2 N solution of H_2SO_4?

First determine the total volume of the new solution by using the relationship between solution conentration and volume:

$$(N_1)(V_1) = (N_2)(V_2)$$
$$(1.4\ N)(100\ \text{mL}) = (1.2\ \text{N})(x\ \text{mL})$$

Solve for the unknown value:

$$\frac{(1.4)(100)}{1.2} = x\ \text{mL}$$

$$116.67 = x\ \text{mL}$$

Therefore, the total volume of the new solution will be 116.67 mL. Since the volume of the original solution is 100 mL, 16.67 mL of water (that is, 116.67 mL − 100 mL) need to be added to obtain the 1.2 N solution:

$$\begin{matrix}100\ \text{mL} \\ \text{of 1.4}\ N\ \text{solution}\end{matrix} + \begin{matrix}16.67\ \text{mL} \\ \text{of water}\end{matrix} = \begin{matrix}116.67\ \text{mL} \\ \text{of 1.2}\ N\ \text{solution}\end{matrix}$$

Example 22

How much water should be added to 80 mL of a 2.5 N solution of H_2CO_3 to obtain a 1.8 N solution of H_2CO_3?

Use the equation relating solution concentration and volume to calculate the total volume of the new solution:

$$(N_1)(V_1) = (N_2)(V_2)$$
$$(2.5\ N)(80\ \text{mL}) = (1.8\ N)(x\ \text{mL})$$

Solve for the unknown value:

$$\frac{(2.5)(80)}{1.8} = x\ \text{mL}$$

$$111.11 = x\ \text{mL}$$

Therefore, the total volume of the new solution will be 111.11 mL. Since the volume of the original solution is 80 mL, 31.11 mL of water (that is, 111.11 mL − 80 mL) needs to be added to obtain the 1.8 N solution:

$$\begin{array}{c}80 \text{ mL} \\ \text{of 2.5 } N \text{ solution}\end{array} + \begin{array}{c}31.11 \text{ mL} \\ \text{of water}\end{array} = \begin{array}{c}111.11 \text{ mL} \\ \text{of 1.8 } N \text{ solution}\end{array}$$

Reaction

A similar equation can be used for calculations involving reactions between samples of two solutions, as illustrated in the following example.

Example 23

Thirty-two mL of a 0.1 N solution of HCl is required to react with (neutralize) 30 mL of a certain base solution. What is the normality of the base solution?

$$(N_1)(V_1) = (N_2)(V_2)$$

$$(0.1 \ N)(32 \text{ mL}) = (x \ N)(30 \text{ mL})$$

$$\frac{(0.1)(32)}{30} = x \text{ normality}$$

$$0.11 = x \text{ normality}$$

Review Questions

1. If 38 lb of chemical are added to 290 lb of water, what is the percent strength of the solution?

2. In 180 gal of solution there are 60 lb of a chemical. What is the percent strength of this solution? Assume the solution has the same density as water—8.34 lbs/gal.

3. You are asked to prepare 40 gal of a 3 percent strength solution. How many pounds of chemical will need to be dissolved in the solution?

4. What volumes of a 2 percent and 5 percent solution must be mixed to make 250 gal of a 2.5 percent solution?

5. How many pounds of a 6 percent solution and water should be mixed to obtain 150 lb of a 5 percent solution?

6. If 93 g of calcium hydroxide [$Ca(OH)_2$] are used in making up a solution, how many moles have been used? (Use the Periodic Table given in Appendix B for atomic weights.)

7. If 0.7 moles of H_2SO_4 are dissolved in 820 mL of solution, what is the molarity of the solution?

8. If 75 g of NaOH are dissolved in 700 mL of solution what is the molarity of the solution? The molecular weight of NaOH is 40.00.

9. You are asked to make up 600 mL of a 1.8 M solution of Na_2CO_3. (a) How many moles of Na_2CO_3 will have to be dissolved in the 600 mL solution? (b) How many grams of Na_2CO_3 does this equal? The molecular weight of Na_2CO_3 is 105.99.

10. If 70 g of NaOH are used in making up a solution, how many equivalent weights of NaOH are used in making up the solution? Use 40.00 g for the equivalent weight of NaOH.

11. If 3 equivalent weights of H_2CO_3 have been used in making up 2.5 L of solution, what is the normality of the solution?

12. If 120 g of NaCl are used in making up 900 mL of solution, what is the normality of the solution? Use 58.44 g for the equivalent weight of NaCl.

13. The water at your treatment plant was analyzed for nitrates. If the nitrate concentration is reported to be 38 mg/L as nitrate, what is the concentration as nitrogen? The atomic weight of nitrogen is 14.01 and the molecular weight of nitrate is 62.01.

14. Laboratory test results indicate that the magnesium concentration of a water sample is 70 mg/L as magnesium. What is the magnesium concentration expressed as $CaCO_3$? Use 12.16 g for the equivalent weight of magnesium and use 50.00 g for the equivalent weight of calcium.

15. How much water should be added to 150 mL of a 0.5 N solution of H_2SO_4 to produce a 0.2 N solution of H_2SO_4?

16. It requires 43 mL of a 1.5 N solution of NaOH to react with (neutralize) 25 mL of a certain acid solution. What is the normality of the acid solution?

Summary Answers

1. 11.6 percent strength solution

2. 4 percent strength solution

3. 10 lb chemical

4. 41.67 gal of the 5 percent solution; 208.33 gal of the 2 percent solution

5. 125 lb of the 6 percent solution; 25 lb of water

6. 1.26 moles

7. 0.85 M solution

8. 2.64 M solution

9. 1.08 moles Na_2CO_3, 114.47 g Na_2CO_3

10. 1.75 equivalent weights of NaOH

11. 1.2 N solution

12. 2.28 N solution

13. 8.59 mg/L nitrate (as nitrogen)

14. 287.83 mg/L magnesium (as $CaCO_3$)

15. 225 mL

16. 2.58 N

Detailed Answers

1.

$$\frac{\text{Percent}}{\text{Strength}} = \frac{\text{Weight of solute}}{\text{Weight of solution}} \times 100$$

$$= \frac{38 \text{ lb chemical}}{(38 \text{ lbs } + 290 \text{ lbs}) \text{ solution}} \times 100$$

$$= \frac{38 \text{ lb chemical}}{328 \text{ lb solution}} \times 100$$

$$= 0.116 \times 100$$

$$= 11.6\% \text{ strength solution}$$

2. First convert 180 gal to pounds:

$$(180 \text{ gal})(8.34 \text{ lb/gal}) = 1501 \text{ lb solution}$$

Now calculate the percent strength:

$$\frac{\text{Percent}}{\text{Strength}} = \frac{\text{Weight of solute}}{\text{Weight of solution}} \times 100$$

$$= \frac{60 \text{ lb chemical}}{1501 \text{ lb solution}} \times 100$$

$$= 0.04 \times 100$$

$$= 4\% \text{ strength solution}$$

3. Use the equation for percent strength:

$$\frac{\text{Percent}}{\text{Strength}} = \frac{\text{Weight of solute}}{\text{Weight of solution}} \times 100$$

Fill in the given information:

$$3 = \frac{x \text{ lbs chemical}}{(40)(8.34) \text{ lb solution}} \times 100$$

Solve for the unknown value:

$$3 = \frac{x}{(40)(8.34)} \times 100$$

Restate this as:

$$\frac{3}{1} = \frac{(x)(100)}{(40)(8.34)}$$

$$\frac{(40)(8.34)(3)}{1} = (x)(100)$$

$$\frac{(40)(8.34)(3)}{100} = x$$

$$10 \text{ lb} = x$$
chemical

4. Use the rectangle method to solve the problem:

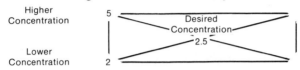

Complete the right side of the rectangle:

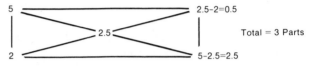

The circled numbers of the right side of rectangle indicate the volume *ratios* by which the two solutions are to be mixed. There are a total of three parts to the new solution. The numbers indicate that 0.5 parts out of three (0.5/3) come from the 5 percent solution and 2.5 parts out of the three (2.5/3) come from the 2 percent solution.

The number of gallons of each of the two solutions which must be used in making a 2.5 percent solution are

$$\left(\frac{0.5}{3}\right)(250 \text{ gal}) = 41.67 \text{ gal of } 5\% \text{ solution}$$

$$\left(\frac{2.5}{3}\right)(250 \text{ gal}) = 208.33 \text{ gal of } 2\% \text{ solution}$$

5. Use the rectangle method to solve the problem. Water is considered a 0 percent solution; therefore, the lower concentration used in the problem will be 0 percent:

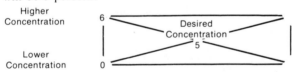

Now complete the right side of the rectangle:

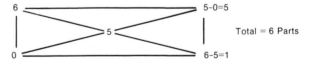

As indicated by the circled numbers on the right side of the rectangle, there is a total of six parts to the new solution. The numbers show that five parts out of six (5/6) is the ratio by which the 6% solution should be added; and one part out of six (1/6) is the ratio by which water (the 0 percent solution) should be added.

The weights of each of the two solutions which are needed to make the 5 percent solution are

$$\left(\frac{5}{6}\right)(150 \text{ lbs}) = 125 \text{ lb of } 6\% \text{ solution}$$

$$\left(\frac{1}{6}\right)(150 \text{ lbs}) = 25 \text{ lb of water}$$

6. First determine the molecular weight of $Ca(OH)_2$:

	No. of atoms		Atomic weight		Total weight
Calcium (Ca)	1	×	40.08	=	40.08
Oxygen (O)	2	×	16.00	=	32.00
Hydrogen (H)	2	×	1.01	=	2.02

Molecular weight of $Ca(OH)_2$ =   74.10

Now calculate number of moles of $Ca(OH)_2$ used in making the solution:

$$\text{Moles} = \frac{\text{Total weight}}{\text{Molecular weight}}$$

$$= \frac{93}{74.10}$$

$$= 1.26 \text{ moles of } Ca(OH)_2$$

7. In this problem, the number of moles in the solution has already been determined. therefore, the molarity of the solution is

$$\text{Molarity} = \frac{\text{Moles of solute}}{\text{Litres of solution}}$$

$$= \frac{0.7 \text{ moles}}{0.820 \text{ L solution}}$$

$$= 0.85 \text{ molarity}$$

$$= 0.85 \ M \text{ solution}$$

8. First determine the number of moles of NaOH in the solution:

$$\frac{\text{Number}}{\text{of Moles}} = \frac{\text{Total weight}}{\text{Molecular weight}}$$

$$= \frac{74}{40.00}$$

$$= 1.85 \text{ moles}$$

Now calculate the molarity of the solution:

$$\text{Molarity} = \frac{\text{Moles of solute}}{\text{Litres of solution}}$$

$$= \frac{1.85 \text{ moles}}{0.700 \text{ L solution}}$$

$$= 2.64 \text{ molarity}$$

$$= 2.64 \text{ } M \text{ solution}$$

9. (a) Use the molarity equation and fill in the given information:

$$\text{Molarity} = \frac{\text{Moles of solute}}{\text{Litres of solution}}$$

$$1.8 \text{ } M = \frac{x \text{ moles Na}_2\text{CO}_3}{0.600 \text{ L solution}}$$

Then solve for the unknown value:

$$(0.600)(1.8) = x$$

$$1.08 \text{ moles Na}_2\text{CO}_3 = x$$

(b) One mole of Na_2CO_3 weighs 105.99 g. Therefore, 1.08 moles weighs

$$(1.08)(105.99 \text{ g}) = 114.47 \text{ g Na}_2\text{CO}_3$$

10.
$$\frac{\text{Number of}}{\text{Equivalent weights}} = \frac{\text{Total weight}}{\text{Equivalent weight}}$$

$$= \frac{70 \text{ g}}{40.00 \text{ g}}$$

$$= 1.75 \text{ equivalent weights of NaOH}$$

11.
$$\text{Normality} = \frac{\text{Equivalent weights of solute}}{\text{Litres of solution}}$$

$$= \frac{3 \text{ equivalent weights}}{2.5 \text{ L solution}}$$

$$= 1.2 \text{ normality}$$

$$= 1.2 \text{ } N \text{ solution}$$

12. First determine the number of equivalent weights used in the solution:

$$\frac{\text{Number of}}{\text{Equivalent weights}} = \frac{\text{Total weight}}{\text{Equivalent weight}}$$

$$= \frac{120 \text{ g}}{58.44 \text{ g}}$$

$$= 2.05 \text{ equivalent weights}$$

Now calculate the normality of the solution:

$$\text{Normality} = \frac{\text{Equivalent weights of solute}}{\text{Litres of solution}}$$

$$= \frac{2.05 \text{ equivalent weights}}{0.900 \text{ L solution}}$$

$$= 2.28 \text{ normality}$$

$$= 2.28 \ N \text{ solution}$$

13. To express the nitrate concentration *as nitrogen*, multiply the 38 mg/L concentration by the ratio of the molecular weights:

$$\left(\frac{\text{Molecular weight of new measurement}}{\text{Molecular weight of old measurement}}\right) (\text{old concentration}) = (\text{new concentration})$$

$$\left(\frac{\text{Molecular weight of N}}{\text{Molecular weight of NO}_3^-}\right) (\text{concentration as NO}_3^-) = (\text{concentration as N})$$

$$\left(\frac{14.01}{62.01}\right) (38 \text{ mg/L as NO}_3^-) = (x \text{ mg/L as N})$$

$$= 8.59 \text{ mg/L as N}$$

14. To express the hardness concentration *as CaCO₃*, multiply the 70 mg/L concentration by the ratio of the equivalent weights:

$$\left(\frac{\text{Equivalent weight of new measure}}{\text{Equivalent weight of old measure}}\right) (\text{old concentration}) = (\text{new concentration})$$

$$\left(\frac{\text{Equivalent weight of CaCO}_3}{\text{Equivalent weight of Mg}}\right) (\text{concentration as Mg}) = (\text{concentration as CaCO}_3)$$

$$\left(\frac{50.00 \text{ g}}{12.16 \text{ g}}\right) (70 \text{ mg/L as Mg}) = (x \text{ mg/L as CaCO}_3)$$

$$= 287.83 \text{ mg/L as CaCO}_3$$

15. Use the equation that relates solution concentration and volume to calculate the total volume of the new solution:

$$(N_1)(V_1) = (N_2)(V_2)$$
$$(0.5\ N)(150\ \text{mL}) = (0.2\ N)(x\ \text{mL})$$
$$\frac{(0.5)(150)}{0.2} = x\ \text{mL}$$
$$375 = x$$

Therefore, the total volume of the new solution will be 375 mL. Since the volume of the original solution is 150 mL, 225 mL of water must be added to obtain the 0.2 N solution:

$$\begin{array}{c}\text{150 mL}\\ \text{of 0.5 } N \text{ solution}\end{array} + \begin{array}{c}\text{225 mL}\\ \text{water}\end{array} = \begin{array}{c}\text{375 mL}\\ \text{of 0.2 } N \text{ solution}\end{array}$$

16.
$$(N_1)(V_1) = (N_2)(V_2)$$
$$(1.5\ N)(43\ \text{mL}) = (x\ N)(25\ \text{mL})$$
$$\frac{(1.5)(43)}{25} = x\ \text{normality}$$
$$2.58 = x\ \text{normality}$$

Chemistry 5

Acids, Bases, and Salts

Inorganic compounds (compounds generally not containing carbon) can be classified into three main groups: (1) acids, (2) bases, and (3) salts. These three terms are commonly used throughout chemistry and it is important that you understand the basic features that distinguish them. The following definitions are adequate for most water treatment chemistry; however, somewhat different definitions may be used in advanced work.

C5-1. Acids

An ACID is any substance that releases hydrogen ions (H^+) when it is mixed into water. For example, shortly after sulfuric acid (H_2SO_4) is mixed into water, many of the H_2SO_4 molecules DISSOCIATE (come apart), forming H^+ and SO_4^- ions. The release of H^+ ions indicates that H_2SO_4 is an acid.

Acids that dissociate readily are known as *strong acids*. Most of the molecules of a strong acid dissociate when mixed into water, releasing a large concentration of hydrogen ions. Examples of strong acids are sulfuric (H_2SO_4), hydrochloric (HCl), and nitric (HNO_3).

Acids that dissociate poorly are known as *weak acids*. They release very few hydrogen ions in water. Examples include carbonic acid (H_2CO_3), which is the acid found in soft drinks; and hydrogen sulfide (H_2S), the compound responsible for the rotten-egg odor found naturally in some ground waters and certain deep surface waters. The following four equations are examples of how acids dissociate when mixed into water:

$$HCl \rightarrow H^+ + Cl^-$$ (a strong acid: generally, more than 99% of the molecules dissociate in water)

$$H_2SO_4 \rightarrow 2H^+ + SO_4^{-2}$$ (a strong acid: generally, more than 99% of the molecules dissociate in water)

$$H_2S \rightarrow 2H^+ + S^{-2}$$ (a weak acid: generally, less than 0.1% of the molecules dissociate in water)

$$H_2CO_3 \rightarrow 2H^+ + CO_3^{-2}$$ (a weak acid: generally, less than 0.1% of the molecules dissociate in water)

Solutions that contain significant numbers of H^+ ions are called ACIDIC. Three other features that distinguish acids from bases and salts are:

- Acids change the color of chemical color indicators:
  —Acids turn litmus paper red.
  —Acids turn phenolphthalein colorless.
  —Acids turn methyl orange to red.

- Acids neutralize bases, resulting in the formation of a salt and water.

- Acids found naturally in foods give the foods a sour taste. The sour flavor of citrus fruits is caused by citric acid. CAUTION: *Tasting the acids found in laboratories and water treatment plants can be dangerous, even fatal.*

C5-2. Bases

A BASE is any substance that produces hydroxyl ions (OH^-) when it dissociates in water. Lime [$Ca(OH)_2$], caustic soda [sodium hydroxide, or NaOH], and common household ammonia [NH_4OH] are familiar examples of bases. Strong bases are those that dissociate readily, releasing a large concentration of hydroxyl ions. NaOH is an example of a very strong, caustic base. Weak bases, such as $Ca(OH)_2$ and NH_4OH, dissociate poorly, releasing few OH^- ions. The following equations are examples of how acids dissociate when mixed into water:

$$NaOH \rightarrow Na^+ + OH^-$$ (a strong base: generally, more than 99% of the molecules dissociate in water)

$$KOH \rightarrow K^+ + OH^-$$ (a strong base: generally, more than 99% of the molecules dissociate in water)

$$Ca(OH)_2 \rightarrow Ca^{+2} + 2OH^-$$ (a relatively weak base: generally, less than 15% of the molecules dissociate in water)

$$NH_4OH \rightarrow NH_4^+ + OH^-$$ (a weak base: generally, less than 0.5% of the molecules dissociate in water)

Solutions that contain significant numbers of OH^- ions are called BASIC SOLUTIONS or ALKALINE SOLUTIONS. The term "alkaline" should not be confused with the term "alkalinity," which has a special meaning in water treatment. Three other features that distinguish bases from acids and salts are:

- Bases change the color of chemical color indicators:
    - —Bases turn litmus paper blue.
    - —Bases turn phenolphthalein red.
    - —Bases turn methyl orange to yellow.

- Bases neutralize acids, resulting in the formation of a salt and water.

- Bases found naturally in foods give the foods a bitter taste. Baking soda and milk of magnesium both taste bitter because they contain basic compounds. CAUTION: *Tasting the bases found in laboratories and water treatment plants can be dangerous, even fatal.*

C5-3.   Salts

SALTS are compounds resulting from an acid–base mixture. The process of mixing an acid with a base to form a salt is called NEUTRALIZATION. Calcium sulfate ($CaSO_4$), for example, is a salt formed by the following acid–base neutralization:

$$H_2SO_4 + Ca(OH)_2 \rightarrow CaSO_4 + H_2O$$

Another example is sodium chloride (NaCl):

$$HCl + NaOH \rightarrow NaCl + H_2O$$

Notice that each acid–base reaction results in a salt plus water. Salts generally have no effect on color indicators. When occurring naturally in foods, salts taste salty; however, like all chemicals found in the laboratory or water treatment plant, *salts may be poisonous, and tasting them could be dangerous or fatal.*

C5-4.   pH

Solutions range from very acidic (having a high concentration of H^+ ions) to very basic (having a high concentration of OH^- ions). When there are exactly as many OH^- ions as H^+ ions, the solution is neutral—neither acidic nor basic—and each OH^- can combine with an H^+ to form a molecule of H_2O. Pure water is neutral, and most salt solutions are neutral or very nearly so.

The pH of a solution is a measurement of how acidic or basic the solution is. The pH scale (Figure 8) runs from 0 (most acidic) to 14 (most basic). Pure water has a pH of 7, the center of the range, neither acidic nor basic.

High Concentration of H^+ Ions	H^+ and OH^- Ions in Balance	High Concentration of OH^- Ions

$0 — 1 — 2 — 3 — 4 — 5 — 6 — 7 — 8 — 9 — 10 — 11 — 12 — 13 — 14$

Pure Acid	Neutral	Pure Base

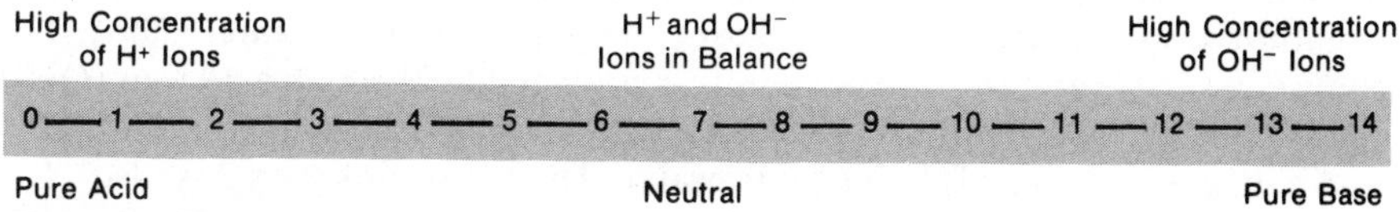

Figure 8.   The pH scale

For each treatment process, there is a pH at which the operation is most effective. If the pH of the water is too low (the water is too acidic) for an operation to be effective, then the pH can be increased by the addition of a base, such as lime [$Ca(OH)_2$]. The OH^- ions released by the base will combine with some of the H^+ ions of the acidic water, forming H_2O molecules and lowering the concentration of H^+ ions.

Similarly, if the pH of water is too high (the water is too basic), then the pH can be lowered by the addition of an acid. In water treatment, pH is often lowered by bubbling carbon dioxide (CO_2) gas through the water. The CO_2 reacts with water to form carbonic acid (H_2CO_3), and the acid dissociates into $2H^+ + CO_3^{-2}$. The H^+ ions then combine with some of the OH^- ions of the basic water, forming H_2O molecules and lowering the concentrations of OH^- ions.

C5-5.   Alkalinity

When acid is added to water that has a high concentration of OH^- ions (basic water, high in pH), the H^+ ions released by the acid combine with the OH^- ions in the water to form H_2O. As long as the water contains OH^- ions, those ions will NEUTRALIZE the added acid and the water will remain basic. After enough acid has been added to combine with all the OH^- ions, addition of more acid will give the water a high concentration of H^+ ions—the water will become acidic.

However, if acid is added to water containing carbonate (CO_3^{-2}) ions in addition to OH^- ions, then some of the H^+ ions released by the acid will combine with the OH^- to form H_2O, and some will combine with the CO_3^{-2} ions to form HCO_3^- (bicarbonate). As more acid is added, the H^+ ions released will combine with the HCO_3^- ions to form H_2CO_3 (carbonic acid). A concentration of H^+ ions will begin to accumulate (the water will become acidic) only after enough acid has been added to convert all the OH^+ to H_2O and to convert all the CO_3^{-2} to HCO_3^- and then to H_2CO_3.

The CO_3^{-2} and HCO_3^- ions in the water increase the capacity of the water to neutralize (or BUFFER) an acid. ALKALINITY is a measurement of a water's capacity to neutralize an acid, whether the neutralization is the result of OH^-, CO_3^{-2}, HCO_3^-, or other negative ions.

In water treatment, OH^-, CO_3^{-2}, and HCO_3^- are the ions causing most of the alkalinity. Alkalinity caused by OH^- is called HYDROXYL ALKALINITY; if caused by CO_3^{-2} it is called CARBONATE ALKALINITY; and if caused by HCO_3^- it is called BICARBONATE ALKALINITY. The combined effect of all three types is reported by a lab as the TOTAL ALKALINITY.

The experiment illustrated in Figure 9 demonstrates the relationship between alkalinity and pH. In Step 1, equal volumes of two solutions are prepared. Solution 1 is made up by mixing a base [lime, $Ca(OH)_2$] into water. The pH of Solution 1 is 11, and the only alkalinity is provided by the hydroxyl ions (OH^-) released by the base. Solution 2 is made up by mixing the same base [$Ca(OH)_2$] and calcium carbonate [$CaCO_3$] into water. The pH of Solution 2 is also 11— that is, Solution 1 and Solution 2 have the same concentration of OH^- ions.

However, Solution 2 has a higher alkalinity than Solution 1, because the alkalinity of Solution 2 includes carbonate ions (CO_3^{-2} released by the calcium carbonate) in addition to the hydroxyl ions (OH^- released by the lime).

In Step 2, sulfuric acid (H_2SO_4) is slowly added to Solution 1 until the pH drops to 7, as indicated by the pH meter. The pH of 7 means that enough H^+ ions have been added (released by the acid) to combine with all the OH^- ions originally in the basic solution. The same volume of acid is added to Solution 2, but the pH meter still indicates a basic solution: pH greater than 7. The carbonate ions (CO_3^{-2}) in solution 2 have combined with some of the H^+ ions released by the acid; as a result, there are not yet enough H^+ ions in Solution 2 to balance the original concentration of OH^- ions.

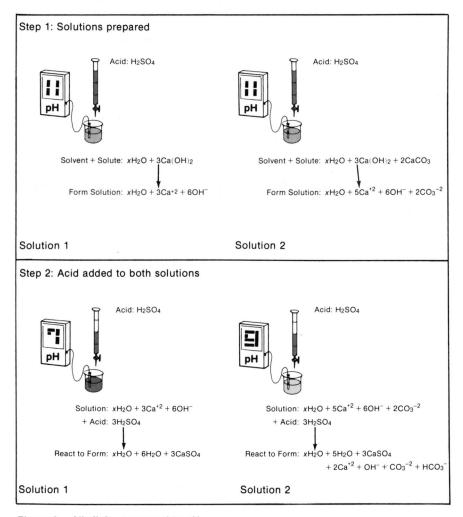

Figure 9.   Alkalinity compared to pH

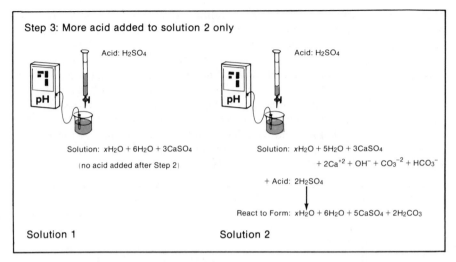

Figure 9 (Continued)

To complete the experiment in Step 3, more acid is added to Solution 2 until the pH drops to a neutral 7. The solutions had the same pH at the beginning of the experiment, and they have the same pH at the end. But Solution 2, which had the higher alkalinity, was capable of neutralizing a greater volume of acid than Solution 1. (Note that the pH meter readings and equations in Figure 9 indicate a slightly simpler behavior of the chemicals than actually would occur in a laboratory.)

Review Questions

1. What are the three main groups of inorganic compounds?

2. List four features that distinguish acids.

3. List four features that distinguish bases.

4. Define salts.

5. What characteristic distinguishes a
   (a) strong acid from a weak acid?
   (b) strong base from a weak base?

6. Write the chemical formulas for the dissociation of one acid and one base.

7. (a) What is the range of the pH scale?
   (b) What range of pH indicates an acidic solution?
   (c) What range of pH indicates a basic solution?
   (d) What is the pH of pure water?

8. Define *alkalinity*.

9. What are the three types of alkalinity most commonly found in water treatment and what ion causes each type?

Summary Answers

(See Detailed Answers.)

Detailed Answers

1. Acids, bases, and salts.

2. Acids (1) releases H^+ ions on dissociation, (2) give a sour taste to foods, (3) turn litmus paper red, and (4) neutralize bases to form a salt and water.

3. Bases (1) release OH^- ions on dissociation, (2) give a bitter taste to foods, (3) turn litmus paper blue, and (4) neutralize acids to form a salt and water.

4. Compounds resulting from an acid–base neutralization reaction.

5. (a) On dissociation a strong acid releases a large concentration of H^+ ions. Weak acids dissociate poorly, releasing very few H^+ ions.

   (b) On dissociation a strong base releases a large concentration of OH^- ions. Weak bases dissociate poorly, releasing very few OH^- ions. Weak bases dissociate poorly, releasing very few OH^- ions.

6. Examples include:

$$HCl \; \rightarrow \; H^+ \; + \; Cl^- \quad \text{(an acid)}$$
$$NaOH \; \rightarrow \; Na^+ \; + \; OH^- \quad \text{(a base)}$$

   See Section C5-3 for other examples.

7. (a) pH ranges from 0 to 14.

   (b) Any pH less than 7 indicates an acidic solution.

   (c) Any pH above 7 indicates a basic solution.

   (d) The pH of pure water is 7.

8. Alkalinity is a measurement of a water's capacity to neutralize acid.

9. The types of alkalinity most commonly encountered in water treatment are (1) hydroxyl alkalinity, caused by the OH^- ion, (2) carbonate alkalinity, caused by the CO_3^{-2} ion, and (3) bicarbonate alkalinity, caused by the HCO_3^{-2} ion.

Chemistry 6

Chemistry of Treatment Processes

In order to operate a water treatment plant successfully, you will need to understand some of the reactions involved when chemicals are added to water, or when constituents are removed from water. The following material discusses the basic chemistry of

- Taste and odor removal
- Coagulation
- Iron and manganese removal
- Lime–soda ash softening
- Recarbonation
- Ion exchange softening
- Scale and corrosion control (stabilization)
- Chlorination

C6-1.   Taste and Odor Removal

The removal of taste and odor may be one of the more difficult problems you will confront. This is due to three important factors. (1) There are many different materials that can cause taste and odor. (2) No common treatment process can remove all types of tastes and odors. (3) The tests used to measure taste and odor rely on keenness of one's sense of taste and smell.

Perhaps the best tools available for detecting, preventing, or eliminating taste and odor problems are two laboratory tests: the threshold odor number (TON) test and the jar test. These tests are as valuable in maintaining quality odor-free water as they are in identifying and treating an odor problem. Threshold odor test results indicate if treated water is within the limit of 3 TON, recommended in the EPA's Secondary Drinking Water Regulations. The results identify specific types of odors and are used with the jar test to find specific treatment methods to remove the odor. Records of threshold odor tests are valuable in identifying those odors that consumers seem most sensitive toward.

Physical removal and chemical oxidation are the two methods normally used to remove tastes and odors. Physical methods for taste and odor removal, not discussed in detail in this text, include (1) coagulation followed by filtration and (2) adsorption. Chemical oxidation to remove tastes and odors can be attained using one of several oxidizing agents:

- Oxygen
- Chlorine
- Potassium permanganate
- Chlorine dioxide
- Ozone

In chemical oxidation the undesirable taste and odor materials are converted chemically to compounds that do not cause tastes or odors. In some cases the conversion allows the materials to be removed by precipitation, although such physical removal is not always essential. The effectiveness of each oxidizing agent varies, depending on the type of odor that is present and the way in which the oxidant is applied.

Oxygen

Oxygen, mixed into the water by aeration, can be effective in removing tastes and odors caused by iron, manganese, and hydrogen sulfide. Aeration brings the offending material into direct contact with oxygen in the atmosphere. The materials are then oxidized into stable non-troublesome forms.

Iron and manganese.   The removal of these two constituents is discussed in section C6-3, "Iron and Manganese Removal."

Hydrogen sulfide.   Aeration removes H_2S primarily by the mechanical scrubbing action of air moving through water; however, some oxidation does occur, as shown in Equations 1 and 2.

$$2H_2S + O_2 \xrightarrow{\text{aeration}} 2H_2O + 2S\downarrow \qquad Eq\ 1$$

Hydrogen sulfide   Sulphur

Further

$$2S + 4O_2 \xrightarrow{\text{aeration}} 2SO_4 \qquad Eq\ 2$$

Sulfur   Sulfate ion

Although oxidation decreases the H_2S, oxidation alone does not completely remove or convert all the H_2S. Equation 1 shows elemental sulfur (solid sulfur) as the final product, which gives the aerated water a milky-blue color. To remove the sulfur, the water must be chemically coagulated and filtered. (The "↓" after the "S" indicates that sulfur is a precipitate.) Because H_2S is best removed by scrubbing at about pH 4.5, it is common practice to bubble CO_2 through the water prior to aeration in order to reduce the pH.

Notice that in Equation 2 further aeration converts the sulfur to sulfate ion. This solves the taste and odor problem; however, it leaves the sulfur in the water in sulfate form—sulfate may be converted back to H_2S if the water becomes anaerobic, such as it might in a dead-end or slightly used portion of the distribution system.

Chlorine

Chlorination, performed within the proper limits, can be an effective way of oxidizing many substances that cause taste and odor. For example, algae-caused odors described as fishy, grassy, or septic can be controlled by prechlorination to a free residual of 0.25–5 mg/L. Iron and manganese can be effectively removed by chlorination, as discussed later under "Iron and Manganese Removal." Hydrogen sulfide (H_2S) can also be successfully oxidized with chlorine, as discussed later under "Chlorination."

Chlorine can intensify certain tastes and odors. For example, chlorine will combine with phenols in water, producing extremely objectionable medicinal tastes (phenol is an aromatic organic compound that gives water the characteristic odor of creosote or coal tar). Chlorine will also intensify earthy odors caused by certain types of algae and actinomycetes. (Actinomycetes are a group of organisms usually associated with taste and odor problems. The group has some characteristics of bacteria and some of fungus.)

Chlorine itself can, under certain circumstances, be the cause of chlorine-like tastes and odors. The level of chlorine residual that will cause taste and odor will vary with the characteristics of the raw water.

Studies have identified more than 700 different organic chemicals ("organics") in drinking water throughout the US. Although organics can cause taste and odor problems in water, there is a more important consideration: long-term exposure to certain organics may pose a threat to the public's health, including the risk of cancer. One group of organics, TRIHALOMETHANES, is formed when chlorine is added to water containing humic acid (a natural organic produced by decaying vegetation). To reduce the amount of trihalomethanes formed in drinking water, prechlorination for taste and odor control should be avoided if humic acids are present in the raw water supply, particularly if other taste and odor control methods are available.

Potassium Permanganate

Potassium permanganate ($KMnO_4$), used either alone or in combination with other chemicals, is effective in removing iron and manganese and oxidizing both organic and inorganic materials that cause taste and odor. When added to water containing taste and odor compounds the reaction is

$$2KMnO_4 + H_2O + \text{Taste and Odor Compounds} \rightarrow \qquad Eq\ 3$$

$$\underset{\substack{\text{Manganese} \\ \text{dioxide}}}{2MnO_2\downarrow} + \underset{\substack{\text{Potassium} \\ \text{hydroxide}}}{2KOH} + \underset{\text{Oxygen}}{3O} + \text{T and O Compounds}$$

This reaction produces oxygen in the water (notice the 3 O's in the equation above) which oxidizes the organic and inorganic taste and odor compounds. The manganese dioxide by-product (insoluble and very finely divided) is removed later, through coagulation, settling, and filtration.

Dosages of $KMnO_4$ will vary from 0.5 to 15 mg/L, although dosages in the range of 0.5–2.5 mg/L are usually adequate to oxidize most taste-and-odor-producing chemicals. Potassium permanganate should be added very early in the treatment process, perhaps at the intake in order to allow enough time for oxidation to take place. In most cases, detention times of 1–1.5 hr are adequate, but for some particularly stubborn taste and odor compounds, detention times of 4–5 hr may be needed.

Water takes on a pink color when potassium permanganate is initially added. Then, as it oxidizes to manganese dioxide (MnO_2), the color begins to change, first to yellow, then orange, and finally brown, which is the color of the manganese dioxide precipitate. By the time the water reaches the filters, the pink color should be gone entirely. If not, oxidation will continue after filtration, giving the water an unacceptable yellow to yellow-brown color. This problem of "color break-through" can be controlled by either reducing the $KMnO_4$ dosage or by moving the point of chemical addition farther upstream.

Chlorine Dioxide

Chlorine dioxide (ClO_2) is a a strong oxidizing agent used primarily to control the odors caused by phenolic compounds. It is about 2–2.5 times as powerful as chlorine. Chlorine dioxide is not purchased as a ready-made chemical like most other water treatment chemicals. Instead, it is made at the treatment plant by adding a solution of sodium chlorite ($NaClO_2$) to a concentrated chlorine solution, as follows:

$$2NaClO_2 + Cl_2 \rightarrow 2ClO_2 + 2NaCl \qquad Eq\ 4$$

The resultant solution should have a pale-yellow color. The amount of chlorine and sodium chlorite needed to reach this color usually varies, depending on the strength of the sodium chlorite. The sodium chlorite commonly used is about 80 percent pure. Theoretically, 1.68 lb of sodium chlorite would be mixed with 0.5 lb of chlorine in order to produce 1.0 lb of chlorine dioxide. However, by using more chlorine (at least as much chlorine as sodium chlorite) practice has shown that the reaction occurs faster and the conversion of the chlorite is more complete.

Chlorine dioxide, although stronger than chlorine, is more expensive to use. Experience has shown that chlorine dioxide is most economically and most effectively used *after* chlorination. In this way larger amounts of chlorine can be used to remove the majority of tastes and odors. Then a small follow-up dose of chlorine dioxide is all that is needed to remove phenols and phenolic compounds, such as chlorophenols formed during chlorination. A dosage of 0.2–0.3 mg/L is usually adequate for taste and odor control.

Ozone

Ozone treatment, called ozonation, can be an effective method of taste and odor control, although experience shows that results are quite variable depending on the quality of the water being treated. Although ozonation is not common in this country, it is widely practiced in Europe. Ozone (O_3) is generated at the treatment plant by releasing an electrical spark into a stream of oxygen under controlled circumstances. Although ozone is a powerful oxidizing agent, the cost to produce it is high.

C6-2.   Coagulation

Coagulation is the gathering together of colloidal and finely divided suspended matter by the addition of a floc-forming chemical or by biological processes. The actual chemical reactions are very complex and are influenced by a variety of factors including

- Type of coagulant used
- Amount of coagulant used
- Type and length of flash mixing
- Type and length of flocculation
- Effectiveness of sedimentation
- Other chemicals used
- Temperature
- pH
- Alkalinity
- Zeta potential
- Raw water quality

Alum

Because it is effective, inexpensive, and easy to apply, aluminum sulfate (usually called alum) is presently the most widely used coagulant in the field of water treatment.

Alum is available in two forms. Filter alum is the name given to solid or dry alum [$Al_2(SO_4)_3 \cdot 14H_2O$]. (The portion of the equation showing " $\cdot$ 14H$_2$O" indicates that a certain amount of water is included chemically along with the alum; in this case, 14 molecules of water are included with each molecule of alum. For simplicity, the water portion of the chemical equation is often left off.) Filter alum is ivory-white and is available in lump, ground, rice, or powdered form. Liquid alum [$Al_2(SO_4)_3 \cdot xH_2O$], is alum already in solution. It is available in three strengths—the strongest being less than half the strength of dry filter alum. Liquid alum can vary in color, depending on strength, from a slight, white irridescent-like color to a yellow-brown.

Depending on the form, filter alum varies in density from 48 to 76 lb/cu ft. Liquid alum has a density of approximately 11 lb/gal, considerably more than water.

When alum is added to water it reacts as follows:

$$Al_2(SO_4)_3 + 3Ca(HCO_3)_2 \rightarrow 2Al(OH)_3\downarrow + 3CaSO_4\downarrow + 6CO_2 \qquad Eq\ 5$$

| Alum | Natural bicarbonate alkalinity | Aluminum hydroxide floc | Calcium sulfate | Carbon dioxide |

$$Al_2(SO_4)_3 + 3Na_2CO_3 + 3H_2O \rightarrow 2Al(OH)_3\downarrow + 3Na_2SO_4 + 3CO_2 \quad Eq\ 6$$

| Alum | Added carbonate alkalinity | Aluminum hydroxide hydroxide | Sodium sulfate | Carbon dioxide |

$$Al_2(SO_4)_3 + 3Ca(OH)_2 \rightarrow 2Al(OH)_3\downarrow + 3CaSO_4\downarrow \qquad Eq\ 7$$

| Alum | Added hydroxide alkalinity | Aluminum hydroxide floc | Calcium sulfate |

Notice in each case that alkalinity is absolutely necessary for the reaction to take place. It does not matter whether the alkalinity is present naturally, as in Equation 5, or whether it is added prior to coagulation, as in Equations 6 and 7. The important point is that there must be enough alkaline substances to react with the alum to form aluminum hydroxide [$Al(OH)_3$], the sticky floc material. About 0.5 mg/L of alkalinity is required for each milligram of alum added per litre. Without an adequate supply of alkalinity the alum will not precipitate (not form aluminum hydroxide floc) and will pass through the filters. If alkalinity is added later, as it might be for corrosion control, the floc will form then and settle out either in the clearwell or in the distribution system, which can cause serious problems.

When alum is added to an alkaline water, as shown in the equation above, coagulation occurs in three steps. First, the positively charged aluminum ions (Al^{+3}) attract the negatively charged particles that cause color and turbidity, and form tiny particles called microflocs. This marks the beginning of coagulation. Second, because many of these microfloc particles are now positively charged, they begin to attract and hold more negatively charged color-causing and turbidity-causing material. These first two coagulation steps occur very quickly (in microseconds).

Third, the microfloc particles grow into easily visible "mature" floc particles. This growth occurs partly by the continual attraction of the color and turbidity materials, partly by the adsorption of virus, bacteria, and algae onto the microfloc, and partly by the random collision of microflocs that cause particles to stick together. Later, during sedimentation, the large floc particles settle rapidly, leaving the water clear.

The alum dosage to use for best results on a particular water will depend on the various factors listed above. In general, alum dosages range from 15 to 100 mg/L.

The effective pH range for alum dosing is between 5.5 and 8.5. It is often desirable to adjust pH within this range (by increasing alkalinity) so that alum will perform at its best. The pH for best performance can be found by using the jar testing technique.

The rate of floc formation varies with temperature. Lower temperature means slower floc formation. Although there is usually no practical way of adjusting water temperature, one can compensate for lower temperatures by increasing the alum dosage.

From time to time raw water quality changes, perhaps as a result of changing seasons, or possibly due to a high runoff of turbid water from a rainstorm. Routine dosage checks using the jar test will help to ensure adequate coagulation/flocculation results.

Ferric Sulfate

Ferric sulfate is a reddish-gray [commercially called Ferric-floc, $Fe_2(SO_4)_3 \cdot 3H_2O$] or grayish-white [commercially called Ferriclear, $Fe(SO_4)_3 \cdot 2H_2O$] granular material. Their densities vary. Ferric-floc has a density of 60–74 lb/cu ft and Ferriclear has a density of 78–90 lb/cu ft.

When ferric sulfate is added to water it reacts as follows:

$$Fe_2(SO_4)_3 + 3Ca(HCO_3)_2 \rightarrow 2Fe(OH)_3\downarrow + 3CaSO_4 + 6CO_2 \qquad Eq\ 8$$

| Ferric sulfate | Natural bicarbonate alkalinity | Ferric hydroxide floc | Calcium sulfate | Carbon dioxide |

The reaction is very similar to alum. Notice that as with alum, ferric sulfate requires alkalinity in the water in order to form the floc particle ferric hydroxide [$Fe(OH)_3$]. If natural alkalinity is not sufficient, then alkaline chemicals (soluble salts containing HCO_3^-, CO_3^{-2}, and OH^- ions) must be added.

Ferric sulfate has several advantages compared to alum. The floc particle [$Fe(OH)_3$] is denser and therefore more easily or quickly removed by sedimentation. Ferric sulfate will react favorably over a much wider range of pH, usually 3.5–9.0, sometimes even greater. However, ferric sulfate can stain equipment; it is difficult to dissolve; its solution is corrosive; and it may react with organics to form soluble iron (Fe^{+2}).[14] (Soluble iron in water causes red water, as well as staining problems and taste.)

The best ferric sulfate dose to use in any coagulation application must be decided on a case-by-case basis using the jar test. However, experience has shown that dosages in water treatment usually run between 5 and 50 mg/L.

C6-3.   Iron and Manganese Removal

Iron and manganese, two troublesome water quality constituents, are found predominantly in ground water supplies, and occasionally in the anaerobic bottom waters of deep lakes. In nature, iron and manganese exist in stable forms known as ferric (Fe^{+3}) and manganic (Mn^{+4}); these forms are insoluble. Under anaerobic conditions, which can develop in ground water aquifers and at the bottom of deep lakes, these are reduced to the soluble forms ferrous (Fe^{+2}) and manganous (Mn^{+2}). The objective of most iron-and-manganese removal processes is to oxidize the reduced forms back into their insoluble forms so they will settle out.

---

[14]Chemistry Section, Chemistry of Treatment Processes (Iron and Manganese Removal).

Chlorine

When iron is present in the form of ferrous bicarbonate $[Fe(HCO_3)_2]$, it is easily treated for removal by chlorine, as shown in the following reactions:

$$2Fe(HCO_3)_2 + Cl_2 + Ca(HCO_3)_2 \rightarrow 2Fe(OH)_3\downarrow + CaCl_2 + 6CO_2 \quad Eq\ 9$$

This reaction, which is almost instantaneous, works in a pH range of 4–10, working best at pH 7. To remove 1 mg/L of iron (Fe^{+2}) requires 0.64 mg/L of chlorine. The resulting precipitate, ferric hydroxide, is easily recognized as a fluffy rust-colored sediment. The calcium bicarbonate $[Ca(HCO_3)_2]$ in the reaction represents the bicarbonate alkalinity of the water. The reaction works using either free or combined chlorine residual.

The reaction with manganese works similarly. The manganese may begin as a salt, such as manganous sulfate $(MnSO_4)$.

$$MnSO_4 + Cl_2 + 4NaOH \rightarrow MnO_2 + 2NaCl + Na_2SO_4 + 2H_2O \quad Eq\ 10$$

Then, with the addition of chlorine and sodium hydroxide, oxidized manganese produces the precipitate manganese dioxide (MnO_2). In order for the reaction to work, 1.3 mg/L of free available chlorine must be added per 1 mg/L of manganese to be removed. Chloramines (combined chlorine residuals) have little effect on manganese.

Sodium hydroxide, NaOH, causes the hydroxyl alkalinity in the sample. The reaction works best in the pH range 6–10. The speed with which manganese oxidizes varies from a few minutes at pH 10, to 2–3 hr at pH 8, to as much as 12 hr at pH 6.

Aeration

Both ferrous iron (Fe^{+2}) and manganous manganese (Mn^{+2}) can be removed by aeration.

$$4Fe(HCO_3)_2 + 10H_2O + O_2 \xrightarrow{\text{aeration}} 4Fe(OH)_3\downarrow + 8H_2CO_3 \quad Eq\ 11$$
$$\text{Ferrous} \qquad\qquad\qquad\qquad \text{Ferric}$$
$$\text{bicarbonate} \qquad\qquad\qquad\quad \text{hydroxide}$$

$$2MnSO_4 + O_2 + 4NaOH \xrightarrow{\text{aeration}} 2MnO_2\downarrow + 2Na_2SO_4 + 2H_2O \quad Eq\ 12$$
$$\text{Manganous} \quad \text{Sodium} \qquad \text{Manganese} \quad \text{Sodium}$$
$$\text{sulfate} \qquad\ \text{hydroxide} \qquad \text{dioxide} \qquad\ \text{sulfate}$$

The removal of iron is actually a two-step reaction. First, the soluble ferrous bicarbonate $[Fe(HCO_3)_2]$ is converted to the less soluble form, ferrous hydroxide $[Fe(OH)_2]$. Then, under further aeration, the ferrous hydroxide is converted to the insoluble form, ferrric hydroxide $[Fe(OH)_3]$, which will filter out or settle out of solution as a fluffy rust-colored sludge. The reaction works best in the pH range 7.5–8.0. It will take about 15 min to complete. To remove 1 mg/L of iron (Fe^{+2}) requires about 0.14 mg/L of oxygen.

The soluble salt of manganese, in this case manganous sulfate $(MnSO_4)$, is oxidized to the insoluble form managanese dioxide (MnO_2), which will filter out or settle out of solution. The reaction works best at a high pH, normally greater than 10. Sodium hydroxide (NaOH), shown as part of the reaction, raises the pH to the desired level and provides the hydroxide alkalinity (hydroxyl ions, OH^-) necessary to raise the pH for the reaction. To remove 1 mg/L of manganese (Mn^{+2}) requires about 0.27 mg/L of oxygen. The manganese removal reaction takes place in about 15 min.

Potassium Permanganate

Potassium permanganate ($KMnO_4$) is a powerful oxidizing agent and can be used successfully to remove both iron and manganese as shown by the following equations. This form of chemical treatment is preferable for manganese removal.

$$Fe(HCO_3)_2 + KMnO_4 + H_2O + 2H^+ \rightarrow \qquad\qquad Eq\ 13$$
$$MnO_2 + Fe(OH)_3 + KHCO_3 + H_2CO_3$$

$$3Mn(HCO_3) + 2KMnO_4 + 2H_2O \rightarrow \qquad\qquad Eq\ 14$$
$$5MnO_2\downarrow + 2KHCO_3 + 4H_2CO_3$$

$$3MnSO_4 + 2KMnO_4 + 2H_2O \rightarrow \qquad\qquad Eq\ 15$$
$$5MnO_2\downarrow + K_2SO_4 + 2H_2SO_4$$

Experience has shown that about 0.6 mg/L of $KMnO_4$ is adequate to remove 1 mg/L of iron (Fe^{+2}). Similarly, it takes about 2.5 mg/L of $KMnO_4$ to remove 1 mg/L of manganese (Mn^{+2}).

Softening

The lime-softening process operates normally at a pH in the range 10–11. Contingent removal of iron and manganese occurs within this range. Iron is eliminated in the form of ferrous hydroxide $Fe(OH)_2$, instead of the familiar ferric hydroxide $Fe(OH)_3$, shown in previous reactions.

C6-4.   Lime–Soda Ash Softening

The two ions most commonly associated with HARDNESS in water are calcium (Ca^{+2}) and magnesium (Mg^{+2}). Although aluminum, strontium, iron, manganese, and zinc ions can also cause hardness, they are not usually present in large enough concentrations to produce a hardness problem. Chemical precipitation is one of the more common methods used to soften water. The chemicals normally used are lime [calcium hydroxide, $Ca(OH)_2$] and soda ash [sodium carbonate, Na_2CO_3].

There are two types of hardness: (1) CARBONATE HARDNESS, caused primarily by calcium bicarbonate, and (2) NONCARBONATE HARDNESS, caused by the salts of calcium and magnesium, such as calcium sulfate [$CaSO_4$], calcium chloride [$CaCl_2$], magnesium chloride [$MgCl_2$], and magnesium sulfate [$MgSO_4$]. Lime is used to remove the chemicals that cause carbonate hardness. Soda ash is used to remove the chemicals that cause noncarbonate hardness.

Carbonate Hardness

When treating for carbonate hardness, lime is the only softening chemical needed, as illustrated in the following reactions:

To remove calcium bicarbonate:

$$Ca(HCO_3)_2 + Ca(OH)_2 \rightarrow 2CaCO_3\downarrow + 2H_2O \qquad\qquad Eq\ 16$$

Calcium   Lime   Calcium   Water
bicarbonate   carbonate

To remove magnesium bicarbonate:

$$Mg(HCO_3)_2 + Ca(OH)_2 \rightarrow CaCO_3\downarrow + MgCO_3 + 2H_2O \quad Eq\ 17$$

Magnesium   Lime   Calcium   Magnesium   Water
bicarbonate   carbonate   carbonate

Then,

$$MgCO_3 + Ca(OH)_2 \rightarrow CaCO_3\downarrow + Mg(OH)_2\downarrow \qquad\qquad Eq\ 18$$

Magnesium   Lime   Calcium   Magnesium
carbonate   carbonate   hydroxide

It takes two separate reactions and twice the lime needed for calcium bicarbonate to remove the magnesium bicarbonate.[15] In Equation 16, lime reacts with calcium bicarbonate to form calcium carbonate ($CaCO_3$). Calcium carbonate is relatively insoluble and precipitates.

In Equation 17, lime reacts with magnesium bicarbonate to form calcium carbonate, which precipitates, and magnesium carbonate, which does not. In addition to being soluble, magnesium carbonate is a form of carbonate hardness. Therefore, the same amount of lime added to Equation 17 is called for in Equation 18. Thus, twice the lime required to remove magnesium bicarbonate is necessary to remove calcium bicarbonate. In Equation 18, the additional lime reacts with magnesium carbonate to form calcium carbonate [$CaCO_3$] and magnesium hydroxide [$Mg(OH)_2$], both relatively insoluble materials that will settle out.

If the water originally had no noncarbonate hardness, then further softening is not needed. However, because $CaCO_3$ and $Mg(OH)_2$ are very slightly soluble, a small amount of hardness remains, usually at least 35 mg/L.

Noncarbonate Hardness

To remove noncarbonate hardness, soda ash must be added to remove the noncarbonate calcium compounds, and soda ash together with lime must be added to remove the noncarbonate magnesium compounds.

To remove calcium noncarbonate hardness:

$$CaSO_4 + Na_2CO_3 \rightarrow CaCO_3\downarrow + Na_2SO_4 \qquad\qquad Eq\ 19$$

Calcium   Soda   Calcium   Sodium
sulfate   ash   carbonate   sulfate

$$CaCl_2 + Na_2CO_3 \rightarrow CaCO_3\downarrow + 2NaCl \qquad\qquad Eq\ 20$$

Calcium   Soda   Calcium   Salt
chloride   ash   carbonate

In both cases the calcium noncarbonate hardness is removed by soda ash. The calcium sulfate and the calcium chloride acquire CO_3 from soda ash and

---

[15]Chemistry Section, Chemical Dosage Problems (Lime–Soda Ash Softening Calculations).

precipitate as $CaCO_3$. The compounds that remain after softening, Na_2SO_4 in Equation 19 and $NaCl$ in Equation 20, are salts and do not cause hardness.

To remove magnesium noncarbonate hardness:

$$MgCl_2 \quad + \quad Ca(OH)_2 \quad \rightarrow \quad Mg(OH)_2\downarrow \quad + \quad CaCl_2 \qquad Eq\ 21$$

Magnesium chloride	Lime	Magnesium hydroxide	Calcium chloride, which must also be removed

$$CaCl_2 \quad + \quad Na_2CO_3 \quad \rightarrow \quad CaCO_3\downarrow \quad + \quad 2NaCl \qquad Eq\ 22$$

Calcium chloride from Eq 21	Soda ash	Calcium carbonate	Salt

$$MgSO_4 \quad + \quad Ca(OH)_2 \quad \rightarrow \quad Mg(OH)_2\downarrow \quad + \quad CaSO_4 \qquad Eq\ 23$$

Magnesium sulfate	Lime	Magnesium hydroxide	Calcium sulfate, which must also be removed

$$CaSO_4 \quad + \quad Na_2CO_3 \quad \rightarrow \quad CaCO_3\downarrow \quad + \quad Na_2SO_4 \qquad Eq\ 24$$

Calcium sulfate from Eq 23	Soda ash	Calcium carbonate	Sodium sulfate

The removal of magnesium noncarbonate hardness with lime forms calcium noncarbonate hardness, which must then be removed with soda ash.

The final reaction related to softening involves carbon dioxide (CO_2), a gas that is found in dissolved form in most natural waters. Unless CO_2 is removed prior to softening (for example, by aeration), it will consume some of the lime added.

$$CO_2 \quad + \quad Ca(OH)_2 \quad \rightarrow \quad CaCO_3\downarrow \quad + \quad H_2O \qquad Eq\ 25$$

Carbon dioxide	Lime	Calcium carbonate	Water

If CO_2 is present, then enough lime must be added at the beginning of softening to allow this reaction to take place and still leave enough lime to complete the softening reactions. Whenever CO_2 is present, the CO_2 reaction occurs before the softening reactions take place.

Chemical softening takes place at a high pH. To precipitate calcium carbonate ($CaCO_3$), a pH of about 9.4 is necessary; the precipitation of magnesium hydroxide requires a pH of 10.6. In both cases the necessary pH is achieved by adding the proper amount of lime.

C6-5. Recarbonation

RECARBONATION is the reintroduction of carbon dioxide into the water either during or after lime–soda ash softening. When hard water is treated by conventional lime softening, the water becomes supersaturated with calcium carbonate and may have a pH of 10.4 or higher. This very fine suspended calcium carbonate can deposit on filter media, cementing together the individual media grains (encrustation), and depositing a scale in the distribution system piping (post-precipitation). To prevent these problems, carbon dioxide is bubbled into the water, lowering the pH and removing calcium carbonate as follows:

$$CaCO_3 \quad + \quad CO_2\uparrow \quad + \quad H_2O \quad \rightarrow \quad Ca(HCO_3)_2 \qquad Eq\ 26$$

Calcium	Carbon	Water	Calcium
carbonate	dioxide		bicarbonate
(in suspension)	(gas)		

This type of recarbonation is usually performed after the coagulated and flocculated waters are settled but before they are filtered, thereby preventing the suspended $CaCO_3$ from being carried out of the sedimentation basin and cementing the filter media. (The symbol "↑" after "CO_2" indicates that carbon dioxide is a gas.)

When the excess-lime technique is used to remove magnesium, a considerable amount of lime remains in the water. This creates a water that is undesirably caustic and high in pH. Carbon dioxide introduced into the water reacts as follows:

$$Ca(OH)_2 \quad + \quad CO_2\uparrow \quad \rightarrow \quad CaCO_3\downarrow \quad + \quad H_2O \qquad Eq\ 27$$

Excess	Carbon	Calcium	Water
lime	dioxide	carbonate	
	(gas)	(precipitate)	

This form of recarbonation is performed after coagulation and flocculation but before final settling. Carbon dioxide reacts with the excess lime, removing the cause of the caustic, high pH condition and, incidentally, removing the calcium which added to the hardness. The product, calcium carbonate, is removed by the filtration process.

It is important to select the correct carbon dioxide dosage.[16] If too much CO_2 is added, the following can happen:

$$Ca(OH)_2 \quad + \quad 2CO_2\uparrow \quad \rightarrow \quad Ca(HCO_3)_2 \qquad Eq\ 28$$

Excess	Carbon	Calcium
lime	dioxide	bicarbonate
	(gas)	

Notice that the excess lime combines with the excessive CO_2 dosage to form carbonate hardness. In the excess-lime method, there is too much calcium hydroxide present and this reaction would significantly increase water hardness. In the methods using lower lime doses, the conversion to calcium bicarbonate may not significantly increase water hardness.

C6.6.   Ion Exchange Softening

The ion exchange process of water softening uses the properties of certain materials (termed CATION EXCHANGE MATERIALS) to exchange the hardness-

---

[16]Chemistry Section, Chemical Dosage Problems (Recarbonation Calculations).

causing cations of calcium and magnesium for nonhardness-causing cations of sodium. The most common cation exchange materials are synthetic polystyrene resins. Each resin particle is a BB-sized, transparent, amber-colored sphere. Each of these insoluble resin spheres contains sodium ions, which are released into the water in exchange for hardness ions of calcium and magnesium. When properly operated, ion exchange is completely effective in removing all hardness, carbonate or noncarbonate.

The two reactions involved in the cation exchange softening process are

$$Ca^{+2} \ + \ Na_2X \ \rightarrow \ CaX \ + \qquad 2Na^+ \qquad\qquad Eq\ 29$$

Hardness cation	Cation exchange resin	Spent resin	Sodium released to treated water in exchange for calcium	

$$Mg^{+2} \ + \ Na_2X \ \rightarrow \ MgX \ + \qquad 2Na^+ \qquad\qquad Eq\ 30$$

Hardness cation	Cation exchange resin	Spent resin	Sodium released to treated water in exchange for magnesium	

The letter "X" is used to represent the exchange resin. Before softening, the resin with the nonhardness cations of sodium appears in the equations as Na_2X. Although this is not a chemical compound, it does behave somewhat like one. The sodium cations ("Na_2" in the resin, "$2Na^+$" after release) are released into the water just as the sodium in Na_2SO_4 would be released when that compound is dissolved in water. Although the resin "X" acts similarly to an anion such as SO_4^{-2}, the resin is actually an insoluble organic material that does not react chemically as SO_4^{-2} would. Instead, it functions more like a "parking lot" for exchangeable cations. The terms CaX and MgX represent the same resin after the exchange has been made.

As shown in the equations, calcium and magnesium hardness ions are removed from the water onto the surface of the resin. In exchange, the resin releases sodium ions. Note that one hardness ion (Mg^{+2} or Ca^{+2}) with a charge of $+2$ is exchanged for two sodium ions, each having a charge of $+1$, a total charge of $+2$. Hence, a $+2$ charge is exchanged for a $+2$ charge, an electrically equivalent exchange.

Equations 29 and 30 can be expanded to show the reactions of the hardness-causing compounds (similar to the equations shown for lime–soda ash softening):

$$Ca(HCO_3)_2 \ + \ Na_2X \ \rightarrow \ CaX \ + \ 2NaHCO_3 \qquad\qquad Eq\ 31$$

$$CaSO_4 \ + \ Na_2X \ \rightarrow \ CaX \ + \ Na_2SO_4 \qquad\qquad Eq\ 32$$

$$CaCl_2 \ + \ Na_2X \ \rightarrow \ CaX \ + \ 2NaCl \qquad\qquad Eq\ 33$$

$$Mg(HCO_3)_2 \ + \ Na_2X \ \rightarrow \ MgX \ + \ 2NaHCO_3 \qquad\qquad Eq\ 34$$

$$MgSO_4 \ + \ Na_2X \ \rightarrow \ MgX \ + \ Na_2SO_4 \qquad\qquad Eq\ 35$$

$$MgCl_2 \ + \ Na_2X \ \rightarrow \ MgX \ + \ 2NaCl \qquad\qquad Eq\ 36$$

In each reaction, the anion originally associated with the hardness cation stays in the softened water; after softening, these anions are associated with the sodium cations released by the resin. Hence, the softened water contains sodium bicarbonate ($NaHCO_3$), sodium sulfate (Na_2SO_4) and sodium chloride ($NaCl$). These compounds do not cause hardness and are present in such small concentrations that they do not cause tastes. Unlike lime–soda ash softening, ion exchange softening operates the same for carbonate and noncarbonate hardness. Both are removed with the same exchange reactions.

After most of the sodium ions are removed from the exchange resin in the softening process, the resin must be REGENERATED in order to restore its softening capacity. That is, the exchange process must be reversed, with the hardness cations of calcium and magnesium being forced out of the resin and replaced by cations of sodium. This reverse exchange is achieved by passing a strong brine solution (a concentrated solution of common table salt) through the resin bed. The two ion exchange regeneration reactions are shown below:

$$CaX + 2NaCl \rightarrow CaCl_2 + Na_2X \qquad\qquad Eq\ 37$$

$$MgX + 2NaCl \rightarrow MgCl_2 + Na_2X \qquad\qquad Eq\ 38$$

When sodium is taken back into the exchange resin, the resin is again ready to be used for softening. The calcium and magnesium, released during regeneration, are carried to disposal by the spent brine solution.

Properly maintained and operated, cation exchange removes *all* hardness. Since water of zero hardness is corrosive, the final step in ion exchange softening is to mix a portion of the unsoftened water with the softened effluent to provide water that is still relatively soft, but that contains enough hardness to be noncorrosive (stable).

C6-7.   Scaling and Corrosion Control

Scaling and corrosion are closely related problems in water treatment. They may be thought of as being at opposite ends of a hypothetical stability scale, as shown in Figure 10.

The objective of scale and corrosion control is to stabilize the water, thus preventing both scale formation and corrosion. The stable range is relatively narrow, requiring careful monitoring during treatment in order to avoid under- or over-shooting the stable range.

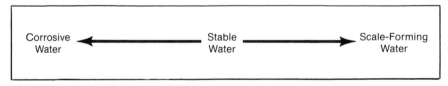

Figure 10.   Hypothetical Stability Scale

Scaling and Corrosion

Scale is the familiar chalky-white deposit frequently found at the bottom of a tea kettle. It is caused by carbonate and noncarbonate hardness constituents in water.

The exact combination of pH and alkalinity that will result in scale (or corrosion) depends on the overall chemical characteristics of the water. Under conditions of pH 7.0–9.5, with corresponding alkalinities of 300–15 mg/L, calcium carbonate $CaCO_3$ will precipitate and form a scale on interior surfaces of pipes and tanks. One of two chemical reactions will occur:

$$Ca^{+2} + CO_3^{-2} \rightarrow CaCO_3\downarrow \qquad\qquad Eq\ 39$$

or

$$Ca^{+2} + 2HCO_3^- \rightarrow CaCO_3\downarrow + CO_2 + H_2O \qquad\qquad Eq\ 40$$

In controlled amounts this scale is beneficial, forming a protective coating inside pipelines and tanks. However, excessive scaling can reduce the capacity of pipelines and the efficiency of heat transfer in boilers.

Corrosion is the oxidation of unprotected metal surfaces. In water treatment, a primary concern is the corrosion of iron and its alloys. Corrosion of iron and steel products is easily identified by the familiar red rust that forms. Iron is an important element in many metallic pipe materials and process equipment. It exists naturally as iron ore, in stable forms such as hematite (Fe_2O_3), magnatite (Fe_3O_4), iron pyrite (FeS_2), and siderite ($FeCO_3$). Smelting converts this ore to elemental iron, which is then used in the manufacture of pipeline materials and treatment equipment. Elemental iron is unstable and has a strong tendency to return by the oxidation or corrosion process to the more stable ore forms noted above. Advanced cases of iron corrosion create the problem of "red water."

There are various theories on how iron corrosion occurs and several factors known to affect corrosion. The following simplified discussion of corrosion chemistry highlights the influence of pH, alkalinity, dissolved oxygen, and carbon dioxide in the corrosion process. Remember that corrosion is a complex process that can be influenced by other factors as well. For example iron bacteria (*Crenothrix* and *Leptothrix*) and sulfate-reducing bacteria can be major causes of corrosion. Increases in water temperature and velocity of flow can accelerate corrosion. The softening process can convert a noncorrosive water to a corrosive one. However, certain constituents in water, such as silica, are believed to protect exposed metal surfaces from corrosion.

When iron corrodes, it is converted from elemental iron (Fe^0) to ferrous ion (Fe^{+2}).

$$Fe^0 \rightarrow Fe^{+2} + 2\ electrons \qquad\qquad Eq\ 41$$

The electrons that come from the elemental iron build up on metal surfaces and inhibit corrosion. If the water has a very low pH (lower than that of potable water), then hydrogen ions (H^+) in solution will react with electrons to form hydrogen gas ($H_2\uparrow$).

$$2H^+ + 2\ electrons \rightarrow 2H_2\uparrow \qquad\qquad Eq\ 42$$

The hydrogen gas coats the metal surface and could reduce corrosion; however, the coating is removed, partly by the scrubbing action of moving water and partly by combination with oxygen (O_2) normally dissolved in the water:

$$2H_2 + O_2 \rightarrow 2H_2O \qquad\qquad Eq\ 43$$

The metal surface is exposed again and corrosion continues. Failure to protect the metal surface or remove corrosion-causing elements will result in destruction of pipes or equipment.

Once the reaction in Equation 41 has occurred, subsequent reactions depend on the chemical characteristics of the water. If water is low in pH, low in alkalinity, and contains dissolved oxygen, then the ferrous ion reacts with water to form ferrous hydroxide.

$$Fe^{+2} + 2H_2O \rightarrow Fe(OH)_2 + H_2\uparrow \qquad\qquad Eq\ 44$$

The insoluble ferrous hydroxide immediately reacts with CO_2 present in low alkalinity water to form soluble ferrous bicarbonate.

$$Fe(OH)_2 + 2H_2CO_3 \rightarrow Fe(HCO_3)_2 + 2H_2O \qquad\qquad Eq\ 45$$

Since ferrous bicarbonate is soluble, it detaches from the metal surface and is mixed throughout the water. The dissolved oxygen in the water then reacts with the ferrous bicarbonate to form insoluble ferric hydroxide.

$$4Fe(HCO_3)_2 + 10H_2O + O_2 \rightarrow Fe(OH)_3 + 8H_2CO_3 \qquad Eq\ 46$$

Because this reaction occurs throughout water, ferric hydroxide does not form a protective coating on the metal surface. The ferric hydroxide appears as suspended particles that cause red water.

If water begins with a higher pH and alkalinity (where CO_2 is not present), then the corrosion reaction can be controlled. The ferrous ion shown in Equation 41 combines with the hydroxyl alkalinity that is present naturally or induced by lime treatment, forming an insoluble film of ferrous hydroxide on the metal surface.

$$Fe^{+2} + 2(OH^-) \rightarrow Fe(OH)_2 \qquad\qquad Eq\ 47$$

If dissolved oxygen is present in the water, it will react with the ferrous hydroxide to form an insoluble ferric hydroxide coating.

$$4Fe(OH)_2 + 2H_2O + O_2 \rightarrow 4Fe(OH)_3 \qquad\qquad Eq\ 48$$

Both ferrous and ferric hydroxide are somewhat porous and, although their coatings retard corrosion, they cannot fully protect the pipe. However, the same high pH and alkalinity conditions that cause the rust coating to form also favor the formation of a calcium carbonate coating. Together these coatings protect the pipe from further corrosion.

Chemical Methods for Scale and Corrosion Control

Table 4 shows common methods used to control scale and corrosion. The following paragraphs contain brief discussions of each method. Lime, soda ash, and caustic soda are typically used to raise pH and alkalinity. Carbon dioxide and sulfuric acid are used to lower pH and alkalinity. However, in situations where stabilization is achieved by pH and alkalinity adjustment, the effects of all chemicals used in water treatment must be taken into account. Alum and ferric sulfate (discussed under "Coagulation") lower pH and alkalinity, as do chlorine (discussed under "Chlorination") and fluosilicic acid (used in fluoridation).

Table 4.   Scale and Corrosion Control Methods

Method	For Control of: Scale	Corrosion
pH/Alkalinity adjustment with lime	■	■
Chelation	■	
Sequestering	■	
Controlled CaCO₃ scaling		■
Other protective chemical coatings		■
Softening	■	

pH/alkalinity adjustment with lime.   For each milligram per litre of lime added, approximately 0.56 mg/L of carbon dioxide is removed. Carbon dioxide is in the form of carbonic acid (H_2CO_3) when dissolved in water. The following equation indicates the chemical reaction that takes place:

$$H_2CO_3 + Ca(OH)_2 \rightarrow CaCO_3 + 2H_2O \qquad\qquad Eq\ 49$$

As carbon dioxide (carbonic acid) is removed, pH increases.

For each milligram per litre of lime added, the alkalinity of the treated water will increase by about 1.28 mg/L. The reaction is:

$$Ca(HCO_3)_2 + Ca(OH)_2 \rightarrow 2CaCO_3 + 2H_2O \qquad\qquad Eq\ 50$$

If the lime dose is too high, excessive scale will form. If it is too low, the water will be corrosive. Langlier's Calcium Carbonate Saturation Index is one technique used to determine the tendency of water to scale or corrode piping and tanks. This method is based on the assumption that every water has a particular pH value where the water will neither deposit scale nor cause corrosion. This stable condition is termed saturation. The pH value, called saturation pH and abbreviated pH_s, varies depending on calcium hardness, alkalinity, and temperature. Once the pH_s is calculated, the LANGLIER INDEX (also called the SATURATION INDEX, S.I.) is found as follows:

$$\text{Langlier Index} = pH - pH_s$$

If the actual pH of the water is less than the calculated pH_s then the water has a negative Langlier Index and may be corrosive. If the actual pH is greater than the calculated pH_s, then the Langlier Index is positive and the water is likely to form scale. In either case the water is unstable. The greater the difference between pH and pH_s, the stronger the tendency for the water to either form scale or cause corrosion. Thus, water with a Langlier Index of +0.4 has a stronger scaling tendency than one with an index of +0.1. Similarly, water with a Langlier Index of −0.4 has a stronger corrosion tendency than one with an index of −0.2. If the pH and pH_s are equal, then the Langlier Index is zero and water is stable. Neither scale formation nor corrosion should occur.

The value of pH_s can be calculated mathematically, but it is simpler to use a graph as illustrated in Figure 11. To find pH_s, calcium hardness in milligrams per litre as $CaCO_3$, alkalinity in milligrams per litre as $CaCO_3$, pH and water

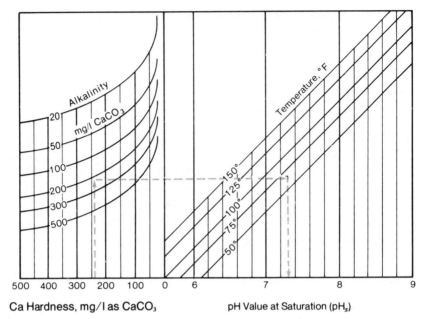

Ca Hardness, mg/l as CaCO₃   pH Value at Saturation (pH$_s$)

Reprinted with permission of Nalco Chemical Company, 2901 Butterfield Road, Oak Brook, IL 60521 from Water: The Universal Solvent; copyright 1977 and the Nalco Water Handbook to be published in 1979 by McGraw-Hill Book Company.

Figure 11.   Determination of pHs for the Langelier Index, Given Hardness, Alkalinity, and Temperature (assuming average TDS of about 500 mg/L)

temperature must be determined. The following information will be used to demonstrate reading the graph:

$$\text{Calcium hardness} = 240 \text{ mg/L as CaCO}_3$$
$$\text{Alkalinity} = 200 \text{ mg/L as CaCO}_3$$
$$\text{pH} = 6.8$$
$$\text{Water temperature} = 70°F$$

Enter the graph at a calcium hardness of 240 mg/L, as shown. Proceed upward to the 200-mg/L alkalinity curve. Travel across to locate the temperature of 70°F (estimate the temperature location between the 50°F and 75°F lines). Move downward from the temperature point to the bottom of the graph and identify the pH$_s$ value as 7.3. Finally, calculate the Langlier Index as follows:

$$\text{Langlier Index} = \text{pH} - \text{pH}_s$$
$$= 6.8 - 7.3$$
$$= -0.5$$

Since the index is negative, the water is corrosive.

Chelation.   CHELATION is a chemical treatment process used to control scale formation. The chemical added is known as a chelating agent. It is a water-soluble compound that captures scale-causing ions in solution, preventing precipitation and scale formation. There are several natural organic materials in water that have chelating ability, including humic acid and lignin. When added to water, the chelating agent reacts with calcium ions to keep them in solution and prevent the formation of calcium carbonate scale. In Equation 51, the "Y" represents the chelating agent.

$$2Ca^{+2} + Na_4Y \rightarrow Ca_2Y + 4Na^+ \qquad Eq\ 51$$

Sequestration.   Sequestration is a chemical addition treatment process that controls scale. The chemical added to the water SEQUESTERS, or holds in solution, scale-causing ions such as calcium, iron, and manganese; thus preventing them from precipitating and forming scale. Any one of several polyphosphates may be used in this process. The most commonly used is sodium hexametaphosphate, $(NaPO_3)_6$. A common dosage for scale prevention is approximately 0.5 mg/L.

Controlled $CaCO_3$ scaling.   One commonly practiced form of corrosion management is controlled $CaCO_3$ scaling. As stated at the beginning of this section, there are three conditions water can have relative to scale formation and corrosion. Water may be corrosive, stable, or scale-forming. By carefully controlling the pH/alkalinity adjustment, the condition of water can be altered so that it is slightly scale-forming. Eventually the scale will build up beyond the desired thickness. Adjusting pH and alkalinity to a point that is slightly corrosive will then dissolve the excess scale.

Controlled scaling requires careful and continuous laboratory monitoring, since the slightest change in the quality of water may require a change in the amount of lime needed. It is also necessary to monitor the thickness of the $CaCO_3$ coating developed in the distribution system pipeline. The coating should be thick enough to prevent corrosion without obstructing the flow of water.

Other chemical protective coatings.   There are two other chemicals used to create protective coatings in pipelines: (1) polyphosphates and (2) sodium silicate. Polyphosphates include sodium hexametaphosphate $[(NaPO_3)_6]$, sold under the trade name "Calgon"; sodium pyrophosphate $[Na_4P_2O_7]$, sold under the trade name "Nalco"; and a group known as "bimetallic glassy phosphates." After being fed into the water, polyphosphates form a phosphate film on interior metal surfaces protecting them from corrosion. Polyphosphates are also effective as sequestering agents for preventing calcium carbonate scale and for stabilizing dissolved iron and manganese. Dosages of 5–10 mg/L are recommended when treatment is initiated. After one to two months protective film is established. The dosages are then reduced and maintained at approximately 1 mg/L.

Sodium silicate $(Na_2Si_4O_9)$, or water-glass, can also be used to control corrosion in water systems. Sodium silicate combines with calcium to form a hard, dense calcium silicate film $(CaSiO_3)$. Dosages vary widely depending on water quality.

Softening.   A major problem caused by water hardness is scale formation. Hard waters form calcium carbonate scales in pipelines and boilers. These scales reduce pipeline capacities, lower boiler heat transfer efficiencies, and cause heat exchange tube failures because excessive heating is required to overcome the insulating effects of scale. The results are higher pumping and maintenance costs, longer repair time, shortened equipment life, and higher fuel and power costs.

There are four major water characteristics that control hardness: the tendency to form scale is greater when the amount of hardness, alkalinity, pH, or temperature is increased. Figures 12, 13, and 14 demonstrate how these characteristics are interrelated. The graphs in the figures were determined for one water sample—they should not be applied in general, but they do illustrate the interaction between the indicated variables.

The ten chemical reactions involved in water softening were presented earlier, under "Lime–Ash Softening." Refer to that material to review hardness removal using lime and soda ash.

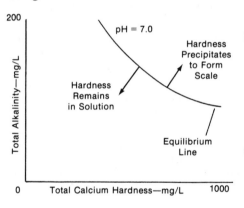

Figure 12. Total Alkalinity vs. Total Calcium Hardness at 77° F—Data Computed for One Water Sample, Not Generally Applicable

Figure 13. The Effect of Total Alkalinity and pH on the Amount of Total Calcium Hardness That Can Be Kept in Solution—Data Calculated for One Water Sample, Not Generally Applicable

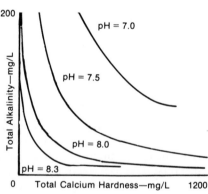

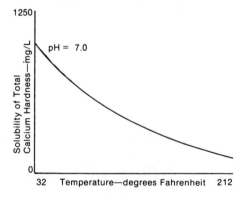

Figure 14. The Effect of Temperature on the Amount of Calcium That Will Stay in Solution— Data Calculated for One Water Sample, Not Generally Applicable

Courtesy of Johnson Controls, Inc.

C6-8.   Chlorination

When chlorine is added to water it reacts to produce various compounds. Some of these compounds are effective disinfectants whereas others represent the end product of a reaction that has removed an undesirable constitutent from the water.

Reaction in Pure Water

When chlorine is added to pure water it reacts as follows:

$$Cl_2 \quad + \quad H_2O \quad \rightarrow \quad HOCl \quad + \quad HCl \qquad Eq\ 52$$

Chlorine   Water   Hypochlorous   Hydrochloric
                                   acid   acid

In this reaction the chlorine combined with water to produce hypochlorous acid ($HOCl$). This is one of the two free available chlorine residual forms. Due to the ease with which $HOCl$ penetrates into and kills bacteria it is the most effective form of chlorine for disinfection. However some of the $HOCl$ (a weak acid) dissociates as follows:

$$HOCl \quad \rightarrow \quad H^+ \quad + \quad OCl^- \qquad Eq\ 53$$

Hypochlorous   Hydrogen   Hypochlorite
acid   ion

As shown by Equation 53, this dissociation produces hydrogen (which neutralizes alkalinity or lowers pH) and hypochlorite ion (OCl^-), the second type of free available chlorine residual. The OCl^- is a relatively poor disinfectant compared with $HOCl$, primarily because of its inability to penetrate into the bacteria.

Equation 54 shows what happens to the hydrochloric acid (a strong acid) formed in the first reaction.

$$HCl \quad \rightarrow \quad H^+ \quad + \quad Cl^- \qquad Eq\ 54$$

Hydrochloric   Hydrogen   Chloride
acid   ion

Notice it also dissociates, forming hydrogen (which neutralizes alkalinity or lowers pH) and chloride ion, one of the same ions formed when common table salt is dissolved in water. Neither the hydrogen nor the chloride ion act as disinfectants.

The effectiveness of chlorination is based on five important factors.

- pH
- concentration
- temperature
- other substances in water
- contact time

The pH strongly influences the ratio of $HOCl$ to OCl^-. As shown in Figure 15, low pH values favor the formation of $HOCl$, the more effective free residual (Equation 52), while high pH values favor the formation of OCl^-, the less effective free residual form (Equation 53). As pH increases from 7.0 to 10.7, the OCl^- form begins to predominate and the time required for the free residual to effectively disinfect increases. The added time is barely detectable in the pH range 7.0–8.5 but is markedly longer for a pH greater than 8.5.

Temperature has two influences.   Very high temperatures (for example, boiling) speed the killing of organisms. However, within the range of temperatures normally found in water, the lower the temperature the more effective the chlorination. There are two reasons. First, as shown in Figure 15, lower temperatures favor the formation of HOCl, which is more effective than OCl$^-$. Second, lower temperatures allow chlorine residuals to persist whereas higher temperatures promote quick dissipation.

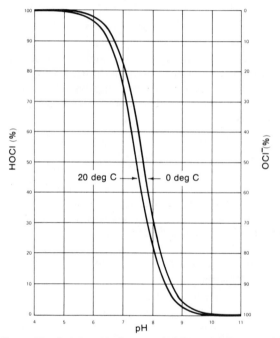

Figure 15.   Relationship Between HOCl, and OCl⁻, and pH

The destruction of organisms (kill) is directly related to the contact time t and concentration of chlorine c. For example, to accomplish a given kill in a given period of contact time, you might need a certain concentration of chlorine. Providing other conditions remain constant (such as pH and temperature), if the contact time is increased, less chlorine concentration is needed to accomplish the same kill; as dosage concentrations are increased the contact time needed can be decreased.

Equations 52–54 represent what happens when chlorine is added to pure water. However, the water treated at a treatment plant is far from pure. The inorganic and organic materials in raw water supplies can and do take part in the reaction with chlorine. Some of the most common materials that react with chlorine in water include

- Ammonia (NH_3)
- Iron (Fe)
- Manganese (Mn)
- Hydrogen sulfide (H_2S)
- Dissolved organic materials

Reaction with Ammonia

One of the most common reactions of chlorine in raw water involves ammonia. Ammonia (NH_3) is an inorganic compound occurring naturally due to decaying vegetation or artificially from domestic and industrial wastewater discharges. Chlorine reacts with ammonia to form chloramines, compounds containing both nitrogen and chloride ions. As shown in Equations 55–57, chloramines are formed in three successive steps:

$$NH_3 \quad + \quad HOCl \quad \rightarrow \quad NH_2Cl \quad + H_2O \qquad Eq\ 55$$

Ammonia　　Hypochlorous　　Monochloramine
　　　　　　acid

$$NH_2Cl \quad + \quad HOCl \quad \rightarrow \quad NHCl_2 \quad + H_2O \qquad Eq\ 56$$

Monochloramine　　Hypochlorous　　Dichloramine
　　　　　　　acid

$$NHCl_2 \quad + \quad HOCl \quad \rightarrow \quad NCl_3 \quad + H_2O \qquad Eq\ 57$$

Dichloramine　　Hypochlorous　　Trichloramine
　　　　　　acid

Whether one chloramine compound or a combination is formed depends on the pH of water and on the presence of enough ammonia. Monochloramine and dichloramine are effective disinfecting agents, but they are commonly not as effective as free chlorine.

Combined available chlorine is a less active oxidizing agent than free available chlorine. When water is chlorinated, free available chlorine reacts rapidly with any oxidizable substance. If water contains natural or added ammonia, then the free available chlorine reacts to form combined available chlorine and the speed of further reactions with oxidizable substances is slowed considerably. The low oxidation potential of combined available chlorine, as compared with free available chlorine, also accounts for slower bactericidal (bacteria killing) action. Under favorable conditions, equivalent bactericidal action is obtained by using approximately 25 times the combined available chlorine residual as free available chlorine. To obtain equivalent bactericidal action with equal amounts of combined available chlorine residual and free available chlorine residual, a contact period approximately 100 times longer is required.

Chloramines are considerably less effective than HOCl as disinfecting agents. Table 5 estimates the effectiveness of the four residual types as compared with HOCl. These numbers are estimates since no method is available to accurately compare effectiveness. Effectiveness can change dramatically, depending on the water's characteristics.

Table 5. **Estimated Effectiveness of Residual Types**

Type	Chemical Abbreviation	Estimated Effectiveness Compared to HOCl
Hypochlorous acid	HOCl	1
Hypochlorite ion	OCl⁻	1/100
Trichloramine*	NCl₃	†
Dichloramine	NHCl₂	1/80
Monochloramine	NH₂Cl	1/150

*Commonly called nitrogen trichloride
†No estimate; possibly more effective than dichloramine.

It is important to place these estimates in proper perspective. Initially, it would appear pointless to use any residual except HOCl. However, from practical experience it is known that chloramines, particularly mono-chloramines, do an acceptable job of disinfection if given enough time. Therefore, contact time and concentration are important factors to consider when using monochloramine or combined residual.

Reaction With Iron

Iron is an undesirable element in water easily removed by chlorination. Iron is often found in ground water supplies, usually in the form of ferrous bicarbonate [$Fe(HCO_3)_2$]. When chlorine is added, the reaction is

$$2Fe(HCO_3)_2 \; + \quad Cl_2 \quad + \; Ca(HCO_3)_2 \; \rightarrow \qquad\qquad Eq\ 58$$

Ferrous   Chlorine   Calcium
bicarbonate   bicarbonate

$$2Fe(OH)_3\!\downarrow \; + \quad CaCl_2 \; + \; 6CO_2$$

Ferric   Calcium   Carbon
hydroxide   chloride   dioxide

Ferric hydroxide [$Fe(OH_3)$] precipitates almost immediately, forming a fluffy, rust-colored sludge. The calcium bicarbonate [$Ca(HCO_3)_2$] in Equation 58 represents alkalinity in the water. Iron may be removed using either the free or combined forms of chlorine residual. Each milligram per litre of iron to be removed requires approximately 0.64 mg/L of chlorine.

Chlorine Reactions With Manganese

Manganese, like iron, is an undesirable constituent often found in ground waters and causes similar problems. Just as iron causes red water, manganese may produce brown or black water.

Chlorine reacts with manganese as shown in Equation 59.

$$MnSO_4 \quad + \quad Cl_2 \quad + 4NaOH \quad \rightarrow \qquad\qquad Eq\ 59$$

Manganous   Chlorine   Sodium
sulfate   hydroxide

$$MnO_2\!\downarrow \quad + \quad 2NaCl \; + \; Na_2SO_4 \; + \; 2H_2O$$

Manganese   Sodium   Sodium   Water
dioxide   chloride   sulfate

The manganese in ground water is normally in the form of a soluble salt, such as manganous sulfate ($MnSO_4$). Chlorination forms the precipitate manganese dioxide (MnO_2), which effectively removes the manganese. The time required to complete the reaction is usually 2 hr.

Each milligram per litre of manganese removed requires 1.3 mg/L of free available chlorine. Unlike iron, manganese is not affected by combined forms of chlorine.

Chlorine Reaction With Hydrogen Sulfide

Chlorine will also react with hydrogen sulfide (H_2S). Rarely found in surface water, H_2S is frequently found in ground-water supplies. In concentrations as low as 0.05 mg/L, H_2S can give water an unpleasant taste. At slightly higher, yet still small concentrations of 0.5 mg/L, the characteristic odor of rotten eggs is noticeable. Hydrogen sulfide can be *fatal* in a few minutes, if inhaled in concentrations equal to or greater than 0.1–0.2 percent by volume in air. At very large concentrations, beginning at 4.3 percent by volume in air, H_2S is flammable. This is one of the more troublesome, obnoxious, and deadly gases that may be encountered in water treatment.

When chlorine is used to remove H_2S, one of two reactions can occur, depending on the chlorine dosage:

$$Cl_2 \;+\; H_2S \;\rightarrow\; 2HCl \;+\; S \qquad\qquad Eq\ 60$$

| Chlorine | Hydrogen sulfide | Hydrochloric acid | Sulphur |

or

$$4Cl_2 \;+\; H_2S \;+\; 4H_2O \;\rightarrow\; 8HCl \;+\; H_2SO_4 \qquad Eq\ 61$$

| Chlorine | Hydrogen sulfide | Water | Hydrochloric acid | Sulfuric acid |

In Equation 60, the reaction is instantaneous and occurs at high pH (the reaction is only about half complete at pH 10). It takes approximately 2.2 mg/L of chlorine to convert 1 mg/L of H_2S to sulfur (S). The sulfur formed by this reaction is a finely divided colloidal-type particle that causes milky-blue turbidity. This turbidity is removed by coagulation and filtration.

In Equation 61, the operationally preferred reaction occurs. Chlorine converts the H_2S to sulfuric acid. The H_2SO_4 and the HCl dissociate into hydrogen, chloride, and sulfate ions. Since the initial concentration of H_2S was slight, these ions were not produced in amounts great enough to cause problems. The reaction in Equation 61 requires approximately 8.9 mg/L of chlorine for each milligram per litre of H_2S to be removed.

Equation 61, like Equation 60, is pH-dependent. If the pH is less than 6.4, all sulfides are converted to sulfates. At a pH of about 7.0, 70 percent of the sulfides change to sulfates, and the remaining 30 percent change to elemental sulfur. In the pH range of about 9.0–10.0, 50 percent of the sulfides are oxidized to sulfates, and the remaining 50 percent oxidize to elemental sulfur.

Sources of Chlorine

There are three types of materials commonly used as a source of chlorine: (1) gaseous chlorine, (2) calcium hypochlorite, and (3) sodium hypochlorite.

Liquid chlorine is a compressed amber-colored gas containing 99.5 percent pure chlorine. At room temperature and pressure, 1 cu ft of liquid will expand to approximately 500 cu ft of gas. Chlorine gas is greenish-yellow in color and visible at high concentrations. It is highly toxic even at concentrations as low as 0.1 percent by volume. As demonstrated in Equation 62, plain chlorine mixed with water produces hypochlorous acid.

$$Cl_2 \quad + \quad H_2O \quad \rightarrow \quad HOCl \quad + \quad HCl \qquad Eq\ 62$$

Chlorine	Water	Hypochlorous	Hydrochloric
gas or		acid	acid
liquid			

Chlorine liquid or gas is neither explosive nor flammable, but it will support combustion. The liquid changes easily to a gas at normal temperatures and pressures. If chlorine remains dry, it will not corrode metal. However, mixed with some moisture it is extremely corrosive. Chlorine liquid is approximately 1.5 times the weight of water. Gas is approximately 2.5 times the weight of air.

Calcium hypochlorite, $Ca(OCl)_2$, is a dry, white or yellow-white, granular material also available in tablets weighing about 0.01 lb. The granular material contains 65 percent available chlorine by weight. This means that when 1 lb of calcium hypochlorite is added to water only 0.65 lb of chlorine is added. Or, in order to add 1 lb of chlorine, 1.54 lb of $Ca(OCl)_2$ must be added.

When added to water $Ca(OCl)_2$ reacts as follows:

$$Ca(OCl)_2 \quad + \quad 2H_2O \quad \rightarrow \quad 2HOCl \quad + \quad Ca(OH)_2 \qquad Eq\ 63$$

| Calcium | Water | Hypochlorous | Lime |
| hypochlorite | | acid | |

Notice that HOCl is produced, just as it was when pure chlorine was added to water. If there are other materials in the water (Fe, Mn, or H_2S), the HOCl will react with them as described previously.

Calcium hypochlorite should be stored carefully to avoid contact with easily oxidized organic material because this type of chlorine can cause fires when brought into contact with many types of organic compounds.

Sodium hypochlorite (NaOCl) is a clear, greenish-yellow liquid chlorine solution normally used in bleaching. Normal household bleach is an example of sodium hypochlorite. It contains 5 percent available chlorine, which is equivalent to 0.42 lb/gal. Commercial bleaches are stronger, containing 9–15 percent available chlorine. Table 6 lists the weight of available chlorine in various strengths of NaOCl solution.

Table 6.   Available Chlorine in NaOCl Solution

Percent Available Chlorine	Available Chlorine lb/gal
10.0	0.833
12.5	1.04
15.0	1.25

Sodium hypochlorite reacts with water to produce the desired HOCl as follows:

$$NaOCl \; + \; H_2O \; \rightarrow \quad HOCl \quad + \quad NaOH \qquad \textit{Eq 64}$$

| Sodium hypochlorite | Water | Hypochlorous acid | Sodium hydroxide |

Sodium hypochlorite solution can be purchased in 5-gal rubber-lined steel drums, and in railroad tank cars. Because it is popular for use in small water systems, it is often purchased in refillable gallon plastic jugs packaged four to a box. Unlike calcium hypochlorite, there is no fire hazard connected with NaOCl storage. It is quite corrosive, however, and should be separated from equipment susceptible to corrosion damage.

Of the three types of chlorine discussed above, gaseous chlorine is more commonly used. Sodium hypochlorite is preferred for small systems, because it is easy to handle and to meter without risk of solid residues clogging pipes and equipment. Since NaOCl is in liquid form, it may be quite expensive to ship to remote areas. Therefore, where transportation costs are significant, dry granular calcium hypochlorite is usually selected.

Review Questions

1. What two forms of free available chlorine residual are produced when chlorine is added to pure water? Give names and chemical abbreviations.

2. Give the names and chemical abbreviations for the three common types of combined chlorine residual.

3. List two differences between free available chlorine residual and combined chlorine residual.

4. List three troublesome chemical characteristics of water which can be removed by chlorination.

5. What materials may be used as sources of chlorine?

6. What is the most active disinfecting compound produced when a hypochlorite is mixed with water?

7. List three methods for iron and manganese removal.

8. List three methods of oxidizing taste- and odor-causing compounds.

9. What pH ranges are most effective for iron and manganese removal by aeration?

10. What range of $KMnO_4$ dosage is effective for most taste and odor removal?

11. What is the most commonly used chemical for coagulation?

12. What is the common range of dosages and the pH operating range for ferric sulfate?

13. What is the most effective pH range for aluminum sulfate?

14. List two advantages and two disadvantages of using ferric sulfate compared with aluminum sulfate.

15. What are the two chemicals used in chemical softening? Give their names and chemical formulas.

16. What are the two general types of hardness? Give examples of each.

17. Which type of hardness is treated with soda ash?

18. At what pH values do $CaCO_3$ and $Mg(OH)_2$ precipitate?

19. List four chemical methods used for scale control and three methods used for corrosion control.

20. What does a Langlier Index of $+2.1$ mean?

21. What does a Langlier Index of -0.9 mean?

22. What does a Langlier Index of zero mean?

23. What is chelation and how does it work?

24. How is controlled $CaCO_3$ scaling used in scale and corrosion control?

25. Softening is one technique used to control scaling. List the four major water characteristics that control the tendency for hardness to precipitate and form scale.

26. Define *ion exchange*.

27. Name and describe the most commonly used ion exchange resin.

28. How many sodium ions are needed to exchange with one calcium or magnesium ion?

29. How are cation ion exchange resins regenerated?

30. What are the lowest practical hardnesses achieved by lime–soda ash and ion exchange softening?

31. Is it desirable to produce zero hardness water? Why?

32. What is recarbonation and what purpose does it serve?

33. Where in the plant is recarbonation commonly practiced?

34. Describe what occurs if too much CO_2 is added to remove excess lime.

Summary Answers

(See Detailed Answers.)

Detailed Answers

1. Hypochlorous acid, HOCl; and hypochlorite ion, OCl^-.

2. Monochloramine, NH_2Cl; Dichloramine, $NHCl_2$; Trichloramine, NCl_3

3. Free available chlorine residual is more active (a better oxidizer) and it tends to dissipate more readily. Combined chlorine residual is slower-acting and the residual tends to last longer.

4. Iron (Fe), manganese (Mn), and hydrogen sulfide (H_2S).

5. Gaseous chlorine, (Cl_2); calcium hypochlorite, $Ca(OCl)_2$; and sodium hypochlorite ($NaOCl$).

6. Hypochlorous acid, HOCl; however, OCl^- can also be produced although it is less active (less effective).

7. Chlorination, aeration, oxidation with potassium permanganate, and softening.

8. Aeration, chlorination, and potassium permanganate.

9. Fe: pH 7.5–8.0; Mn: above pH 10.

10. 0.5–15 mg/L, usually 0.5–2.5 mg/L.

11. Aluminum sulfate, $Al_2(SO_4)_3$.

12. Dosage range: 15 mg/L–150 mg/L; pH range: 5.5–8.5.

13. pH range: 5.5–8.5 and higher.

14. Advantages: denser, more rapidly settling floc and wider workable pH range.
    Disadvantages: corrosive, stains equipment, problems in dissolving, and can produce soluble iron (Fe^{+2}).

15. Lime: $Ca(OH)_2$; Soda Ash: Na_2CO_3.

16. Carbonate hardness: $Ca(HCO_3)_2$, $Mg(HCO_3)_2$, and $MgCO_3$; non-carbonate hardness: $CaSO_4$, $CaCl_2$, $MgSO_4$, and $MgCl_2$.

17. Noncarbonate hardness.

18. $CaCO_3$: pH 9.4; $Mg(OH)_2$: pH 10.6.

19. For scale control: chelation, sequestering, softening, and pH/alkalinity adjustment with lime. For corrosion control: controlled $CaCO_3$ scaling, other protective chemical coatings, and pH/alkalinity adjustment with lime.

20. The water is scale forming.

21. The water is corrosive.

22. The water is stable.

23. It is a chemical treatment process used to control scaling. The chelating agent holds scale-causing ions in solution so they cannot precipitate out and cause scale.

24. For a period of time, the pH/alkalinity balance is adjusted so that water is slightly scale-forming. Then, for a period of time, the pH/alkalinity balance is readjusted so the water is slightly corrosive; enough to slowly dissolve the $CaCO_3$ scale previously formed. This cycle, repeated routinely, prevents excessive scale and helps to control corrosion.

25. Amount of hardness, alkalinity, pH, and temperature.

26. The reversible exchange of ions between the water being treated and the exchange material.

27. Polystyrene resins: BB-sized, clear; amber-colored spheres.

28. Two, an electrically equivalent amount.

29. By backwashing with a strong salt or brine solution.

30. 35 mg/L–85 mg/L for lime–soda ash; 0 mg/L for ion exchange.

31. No. Zero-hardness water is corrosive.

32. The introduction of CO_2 back into the water. This reduces high pH conditions, stabilizing the water and preventing filter encrustation and after-precipitation in the mains.

33. After coagulation/flocculation/sedimentation and before filtration, or before final settling, or before leaving the plant, depending on the reason for recarbonation.

34. The reaction forms $Ca(HCO_3)_2$ which significantly increases water hardness.

Chemistry 7

Chemical Dosage Problems

One of the more common uses of mathematics in water treatment practices is chemical dosage calculations. As a basic or intermediate level operator, there are generally seven types of dosage calculations that you may be required to perform:

- Milligrams-per-litre to pounds-per-day conversions
- Milligrams-per-litre to percent conversions
- Feed rate conversions
- Chlorine dosage/demand/residual calculations
- Percent strength calculations
- Solution dilution calculations
- Reading nomographs

Percent-strength calculations,[17] solution–dilution calculations,[18] and reading nomographs[19] are discussed in other sections of this book. The remaining four types of calculations are explained in the material that follows.

In addition to the general types of calculations, the operator may be required to perform calculations dealing with specific treatment processes. The last part of this chapter deals with the mathematics specifically needed for lime–soda ash softening, ion exchange softening, recarbonization, and fluoridation.

C7-1. Milligrams-per-Litre to Pounds-per-Day Conversions

The formula for converting milligrams per litre to pounds per day is derived from the formula for converting parts per million (ppm) to pounds per day, which is as follows.

---

[17]Chemistry Section, Solutions (Percent Strength).
[18]Chemistry Section, Solutions (Percent Strength—Dilution Calculations).
[19]Mathematics Section, Graphs and Tables (Graphs—Nomographs).

	Flow	Conversion	Feed
Dosage	rate	factor	rate
(ppm)	(mgd)	(8.34 lb/gal) =	(lb/day)

In the range of 0–2000 mg/L, milligrams per litre are approximately equal to parts per million. For example, 150 mg/L of calcium is approximately equal to 150 ppm of calcium. Therefore, "mg/L" can be substituted for "ppm" in the equation just given. The substitution yields the following equation, which is used to convert between milligrams per litre and pounds per day:

	Flow	Conversion	Feed
Dosage	rate	factor	rate
(mg/L)	(mgd)	(8.34 lb/gal) =	(lb/day)

Converting milligrams per litre to pounds per day is a common water treatment calculation. Therefore, you should *memorize the conversion formula.* The following examples illustrate how the formula is used.

Example 1
The dry alum dosage rate is 12 mg/L at a water treatment plant. The flow rate at the plant is 3 mgd. How many pounds per day of alum are required?

This is a milligrams-per-litre to pounds-per-day conversion problem; therefore, first write the equation that relates the two terms:

$$(mg/L)(mgd)(8.34\ lb/gal) = lb/day$$

Now fill in the information given in the problem and solve for the unknown value:[20]

$$(12\ mg/L)(3\ mgd)(8.34\ lb/gal) = x\ lb/day$$

$$300.24\ lb/day = x\ lb/day$$

Example 2
Fluoride is added at a concentration of 1.5 mg/L. The flow rate at the treatment plant is 2 mgd. How many pounds per day of fluoride are added?

First write the conversion equation:

$$(mg/L)(mgd)\ 8.34\ lb/gal) = lb/day$$

Fill in the information given in the problem and solve for the unknown value:

---

[20]Mathematics Section, Solving for the Unknown Value.

$$(1.5 \text{ mg/L})(2 \text{ mgd})(8.34 \text{ lb/gal}) = x \text{ lb/day}$$
$$25.02 \text{ lb/day} = x$$

Example 3

The chlorine dosage rate at a water treatment plant is 2 mg/L. The flow rate at the plant is 700,000 gpd. How many pounds per day of chlorine are required?

First write the conversion formula:

$$(\text{mg/L})(\text{mgd})(8.34 \text{ lb/gal}) = \text{lb/day}$$

Before the information given in the problem can be filled in, 700,000 gpd must be converted to million gallons per day. To do this, locate the position of the "millions comma" and move the decimal to there. For example, in the number 2,400,000, the millions comma is between the 2 and 4. To express 2,400,000 gpd as million gallons per day, replace the millions comma with a decimal point:

$$2,400,000 \text{ gpd} = 2.4 \text{ mgd}$$

In this example, to convert 700,000 gpd to million gallons per day:

,700,000 gpd

↑ millions comma

$$700,000 \text{ gpd} = 0.7 \text{ mgd}$$

Now write the information into the equation and solve for the unknown value:

$$(2 \text{ mg/L})(0.7 \text{ mgd})(8.34 \text{ lb/gal}) = x \text{ lb/day}$$
$$11.68 \text{ lb/day} = x$$

Example 4

A pump discharges 400 gpm. What chlorine feed rate (pounds-per-day) is required to provide a dosage of 2.5 mg/L?

This problem involves *flow* rate (gallons per minute), concentration (milligrams per litre), and feed rate (pounds per day). First, write the conversion equation:

$$(\text{mg/L})(\text{Flow mgd})(8.34 \text{ lb/gal}) = \text{lb/day}$$

Before the information given in the problem can be filled in, the flow rate in gallons per minute will have to be converted to flow rate in million gallons per day:

$$(400 \text{ gpm})(1440 \text{ min/day}) = 576,000 \text{ gpd}$$
$$= 0.576 \text{ mgd}$$

Now write the information into the equation:

$$(2.5 \text{ mg/L})(0.576 \text{ mgd})(8.34 \text{ lb/gal}) = 12.01 \text{ lb/day}$$

In the four preceding examples the unknown value was always pounds per day. In the equation, however, any one of the three variables may be the unknown value:

$$(\text{mg/L})(\text{mgd})(8.34 \text{ lb/gal}) = \text{lb/day}$$
$$\text{variables}$$

Occasionally, for example, you may know how many pounds per day of chemicals are added and what the plant flow rate is, and need to know what chemical concentration in milligrams per litre this represents.

The next two examples illustrate the procedure used.

Example 5

In a treatment plant, 250 lb/day of dry alum is added to a flow of 1,550,000 gpd. What is this dosage expressed as in milligrams per litre?

From the statement of the problem you can see that this is a milligrams-per-litre and pounds-per-day problem. So, as in previous examples, first write the conversion equation that relates the two terms:

$$(\text{mg/L})(\text{mgd})(8.34 \text{ lb/gal}) = \text{lb/day}$$

Next, the flow rate in gallons per day must be converted to million gallons per day:

1,550,000 gpd

millions comma

$$1,550,000 \text{ gpd} = 1.55 \text{ mgd}$$

Now complete the problem by filling in the information given and solving for the unknown value:

$$(x \text{ mg/L})(1.55 \text{ mgd})(8.34 \text{ lb/gal}) = 250 \text{ lb/day}$$

$$x = \frac{250}{(1.55)(8.34)}$$

$$x = 19.34 \text{ mg/L}$$

Example 6

On one day at a treatment plant the coagulation process was operated for 15 hr. The average flow during the period was 200 gpm; 30 lb of alum were fed into the water. What was the alum dosage in milligrams per litre during that period?

The problem involves flow rate, concentration, and feed rate. Write the equation that relates the terms:

$$(mg/L)(mgd)(8.34\ lb/gal) = lb/day$$

Before the information given in the problem can be written into the equation, the flow rate must be expressed in million gallons per day. The flow rate was not continuous during the 24 hours; therefore, the usual gallons-per-minute to million-gallons-per-day conversion cannot be used. Because the 200 gpm flow rate continued for only 15 hr, the total flow for the day was

$$(200\ gpm)(60\ min/hr)(15\ hr) = 180,000\ gal\ flow$$
$$during\ the\ day$$

$$= 0.18\ mgd$$

Now write the information into the equation and solve for the unknown:

$$(x\ mg/L)(0.18\ mgd)(8.34\ lb/gal) = 30\ lb/day$$

$$x = \frac{30}{(0.18)(8.34)}$$

$$x = 19.98\ mg/L\ alum$$

In some problems the flow variable might be the unknown value. The following example illustrates the procedure used to solve such a problem:

Example 7

If 100 lb/day of dry alum were fed into the flow at a treatment plant to achieve a chemical dosage of 20 mg/L, what was the flow rate at the plant in million gallons per day?

First write the equation that relates milligrams per litre and pounds per day:

$$(mg/L)(mgd)(8.34\ lb/gal) = lb/day$$

Then fill in the information given and solve for the unknown value:

$$(20\ mg/L)(x\ mgd)(8.34\ lb/gal) = 100\ lb/day$$

$$x = \frac{100}{(20\ (8.34)}$$

$$x = 0.6\ mgd$$

One variation of the mg/L to lb/day problem involves the calculation of hypochlorite dosage. Calcium hypochlorite (usually 65 or 70 percent available chlorine) or sodium hypochlorite (usually 5–15 percent available chlorine) is sometimes used for chlorination instead of chlorine gas (100 percent available chlorine).

To obtain the same disinfecting power as chlorine gas, a greater weight of calcium hypochlorite would be required, and an even greater weight of sodium hypochlorite would be needed. For example, assume that water flowing through a treatment plant required a chlorine dosage of 300 lb/day. If chlorine gas were used for disinfection, the proper dosage rate would be 300 lb/day. If hypochlorites were used, however, about 430 lb/day of calcium hypochlorite or 3000 lb/day of sodium hypochlorite would be required to obtain the same disinfecting power.

In calculating the pounds of hypochlorite required, first calculate the pounds per day of chlorine (100 percent available) required, and then determine the amount of hypochlorite required. The following examples illustrate the calculations.

Example 8

Disinfection at a treatment plant requires 280 lb/day of chlorine. If calcium hypochlorite (65 percent available chlorine) is used, how many pounds per day will be required?

Since there is only 65 percent available chlorine in the calcium hypochlorite compound, *more* than 280 lb/day of calcium hypochlorite will have to be added to the water to obtain the same disinfecting power as 280 lb/day of 100 percent available chlorine. In fact, *65 percent of some number greater than 280* should equal 280:

(65%)(Greater number of lb/day) $= 280$ lb/day

This equation can be restated as:

$$(0.65)(x \text{ lb/day}) = 280 \text{ lb/day}$$

Then solve for the unknown value:

$$x = \frac{280}{0.65}$$

$$x = 430.77 \text{ lb/day}$$
$$\text{calcium hypochlorite}$$
$$\text{required}$$

Example 9

A water supply requires 30 lb/day of chlorine for disinfection. If sodium hypochlorite with 10 percent available chlorine is used for the disinfection, how many pounds per day of sodium hypochlorite are required?

Since there is only 10 percent available chlorine in the sodium hypochlorite, considerably more than 30 lb/day will be required to accomplish the disinfection. In fact, 10 percent of some greater number should equal 30 lb/day:

(10%)(Greater number of lb/day) $= 30$ lb/day

This equation can be restated as

$$(0.10)(x \text{ lb/day}) = 30 \text{ lb/day}$$

Then solve for the unknown value:

$$x = \frac{30}{0.10}$$

$$= 300 \text{ lb/day}$$
sodium hypochlorite
required

Normally, hypochlorite problems as shown in the two examples above are part of a milligrams-per-litre to pounds-per-day calculation. The following example illustrates the combined calculations.

Example 10

How many pounds per day of hypochlorite (70 percent available chlorine) are required for disinfection in a plant where the flow rate is 1.4 mgd and the chlorine dosage is 2.5 mg/L?

First calculate how many pounds per day of 100 percent chlorine are required, then calculate the hypochlorite requirement.

Write the equation that converts milligrams per litre concentration to pounds per day:

$$(\text{mg/L})(\text{mgd})(8.34 \text{ lb/gal}) = \text{lb/day}$$

Fill in the equation with the information given in the problem:

$$(2.5 \text{ mg/L})(1.4 \text{ mgd})(8.34 \text{ lb/gal}) = x \text{ lb/day}$$

$$29.19 \text{ lb/day} = x$$
chlorine required

Since 70-percent hypochlorite is to be used, more than 29.19 lb/day of hypochlorite will be required. In fact, 70 percent of some greater number should equal 29.19 lb/day:

$$(70\%)(\text{Greater number of lb/day}) = 29.19 \text{ lb/day}$$

This equation can be restated as

$$(0.7)(x \text{ lb/day}) = 29.19 \text{ lb/day}$$

Then solve for the unknown value:

$$x = \frac{29.19}{0.7}$$

$$= 41.7 \text{ lb/day}$$
hypochlorite required

Sometimes chlorine is used to disinfect tanks or pipelines. In such cases, a certain concentration must be achieved by the one-time addition of chlorine to a volume of water. The equation used to convert from milligrams per litre to pounds of chlorine required is similar to the equation used to convert to pounds-per-day feed rate; however, the volume of water in the container to be disinfected replaces the flow rate through the plant. Thus, the equation to calculate pounds of chlorine needed, given the disinfection dosage to be achieved, is as follows:

	Volume of	Conversion	Chlorine
Dosage	container	factor	weight
(mg/L)	(mil gal)	(8.34 lb/gal) =	(lb)

The following two examples illustrate how the equation is used.

Example 11

How many pounds of chlorine (100 percent available) are required to disinfect a 150,000-gal tank if the tank is to be disinfected with 50 mg/L of chlorine?

First write the conversion equation:

$$(mg/L)(mil\ gal)(8.34\ lb/gal) = lb$$

Next convert the 150,000-gal tank volume to millions of gallons:

$$150,000\ gal = 0.15\ mil\ gal$$

Then fill in the equation and solve:

$$(50\ mg/L)(0.15\ mil\ gal)(8.34\ lb/gal) = x\ lb$$
$$62.55\ lb = x$$
chlorine required

Example 12

How many pounds of hypochlorite (65 percent available chlorine) are required to disinfect 4000 ft of 24-in. water line if an initial dose of 40 mg/L is required?

This problem is similar to the previous one; however, in this problem the volume[21] of the water line must be calculated:

$$(0.785)(2\ ft)(2\ ft)(4000\ ft)(7.48\ gal/cu\ ft) = 93,949\ gal$$

The volume must be expressed in terms of million gallons:

$$93,949\ gal = 0.094\ mil\ gal$$

Next, use the equation that relates milligrams per litre to weight of chlorine required:

$$(40\ mg/L)(0.094\ mil\ gal)(8.34\ lb/gal) = x\ lb$$
$$31.36\ lb/day = x$$
chlorine required

Since 65-percent hypochlorite is to be used, more than 31.36 lb/day of hypochlorite will be required. In this case, 65 percent of some greater number should equal 31.36 lb/day:

---

[21] Mathematics Section, Volume Measurements.

$$(65\%)(\text{Greater number of lb}) = 31.36 \text{ lb}$$

This equation can be restated as

$$(0.65)(x \text{ lb}) = 31.36 \text{ lb}$$

Then solve for the unknown value:

$$x = \frac{31.36}{0.65}$$

$$x = 48.25 \text{ lb}$$
hypochlorite required

C7-2.   Milligrams-per-Litre to Percent Conversions

A concentration or dosage expressed as milligrams per litre can also be expressed as percent. Milligrams per litre are approximately equal to parts per million, and percent[22] means "parts per hundred."

$$\text{mg/L} = \text{parts per million} = \frac{\text{parts}}{1,000,000}$$

$$\text{percent} = \text{parts per hundred} = \frac{\text{parts}}{100}$$

Because $1,000,000 \div 100 = 10,000$, converting from parts per million (or milligrams per litre) to percent is accomplished by dividing by 10,000. The following box diagram[23] can be used:

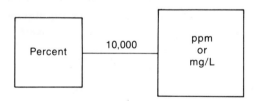

It will help to remember that division by 10,000 can be performed by moving the decimal point to the left four places (four places because 10,000 has four zeros), yielding a much smaller number:

$$31,400 \div 10,000 = 3.1400$$

$$= 3.14$$

Multiplication by 10,000 can be performed by moving the decimal point to the right four places, yielding a much larger number:

$$2.31 \times 10,000 = 23100.$$

$$= 23,100$$

---

[22]Mathematics Section, Percent.
[23]Mathematics Section, Conversions.

The following examples illustrate the use of the box diagram:

Example 13
A chemical is to be dosed at 25 mg/L. Express the dosage as percent.

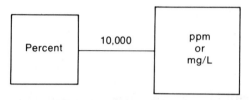

You are moving from a larger to a smaller box, so division by 10,000 is indicated:

$$\frac{25}{10,000} = 0.0025\%$$

Example 14
Express 120 ppm as percent.

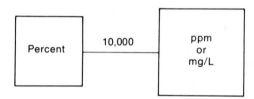

Referring to the box diagram, you are moving from a large box to a small box, so division is indicated:

$$\frac{120}{10,000} = 0.0120\%$$

Example 15
The sludge in a clarifier has a total solids concentration of 2 percent. Express the concentration as milligrams per litre of total solids.

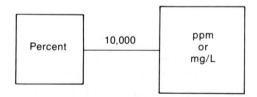

Referring to the box diagram, you are moving from a smaller to a larger box, so multiply by 10,000:

$$(2)(10,000) = 20,000 \text{ mg/L}$$

Table 7 shows the relationship between percent concentration and milligrams-per-litre concentration.

Table 7.   Percent vs. Milligrams-per-Litre Concentration

Percent	mg/L or ppm	Percent	mg/L or ppm
100.000	1,000,000	0.100	1,000
50.000	500,000	0.050	500
10.000	100,000	0.010	100
5.000	50,000	0.005	50
1.000	10,000	0.001	10
0.500	5,000	0.0001	1

C7-3.   Feed Rate Conversions

Some chemical dosage problems involve a conversion of feed rates from one term to another, such as from pounds per day to pounds per hour, or from gallons per day to gallons per hour. Many of these conversions are discussed in the mathematics section.[24] Four types of feed rate conversions not covered in the math section are discussed in the following material:

- gallons-per-hour to gallons-per-day
- gallons-per-hour to pounds-per-day
- pounds-per-hour to pounds-per-day
- pounds-per-hour to pounds-per-day

As with many of the conversions discussed in the math section, the *box method* will be used in making the conversions. The following box diagram will be used:

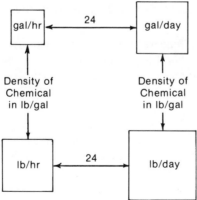

Example 16
The feed rate for a chemical is 230 lb/day. What is the feed rate expressed in pounds per hour?

---

[24]Mathematics Section, Conversions (Flow Rate).

In moving from lb/day to lb/hr, you are moving from a larger box to a smaller box. Therefore, division by 24 is indicated:

$$\frac{230 \text{ lb/day}}{24} = 9.58 \text{ lb/hr feed rate}$$

Example 17

A chemical has a density of 11.58 lb/gal. The desired feed rate for the chemical is 0.6 gal/hr. How many pounds per day is this?

Converting from gal/hr to lb/day, you are moving from a smaller box to a larger box ($gal/hr \rightarrow gal/day$) and then to a still larger box ($gal/day \rightarrow lb/day$). Therefore, multiplication by the density of the chemical and by 24 is indicated:

$$(0.6 \text{ gal/hr}) \left(\begin{array}{c} \text{Density of} \\ \text{chemical} \\ \text{in lb/gal} \end{array} \right) (24) = \begin{array}{c} \text{lb/day} \\ \text{feed rate} \end{array}$$

$$(0.6 \text{ gal/hr})(11.58 \text{ lb/gal})(24) = x \text{ lb/day}$$

$$166.75 \text{ lb/day} = x$$
$$\text{feed rate}$$

C7-4. Chlorine Dosage/Demand/Residual Calculations

The chlorine requirement or chlorine dosage is the sum of the chlorine demand and the desired chlorine residual. This can be expressed mathematically as

Chlorine		Chlorine		Chlorine
dosage	=	demand	+	residual
(mg/L)		(mg/L)		(mg/L)

Example 18

A water is tested and found to have a chlorine demand of 6 mg/L. The desired chlorine residual is 0.2 mg/L. How many pounds of chlorine will be required daily to chlorinate a flow of 8 mgd?

Chlorine		Chlorine		Chlorine
dosage	=	demand	+	residual
(mg/L)		(mg/L)		(mg/L)

Fill in the information given in the problem:

$$\text{Dosage} = 6 \text{ mg/L} + 0.2 \text{ mg/L}$$
$$= 6.2 \text{ mg/L}$$

Now convert the dosage in milligrams per litre to feed rate in pounds per day:

$$(6.2 \text{ mg/L})(8 \text{ mgd})(8.34 \text{ lb/gal}) = 413.66 \text{ lb/day}$$
$$\text{chlorine feed rate}$$

Example 19

The chlorine demand of a water is 5.5 mg/L. A chlorine residual of 0.3 mg/L is desired. How many pounds of chlorine will be required daily for a flow of 28 mgd?

$$
\begin{array}{ccc}
\text{Chlorine} & \text{Chlorine} & \text{Chlorine} \\
\text{dosage} = & \text{demand} + & \text{residual} \\
\text{(mg/L)} & \text{(mg/L)} & \text{(mg/L)}
\end{array}
$$

$$\text{Dosage} = 5.5 \text{ mg/L} + 0.3 \text{ mg/L}$$
$$= 5.8 \text{ mg/L}$$

Convert dosage in milligrams per litre to feed rate in pounds per day:

$$(5.8 \text{ mg/L})(28 \text{ mgd})(8.34 \text{ lb/gal}) = 1354 \text{ lb/day}$$
$$\text{chlorine feed rate}$$

Note that prechlorination usually does not require a residual to be achieved. Setting the residual equal to zero gives

$$
\begin{array}{ccc}
\text{Chlorine} & \text{Chlorine} & \text{Chlorine} \\
\text{dosage} = & \text{demand} + & \text{residual} \\
\text{(mg/L)} & \text{(mg/L)} & \text{(mg/L)}
\end{array}
$$

$$
\begin{array}{ccc}
\text{Chlorine} & \text{Chlorine} & \\
\text{dosage} = & \text{demand} + & 0 \\
\text{(mg/L)} & \text{(mg/L)} &
\end{array}
$$

$$
\begin{array}{cc}
\text{Chlorine} & \text{Chlorine} \\
\text{dosage} = & \text{demand} \\
\text{(mg/L)} & \text{(mg/L)}
\end{array}
$$

Therefore, when no residual is required, as in most prechlorination, the dosage is equal to the chlorine demand of the water.

The equation used to calculate chlorine dosage can also be used to calculate chlorine demand when dosage and residual are given, or to calculate chlorine residual when dosage and demand are given. The following examples illustrate the calculations.

Example 20

The chlorine dosage for a water is 5 mg/L. The chlorine residual after 30 min contact time is 0.6 mg/L. What is the chlorine demand in milligrams per litre?

$$\begin{array}{ccc} \text{Chlorine} & \text{Chlorine} & \text{Chlorine} \\ \text{dosage} = & \text{demand} + & \text{residual} \\ \text{(mg/L)} & \text{(mg/L)} & \text{(mg/L)} \end{array}$$

$$5 \text{ mg/L} = x \text{ mg/L} + 0.6 \text{ mg/L}$$

Now solve for the unknown value:

$$5 \text{ mg/L} - 0.6 \text{ mg/L} = x \text{ mg/L}$$
$$4.4 \text{ mg/L} = x$$
$$\text{demand}$$

To check this:

$$5 \text{ mg/L} = 4.4 \text{ mg/L} + 0.6 \text{ mg/L}$$

Example 21

The chlorine dosage of a water is 7.5 mg/L, and the chlorine demand is 7.1 mg/L. What is the chlorine residual?

$$\begin{array}{ccc} \text{Chlorine} & \text{Chlorine} & \text{Chlorine} \\ \text{dosage} = & \text{demand} + & \text{residual} \\ \text{(mg/L)} & \text{(mg/L)} & \text{(mg/L)} \end{array}$$

Fill in the given information and solve for the unknown value:

$$7.5 \text{ mg/L} = 7.1 \text{ mg/L} + x \text{ mg/L}$$
$$7.5 \text{ mg/L} - 7.1 \text{ mg/L} = x \text{ mg/L}$$
$$0.4 \text{ mg/L} = x$$

To check this:

$$7.5 \text{ mg/L} = 7.1 \text{ mg/L} + 0.4 \text{ mg/L}$$

Sometimes the dosage may not be given in milligrams per litre, but the feed rate setting of the chlorinator and the daily flow rate through the plant will be known. In such cases, feed rate and flow rate should be used to calculate the dosage in milligrams per litre (following the procedure shown previously in examples 5 and 6). The dosage in milligrams per litre can then be used in the dosage/demand/residual equation, as in examples 20 and 21.

C7-5. Lime–Soda Ash Softening Calculations

There are two methods presented for calculating lime and soda ash dosages:

• A conventional method

• A conversion factor method

The conventional method, although much longer, is helpful in understanding the chemical and mathematical relationships involved in softening. The conversion factor method is simpler and quicker, and is the more practical method to use in day-to-day operations.

In both methods of calculation, the lime and soda ash dosages required are dependent on the carbonate and noncarbonate constituents in the water. Although hard water will always require lime, it may not always need soda ash. Lime is used to remove carbonate hardness, and both lime and soda ash are used to remove noncarbonate hardness. The amounts of carbonate and noncarbonate hardness can be found as shown in Table 8.

Table 8.   Finding Carbonate and Noncarbonate Hardness

Laboratory Results TH = Total Hardness TA = Total Alkalinity	Noncarbonate Hardness (Lime and Soda Ash Used)	Carbonate Hardness (Lime Only Used)
1. TH less than TA	0	TH
2. TH = TA	0	TH
3. TH greater than TA	TH-TA	TA

To determine whether lime only or both lime and soda ash are needed, total hardness must be compared to total alkalinity. NOTE: in some areas of the US, the only alkalinity occurring naturally in water is bicarbonate alkalinity; in these areas, many labs report only the bicarbonate alkalinity concentration or only the total alkalinity concentration, assuming that the operator will be aware that, for the water tested, total alkalinity = bicarbonate alkalinity. However, to be certain of the characteristics of the water tested, an operator should request a detailed lab report showing the total alkalinity as well as the bicarbonate, carbonate, and hydroxyl alkalinity.

A total hardness less than or equal to total alkalinity (condition 1 or 2 in Table 8) indicates two important facts about lime and soda ash dosages:

- Since the noncarbonate concentration is zero, only lime treatment will be required. (No soda ash will be required.)

- The lime normally added to treat magnesium noncarbonate hardness (for example, $MgCl_2$, $MgSO_4$) is not required.

The following chemical equation is used to determine lime dosage requirements when conditions 1 or 2 exist:

Lime Dosage $= [CO_2] + [\text{Total Hardness}] + [Mg] + [\text{Excess}]$ *Eq 1*

The brackets, [ ], in the equation mean "the concentration of." Therefore, when total hardness is less than or equal to total alkalinity, the lime dosage equals the sum of the concentrations of carbon dioxide, total hardness, and magnesium, plus any excess lime desired.

In most operating situations, more lime will be needed than is indicated by the chemical equations. This additional amount, called "excess lime," is added either to speed the reactions, or to force them to react more completely. It is common practice to add from 25 to 50 mg/L as $CaCO_3$ of excess lime. An equivalent amount of soda ash also must be added in order to remove the excess calcium ion added with the lime. To remove 1 mg/L of excess lime requires 1.06 mg/L as Na_2CO_3 of excess soda ash.

When total hardness is greater than total alkalinity (condition 3 in Table 8), carbonate and noncarbonate hardness constituents are present in the water. This means that both lime and soda ash will be needed in treating for hardness: lime to remove carbonate hardness constituents, and both lime and soda ash to remove noncarbonate hardness constituents in the water. The following chemical equations are then used to determine lime and soda ash dosage requirements:

$$\text{Lime dosage} = [CO_2] + [HCO_3^-] + [Mg] + [\text{Excess}] \qquad Eq\ 2$$

$$\text{Soda ash dosage} = [\text{Total Hardness}] - [HCO_3^-] + [\text{Excess}] \qquad Eq\ 3$$

The lime dosage equals the sum of the concentrations of carbon dioxide, bicarbonate alkalinity, and magnesium, plus the excess lime desired. The soda ash dosage is equal to the total hardness, minus the bicarbonate alkalinity, plus any excess soda ash desired. Note that when the bicarbonate alkalinity (HCO_3^-) is equal to the total alkalinity (TA), Equation 3 could be written

$$\text{Soda ash dosage} = [\text{Total Hardness}] - [TA] + [\text{Excess}] \qquad Eq\ 3_A$$

And, as shown in Table 8, when the total hardness (TH) is greater than the total alkalinity, TH $-$ TA = Noncarbonate Hardness. Under these conditions, Equation 3 could be written

$$\text{Soda ash dosage} = [\text{Noncarbonate Hardness}] + [\text{Excess}] \qquad Eq\ 3_B$$

The two ions that most commonly cause hardness in water are calcium and magnesium. Although aluminum, strontium, iron, manganese, barium, and zinc ions can also cause hardness, they are not usually present in large enough concentrations to cause a significant problem. Consequently it is usually assumed, without much sacrifice in accuracy, that

$$[\text{Total hardness}] = [Ca] + [Mg] \qquad Eq\ 4$$

When only calcium and magnesium are measured in the laboratory, the total hardness can be found using Equation 4. When the laboratory measures total hardness and calcium, Equation 4 can be used to determine the magnesium concentration.

In some cases, total hardness, calcium, and magnesium are all measured in the laboratory. When this occurs, total hardness should be used for Equations 1 and 3 instead of the sum of calcium and magnesium, since total hardness includes the effects of all hardness-causing ions.

Equations 1–4 are the basis for lime and soda ash dosage determinations using either the conventional or the conversion factor method. Before these four equations can be used, the concentrations of CO_2, Total Hardness, Mg, HCO_3^-, and excess lime must all be expressed in "mg/L as $CaCO_3$." Usually, when a laboratory test is made for CO_2, the results are reported in "mg/L as CO_2." Similarly, a magnesium concentration would be reported in "mg/L as Mg," and a bicarbonate alkalinity would be reported in "mg/L as HCO_3^-," or in "mg/L as $CaCO_3$." Since these concentrations are not expressed in the same units of measure, they cannot be added together. However, each concentration can be converted[25] to "mg/L as $CaCO_3$" (the units commonly used to express hardness). After all laboratory results are converted to these units, they can be combined as in equations 1–4.

---

[25]Chemistry Section, Solutions (Hardness).

The conventional and conversion factor methods are based on the theory that all reactions occur in definite proportions to the compounds involved. More specifically, the reactions occur in definite proportion to the equivalent weights of the reacting ions or compounds. The difference between the two methods is that, in the conversion factor method, some of the conversion calculations (such as conversion from "mg/L as HCO_3^-" to "mg/L as $CaCO_3$") are shortened by using conversion factors in which part of the calculation has been completed.

Conventional Method

Given the total hardness, calcium, magnesium, bicarbonate alkalinity, total alkalinity, and free carbon dioxide concentrations from laboratory test results, the conventional method uses equivalent weights to convert all hardness data to milligrams per litre as $CaCO_3$, and then uses the $CaCO_3$ concentrations to calculate the quantity of lime and soda ash needed. This consists of six steps, which are summarized below and explained in detail in the following examples:

Step 1: Convert all data to concentrations as $CaCO_3$, using the following equivalent weights:

$$Ca = 20$$
$$Mg = 12$$
$$HCO_3^- = 61$$
$$CO_2 = 22$$
$$CaCO_3 = 50$$

For example:

$$5 \text{ mg/L as } CO_2 = (5)\left(\frac{50}{22}\right)$$
$$= 11.36 \text{ mg/L as } CaCO_3$$

Step 2: Determine whether lime only or both lime and soda ash will be required in treating the hardness constituents. (Use Table 8)

Step 3: Calculate the lime dosage or lime and soda ash dosage (milligrams per litre as $CaCO_3$) using the following equations:

When total hardness is less than or equal total alkalinity:

Lime dosage = [CO_2] + [Total Hardness] + [Mg] + [Excess]   *Eq 1*

When total hardness is greater than total alkalinity:

Lime dosage = [CO_2] + [HCO_3^-] + [Mg] + [Excess]   *Eq 2*

Soda ash dosage = [Total Hardness] − [HCO_3^-] + [Excess]   *Eq 3*

or, when noncarbonate hardness is given,

Soda ash dosage = [Noncarbonate Hardness] + [Excess]   *Eq 3B*

Step 4: Convert lime and soda ash dosages expressed as $CaCO_3$ to dosages expressed as CaO and as Na_2CO_3, respectively, using the following equivalent weights:

$$CaCO_3 = 50$$
$$CaO = 28$$
$$Na_2CO_3 = 53$$

For example:

$$227 \text{ mg/L lime as } CaCO_3 = (227)\left(\frac{28}{50}\right) = 127.12 \text{ mg/L as } CaO$$

Step 5: Adjust dosages for the purity of the lime or soda ash used. For example:

$$(88\%)(\text{Actual dosage of } 88\% \text{ pure } CaO) = 127 \text{ mg/L dosage of } 100\% \text{ pure } CaO$$

$$\text{Actual dosage of } 88\% \text{ pure } CaO = 144.5 \text{ mg/L}$$

Step 6: Convert these adjusted dosages to pounds per million gallon or pounds per day. For example:

$$(\text{mg/L})(8.34 \text{ lb/gal}) = \text{lb/mil gal}$$

$$(\text{mg/L})(\text{mgd flow})(8.34 \text{ lb/gal}) = \text{lb/day}$$

The following examples illustrate softening calculations using the conventional method.

Example 22

The laboratory returned the following test results:

Calcium	$= 100$ mg/L as Ca
Total hardness	$= 300$ mg/L as $CaCO_3$
Bicarbonate alkalinity	$= 250$ mg/L as HCO_3^-
Total alkalinity	$= 205$ mg/L as $CaCO_3$
Carbon dioxide	$= 25$ mg/L as CO_2

Express each concentration as $CaCO_3$.

Using the equivalent weight procedure,[26] the conversions can be made as follows:

Calcium: $(100 \text{ mg/L})\left(\frac{50}{20}\right) = 250$ mg/L as $CaCO_3$

Total hardness: (No conversion needed) $= 300$ mg/L as $CaCO_3$

Bicarbonate alkalinity: $(250 \text{ mg/L})\left(\frac{50}{61}\right) = 204.92$ mg/L as $CaCO_3$

Total alkalinity: (No conversion needed) $= 205$ mg/L as $CaCO_3$

Carbon dioxide: $(25 \text{ mg/L})\left(\frac{50}{22}\right) = 56.82$ mg/L as $CaCO_3$

---

[26]Chemistry Section, Solutions (Hardness).

Example 23

The laboratory returned the following test results:

Magnesium	$= 45$ mg/L as Mg
Total hardness	$= 280$ mg/L as $CaCO_3$
Bicarbonate alkalinity	$= 245$ mg/L as HCO_3^-
Total alkalinity	$= 202$ mg/L as $CaCO_3$
Carbon dioxide	$= 20$ mg/L as CO_2

(a) Express each concentration as $CaCO_3$.

(b) According to Step 2 in the softening calculations (and using Table 8), will lime only or lime and soda ash be required in treating the hardness constituents?

(a) Using the equivalent weight procedure, as in Example 22, make the conversions as follows:

Magnesium: $(45 \text{ mg/L})\left(\dfrac{50}{12}\right) = 187.5$ mg/L as $CaCO_3$

Total hardness: (No conversion needed) $= 280$ mg/L as $CaCO_3$

Bicarbonate alkalinity: $(245 \text{ mg/L})\left(\dfrac{50}{61}\right) = 200.82$ mg/L as $CaCO_3$

Total alkalinity: (No conversion needed) $= 202$ mg/L as $CaCO_3$

Carbon dioxide: $(20 \text{ mg/L})\left(\dfrac{50}{22}\right) = 45.45$ mg/L as $CaCO_3$

(b) In this example, total hardness (280 mg/L as $CaCO_3$) is greater than total alkalinity (202 mg/L as $CaCO_3$). As shown in Table 8, the water will have both noncarbonate and carbonate hardness constituents and will, therefore, require both lime and soda ash treatment.

Example 24

The laboratory test results for a water sample were as follows:

Magnesium	$= 32$ mg/L as Mg
Total hardness	$= 345$ mg/L as $CaCO_3$
Bicarbonate alkalinity	$= 156$ mg/L as HCO_3^-
Total alkalinity	$= 128$ mg/L as $CaCO_3$
Carbon dioxide	$= 5$ mg/L as CO_2

(a) Determine whether lime alone or lime and soda ash will be required in treating the hardness constituents.

(b) Calculate the lime dosage or lime and soda ash dosages in (milligrams per litre as $CaCO_3$). (Assume no excess lime or soda ash is to be added.)

(a) The total hardness (345 mg/L as $CaCO_3$) is greater than the total alkalinity (128 mg/L as $CaCO_3$). Therefore, according to Table 8, water will have both carbonate and noncarbonate hardness constituents and will require both lime and soda ash treatment.

(b) Before calculating the lime and soda ash dosages using equations 1–3, express all concentrations in terms of $CaCO_3$:

Magnesium: $(32 \text{ mg/L})\left(\dfrac{50}{12}\right) = 133.33$ mg/L as $CaCO_3$

Total hardness: (No conversion needed) $= 345$ mg/L as $CaCO_3$

Bicarbonate alkalinity: $(156 \text{ mg/L})\left(\dfrac{50}{61}\right) = 127.87$ mg/L as $CaCO_3$

Total alkalinity: (No conversion needed) $= 128$ mg/L as $CaCO_3$

Carbon dioxide: $(5 \text{ mg/L})\left(\dfrac{50}{22}\right) = 11.36$ mg/L as $CaCO_3$

Now use the dosage equations shown in Step 3:

Lime dosage $= [CO_2] + [HCO_3^-] + [Mg] + [\text{Excess}]$ *Eq 2*

$\qquad = 11.36 \text{ mg/L} + 128 \text{ mg/L} + 133.33 \text{ mg/L} + 0$

$\qquad = 272.69$ mg/L as $CaCO_3$

Soda ash dosage $= [\text{Total hardness}] - [HCO_3^-] + [\text{Excess}]$ *Eq 3*

$\qquad = 345 \text{ mg/L} - 128 \text{ mg/L} + 0$

$\qquad = 217$ mg/L as $CaCO_3$

If, instead of a magnesium ion concentration, a calcium ion concentration is given in the test results, then the magnesium ion concentration can still be calculated by using Equation 4 and the Total Hardness. Then the lime dosage can be determined in the usual manner:

Example 25

The laboratory returned the following results of testing:

Calcium	$= 50$ mg/L as Ca
Total hardness	$= 150$ mg/L as $CaCO_3$
Bicarbonate alkalinity	$= 149$ mg/L as HCO_3^-
Total alkalinity	$= 130$ mg/L as $CaCO_3$
Carbon dioxide	$= 10$ mg/L as CO_2

Calculate the lime dosage milligrams per litre as $CaCO_3$ and, if needed, the soda ash dosage. (Assume no excess lime or soda ash to be added.)

Since the total alkalinity (130 mg/L) is less than the total hardness, the water has both carbonate and noncarbonate hardness constituents. Therefore, both lime and soda ash will be required.

Before the equations 1–3 can be used in determining lime

and soda ash dosages, all the concentrations must be expressed in similar terms (as $CaCO_3$).

Use the equivalent weight procedure to make the conversions to "as $CaCO_3$:"

Calcium: $(50 \text{ mg/L})\left(\dfrac{50}{20}\right) = 125 \text{ mg/L as } CaCO_3$

Total hardness: (No conversion needed) $= 150 \text{ mg/L as } CaCO_3$

Bicarbonate alkalinity: $(149 \text{ mg/L})\left(\dfrac{50}{61}\right) = 122.13 \text{ mg/L as } CaCO_3$

Total alkalinity: (No conversion needed) $= 130 \text{ mg/L as } CaCO_3$

Carbon dioxide: $(10 \text{ mg/L})\left(\dfrac{50}{22}\right) = 22.73 \text{ mg/L as } CaCO_3$

Next, determine the concentration of magnesium. Total hardness is basically calcium ion concentration plus magnesium ion concentration. Mathematically this is

$$[\text{Total hardness}] = [\text{Ca}] + [\text{Mg}] \qquad Eq\ 4$$

In this example, two of these factors are known, total hardness and calcium. So, write the known information into the equation:

$$[\text{Total hardness}] = [\text{Ca}] + [\text{Mg}]$$
$$150 \text{ mg/L} = 125 \text{ mg/L} + x$$

Now solve for the unknown value,[27] making sure that all concentrations in the above equation were expressed in terms of the same constituent (as $CaCO_3$). Since calcium represents 125 mg/L of the total 150 mg/L hardness, magnesium represents the balance of the hardness, or:

$$\text{Mg} = 150 \text{ mg/L} - 125 \text{ mg/L}$$
$$= 25 \text{ mg/L as } CaCO_3$$

Keep in mind that this calculation, though accurate enough, is not exact. Since there may be small amounts of other hardness-producing cations than calcium and magnesium, it is likely that if the magnesium ion concentration were measured, it would be slightly less than the 25 mg/L calculated above.

Now the lime and soda ash dosages can be calculated using the equations:

Lime dosage $= [CO_2] + [HCO_3^-] + [\text{Mg}] + [\text{Excess}] \qquad Eq\ 2$
$$= 22.73 \text{ mg/L} + 122.13 \text{ mg/L} + 25 \text{ mg/L} + 0$$
$$= 169.86 \text{ mg/L as } CaCO_3$$

---

[27]Mathematics Section, Solving for the Unknown Value.

$$\text{Soda ash dosage} = [\text{Total hardness}] - [\text{HCO}_3^-] + [\text{Excess}] \qquad Eq \ 3$$
$$= 150 \text{ mg/L} - 122.13 \text{ mg/L} + 0$$
$$= 27.87 \text{ mg/L as CaCO}_3$$

Dosages expressed as $CaCO_3$ are useful in order to simplify and standardize the calculation process. However, these dosages are not useful operationally until they are expressed as lime or as soda ash. Once they have been expressed in terms of their own equivalent weight, the operator can use the dosages to dispense the correct amount of chemical. Examples 26 and 27 illustrate this calculation.

Example 26

Using the lime and soda ash dosages given in Example 25, determine these dosages expressed in terms of quicklime (lime as CaO) and soda ash (Na_2CO_3), respectively.

To solve this problem, make the equivalent weight conversions. This is done by multiplying the dosages by the equivalent weight of the chemical being changed to and dividing that result by the equivalent weight of the chemical being changed from:

$$\text{Lime dosage} = (169.86 \text{ mg/L})\left(\frac{28}{50}\right)$$
$$= 95.12 \text{ mg/L as CaO}$$
$$\text{Soda ash dosage} = (27.87 \text{ mg/L})\left(\frac{53}{50}\right)$$
$$= 29.54 \text{ mg/L as Na}_2\text{CO}_3$$

Example 27

The most current laboratory data sheet shows the following water hardness characteristics:

Calcium	$= 140$ mg/L as $CaCO_3$
Total hardness	$= 180$ mg/L as $CaCO_3$
Bicarbonate alkalinity	$= 220$ mg/L as $CaCO_3$
Total alkalinity	$= 220$ mg/L as $CaCO_3$
Carbon dioxide	$= 6$ mg/L as $CaCO_3$

Find the concentration of quicklime (lime as CaO) and soda ash (Na_2CO_3), if needed. (Assume that no excess lime or soda ash is to be added.)

Since the laboratory has reported the results as $CaCO_3$, the problem can be solved by beginning with Step 2. In this example, total hardness (180 mg/L as $CaCO_3$) is less than total alkalinity (220 mg/L as $CaCO_3$.) This is condition 1 in Table 8, indicating the addition of lime alone. The equation used to calculate lime dosage as $CaCO_3$ utilizes magnesium concentration information, therefore, calculate the magnesium concentration by using total hardness and calcium concentration information:

$$[\text{Total hardness}] = [\text{Ca}] + [\text{Mg}] \qquad Eq\ 4$$

Insert the known information in the equation:

$$180\ \text{mg/L} = 140\ \text{mg/L} + x$$

Solving for the unknown gives a magnesium concentration of

$$180\ \text{mg/L} - 140\ \text{mg/L} = 40\ \text{mg/L as } CaCO_3$$

Because total hardness is less than total alkalinity, Equation 1 should be used in determining lime dosage requirements:

$$\text{Lime dosage} = [CO_2] + [\text{Total hardness}] + [\text{Mg}] + [\text{Excess}]\ Eq\ 1$$
$$= 6\ \text{mg/L} + 180\ \text{mg/L} + 40\ \text{mg/L} + 0$$
$$= 226\ \text{mg/L as } CaCO_3$$

NOTE: if equations 2 and 3 are used with data indicating a total hardness less than the total alkalinity, then the soda ash dosage calculation will result in a zero or negative number, and the calculation must be repeated using Equation 1.

To complete this problem, use the equivalent weight calculation to convert milligrams per litre as $CaCO_3$ to milligrams per litre as CaO:

$$\text{Lime dosage} = (226\ \text{mg/L})\left(\frac{28}{50}\right)$$
$$= 126.56\ \text{mg/L as CaO}$$

Once the milligrams-per-litre concentration of lime and soda ash has been calculated, the information is expressed in the more practical dosage terms of pounds per million gallons or pounds per day, as noted in Step 6 of the chemical softening calculations. The following examples illustrate this calculation.

Example 28

How many pounds of lime per day would be needed to soften the raw water in Example 27 at the rate of 640,000 gpd?

This is a typical milligrams-per-litre to pounds-per-day calculation.[28] To convert 126.56 mg/L CaO to pounds per day insert the known information in the following equation:

$$(\text{mg/L})(\text{mgd flow})(8.34\ \text{lb/gal}) = \text{lb/day}$$

$$(126.56\ \text{mg/L})(0.64\ \text{mgd})(8.34\ \text{lb/gal}) = 675.53\ \text{lb/day CaO}$$

---

[28]Chemistry Section, Chemical Dosage Problems (Milligrams-per-Litre to Pounds-per-day Conversions).

The lime and soda ash used in water softening treatment are not always 100 percent pure (lime is most often not 100 percent pure, whereas soda ash is usually, but not always, 100 percent pure). Suppose for example, a certain powdered quicklime were only 95 percent pure (contained only 95 percent CaO). This would mean that if a dosage of 60 mg/L of quicklime were required, more than 60 mg/L (in fact, 63 mg/L) would actually have to be added to the water to compensate for the impurity of the powder.

To calculate the actual dosage needed of chemical that is less than 100 percent pure, remember that the general equation for percent problems is[29]

$$\text{Percent} = \frac{\text{Part}}{\text{Whole}} \times 100$$

When calculating dosages of chemicals with less than 100 percent purity, the "whole" is the amount of impure, commercial chemical actually measured into the water; the "percent" is the purity of the commercial chemical; and the "part" is the calculated dosage of pure chemical that is required for the treatment process. This is summarized in the following equation:

$$\left(\begin{array}{c} \% \text{ Purity} \\ \text{of chemical} \\ \text{actually used} \end{array} \right) \left(\begin{array}{c} \text{Actual dosage} \\ \text{required for} \\ \text{chemical used} \end{array} \right) = \left(\begin{array}{c} \text{Calculated dosage} \\ \text{of 100\% pure} \\ \text{chemical required} \end{array} \right)$$

Example 29

Assume that 675.53 lb/day CaO is required for treating a certain water, as calculated in the previous example. How many pounds per day of the 90 percent pure quicklime will actually be required?

Keep in mind that the answer to this problem will be slightly more than 675.53 lb/day. To determine the required dosage of the 90 percent pure chemical that is actually used, insert the information in the equation:

(% Purity)(Actual dosage) = (Calculated dosage)

Convert the percent to a decimal:

(0.90)(Actual dosage) = 675.53 lb/day

Then solve for the unknown number:

$$\text{(Actual dosage)} = \frac{675.53}{0.90}$$

$$= 750.58 \text{ lb/day CaO required, 90\% pure}$$

Remember that, whenever the purity of the chemical used is less than 100 percent, the actual dosage will be some greater number than the calculated dosage of 100 percent pure chemical.

---

[29] Mathematics Section, Percent.

Example 30

It has been calculated (see Example 26) that 95.12 mg/L quicklime as CaO and 29.54 mg/L soda ash as Na_2CO_3 are required in treating a certain water. The quicklime to be used is 94 percent pure; the soda ash is 100 percent pure, and the plant flow at the time is 1.6 mgd. How many pounds per day of quicklime and soda ash should be used?

First, calculate the pounds-per-day requirement for the lime and soda ash:

Lime:

$$(mg/L)(mgd\ flow)(8.34\ lb/gal) = lb/day$$

$$(95.12\ mg/L)(1.6\ mgd)(8.34\ lb/gal) = 1269\ lb/day\ CaO\ (100\%\ pure)$$

Soda ash:

$$(mg/L)(mgd\ flow)(8.34\ lb/gal) = lb/day$$

$$(29.54\ mg/L)(1.6\ mgd)(8.34\ lb/gal) = 394.18\ lb/day\ Na_2CO_3\ (100\%\ pure)$$

Now calculate the adjustment to compensate for the impurity of the chemicals:

Lime:

$$(94\%)(Actual\ dosage) = 1269\ lb/day$$

$$(0.94)(Actual\ dosage) = 1269\ lb/day$$

$$Actual\ dosage = \frac{1269}{0.94}$$

$$= 1350\ lb/day\ CaO\ required,\ 94\%\ pure$$

Soda ash:

Since the chemical is 100 percent pure, 394.18 lb/day Na_2CO_3 is the actual dosage required.

It is common practice to add more lime than determined by calculation because it increases the speed and improves the completeness of the reaction. The following two examples illustrate the use of excess lime information in the calculation of lime and soda ash dosages.

Example 31

A raw water sample has the following water quality characteristics:

Calcium	= 152 mg/L as Ca
Magnesium	= 29 mg/L as Mg
Bicarbonate alkalinity	= 317 mg/L as HCO_3^-
Carbon dioxide	= 9 mg/L as CO_2

It has been decided to include an excess lime dose of 35 mg/L as $CaCO_3$. Including this excess, calculate the lime

(89 percent pure) and soda ash (100 percent pure) dosage (a) in milligrams per litre as lime or soda ash; (b) in pounds per million gallons; and (c) in pounds per day. Assume a plant flow of 479,500 gpd.

(a) First, express all data as $CaCO_3$:

Calcium: $(152)\left(\dfrac{50}{20}\right)$ = 380 mg/L as $CaCO_3$

Magnesium: $(29)\left(\dfrac{50}{12}\right)$ = 120.83 mg/L as $CaCO_3$

Bicarbonate alkalinity: $(317)\left(\dfrac{50}{61}\right)$ = 259.84 mg/L as $CaCO_3$

Carbon dioxide: $(9)\left(\dfrac{50}{22}\right)$ = 20.45 mg/L as $CaCO_3$

Now calculate the lime and soda ash requirements:

$$\text{Lime dosage} = [CO_2] + [HCO_3^-] + [Mg] + [\text{Excess}] \qquad Eq\ 2$$
$$= 20.45 + 259.84 + 120.83 + 35$$
$$= 436.12 \text{ mg/L as } CaCO_3$$

$$\text{Soda ash as } CaCO_3 = [\text{Total hardness}] - [HCO_3^-] + [\text{Excess}] \quad Eq\ 3$$
$$[\text{Total hardness}] = [Ca] + [Mg]$$
$$= 380 + 120.83$$
$$= 500.83 \text{ mg/L as } CaCO_3$$
$$\text{Soda ash dosage} = 500.83 - 259.84 + 35$$
$$= 275.99 \text{ mg/L as } CaCO_3$$

Next, express the dosages as lime or soda ash:

$$\text{Lime dosage} = (436.12)\left(\frac{28}{50}\right)$$
$$= 244.23 \text{ mg/L as } CaO \text{ (pure)}$$

$$\text{Soda ash dosage} = (275.99)\left(\frac{53}{50}\right)$$
$$= 292.55 \text{ mg/L as } Na_2CO_3 \text{ (pure)}$$

Finally, adjust for purity:

$$\text{Actual lime dosage} = \frac{244.23}{0.89}$$
$$= 274.42 \text{ mg/L as } CaO, 89\% \text{ pure}$$

No purity adjustment is needed for soda ash, therefore:

$$\text{Soda ash dosage} = 292.55 \text{ mg/L as } Na_2CO_3 \text{ 100\% pure}$$

(b) The pounds per million gallons equivalent to any dosage is found by multiplying the dosage by 8.34 as follows:

$$lb/mil\ gal = (mg/L)(8.34\ lb/gal)$$

$$Lime\ dosage = (274.42\ mg/L)(8.34\ lb/gal)$$

$$= 2288.66\ lb/mil\ gal\ of\ 89\%\ CaO$$

$$Soda\ ash\ dosage = (292.55\ mg/L)(8.34\ lb/gal)$$

$$= 2439.87\ lb/mil\ gal\ of\ 100\%\ Na_2CO_3$$

(c) Similarly, the pounds per day can be found as follows:

$$(mg/L)(mgd\ flow)(8.34\ lb/gal) = lb/day$$

$$Lime\ dosage = (274.42\ mg/L)(0.4795\ mgd)(8.34\ lb/gal)$$

$$= 1097.41\ lb/day\ of\ 89\%\ CaO$$

$$Soda\ ash\ dosage = (292.55)(0.4795\ mgd)(8.34\ lb/gal)$$

$$= 1169.92\ lb/day\ of\ 100\%\ Na_2CO_3$$

Example 32

The test results for a raw water sample were returned from the laboratory as follows:

Total hardness	$= 220\ mg/L$ as $CaCO_3$
Calcium	$= 66\ mg/L$ as Ca
Magnesium	$= 13\ mg/L$ as Mg
Bicarbonate alkalinity	$= 156\ mg/L$ as HCO_3^-
Carbon dioxide	$= 5\ mg/L$ as CO_2

The plant flow on January 4 averaged 1.2 mgd. How many pounds of quicklime (88 percent pure) and soda ash (100 percent pure) should have been used that day? It is common practice at this plant to add 26 mg/L excess lime as $CaCO_3$.

Begin by converting all the constituent concentrations to concentrations as $CaCO_3$.

Total hardness: $= 220\ mg/L$ as $CaCO_3$

Calcium: $(66)\left(\dfrac{50}{20}\right) = 165\ mg/L$ as $CaCO_3$

Magnesium: $(13)\left(\dfrac{50}{12}\right) = 54.17\ mg/L$ as $CaCO_3$

Bicarbonate alkalinity: $(156)\left(\dfrac{50}{61}\right) = 127.87\ mg/L$ as $CaCO_3$

Carbon dioxide: $(5)\left(\dfrac{50}{22}\right) = 11.36\ mg/L$ as $CaCO_3$

Next calculate lime dosage as $CaCO_3$, according to the following equation:

$$\text{Lime dosage} = [CO_2] + [HCO_3^-] + [Mg] + [\text{Excess}] \qquad Eq\ 2$$
$$= 11.36 + 127.87 + 54.17 + 26$$
$$= 219.4 \text{ mg/L as } CaCO_3$$

In a similar manner, calculate the soda ash dosage by this equation:

$$\text{Soda ash dosage} = [\text{Total hardness}] - [HCO_3^-] + [\text{Excess}] \qquad Eq\ 3$$
$$= 220 - 127.87 + 26$$
$$= 118.13 \text{ mg/L as } CaCO_3$$

Notice that total hardness, calcium, and magnesium were all measured in the laboratory. In such cases, it is best to use total hardness in the soda ash calculation since the sum of calcium and magnesium can be slightly smaller (220 mg/L as $CaCO_3$, versus 219.17 mg/L as $CaCO_3$).

The dosages calculated so far must now be expressed in terms of the equivalent weight of the softening chemicals used (as CaO or as Na_2CO_3).

$$\text{Lime dosage} = (219.4)\left(\frac{28}{50}\right)$$
$$= 122.86 \text{ mg/L as } CaO \text{ (pure)}$$
$$\text{Soda ash dosage} = (118.13)\left(\frac{53}{50}\right)$$
$$= 125.22 \text{ mg/L as } Na_2CO_3 \text{ (pure)}$$

These are the concentrations needed if the quicklime and soda ash are 100 percent pure. But the quicklime used is 88 percent pure, therefore, adjust the quicklime dosage by dividing the 100 percent pure dosage by the actual purity to be used. Therefore:

$$\text{Actual lime dosage} = \frac{122.86}{0.88}$$
$$= 139.61 \text{ mg/L as } CaO \text{ 88\% pure}$$

The soda ash used was 100 percent pure so no adjustment need be made.

Finally, based on the average flow rate for the day, given as 1.2 mgd, determine the total pounds of lime and soda ash to be used per day, as follows:

$$\text{Lime dosage} = (139.61 \text{ mg/L})(8.34 \text{ lb/gal})(1.2 \text{ mgd})$$
$$= 1397.22 \text{ lb/day of 88\% pure } CaO$$
$$\text{Soda ash dosage} = (125.22 \text{ mg/L})(8.34 \text{ lb/day})(1.2 \text{ mgd})$$
$$= 1253.20 \text{ lb/day of 100\% pure } Na_2CO_3$$

Conversion Factor Method

All of the equivalent weight conversions required in the conventional method have been combined into single factors shown in Table 9. These factors, multiplied by the concentration of the corresponding constituent, will give the lime or soda ash dosage needed to remove that constituent, either in units of milligrams per litre or pounds per million gallons. Note that the total lime dosage is found by summing the amounts of lime needed to remove carbon dioxide, bicarbonate alkalinity, magnesium, and adding that sum to the excess lime required (if any). The total soda ash dosage is found similarly, by adding the amount of soda ash required to remove noncarbonate hardness plus any excess soda ash required. One final step is required, and that is adjusting the dosage based on the actual purity of the lime and soda ash.

Table 9.   Factors for Converting Constituent Concentrations to Softening Chemical Dosages*

Constituent Concentration	Factors* for Converting to:			
mg/L	Lime (CaO, 100%)		Soda Ash (Na$_2$CO$_3$ 100%)	
	mg/L	lb/MG	mg/L	lb/MG
Carbon dioxide as CO$_2$	1.27	10.63	—	—
Bicarbonate alkalinity as CaCO$_3$	0.56	4.67	—	—
Magnesium as Mg	2.31	19.24	—	—
Excess lime as CaCO$_3$	0.56	4.67	—	—
Noncarbonate hardness as CaCO$_3$	—	—	1.06	8.83
Excess soda ash as CaCO$_3$	—	—	1.06	8.83

*Multiply constituent concentration by "Factor" to obtain softening chemical dosage in units noted.

Example 33

The following test result data was provided by the laboratory:

$$CO_2 = 25 \text{ mg/L as } CO_2$$
$$HCO_3^- = 205 \text{ mg/L as } CaCO_3$$
$$Mg = 9 \text{ mg/L Mg}$$
$$\text{Noncarbonate hardness} = 95 \text{ mg/L as } CaCO_3$$

Assuming no excess lime is added, find the correct dosages in milligrams per litre for lime (90% CaO) and soda ash (99% Na$_2$CO$_3$) required to remove all hardness.

First, check to see that the data is expressed in the same terms noted in Table 9. If the terms disagree, convert to the appropriate terms using the equivalent weight method. Then proceed with the calculations, beginning with the lime calculations:

$$CO_2: (25)(1.27) = 31.75 \text{ mg/L as } CaO$$

$$HCO_3^-:(205)(0.56) = 114.8 \text{ mg/L as } CaO$$

$$Mg: (9)(2.31) = \underline{20.79 \text{ mg/L as } CaO}$$

$$\text{Total lime dosage} = 167.34 \text{ mg/L as } CaO$$

Now adjust the lime dosage for purity.

$$\text{Actual lime dosage} = \frac{167.34}{0.9}$$

$$= 185.93 \text{ mg/L as CaO 90\% pure}$$

Next find the soda ash dosage:

$$\text{Soda ash dosage} = (95)(1.06)$$

$$= 100.7 \text{ mg/L as Na}_2\text{CO}_3$$

Then, adjust for purity:

$$\text{Actual soda ash dosage} = \frac{100.70}{0.99}$$

$$= 101.72 \text{ mg/L as Na}_2\text{CO}_3 \text{ 99\% pure}$$

Example 34

The following laboratory test results show the hardness of the raw water supply:

$$\text{CO}_2 = 20 \text{ mg/L as CaCO}_3$$
$$\text{Total hardness} = 200 \text{ mg/L as CaCO}_3$$
$$\text{Magnesium} = 50 \text{ mg/L as CaCO}_3$$
$$\text{Total alkalinity} = 100 \text{ mg/L as CaCO}_3$$

The objective is to remove all CO_2 and carbonate hardness while reducing magnesium to 5 mg/L as $CaCO_3$ and noncarbonate hardness to 75 mg/L as $CaCO_3$. Calculate the required dosages in pounds per million gallons. The quicklime used is 88 percent pure, and the soda ash is 98 percent pure. No excess lime is added.

First, list the amount to be removed of each constituent:

$$\text{CO}_2: \qquad\qquad = 20 \text{ mg/L as CaCO}_3$$

$$\text{Carbonate hardness:} \quad = 100 \text{ mg/L as CaCO}_3$$

$$\text{Mg: 50 mg/L} - 5 \text{ mg/L} = 45 \text{ mg/L as CaCO}_3$$

$$\text{Noncarbonate hardness:} \quad = ?$$

To determine the noncarbonate hardness refer to Table 8, which indicates that noncarbonate hardness is present only when the total hardness is greater than the total alkalinity. To calculate noncarbonate hardness, subtract the total alkalinity from the total hardness:

$$\text{Noncarbonate hardness} = \text{Total hardness} - \text{Total alkalinity}$$

$$= 200 \text{ mg/L} - 100 \text{ mg/L}$$

$$= 100 \text{ mg/L as CaCO}_3$$

Then the amount to be removed is

$$100 \text{ mg/L} - 75 \text{ mg/L} = 25 \text{ mg/L as CaCO}_3$$

Now convert as needed to the units shown in Table 9:

$$CO_2: \quad (20)\left(\frac{22}{50}\right) = 8.8 \text{ mg/L as CO}_2$$

$$HCO_3^-: \qquad\qquad = 100 \text{ mg/L as CaCO}_3$$

$$Mg: \quad (45)\left(\frac{12}{50}\right) = 10.8 \text{ mg/L as Mg}$$

$$NCH: \qquad\qquad = 25 \text{ mg/L as CaCO}_3$$

Then calculate the required dosages:

$$CO_2: \quad (8.8)(10.63) = 93.54 \text{ lb/MG as CaO}$$
$$HCO_3^-: \quad (100)(4.67) = 467 \text{ lb/MG as CaO}$$
$$Mg: \quad (10.8(19.24) = \underline{207.79 \text{ lb/MG as CaO}}$$

$$\text{Lime dosage} = 768.33 \text{ lb/MG as CaO}$$

$$NCH: \quad (25)(8.83) = \underline{220.75 \text{ lb/MG as Na}_2\text{CO}_3}$$

$$\text{Soda ash dosage} = 220.75 \text{ lb/MG as Na}_2\text{CO}_3$$

Finally, adjust for purity:

$$\text{Actual lime dosage} = \frac{768.33}{0.88}$$

$$= 873.1 \text{ lb/MG as CaO 88\% pure}$$

$$\text{Actual soda ash dosage} = \frac{220.75}{0.98}$$

$$= 225.26 \text{ lb/MG as Na}_2\text{CO}_3 \text{ 98\% pure}$$

Example 35

Based on the following hardness data, calculate the lime and soda ash requirements for complete removal at an average daily flow rate of 1,137,600 gpd. Excess lime and soda ash are added at the rate of 27 mg/L as $CaCO_3$. The soda ash used is 100 percent pure. The quicklime is 90 percent pure. Give results in milligrams per litre.

$$\begin{aligned}
\text{Total hardness} &= 215 \text{ mg/L as CaCO}_3 \\
\text{Mg} &= 15.8 \text{ mg/L as Mg} \\
\text{HCO}_3^- &= 185 \text{ mg/L as CaCO}_3 \\
\text{CO}_2 &= 25.8 \text{ mg/L as CO}_2
\end{aligned}$$

Since all lab data is in the correct units for use with the table, begin with the dosage calculations.

$$CO_2: \ (25.8)(1.27) \ = \ 32.77 \ mg/L \ as \ CaO \ 100\%$$
$$HCO_3^-: \ (185)(0.56) \ = \ 103.6 \ mg/L \ as \ CaO \ 100\%$$
$$Mg: \ (15.8)(2.31) \ = \ 36.5 \ mg/L \ as \ CaO \ 100\%$$
$$Excess \ lime: \ \ (27)(0.56) \ = \ \underline{15.12 \ mg/L \ as \ CaO \ 100\%}$$

$$Lime \ dosage \ = \ 187.99 \ mg/L \ as \ CaO \ 100\%$$

$$Determine \ NCH: \qquad NCH \ = \ TH \ - \ TA$$
$$= \ 215 \ mg/L \ - \ 185 \ mg/L$$
$$= \ 30 \ mg/L \ as \ CaCO_3$$

Then use Table 9:

$$NCH: \ \ (30)(1.06) \ = \ 31.8 \ mg/L \ as \ Na_2CO_3 \ 100\%$$
$$Excess \ soda \ ash: \ \ (27)(1.06) \ = \ \underline{28.62 \ mg/L \ as \ Na_2CO_3 \ 100\%}$$

$$Soda \ ash \ dosage \qquad = \ 60.42 \ mg/L \ as \ Na_2CO_3 \ 100\%$$

Next, adjust for purity as follows:

$$Actual \ lime \ dosage \ = \ \frac{187.99}{0.9}$$

$$= \ 208.88 \ mg/L \ as \ CaO \ 90\% \ pure$$
$$Actual \ soda \ ash \ dosage \ = \ 60.42 \ mg/L \ as \ Na_2CO_3 \ 100\% \ pure$$

C7-6. Recarbonation Calculations

As discussed under "Chemistry of Recarbonation," there are three possible reactions involved when carbon dioxide is used to stabilize a lime-softened water.[30] Equation 26 from "Chemistry of Recarbonation" shows that calcium carbonate, in suspension, will react with carbon dioxide to form calcium bicarbonate. Equations 27 and 28 show that excess lime will also react with carbon dioxide, forming either calcium carbonate precipitate or soluble calcium bicarbonate.

$CaCO_3$	+	$CO_2\uparrow$	+	H_2O	$\rightarrow$	$Ca(HCO_3)_2$	*Eq 26*
Calcium		Carbon		Water		Calcium	*Chemistry 6*
carbonate		dioxide				bicarbonate	
		(gas)				(in suspension)	

$Ca(OH)_2$	+	$CO_2\uparrow$	$\rightarrow$	$CaCO_3\downarrow$	+	H_2O	*Eq 27*
Excess		Carbon		Calcium		Water	*Chemistry 6*
lime		dioxide		carbonate			
		(gas)		(precipitate)			

---

[30]Chemistry Section, Chemistry of Treatment Processes (Chemistry of Recarbonation).

$$Ca(OH)_2 \;+\; 2CO_2\uparrow \;\rightarrow\; Ca(HCO_3^-)_2 \qquad\qquad Eq\ 28$$

Excess lime	Carbon dioxide (gas)	Calcium bicarbonate (in suspension)	*Chemistry 6*

The conditions described by these equations can and do occur together. Consequently, the carbon dioxide dosage needed for recarbonation is the amount needed to reduce calcium carbonate to a desired level and to completely eliminate excess lime. The amount of calcium carbonate present is measured by the alkalinity test and is assumed to be carbonate alkalinity (see *Introduction to Water Quality Analyses,* Physical/Chemical Tests Module—Alkalinity). The amount of excess lime present, also measured by the alkalinity test, is assumed equal to the hydroxide alkalinity.

The correct carbon dioxide dosage is found by adding 0.44 mg/L of CO_2 for every 1 mg/L of carbonate or hydroxide alkalinity expressed as $CaCO_3$. That is, to remove 5 mg/L as $CaCO_3$ of carbonate alkalinity and 5 mg/L as $CaCO_3$ of hydroxide alkalinity will require $(5 + 5)(0.44) = 4.4$ mg/L of CO_2 as CO_2. Consider the following example:

Example 36

The following results of water quality analyses were reported for water in the process of treatment. Calculate the correct CO_2 dosage.

Quality in mg/L as $CaCO_3$

	Raw water quality	After lime-soda treatment	Desired after recarbonation
Bicarbonate alkalinity	250	0	5
Carbonate alkalinity	0	35	30
Hydroxide alkalinity	0	40	0

First, find out how much carbonate and hydroxide alkalinity must be removed. For carbonate alkalinity, subtract the desired concentration from the concentration resulting from the lime-soda treatment:

Carbonate alkalinity to be removed $=$ 35 mg/L $-$ 30 mg/L

$=$ 5 mg/L as $CaCO_3$

Do the same for hydroxide alkalinity:

Hydroxide alkalinity to be removed $=$ 40 mg/L $-$ 0 mg/L

$=$ 40 mg/L as $CaCO_3$

Then add the two alkalinities to find the combined amount to be removed:

Combined alklainity to be removed $=$ 5 mg/L $+$ 40 mg/L

$=$ 45 mg/L as $CaCO_3$

Finally, find the carbon dioxide dose as follows:

$$CO_2 \text{ dose} = (\text{Alkalinity to be removed})(0.44)$$
$$= (45 \text{ mg/L})(0.44)$$
$$= 19.8 \text{ mg/L}$$

C7-7. Ion Exchange Softening Calculations

The operation of an ion exchange water softener (more correctly called a cation exchange water softener) involves four types of calculations:

- Hardness removal capacity
- Volume of water softened per cycle
- Length of softening run or cycle
- Regeneration salt requirement

All ion exchange materials (the zeolite types and the resin types) have maximum exchange capacities. The usual capacity ranges are summarized below.

Exchange Material	Exchange Capacity (grains per cubic foot)
Zeolites	2,800–11,000
Resins	11,000–35,000

The use of the unit "grain" is very common in this branch of water treatment. It is used in such terms as grains per cubic foot and grains per gallon (gpg). The following conversion factors will be useful in converting these somewhat old-fashioned units to more familiar units.

$$1 \text{ grain} = 64.80 \text{ mg}$$
$$1 \text{ gpg} = 17.12 \text{ mg/L}$$
$$1 \text{ gpg} = 142.86 \text{ lb/MG}$$

Example 37
A treatment unit contains 20 cu ft of cation exchange material which has a rated removal capacity of 20,000 grains per cubic foot. What is the total hardness removal capacity? Give answer in grains.

Since each cubic foot of material can remove 20,000 grains of hardness and there are 20 cu ft of material, the answer can be found in one multiplication:

$$\text{Total removal capacity} = (20,000 \text{ grain/cu ft})(20 \text{ cu ft})$$
$$= 400,000 \text{ grains removed per cycle}$$

Example 38
The unit in Example 37 treats water with a hardness of 253 mg/L. How many gallons of water can be softened before the exchange material must be regenerated?

First, convert hardness from milligrams per litre to grains per gallon in order to be consistent with the units of removal capacity (grains per cycle) found in Example 37:

$$gpg = \frac{mg/L}{17.12}$$

$$Hardness = \frac{253 \ mg/L}{17.12}$$

$$= 14.78 \ grains \ per \ gallon$$

Now, find the gallons of water softened by dividing the removal capacity per cycle by the hardness as follows:

$$Volume \ of \ water \ softened = \frac{Removal \ capacity \ per \ cycle}{Hardness}$$

$$Volume \ of \ water \ softened = \frac{400,000 \ grains}{14.78 \ gpg}$$

$$= 27,063 \ gal \ per \ cycle$$

Example 39

The area of the exchanger material bed is 5 sq ft and the softening loading rate is 6 gpm per sq ft. Find the volume of water softened every minute and calculate the length of the softening run cycle, assuming the volume of softened water is 27,063 gal before regeneration is needed.

The volume of water softened each minute is found by multiplying the bed area times the loading rate:

$$Volume \ per \ minute = (Bed \ area)(Loading \ rate)$$

$$= (5 \ sq \ ft)(6 \ gpm/sq \ ft)$$

$$= 30 \ gpm$$

Then, divide the total volume treated per cycle by the volume per minute to find the run length.

$$Run \ length = \frac{Volume \ per \ cycle}{Volume \ per \ minute}$$

$$Run \ length = \frac{27,063 \ gal}{30 \ gpm}$$

$$= 902.1 \ min, \ or \ \frac{902.1}{60} = 15.04 \ hr$$

Example 40

The exchange material in Example 37 can be regenerated by adding 0.35 lb of salt for every 1000 grains of hardness removed. How many pounds of salt are required for this regeneration?

As found in Example 37, the exchange material can remove 400,000 grains of hardness. After the exchange capacity has been exhausted, it takes 0.35 lb of salt to restore each 1000 grains of removal capacity, therefore:

Salt requirement = (Removal Capacity)(Regeneration Requirement)

$$= \text{(Removal Capacity)} \left(\frac{\text{Salt requirement in lb}}{1000 \text{ grains of hardness}} \right)$$

$$\text{Salt requirement} = (400,000 \text{ grain}) \left(\frac{0.35 \text{ lb}}{1000 \text{ grain}} \right)$$

$$= (400)(0.35)$$

$$= 140 \text{ lb salt}$$

Example 41

The salt required for exchange material regeneration is often prepared in advance as a solution of salt and water and held in storage for use as needed. The solution used in Example 40 contains 0.5 lb of salt/gal (equivalent to a 6 percent salt solution). Express the answer in gallons of salt solution.

To make the conversion, divide pounds of salt by pounds of salt per gallon:

$$\text{Gal of salt solution} = \frac{\text{lb of salt needed}}{\text{lb of salt/gal of solution}}$$

$$\text{Gal salt solution} = \frac{140 \text{ lb}}{0.5 \text{ lb/gal}}$$

$$= 280 \text{ gal}$$

Example 42

One unit of an ion exchange water softener uses polysterene resin and has these operating characteristics in a certain plant:

Bed depth: 24 in.
Loading rate: 5 gpm/sq ft
Brine solution strength: 12.5% (1.24 lb/gal)
Surface area: 7.5 sq ft
Volume of exchange material: 15 cu ft
Rated removal capacity: 30,000 grains/cu ft
Average water hardness: 180 mg/L
Salt regeneration requirement: 0.35 lb/1000 grains

Calculate the following:

(a) Total removal capacity
(b) Volume treated per cycle

(c) Softening run length
(d) Pounds of salt needed for regeneration
(e) Gallons of salt solution needed for regeneration

(a)

$$\begin{aligned}
\text{Total removal capacity} &= \text{(grains per cu ft)(cu ft resin)} \\
&= \text{(30,000 grain/cu ft)(15 cu ft)} \\
&= \text{450,000 grains removed per cycle}
\end{aligned}$$

(b)

$$\begin{aligned}
\text{Hardness in gpg} &= \frac{180 \text{ mg/L}}{17.12} \\
&= \frac{180}{17.12} \\
&= 10.51 \text{ gpg}
\end{aligned}$$

$$\begin{aligned}
\text{Volume treated} &= \frac{\text{Removal Capacity per Cycle}}{\text{Hardness}} \\
&= \frac{450,000 \text{ grain}}{10.51 \text{ gpg}} \\
&= 42,816 \text{ gal per cycle}
\end{aligned}$$

(c)

$$\begin{aligned}
\text{Volume per minute} &= \text{(Bed area)(Loading rate)} \\
&= \text{(7.5 sq ft)(5 gpm/sq ft)} \\
&= 37.5 \text{ gpm}
\end{aligned}$$

$$\begin{aligned}
\text{Run length} &= \frac{\text{Volume Treated per Cycle}}{\text{Volume Treated per minute}} \\
&= \frac{42,816 \text{ gal}}{37.5 \text{ gpm}} \\
&= 1142 \text{ min, or 19 hr}
\end{aligned}$$

(d)

$$\begin{aligned}
\text{Pounds salt} &= \text{(Removal capacity)(Regeneration requirement)} \\
&= \text{(450,000 grain)}\left(\frac{0.35 \text{ lb}}{1000 \text{ grain}}\right) \\
&= 157.5 \text{ lb salt}
\end{aligned}$$

(e)

$$\begin{aligned}
\text{Gallons of salt solution} &= \frac{\text{(lb salt needed)}}{\text{(lb salt/gal of solution)}} \\
&= \frac{157.5 \text{ lb}}{1.24 \text{ lb/gal}} \\
&= 127.02 \text{ gal of solution}
\end{aligned}$$

C7-8. Fluoridation Calculations

There are three chemical compounds commonly used in water fluoridation:

- Sodium fluoride, NaF
- Sodium silicofluoride, Na_2SiF_6
- Fluosilicic acid, H_2SiF_6

Since each compound is used for its fluoride ion content, it is important to determine the exact amount of fluoride ion available in a given fluoride compound. This is determined by two factors:

- The weight of fluoride relative to the total weight of the compound
- The purity of the compound

Example 43

Find the percentage of fluoride ion contained in sodium fluoride, NaF.[31]

Find the molecular weight of sodium fluoride by adding the atomic weights of the individual elements:

$$\begin{array}{rl} Na: & 23 \\ F: & \underline{19} \\ & 42 \text{ molecular weight of NaF} \end{array}$$

Then divide the weight of the fluoride ion by the total molecular weight of the compound and multiply by 100, as follows:

$$\text{Percent F} = \left(\frac{19}{42}\right)(100)$$

$$= 45\% \text{ fluoride ion (rounded from 45.24)}$$

Example 44

The commercial strength of the sodium fluoride in Example 43 is between 95 and 98 percent. That is, each 100-lb bag of chemical compound delivered to the plant contains 95–98 lb of sodium fluoride and 2–5 lb of impurities. How much fluoride ion is contained in each 100-lb bag?

From Example 43 you know that if it contained pure sodium fluoride, each 100-lb bag would have 45 percent or $(0.45)(100) = 45$ lb fluoride ion. However, the bags contain slightly impure NaF. Therefore, each bag holds only 95–98 percent of what it would if the NaF were 100 percent pure.

---

[31]Chemistry Section, Valence, Chemical Formulas, and Chemical Equations (Chemical Formulas and Equations—Determining percent by Weight of Elements in a Compound).

or from $(45)(0.95) = 42.75$ lb of F (43%)

to $(45)(0.98) = 44.1$ lb of F (44%)

Example 45

Find the percentage of fluoride ion contained in sodium silicofluoride (Na_2SiF_6).

Find the molecular weight of sodium silicofluoride by adding the atomic weights of the individual elements, as follows:

Na_2: $(2)(23) = 46$
Si: $= 28$
F_6: $(6)(19) = 114$
             $\overline{}$
188 molecular weight of Na_2SiF_6

Then divide the total weight of the fluoride ion by the molecular weight of the compound and multiply by 100, as follows:

$$\text{Percent F} = \left(\frac{114}{188}\right)(100)$$

$$= 61\% \text{ fluoride ion (rounded from 60.64)}$$

Example 46

The commercial strength of the sodium silicofluoride in Example 45 is normally 98.5 percent. That is, each 100-lb bag contains 98.5 lb of the compound and 1.5 lb of impurities. How much fluoride ion does each 100-lb bag contain?

From Example 45 you know that if it contained pure sodium silicofluoride, each bag would hold 61 percent or 61 lb of fluoride ion. However, the bags contain slightly impure Na_2SiF_6, with only 98.5 percent of the weight of each bag being composed of Na_2SiF_6. Therefore, the actual weight of fluoride ion in a 100-lb bag is:

$(98.5\%)(61 \text{ lb}) = \text{weight of F in 100-lb bag}$

$(0.985)(61 \text{ lb}) = \text{weight of F in 100-lb bag}$

$60.09 \text{ lb} = \text{weight of F in 100-lb bag}$

That is, 60 lb (rounded from 60.09 lb)—or 60%—of every 100-lb bag is fluoride ion.

Example 47

Find the percentage of fluoride ion contained in fluosilicic acid, H_2SiF_6.

Find molecular composition of fluosilicic acid, by weight, by adding the atomic weights of the individual elements:

$$H_2: \quad (2)(1) = \quad 2$$
$$Si: \qquad\quad = \quad 28$$
$$F_6: \quad (6)(19) = \underline{114}$$

144 molecular weight of H_2SiF_6

Then divide the total weight of the fluoride ion by the molecular weight of the compound and multiply by 100 as follows:

$$\text{Percent F} = \left(\frac{114}{144}\right)(100)$$

$$= 79\% \text{ fluoride ion (rounded from 79.17)}$$

Example 48

The commercial strength of the fluosilicic acid in Example 47 is quite variable, ranging from 20 to 35 percent pure. That is, each pound of commercial acid contains 0.2 to 0.35 lb of pure acid. The remainder of the liquid, 0.65 to 0.8 lb/lb, consists of impurities and water. The density of the commercial acid is about 10.5 lb/gal. How much fluoride ion is contained in each gallon of acid?

From Example 47 you know that if the fluosilicic acid were 100 percent pure, each pound would contain 79 percent or 0.79 pounds of fluoride ion. However, commercial fluosilicic acid is very dilute, containing only 20 to 35 percent pure acid. To determine the weight of fluoride ion contained in 1 lb of commercial (dilute) acid, multiply the *percentage of fluoride ion per pound of pure acid* times the *percentage of pure acid per pound of commercial acid:*

Weight of fluoride ion in 1 lb commercial acid:

from $(0.79)(0.20) = 0.16$ lb or 16% pure fluoride ion

to $(0.79)(0.35) = 0.28$ lb or 28% pure fluoride ion

Remember that the acid has a density of 10.5 lb/gal. Each pound of acid contains from 0.16 to 0.28 lb fluoride ion; therefore, 1 gal of acid contains 10.5 times that amount:

Weight of fluoride ion in 1 gal commercial acid:

from $(10.5)(0.16 \text{ lb}) = 1.68$ lb

to $(10.5)(0.28 \text{ lb}) \quad = 2.94$ lb

Example 49

A water treatment plant fluoridates by direct addition of commercial fluosilicic acid. The strength of the acid is 25 percent and the density is 10.5 lb/gal. The treated water flow rate is 1 mgd. What is the dosage rate needed in order to fluoridate at 1 milligram per litre?

Example 47 shows the fluoride ion content of pure fluosilicic acid to be 79 percent. To adjust for the dilution, multiply the ion content by the strength of the commercial acid, as in example 48.

$(0.79)(0.25) = 0.197$ lb fluoride ion per lb of commercial acid

The acid density is 10.5 lb/gal; therefore, find the weight of fluoride ion per gallon of commercial acid as follows:

$(10.5 \text{ lb}/\text{gal})(0.197 \text{ lb F}/\text{lb commercial acid}) =$
$$2.07 \text{ lb F}/\text{gal commercial acid}$$

With each gallon of commercial acid added to the water, 2.07 lb of fluoride ion is added.

The desired fluoride dose of 1 mg/L is equivalent to

$(1 \text{ mg}/\text{L})(8.34 \text{ lb}/\text{gal})(1 \text{ mgd}) = 8.34 \text{ lb}/\text{mil gal}$

Since the flow rate is 1 mgd, the weight of pure fluoride ion needed is 8.34 lb/day. The number of gallons of commercial acid needed to achieve this dosage is

$$\frac{8.34 \text{ lb}/\text{day}}{2.07 \text{ lb F}/\text{gal commercial acid}} = 4.03 \text{ gal commercial acid}/\text{day}$$

Example 50

How many pounds of NaF are required to fluoridate 1 mgd to a level of 1 mg/L F? The NaF used is 95 percent pure.

The fluoride ion content of 95 percent pure NaF is 43 percent (see Example 44). Since the flow rate is 1 mgd, the desired dosage of 1 mg/L is equivalent to 8.34 lb/mil gal, or 8.34 lb/day. Therefore, to solve the problem determine the amount of NaF needed to achieve an 8.34 lb/day dosage of fluoride ion. Note that, since only about half (43 percent) of each pound of NaF is fluoride ion, to get 8.34 lb of F will require the addition of a little more than 16 lb of NaF. More precisely:

$$(43\%)(\text{lb}/\text{day NaF, } 95\%) = 8.34 \text{ lb F}/\text{day}$$
$$(0.43)(\text{lb}/\text{day NaF, } 95\%) = 8.34 \text{ lb F}/\text{day}$$
$$(\text{lb}/\text{day NaF, } 95\%) = \frac{8.34}{0.43}$$
$$= 19.4 \text{ lb}/\text{day NaF, } 95\%$$

Example 51

An important feature of NaF is its relatively constant solubility. The strength of an NaF solution is consistently 4 percent, by weight, within the range of temperatures common in water treatment ($0\text{–}25^0\text{C}$). How much of such a solution would have to be added each day to meet the dosage requirement in Example 50? What would this dosage requirement be in gallons per minute?

Based on the answer to Example 50, you know that the desired dosage is 19.4 lb NaF/day. Because of NaF's relatively constant solubility, 19.4 lb must be 4 percent of the total weight of chemical solution added. Therefore

$$(4\%)(\text{Solution added}) = 19.4 \text{ lb}$$

$$\text{Solution added} = \frac{19.4}{0.04}$$

$$= 485 \text{ lb/day NaF solution}$$

The density of a 4 percent NaF solution is about 8.6 lb/gal. So the dosage rate of 4 percent NaF solution is

$$\frac{485 \text{ lb}}{8.6 \text{ lb/gal}} = 56.4 \text{ gpd of } 4\% \text{ NaF}$$

$$= 2.35 \text{ gph of } 4\% \text{ NaF}$$

$$= 0.039 \text{ gpm of } 4\% \text{ NaF}$$

Example 52

A shipment of 98.5 percent pure sodium silicofluoride has an average fluoride content of 60 percent (see Example 46). How many pounds of Na_2SiF_6 should be added daily to treat a 1 mgd volume to a level of 1 mg/L?

Follow the same procedure described in Example 50:

$$(60\%)(\text{Desired dosage}) = 8.34 \text{ lb F/day}$$

$$(0.60)(\text{Desired dosage}) = 8.34 \text{ lb F/day}$$

$$\text{Desired dosage} = \frac{8.34 \text{ lb F/day}}{0.60}$$

$$= 13.91 \text{ lb } Na_2SiF_6/\text{day}$$

Example 53

The Na_2SiF_6 from Example 52 is mixed with water to make a solution of 0.5 percent by weight. How much solution must be added daily to treat 1 mgd to the level of 1 mg/L?

Follow the same procedure described in Example 51:

$$(0.5\%)(\text{Solution added}) = 13.9 \text{ lb}$$

$$= \frac{13.9 \text{ lb}}{0.005}$$

$$= 2780 \text{ lb/day } Na_2SiF_6, \\ 0.5\% \text{ solution}$$

Such a weak solution would be about the same density as water, 8.34 lb/gal. Therefore, the dosage rate of the 0.5 percent solution in gallons per day is

$$\frac{2780 \text{ lb}}{8.34 \text{ lb/gal}} = 333.33 \text{ gpd of } 0.5\% \text{ Na}_2\text{SiF}_6$$

$$= 13.89 \text{ gph}$$

$$= 0.23 \text{ gpm}$$

Example 54

The natural fluoride ion concentration of a water supply is 0.2 mg/L. How many pounds of sodium silicofluoride must be used each day to treat 0.5 mgd of water to a level of 1 mg/L? (The sodium silicofluoride is typically 98.5 percent pure.)

To raise the fluoride level from 0.2 mg/L to 1 mg/L, increase the fluoride ion concentration by 0.8 mg/L, that is:

1 mg/L − 0.2 mg/L = 0.8 mg/L required increase in
fluoride ion concentration

Determine dosage in pounds per day:

Required dosage
pure fluoride ion = (0.8 mg/L)(0.5 mgd)(8.34 lb/day)

$$= 3.34 \text{ lb/day}$$

So, 3.34 lb of 100 percent pure fluoride ion should be added every day. However, 98.5 percent Na_2SiF_6 contains only 60 percent of the total weight of pure fluoride ion that must be added (see Example 46). Therefore, to compute the proper dosage:

$$(60\%)(\text{lb/day Na}_2\text{SiF}_6, 98.5\%) = 3.34 \text{ lb F/day}$$

$$(0.60)(\text{lb/day Na}_2\text{SiF}_6, 98.5\%) = 3.34 \text{ lb F/day}$$

$$\text{lb/day of Na}_2\text{SiF}_6 = \frac{3.34}{0.60}$$

$$= 5.57 \text{ lb/day}$$

Example 55

How many gallons of water are required to dissolve 5.57 lb of Na_2SiF_6, at a temperature of 15.6° C? The solubility of Na_2SiF_6 is 0.62 gm/100 mL at 15.6° C.

First convert to English units of measure:[32]

---

[32]Mathematics Section, Conversions (Metric System Conversions).

$$\frac{0.62 \text{ gm}}{100 \text{ mL}} = \frac{0.62 \text{ gm}}{100 \text{ mL}} \times \frac{3785 \text{ mL}}{\text{gal}} \times \frac{1 \text{ lb}}{454 \text{ gm}}$$

$$= 0.0525 \text{ lb/gal}$$

This means that each gallon of water dissolves 0.0525 lb of Na_2SiF_6. Therefore, to determine how many gallons are required to dissolve 5.57 lb of Na_2SiF_6, you need to know how many units of 0.0525 lb there are in 5.57 lb. Stated mathematically this is:

$$\text{Solution water required} = \frac{5.57 \text{ lb}}{0.0525 \text{ lb/gal}}$$

$$= 106.1 \text{ gal}$$

Example 56

What rate should the chemical metering pump be set to pump the fluoride solution of Example 55 into the water system? The treatment plant flow rate is constant at a rate of 0.5 mgd. Give answer in gallons per hour.

Since the 106.1 gal supply of fluoride solution is a 1-day supply, and since the plant flow rate is constant, the fluoride solution must be added uniformly throughout the 24 hour period:

$$\text{Solution feed setting} = \frac{106.1 \text{ gal}}{24 \text{ hr}}$$

$$= 4.42 \text{ gph}$$

Review Questions

Milligrams per Litre to Pounds per Day

1. The dry alum feed rate is 380 lb/day. The alum is added to a plant flow of 3,250,000 gpd. What is the alum dosage expressed in milligrams per litre?

2. At a water treatment plant, chlorine is fed at the rate of 3.5 mg/L. The flow at the plant is 4.5 mgd. How many pounds per day of chlorine are added?

3. Fluoride is dosed at a rate of 2.2 mg/L with a plant flow of 750,000 gpd. How many pounds per day of fluoride are required?

4. On one day the coagulation process at a plant was operated for 10 hr. During that period the flow rate through the coagulation process was 250 gpm. If 45 lb of alum were fed into the water, what was the alum dosage in milligrams per litre during the period?

5. To disinfect a water supply requires 25 lb/day of chlorine. If a commercially available sodium hypochlorite (15 percent available chlorine) is to be used for disinfection, how many pounds per day of sodium hypochlorite will be required?

6. How many pounds of hypochlorite (65 percent available chlorine) are required to disinfect 1500 ft of 8-in. water line if an initial dose of 45 mg/L is required?

Milligrams per Litre to Percent

7. Express 150 mg/L as percent.

8. What is 35,000 mg/L expressed as a percent concentration?

9. A sludge has a total solids concentration of 3.25 percent. What is the concentration expressed as milligrams per litre of solids?

Feed Rate Conversions

10. The feed rate for a chemical is 180 lb/day. What is the feed rate in pounds per hour?

11. The desired feed rate is 28 gal/day. What is the feed rate in pounds per hour?

Chlorine Dosage/Demand/Residual Problems

12. A water is tested and found to have a chlorine demand of 4.7 mg/L. The desired chlorine residual is 0.3 mg/L. How many pounds of chlorine will be required daily to chlorinate a flow of 3.5 mgd?

13. The water at a plant is dosed with 7.2 mg/L chlorine. The chlorine residual after 30 min contact time is found to be 0.4 mg/L. What is the chlorine demand expressed in mg/L?

Lime–Soda Ash Softening Calculations

14. Express the following laboratory test results as $CaCO_3$:

Calcium	$= 120$ mg/L as Ca
Magnesium	$= 30$ mg/L as Mg
Bicarbonate alkalinity	$= 255$ mg/L as HCO_3
Total alkalinity	$= 215$ mg/L as $CaCO_3$
Carbon dioxide	$= 15$ mg/L as CO_2

15. The laboratory returned the results listed below. Determine how much lime is needed and, if required, how much soda ash is needed to soften this water. Assume no excess lime or soda ash is to be added. Express results as $CaCO_3$. Use the "conventional" method to solve the problem.

Calcium	$= 60$ mg/L as Ca
Total hardness	$= 175$ mg/L as $CaCO_3$
Bicarbonate alkalinity	$= 160$ mg/L as HCO_3
Total alkalinity	$= 140$ mg/L as $CaCO_3$
Carbon dioxide	$= 10$ mg/L as CO_2

16. Express the answers to Question 15 as lime and as soda ash.

17. Using 94-percent pure quicklime and 100-percent soda ash, how many pounds per day of chemicals will be needed in Question 16 if the flow rate is 2 mgd?

18. A raw water sample tested as follows:

Total hardness	$= 238$ mg/L as $CaCO_3$
Calcium	$= 70$ mg/L as Ca
Magnesium	$= 15$ mg/L as Mg
Bicarbonate alkalinity	$= 160$ mg/L as HCO_3
Carbon dioxide	$= 9$ mg/L as CO_2

    On the day the sample was drawn the plant flow averaged 0.92 mgd. How many pounds of 89-percent pure quicklime (as CaO) and 100-percent pure soda ash (as Na_2CO_3) should have been used that day if an excess of 19 mg/L lime was added? Use the "conventional" method to solve the problem.

19. Based on the following hardness data, calculate the lime and soda ash requirements (pounds per day) for complete removal at an average daily flow rate of 2.18 mgd. Excess lime is added at the rate of 21 mg/L as $CaCO_3$. The soda ash used is 100-percent pure. The quicklime is 90-percent pure. Use the "conversion factor" method and Table 9.

$$Total\ hardness\ =\ 192\ mg/L\ as\ CaCO_3$$
$$Mg\ =\ 12\ mg/L\ as\ Mg$$
$$HCO_3^-\ =\ 167\ mg/L\ as\ CaCO_3$$
$$CO_2\ =\ 18\ mg/L\ as\ CO_2$$
$$Total\ alkalinity\ =\ 167\ mg/L\ as\ CaCO_3$$

Recarbonation Calculations

20. The following results of water quality analyses were reported for treated water. Calculate the correct CO_2 dosage. Results of the analyses are given in "mg/L as $CaCO_3$."

	Raw Water Quality	After Lime–Soda Treatment	Desired After Recarbonation
Bicarbonate alkalinity	250	0	5
Carbonate alkalinity	0	35	20
Hydroxide alkalinity	0	50	0

Ion Exchange Calculations

21. An ion exchange water softener unit holds cationic polystyrene resin and has these operating features:

Bed depth:	26 in.
Loading rate:	4.8 gpm/sq ft
Brine strength:	12.9% (1.24 lb/gal)
Surface area:	4.1 sq ft
Volume of exchange material:	8.88 cu ft
Rated removal capacity:	30,000 grains/cu ft
Average water hardness:	142 mg/L
Salt regeneration requirement:	0.35 lb salt/1000 grains of hardness removed

Calculate the following:

(a) Total removal capacity
(b) Volume treated per cycle
(c) Softening run length in hours
(d) Pounds of salt needed for regeneration
(e) Gallons of brine needed for regeneration

Fluoridation Calculations

22. Calculate the percentage of fluoride by weight in each of the three commonly used fluoride compounds—sodium fluoride (NaF), sodium

silicofluoride (Na_2SiF_6), and fluosilicic acid (H_2SiF_6). Use the following atomic weights: sodium = 23; fluorine = 19; silica = 28; and hydrogen = 1.

23. The purity of a 100-lb bag of NaF is labeled "96 percent." How many pounds of fluoride ion are contained in the bag of NaF? (Use NaF information calculated in Problem 22.)

24. Fluoridation is accomplished by direct addition of fluosilicic acid, as delivered. The strength of the acid is 32 percent and the density is 11.1 lb/gal. How many gallons of acid are needed daily to fluoridate 0.16 mgd to a level of 1 mg/L? (Use H_2SiF_6 data calculated in Problem 22.)

25. A water is fluoridated using a saturated solution of NaF, 95-percent pure. The density of the solution is about 8.6 lb/gal. How many gallons per day of the solution would be needed to fluoridate 0.16 mgd to a level of 1 mg/L?

26. How much sodium silicofluoride (98.5% pure) is needed to treat a 6.4 mgd flow to a level of 1.2 mg/L if the natural fluoride concentration is 0.3 mg/L?

27. (a) How much water at 15.6°C is needed to dissolve 80 lb of Na_2SiF_6 and (b) what gallons-per-minute feed rate should be used to deliver this solution uniformly over 24 hr? The solubility of Na_2SiF_6 is 0.62 gm/100 ml at 15.6°C.

Summary Answers

1. 14.02 mg/L chlorine

2. 131.36 lb/day fluoride

3. 13.76 lb/day alum

4. 35.97 lb/day alum

5. 166.67 lb/day sodium hypochlorite

6. 2.31 lb/day hypochlorite

7. 0.015%

8. 3.5%

9. 32,500 mg/L solids

10. 7.5 lb/hr

11. 9.73 lb/hr

12. 145.95 lb/day chlorine

13. 6.8 mg/L demand

14. Ca = 300 mg/L as $CaCO_3$; Mg = 125 mg/L as $CaCO_3$; HCO_3^-= 209.02 mg/L as $CaCO_3$; TA = 215 mg/L as $CaCO_3$; CO_2 = 34.09 mg/L as $CaCO_3$.

15. Lime and soda ash are both required; lime dosage = 178.88 mg/L as $CaCO_3$; soda ash dosage = 43.85 mg/L as $CaCO_3$.

16. Lime = 100.17 mg/L as $CaCO_3$; soda ash = 46.53 mg/L as Na_2CO_3.

17. Lime = 1777 lb/day; soda ash = 776.12 lb/day.

18. 1125 lb/day 89-percent pure CaO; 1024 lb/day 100-percent pure Na_2CO_3.

19. Quicklime = 3149 lb/day CaO 90-percent pure; Soda ash = 886.52 lb/day Na_2CO_3 100-percent pure.

20. 28.6 mg/L.

21. (a) 266,400 grains per cycle; (b) 32,135 gal treated; (c) 27.22 hr; (d) 93.24 lb salt; (e) 75.19 gal of brine.

22. NaF: 45%; Na_2SiF_6: 61%; H_2SiF_6: 79%.

23. 43.2 lb.

24. 0.48 gpd acid.

25. 8.98 gpd of 4-percent NaF solution, 95-percent pure.

26. 79.55 lb/day 98.5-percent pure Na_2SiF_6.

27. 1547 gpd; 1.07 gpm.

Detailed Answers

1. $$(mg/L)(mgd)(8.34 \text{ lb/gal}) = \text{lb/day}$$

Fill in the information given, then solve for x:

$$(x \text{ mg/L})(3.25 \text{ mgd})(8.34 \text{ lb/gal}) = 380 \text{ lb/day}$$

$$x \text{ mg/L} = \frac{380}{(3.25)(8.34)}$$

$$x \text{ mg/L} = 14.02 \text{ mg/L}$$
chlorine dosage

2. $$(mg/L)(mgd)(8.34 \text{ lb/gal}) = \text{lb/day}$$

Fill in the information given, then solve for the unknown value:

$$(3.5 \text{ mg}/\text{L})(4.5 \text{ mgd})(8.34 \text{ lb}/\text{gal}) = x \text{ lb}/\text{day}$$
$$131.36 = x \text{ lb}/\text{day}$$
$$\text{fluoride}$$

3. $(\text{mg}/\text{L})(\text{mgd})(8.34 \text{ lb}/\text{gal})$

Fill in the information given, then solve for the unknown value:

$$(2.2 \text{ mg}/\text{L})(0.75 \text{ mgd})(8.34 \text{ lb}/\text{day}) = x \text{ lb}/\text{day}$$
$$13.76 \text{ lb}/\text{day alum} = x \text{ lb}/\text{day}$$
$$\text{alum}$$

4. $(\text{mg}/\text{L})(\text{Flow mgd})(8.34 \text{ lb}/\text{gal}) = \text{lb}/\text{day}$

Before writing the information into the equation, the flow rate must be expressed in million gallons per day. The total flow for the day was

$$(250 \text{ gpm})(60 \text{ min}/\text{hr})(10 \text{ hr}) = 150{,}000 \text{ gal flow}$$
$$= 0.15 \text{ mgd}$$

Now fill in the information and solve for the unknown:

$$(x \text{ mg}/\text{L})(0.15 \text{ mgd})(8.34 \text{ lb}/\text{gal}) = 45 \text{ lb}/\text{day}$$
$$x = \frac{45}{(0.15)(8.34)}$$
$$x = 35.97 \text{ mg}/\text{L}$$
$$\text{alum dosage}$$

5. Since there is only 15 percent available chlorine in the sodium hypochlorite, considerably more than 25 lb/day of hypochlorite will be required to accomplish the disinfection. In fact, 15 percent of some greater number should equal 25 lb/day:

$$(15\%)(\text{Greater number of lb}/\text{day}) = 25 \text{ lb}/\text{day}$$

This equation can be restated as

$$(0.15)(x \text{ lb}/\text{day}) = 25 \text{ lb}/\text{day}$$

Then solve for the unknown value:

$$x = \frac{25}{0.15}$$
$$x = 166.67 \text{ lb}/\text{day}$$
$$\text{sodium hypochlorite}$$

6. First calculate the volume of the water line in gallons:

$$(0.785)(0.67 \text{ ft})(0.67 \text{ ft})(1500 \text{ ft})(7.48 \text{ gal}/\text{cu ft}) = 3954 \text{ gal}$$

Express the volume as millions of gallons:

$$3954 \text{ gal} = 0.003954 \text{ mil gal}$$

Rounding this gives

$$= 0.004 \text{ mil gal}$$

Calculate the chlorine required (pounds per day) using the equation that relates milligrams per litre and pounds per day:

$$(45 \text{ mg/L})(0.004 \text{ mil gal})(8.34 \text{ lb/gal}) = 1.5 \text{ lb}$$
$$\text{chlorine}$$

Since 65 percent hypochlorite is to be used, more than 1.5 lb/day of hypochlorite will be required. In fact, 65 percent of some greater number should equal 1.5 lb:

$$(65\%)(\text{Greater number of lb}) = 1.5 \text{ lb}$$

This equation can be restated as

$$(0.65)(x \text{ lb}) = 1.5 \text{ lb}$$

Then solve for the unknown value:

$$x = \frac{1.5}{0.65}$$
$$= 2.31 \text{ lb}$$
$$\text{hypochlorite}$$

7. Use the box diagram that relates percent and milligrams per litre. In converting from milligrams per litre to percent, you are moving from a large box to a small box. Therefore, division is indicated.

$$\frac{150}{10,000} = 0.015\%$$

8. Use the box diagram that relates percent and milligrams per litre. In converting from milligrams per litre to percent, you are moving from a large box to a small box. Therefore, division is indicated.

$$\frac{35,000}{10,000} = 3.5\%$$

9. Use the box diagram that relates percent and milligrams per litre. In converting from percent to milligrams per litre, you are moving from a small box to a large box. Therefore, multiplication is indicated.

$$(3.25)(10,000) = 32,500 \text{ mg/L}$$

10. Use the four-box diagram that relates the terms to be converted. In converting from *lb/day* to *lb/hour*, you are moving from a larger box to a smaller box. Therefore, division by 24 is indicated:

$$\frac{180 \text{ lb/day}}{24 \text{ hr/day}} = 7.5 \text{ lb/hour}$$
$$\text{feed rate}$$

11. Use the four-box diagram that relates the terms to be converted. Converting first from *gal/day* to *lb/day*, you are moving from a smaller box to a larger box. Therefore, multiplication by 8.34 is indicated. Then in making the conversion from *lb/day* to *lb/hr*, you are moving from a larger box to a smaller box, indicating division by 24:

$$\frac{(28 \text{ gal/day})(8.34 \text{ lb/gal})}{24 \text{ hr/day}} = 9.73 \text{ lb/hr}$$

12.
$$\begin{array}{ccc} \text{Chlorine} & \text{Chlorine} & \text{Chlorine} \\ \text{dosage} & = \text{demand} + & \text{residual} \\ \text{(mg/L)} & \text{(mg/L)} & \text{(mg/L)} \end{array}$$

Use the chlorine dosage/demand/residual equation, and fill in the information given in the problem:

$$\text{Dosage} = 4.7 \text{ mg/L} + 0.3 \text{ mg/L}$$
$$= 5.0 \text{ mg/L}$$

Converting the milligrams per litre dosage to pounds per day:

$$(5.0 \text{ mg/L})(3.5 \text{ mgd})(8.34 \text{ lb/gal}) = 145.95 \text{ lb/day chlorine}$$

13. Fill the information given in the problem into the chlorine dosage/demand/residual equation:

$$7.2 \text{ mg/L} = x \text{ mg/L} + 0.4 \text{ mg/L}$$

Solve for the unknown value:

$$7.2 \text{ mg/L} - 0.4 \text{ mg/L} = x \text{ mg/L}$$
$$6.8 \text{ mg/L} = x \text{ mg/L}$$

To check this:

$$7.2 \text{ mg/L} = 6.8 \text{ mg/L} + 0.4 \text{ mg/L}$$

14.

Calcium: $(120 \text{ mg/L})\left(\frac{50}{20}\right) = 300 \text{ mg/L as } CaCO_3$

Magnesium: $(30 \text{ mg/L})\left(\frac{50}{12}\right) = 125 \text{ mg/L as } CaCO_3$

Bicarbonate alkalinity: $(255 \text{ mg/L})\left(\frac{50}{61}\right) = 209.02 \text{ mg/L as } CaCO_3$

Total alkalinity: No conversion needed $= 215 \text{ mg/L as } CaCO_3$

Carbon dioxide: $(15 \text{ mg/L})\left(\frac{50}{22}\right) = 34.09 \text{ mg/L as } CaCO_3$

15. Since the total alkalinity (140 mg/L as $CaCO_3$) is less than the total hardness (175 mg/L as $CaCO_3$), the water should have both carbonate and noncarbonate hardness. (See Table 8.) Lime and soda ash will be needed. First, convert all lab results to concentrations as $CaCO_3$:

Calcium: $\qquad\qquad\qquad (60 \text{ mg/L})\left(\dfrac{50}{20}\right) = 150 \text{ mg/L}$

Total hardness: $\qquad$ (no conversion needed) $= 175 \text{ mg/L}$

Bicarbonate alkalinity: $\qquad (160 \text{ mg/L})\left(\dfrac{50}{61}\right) = 131.15 \text{ mg/L}$

Total alkalinity: $\qquad$ (no conversion needed) $= 140 \text{ mg/L}$

Carbon dioxide: $\qquad\qquad (10 \text{ mg/L})\left(\dfrac{50}{22}\right) = 22.73 \text{ mg/L}$

Then find the magnesium concentration:
$$[\text{Total hardness}] = [\text{Ca}] + [\text{Mg}] \qquad\qquad Eq\ 4$$
which may be restated as:
$$\begin{aligned}[\text{Mg}] &= [\text{Total hardness}] - [\text{Ca}] \\ &= 175 \text{ mg/L} - 150 \text{ mg/L} \\ &= 25 \text{ mg/L as } CaCO_3\end{aligned}$$

Next, calculate the lime dosage and soda ash dosage:
$$\begin{aligned}\text{Lime dosage} &= [CO_2] + [HCO_3^-] + [\text{Mg}] + [\text{Excess}] \qquad Eq\ 2\\ &= 22.73 \text{ mg/L} + 131.15 \text{ mg/L} + 25 \text{ mg/L} + 0 \\ &= 178.88 \text{ mg/L as } CaCO_3\end{aligned}$$

$$\begin{aligned}\text{Soda ash dosage} &= [\text{Total hardness}] - [HCO_3^-] + [\text{Excess}] \quad Eq\ 3\\ &= 175 \text{ mg/L} - 131.15 \text{ mg/L} + 0 \\ &= 43.85 \text{ mg/L as } CaCO_3\end{aligned}$$

16.
$$\begin{aligned}\text{Lime dosage} &= (178.88 \text{ mg/L})\left(\frac{28}{50}\right) \\ &= 100.17 \text{ mg/L as CaO}\end{aligned}$$

$$\begin{aligned}\text{Soda ash dosage} &= (43.9 \text{ mg/L})\left(\frac{53}{50}\right) \\ &= 46.53 \text{ mg/L as } Na_2CO_3\end{aligned}$$

17. $(mg/L)(mgd\ flow)(8.34\ lb/gal) = lb/day$

Therefore the lime and soda ash dosages are

$$(100.17\ mg/L)(2\ mgd)(8.34\ lb/gal) = lb/day\ lime$$
$$= 1671\ lb/day\ lime$$
$$\text{if } 100\%\ \text{pure}$$

$$(46.53\ mg/L)(2\ mgd)(8.34\ lb/gal) = lb/day\ soda\ ash$$
$$= 776.12\ lb/day$$
$$\text{if } 100\%\ \text{pure}$$

Now adjustments for purity must be made. Since the lime is only 94-percent pure, an adjustment must be made. More than 1670 lb/day lime will be required to accomplish the treatment:

$$(\%\ Purity)(Actual\ Dosage) = (Calculated\ Dosage)$$
$$(0.94)(Actual\ dosage) = 1671\ lb/day$$
$$Actual\ dosage = \frac{1671}{0.94}$$
$$= 1777\ lb/day\ of\ 94\%\ lime$$

In this problem soda ash is assumed to be 100-percent pure, so the soda ash dosage needed is 776.12 lb/day.

18. Begin by converting all the constituent concentrations to concentrations as $CaCO_3$:

Total hardness: (No conversion) $= 238\ mg/L$ as $CaCO_3$

Calcium: $(70)\left(\dfrac{50}{20}\right) = 175\ mg/L$ as $CaCO_3$

Magnesium: $(15)\left(\dfrac{50}{12}\right) = 62.5\ mg/L$ as $CaCO_3$

Bicarbonate alkalinity: $(160)\left(\dfrac{50}{61}\right) = 131.15\ mg/L$ as $CaCO_3$

Carbon dioxide: $(9)\left(\dfrac{50}{22}\right) = 20.45\ mg/L$ as $CaCO_3$

Next, calculate lime and soda ash dosages:

$$Lime\ dosage = [CO_2] + [HCO_3^-] + [Mg] + [Excess] \qquad Eq\ 2$$
$$= 20.45 + 131.15 + 62.5 + 19$$
$$= 233.1\ mg/L\ as\ CaCO_3$$

$$Soda\ ash\ dosage = [Total\ Hardness] - [HCO_3^-] + [Excess] \qquad Eq\ 3$$
$$= 238 - 131.15 + 19$$
$$= 125.85\ mg/L\ as\ CaCO_3$$

Now express the dosages in terms of the equivalent weight of the softening chemicals used (that is, as CaO or as Na_2CO_3).

$$\text{Lime dosage} = (233.1)\left(\frac{28}{50}\right)$$

$$= 130.54 \text{ mg/L as } CaO$$
$$(100\% \text{ pure})$$

$$\text{Soda ash dosage} = (125.85)\left(\frac{53}{50}\right)$$

$$= 133.4 \text{ mg/L as } Na_2CO_3$$
$$(100\% \text{ pure})$$

Now the adjustment for chemical purity must be made. Since the purity of the quicklime is not 100 percent but only 89 percent, more than 130.1 mg/L quicklime will be required. To calculate this larger, actual dosage:

$$(0.89)(\text{Actual dosage}) = 130.54 \text{ mg/L}$$

$$\text{Actual dosage} = \frac{130.1}{0.89}$$

$$= 146.67 \text{ mg/L as } CaO,$$
$$89\% \text{ pure}$$

In this problem it is assumed that the soda ash is 100-percent pure. Therefore, the actual dosage required is 133.4 mg/L as Na_2CO_3. Finally, based on the 0.92 mgd average flow rate for the day, the total pounds of lime and soda ash used per day are:

$$\text{lb/day Lime dosage} = (146.67 \text{ mg/L})(0.92 \text{ mgd})(8.34 \text{ lb/gal})$$

$$= 1125 \text{ lb/day } CaO, 89\% \text{ pure}$$

$$\text{lb/day Soda ash dosage} = (133.4 \text{ mg/L})(0.92 \text{ mgd})(8.34 \text{ lb/gal})$$

$$= 1024 \text{ lb/day } Na_2CO_3, 100\% \text{ } Na_2CO_3$$

19. Since all lab data are in the correct units for use with the table, begin with the dosage calculations. Using the "mg/L" conversions:

$$CO_2: \quad (18)(1.27) = \quad 22.86 \text{ mg/L as } CaO \text{ 100\%}$$
$$HCO_3: \quad (167)(0.56) = \quad 93.52 \text{ mg/L as } CaO \text{ 100\%}$$
$$Mg: \quad (12)(2.31) = \quad 27.72 \text{ mg/L as } CaO \text{ 100\%}$$
$$\text{Excess lime:} \quad (21)(0.56) = \quad \underline{11.76 \text{ mg/L as } CaO \text{ 100\%}}$$
$$\text{Lime dosage} = 155.86 \text{ mg/L as } CaO \text{ 100\%}$$

Since the quicklime is only 90-percent pure, the dosage must be adjusted to account for purity. More than 155.86 mg/L will be required for proper treatment. To calculate the larger, actual dosage:

$$(0.90)(\text{Actual dosage}) = 155.86 \text{ mg/L}$$
$$\text{Actual dosage} = \frac{155.86}{0.90}$$
$$= 173.18 \text{ mg/L as CaO 90\% pure}$$

Calculate the soda ash dosage:

NCH: $(25)(1.06) = 26.5$ mg/L as Na_2CO_3 100%

Excess soda ash: $(21)(1.06) = 22.26$ mg/L as Na_2CO_3 100%

Soda ash dosage $= 48.76$ mg/L as Na_2CO_3 100%

No adjustment need be made for purity since the soda ash is assumed to be 100-percent pure.

Therefore, the lime and soda ash dosages in pounds per day are

$$(173.18 \text{ mg/L})(2.18 \text{ mgd})(8.34 \text{ lb/gal}) = 3149 \text{ lb/day CaO,} \quad 90\% \text{ pure}$$

$$(48.76 \text{ mg/L})(2.18 \text{ mgd})(8.34 \text{ lb/gal}) = 886.52 \text{ lb/day } Na_2CO_3, \quad 100\% \text{ pure}$$

20. The first step is to find out how much carbonate and hydroxide alkalinity must be removed:

Carbonate alkalinity to be removed $= 35$ mg/L $- 20$ mg/L
$$= 15 \text{ mg/L as CaCO}_3$$

Hydroxide alkalinity to be removed $= 50$ mg/L $- 0$ mg/L
$$= 50 \text{ mg/L as CaCO}_3$$

Then add the two alkalinities to find the combined amount needed to be removed:

Combined alkalinity to be removed $= 15$ mg/L $+ 50$ mg/L
$$= 65 \text{ mg/L as CaCO}_3$$

Finally, find the carbon dioxide dose as follows:

CO_2 dose $=$ (Alkalinity to be removed)(0.44)
$$= (65 \text{ mg/L})(0.44)$$
$$= 28.6$$

21. (a)
Total removal capacity $=$ (grains/cu ft)(cu ft resin)
$$= (30,000)(8.88)$$
$$= 266,400 \text{ grains removed per cycle}$$

(b) $\quad \dfrac{\text{Hardness in}}{\text{grains per gallon}} = \dfrac{\text{mg/L}}{17.12}$

$$= \dfrac{142}{17.12}$$

$$= 8.29 \text{ gpg}$$

Volume treated $= \dfrac{\text{Removal capacity per cycle}}{\text{Hardness}}$

$$= \dfrac{266,400}{8.29}$$

$$= 32,135 \text{ gal per cycle}$$

(c) First calculate the volume treated each minute:

Volume per minute $=$ (Bed area)(loading rate)

$$= (4.1)(4.8)$$

$$= 19.68 \text{ gpm}$$

Since the total cycle treats 32,135 gal, the length of the cycle can be calculated as:

Run length $= \dfrac{\text{Volume treated per cycle}}{\text{Volume treated per minute}}$

$$= \dfrac{32,135 \text{ gal}}{19.68 \text{ gpm}}$$

$$= 1633 \text{ min, or } 27.22 \text{ hr}$$

(d) The salt dosage is based on the removal capacity and the regeneration requirement:

Pounds salt $=$ (Removal capacity)(Regeneration requirement)

$$= (266,400)\left(\dfrac{0.35}{1000}\right)$$

$$= 93.24 \text{ lb salt}$$

(e) Gallons of brine $= \dfrac{(\text{lb salt needed})}{(\text{lb salt/gal of solution})}$

$$= \dfrac{93.24}{1.24}$$

$$= 75.19 \text{ gal of brine}$$

22. (See Examples 43, 45, and 47 for calculations.)

23. As calculated in Problem 22, a bag of 100-percent pure NaF would contain 45 percent fluoride ion (or in the case of a 100-lb bag of NaF, 45 lb of fluoride ion). In this problem, however, the bag only contains 96-percent NaF; therefore, it would contain slightly less than 45.24 lb NaF:

$$(45 \text{ lb F})(0.96) = 43.2 \text{ lb F per } 100 \text{ lb of } 95\% \text{ NaF}$$

24. Using the information calculated in Problem 22, a 100-percent purity fluosilicic acid contains 79 percent fluoride ion. Since the commercial acid in this problem is only 32-percent pure, every pound of commercial acid contains:

$$(79\%)(32\%)(1 \text{ lb commercial acid}) = x \text{ lb pure fluoride}$$

$$(0.79)(0.32)(1 \text{ lb}) = 0.25 \text{ lb of fluoride}$$

Since 1 lb of acid contains 0.25 lb F, 11.1 lb of acid would contain 11.1 times that amount:

$$(11.1)(0.25 \text{ lb F/lb acid}) = 2.78 \text{ lb F/gal acid}$$

This means that with each gallon of acid added to the water, 2.78 lb of fluoride ion is added.

The desired fluoride ion dose, 1 mg/L, is equivalent to:

$$(1 \text{ mg/L})(0.16 \text{ mgd})(8.34 \text{ lb/gal}) = 1.33 \text{ lb F/day}$$

Since there is 2.78 lb F in each gallon of acid, to achieve the proper dosage (1.33 lb F), only about a half gallon of acid will be needed. More precisely:

$$\frac{1.33 \text{ lb F/day}}{2.78 \text{ lb F/gal acid}} = 0.48 \text{ gal acid/day}$$

25. To determine the gallons per day of saturated solution of NaF needed, you will have to calculate the following:

    (a) pounds per day fluoride ion needed
    (b) pounds per day NaF needed
    (c) pounds per day solution needed
    (d) gallons per day solution needed

(a) A dosage of 1 mg/L fluoride ion is required. Calculate pounds of fluoride ion required per day as follows:

$$(1 \text{ mg/L})(0.16 \text{ mgd})(8.34 \text{ lb/gal}) = \text{lb/day fluoride ion}$$

$$= 1.33 \text{ lb/day fluoride ion}$$

(b) As calculated in Problem 22, a 100-percent pure compound of NaF contains only 45 percent fluoride ion. Since at 95-percent purity there is less fluoride in the compound, the adjusted percentage should be less than 45.24 percent:

$$(0.95)(45\%) = 42.75\% \text{ fluoride ion}$$

Since each lb of NaF contains 0.43 lb F, the relationship between NaF and fluoride ion content can be expressed mathematically as:

$$(0.43)(x \text{ lb/day NaF}) = 1.33 \text{ lb/day F}$$

$$x = \frac{1.33}{0.43}$$

$$= 3.09 \text{ lb/day NaF}$$

(c) A saturated NaF solution contains only 4 percent NaF by weight. Using this information, the total pounds per day of solution can be calculated:

$$(0.04)(x \text{ lb/day solution}) = 3.09 \text{ lb/day NaF}$$

$$x = \frac{3.09}{0.04}$$

$$= 77.25 \text{ lb/day solution}$$

(d) Using the appropriate conversion factor, convert the pounds per day of solution to gallons per day:

$$\frac{77.25 \text{ lb/day}}{8.6 \text{ lb/gal}} = 8.98 \text{ gpd}$$

Therefore, dosing with a saturated 95-percent NaF solution would require about 9 gpd.

26. To raise the fluoride concentration to 1.2 mg/L, you must add

$$1.2 \text{ mg/L} - 0.3 \text{ mg/L} = 0.9 \text{ mg/L}$$

Then, to calculate pounds of fluoride needed per day,

$$\text{lb/day F needed} = (0.9 \text{ mg/L})(6.4 \text{ mgd})(8.34 \text{ lb/gal})$$

$$= 48.04 \text{ lb/day F}$$

Since Na_2SiF_6 is not 100 percent fluoride ion, and the commercial compound is not 100-percent pure Na_2SiF_6, adjustments must be made. From Answer 22, pure Na_2SiF_6 contains 61% F. Therefore:

$$\text{lb/day} = \frac{48.04}{0.61}$$

$$= 78.75 \text{ lb } Na_2SiF_6 \text{ (100\% pure)}$$

Since the Na_2SiF_6 is only 98.5-percent pure:

$$\text{lb/day} = \frac{78.75 \text{ lb}}{0.99}$$

$$= 79.55 \text{ lb/day } Na_2SiF_6 \text{ (98.5\% pure)}$$

27. (a) Solubility $= 0.62 \text{ gm/100 mL}$ at $15.6° \text{ C}$, which in units of pounds per gallon is

$$\text{Solubility in lb/gal} = \left(\frac{0.62 \text{ gm}}{100 \text{ mL}}\right)\left(\frac{3785 \text{ mL}}{\text{gal}}\right)\left(\frac{1 \text{ lb}}{454 \text{ gm}}\right)$$

$$= 0.0517 \text{ lb/gal solubility}$$

This means that 1 gal of water is required to dissolve each unit of 0.0517 lb Na_2SiF_6. To determine the total number of gallons required to dissolve 80 lb of Na_2SiF_6, you must determine how many units of 0.0517 lb there are in 80 lb:

$$\frac{80 \text{ lb}}{0.0517 \text{ lb}} = 1547$$

Because there are 1547 units of 0.0517 (each one representing 1 gal of water needed), the total number of gallons of water needed to dissolve 80 lb of Na_2SiF_6 is 1547 gal.

(b) To evenly feed 1547 gallons over one day, the hourly chemical feed pump pumping rate should be

$$\frac{1547 \text{ gpd}}{1440 \text{ min/day}} = 1.07 \text{ gpm}$$

Electricity

Electricity 1

Electricity, Magnetism, and Electrical Measurements

E1-1. Electricity and Magnetism

All matter is composed of atoms, and the basic particles that make up atoms are protons, neutrons, and electrons.[1] Because electrons are held to atoms by a relatively weak force, the outer electrons of many atoms can easily be removed or transferred to other atoms. This movement may be the result of chemical action, or it may be due to electric or magnetic forces. The study of the movement and effect of large numbers of electrons is the study of electricity.

For purposes of explanation, electricity is often classified as either static or dynamic. Both forms of electricity are composed of large numbers of electrons, and both forms interact with magnetism. The study of the interaction is called electromagnetics.

Static Electricity

Static electricity refers to a state in which electrons have accumulated but are not flowing from one position to another. Static electricity is often referred to as electricity at rest. This does not mean that it is not ready to flow once given the opportunity. An example of this phenomenon is often experienced when one walks across a dry carpet and then touches a doorknob—a spark at the fingertips is likely noticed and a shock is usually felt. Static electricity is often caused by friction between two bodies as by rotating machinery—conveyor belts and the like—moving through dry air. Static electricity is prevented from building up by properly bonding equipment to ground or earth.

Dynamic Electricity

Dynamic electricity is electricity in motion. Motion can be of two types—one in which the electric current flows continuously in one direction (direct current),

---

[1]Chemistry Section, The Structure of Matter.

and a second in which the electric current reverses its direction of flow in a periodic manner (alternating current).

Direct current. Current that flows continuously in one direction is referred to as direct current (DC). In this case, the voltage remains at a fixed potential across the path through which the current is flowing. Direct current is developed by batteries and can also be developed by rotating-type DC generators.

Alternating current. Electric current that reverses its direction in a periodic manner—rising from zero to maximum strength, returning to zero, and then going through similar variations of strength in the opposite direction—is referred to as alternating current (AC). In this case, the voltage across the circuit varies in potential force in a periodic manner similar to the variations in current. Alternating current is generated by rotating-type AC generators.

Induced current. A change in current in one electrical circuit will induce a voltage in a nearby conductor. If the conductor is of such configuration that a closed path is formed, the induced voltage causes an induced current to flow. This phenomenon is the basis for explaining the property of a transformer. Since changes of current are a requirement, it is obvious that DC circuits cannot use transformers, whereas AC circuits are adaptable to the use of transformers. The function of a transformer is of great value because the voltage of the induced current may be increased or diminished to any extent, depending upon the requirements of the relation and proximity of the electrical circuits.

Electromagnetics

The behavior of electricity is determined by two forces: electric and magnetic. A familiar example of the electric force is the force that attracts dust to phonograph records in dry climates. The magnetic force holds a horseshoe magnet to a piece of iron and aligns a compass with magnetic north. Wherever there are electrons in motion, both forces are created. These electromagnetic forces act on the electrons themselves and on the matter through which they move. Under certain conditions, the motion of electrons sends electromagnetic waves of energy over great distances.

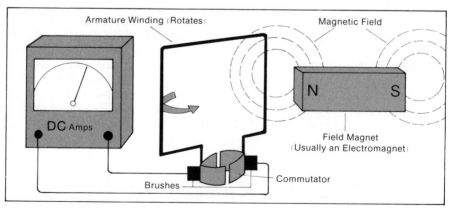

Figure 1. In a Generator, Current is Produced When a Coil of Wire Moves Through a Magnetic Field.

Electromagnetic energy is responsible for the transmission of a source of power. Because of electromagnetic energy, power can be transmitted across the air gap between the stator and rotor of a motor. Radio waves are electromagnetic. Radar is electromagnetic. So is light, which unlike most forms of electromagnetic waves, can be seen. The doorbell is an electromagnetic device; so is the solenoid valve on your dishwasher. In fact, much of the electrical equipment with which you are familiar is electromagnetic; your clock motor, the alternator on your car, and the electric typewriter use electromagnetics.

You will find it worthwhile to make your own list of equipment and appliances which are all-electric and which are electromagnetic. You will be surprised at how much you already know about this subject.

In fact, because electricity is such a part of everyday life, even persons who have not studied electricity know more about it today than did early scientific observers. Electricity is often taken for granted, but it should be respected. Every operator should become familiar with safety rules and practices pertaining to electricity.

E1-2. Electrical Measurements and Equipment

The following paragraphs introduce the basic terminology used to describe electricity and electrical equipment. To help the reader understand the new concepts, the behavior of electricity is compared to the behavior of water. A more detailed and more technical description of the terms introduced is given in the next chapter.

Molecule of liquid—Electron of electricity. The smallest unit of liquid is a molecule, whereas the smallest unit of electricity is an electron.

Flow rate (gpm)—Current (A). The rate at which water flows through a pipe is expressed as gallons per minute (gpm)—it could be expressed as molecules per minute, but the numbers involved would be inconveniently large. Similarly, the rate at which electricity flows through a conductor (called current) could be expressed as electrons per second, but the numbers involved would be too large to be practical. Therefore, the unit of electrical current commonly used is the ampere (A), which represents a flow rate of about 6,300,000,000,000,000,000 electrons per second. The flow rate of liquid expressed in gallons per minute is, therefore, analogous to the flow rate (current) of electricity expressed in amperes.

Pressure (psi)—Potential (V). Liquid flow requires a certain head or pressure. The liquid will tend to flow from the high pressure to the low pressure. Electrical potential is similar to head or pressure. Electric current will always tend to flow from high potential to low potential. Electric potential is expressed in terms of volts (V). Whereas liquid pressure is generally expressed in terms of pounds per square inch with reference to atmospheric pressure, electrical potential is generally expressed in terms of voltage with respect to ground or earth, with the general assumption that ground or earth is at zero potential.

Pressure drop—Voltage drop. The flow of liquid through a pipe is accompanied by a pressure drop as a result of the friction of the pipe. This pressure drop is often referred to as friction head. Similarly, the flow of electricity through a wire is accompanied by a voltage drop as a result of the resistance of the wire.

Friction—Resistance. The flow of water through a pipe is limited by the amount of friction in the pipe. Similarly, the flow of electricity through a wire is limited by resistance.

Pump—Generator. A pump uses the energy of its prime mover, perhaps a gasoline engine, to move water; the pump creates pressure and flow. Similarly, a generator, powered by a prime mover such as a gasoline engine or a water turbine, causes electricity to flow through the conductor; the generator creates voltage and current.

Turbine—Motor. A water-driven turbine, like those found in hydroelectric power stations, takes energy from flowing water and uses it to turn the output shaft. An electric motor uses the energy of an electric current to turn the motor shaft.

Turbine-driven pump—Motor-driven generator. A turbine running off a high pressure, low flow-rate stream of water (or of hydraulic fluid) can be used to drive a pump to move water at low pressure, but at a high flow-rate. (Such an arrangement might be useful for dewatering, for example, although it is not common.) Similarly, a motor operating from a high voltage, low current source can be used to run a generator having a low voltage, high current output. Note that both the water system and the electrical system could be reversed, taking a low pressure (voltage), high flow-rate (current) source and creating a high pressure (voltage), low flow-rate (current) output. The motor driven generator (also called a dynamotor) is sometimes used to convert DC battery power to a higher voltage, lower current AC power.

(Turbine-driven pump)—AC transformer. For most electrical applications, the motor-driven generator is replaced with a transformer, which does exactly the same thing—it transforms low voltage, high current electricity into high voltage, low current electricity; or it can transform high voltage, low current into low voltage, high current. The transformer will only work with alternating current (AC). It is a very simple device which uses the electromagnetic properties of electricity.

Reservoir—Storage battery. Liquids can be stored in tanks and reservoirs, whereas direct-current electricity can be stored in batteries and capacitors.

Flooding—Short circuits. The washout of a dam or the break of a water main can cause various degrees of flooding and damage, depending upon the pressure and, more importantly, upon the quantity of water that is released. Similarly, an electric fault current, referred to as a short circuit, can cause excessive damage, depending upon the voltage and, more importantly, upon the quantity of electric energy that is released. This latter item, the quantity of electric energy that might be released, is referred to as the short-circuit capability of the power system. It is important to recognize that this item, the short-circuit capability, determines the physical size and ruggedness of the electrical switching equipment required at each particular plant. In other words, for nearly identical plants, the electrical equipment may be quite different in size because of the different short-circuit capabilities of the power system at the plant site.

Instantaneous transmission of pressure—Electricity. When we let water into a pipe that is already filled, the water that we let into the pipe at one end is not the same as that which promptly rushes out of the other end. The water we let in pushes the water already in the pipe out ahead of it. The pressure, however, is transmitted almost instantaneously. The action of electricity is much the same. The electric current in a wire travels at high speeds, very close that that of light's. When a light switch is turned on, the power starts the electrons on their way.

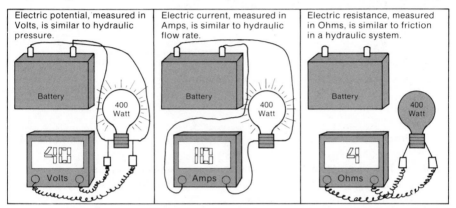

| Electric potential, measured in Volts, is similar to hydraulic pressure. | Electric current, measured in Amps, is similar to hydraulic flow rate. | Electric resistance, measured in Ohms, is similar to friction in a hydraulic system. |

Figure 2. **Ohm's Law States That Voltage Drop _E_ Across an Electrical Component Equals Current _I_ Flowing Through the Component Multiplied by Resistance _R_ of the Component: _E_ = _IR_.**

Review Questions

1. The study of electricity is the study of _____.

2. What is static electricity?

3. What is dynamic electricity?

4. What are the two types of force that determine the movement of electrons?

5. Define _direct current_ and _alternating current_.

6. Give three examples of electromagnetic waves.

7. What are the units of measurement of electrical potential and electrical current?

8. For each of the following hydraulic concepts or mechanisms, list the comparable electrical concept or device:

   (a) Pressure   (c) Friction   (e) Reservoir   (g) Flooding

   (b) Flow rate   (d) Pump   (f) Turbine-driven pump

Detailed Answers

1. The behavior and effect of the motion of large numbers of electrons.

2. Static electricity refers to a state in which electrons have accumulated but are not flowing from one position to another.

3. Dynamic electricity is electricity in motion.

4. Electric, magnetic.

5. Direct current flows continuously in one direction. Alternating current reverses its direction in a periodic manner.

6. Radio waves, radar, light.

7. Potential is measured in volts (V). Current is measured in amperes (A).

8. (a) Potential or volts, (b) current or amperes, (c) resistance, (d) generator, (e) storage battery, (f) motor-driven generation or—for AC only—transformer, (g) short circuits.

Electricity 2

Electrical Quantities and Terms

The following is a glossary of some of the more frequently used units of measurement of electricity, specifically with regard to power systems. Many of these are commonly used in describing a system, a piece of electrical equipment, or an amount of electricity required or used. You should become familiar with their usage. The terms defined here are also listed in the glossary at the back of the book.

AC. An abbreviation for alternating current.

Alternating current. This is the term applied to an electric current that is continually reversing its direction of flow at regular intervals. During each interval that it flows in one direction it starts at zero, rises to maximum flow, then falls back to zero flow. It then starts from this zero flow, rises to a maximum flow, but in the opposite direction, then again falls back to zero flow. Each repetition from zero to a maximum in one direction and then through zero to a maximum in the other direction and back to zero, is called a complete cycle. The number of cycles made in one second are now called hertz (Hz). Thus, a 60-cycle AC power system is a 60-Hz AC power system. The change to the word hertz is place of cycles-per-second has occurred only recently. There is an effort being made to give all electrical terms the name of some person who has contributed to our knowledge of electricity; thus cycles-per-second are named for a man named Hertz.

Ampere. This term measures electric current, either AC or DC. One ampere (A) is the amount of current that would flow through a resistance of 1 ohm when the voltage on the circuit is 1 volt (V). Most equipment is rated in amperes. If the equipment is operated above its ampere rating, it is overloaded and may be damaged. For example, a motor will have a nameplate that will state how many amperes it will draw when operating fully loaded. If a motor draws more amperes than its nameplate rating, it may be assumed to be overloaded. Amperes are measured with an ammeter.

Ampere-hour. This is a measure of a quantity of electricity. One ampere flowing for 1 hr is an ampere-hour (amp-hr). This term is used chiefly as a rating of a storage battery, and it simply indicates the amount of energy that a battery can deliver before it needs recharging. A 100-amp-hr rating on a battery means it, for example, can deliver 100 A for 1 hr, 1 A for 100 hr, 2 A for 50 hr, or 4 A for 25 hr.

Apparent power. This is the product of AC volts and AC amperes, and equals volt-amperes (VA). Although this term will not be used frequently, it should be pointed out that this is different from real power. Apparent power is normally expressed as kVA, meaning thousand volt-amperes. Some of the current that flows in a circuit may not produce any power. This may seem strange at first, but let's look at it from another point of view. If you tie a rope on your car and try to pull it down the street by hand, would you walk in front of the car or would you go off to the side and walk down the sidewalk? Obviously, if you went over to the sidewalk, you would have to pull much harder than if you were directly in front of the car. In this case, you would be using more power than was really necessary to move the car forward. The only power that is of any value in this case is the power that moves the car forward; the "power" you spend in pulling the car sideways is of no use and could be considered useless. Apparent power would be the power you actually spend; real power would be that which actually moves the car forward; and unreal power would be that which was spent but which was unnecessary. In electrical systems, the percentage of apparent power that is real power is called the power factor. One hundred kVA at 80 per cent power factor represents 80 kW of real power. You will frequently hear the expression "power factor," and it is usually expressed as a decimal, as 0.8, instead of as a per cent. Apparent power is measured in volt-amperes; real power is measured in watts; and unreal power is measured in vars. Because these units are quite small with respect to the sizes of electrical loads usually found in water-distribution systems, they are generally expressed in thousands. Kilo means thousand. Hence, you will generally find apparent power expressed in kVA (kilovolt-amperes), real power expressed in kW (kilowatts), and unreal power expressed in kvar (kilovars).

Cycle. Cycle applies to alternating current systems and refers to one complete change in direction of current flow from 0 to maximum to 0 in one direction, then to maximum and back to 0 in the opposite direction. See "alternating current ."

Capacitance. Capacitance is a part of the total impedance of an electrical circuit tending to resist the flow of current. It acts somewhat like inertia in that it tries to resist a sudden change of voltage. For example, your car has inertia, and you will recall that if it is standing still, it takes more force to start it moving than it does to keep it moving. In like respect, after a car is moving, it tends to keep moving and cannot easily be stopped instantly. The capacitance of a circuit is one of the reasons for the unreal power defined previously.

Capacitors are sometimes used to prevent a sudden rise in voltage and sometimes are used to prevent a sudden drop in voltage. Capacitors are also used in some cases to isolate devices so that no direct-wire physical connection between them is necessary. Frequently, capacitors are used in conjunction with induction motors to raise the power factor of the electrical system.

Capacitive reactance is measured in ohms.

Demand. This is an expression used frequently in the sale or purchase of electric power. Demand is the amount of electric power expressed in watts (W) or kilowatts (kW) that may be required during a certain time interval. Usually power utilities determine use during a 15-min or a 30-min interval. This is really a measure of how big the power requirement is, not how much power is used. For example, you might have a 2000-hp motor that you start and use only once or twice a year, and then only for a couple of hours each time it is used. Obviously, you wouldn't be buying a lot of electric power during the year for operation of this motor; but it should be recognized that regardless of how seldom the load is used, the power company has to have that amount of power readily available whenever the load is applied to its system. As a result, the power company may make a separate charge for the size of the load in addition to the charge for energy actually consumed. This is called a demand charge. Demand is measured by a demand meter, which actually measures the maximum kilowatts that have been required during any particular demand interval—15 min, for instance. The demand meter has a maximum-reading pointer. It is not uncommon that the maximum demand established in any particular month will then apply, as far as billing is concerned, for the following eleven months. It is obvious why it is of chief concern to the operator to prevent establishing a new maximum demand, since this may increase the price of purchased power for the following year.

This concept of demand is applicable to water-distribution systems. You may have a customer who normally uses less than 20 gpm of water but who might very infrequently want several hundred gpm for a short time. In such an instance, pipelines and supply facilities would have to be sized for the maximum usage, even though the usual quantity were not too different from the average residential service.

DC. An abbreviation for direct current.

Direct current. An electric current that flows continuously in one direction. Direct current is used quite frequently for controls and instrumentation. It can be generated on the site by using an engine-driven DC generator. It most frequently is converted from alternating current by means of an AC motor-driven DC generator, or directly converted by means of a rectifier. The storage battery is a common source of direct current that is kept charged by a battery charger, which is essentially a regulated rectifier. Of course, dry cells (similar to flashlight batteries) can be used to supply DC current, but these are limited to small loads.

Synchronous motors require a comparatively small amount of direct current in addition to their AC-current requirements. Magnetic couplings used as variable-speed drives require both AC and DC current. Most electronic instruments use direct current. For example, a signal of 4-20 milliamperes (mA) DC is commonly used with electronic instruments. This is a very small amount of current. Milli is a prefix meaning thousandth. Because 4-20 thousandths of an ampere DC output is a very small current flow, such an instrument will contain a very small wire size. Direct current from an auxiliary battery supply is frequently used for both power and control of valves where continued operation is necessary should AC power fail. Many of the new electronic, variable-speed, motor controllers—SCR drives, for example—employ direct current. SCR refers to silicon-controlled rectifiers. There are many types of SCR drives on the market for comparatively small motors. Many emergency-lighting systems are designed for operation on direct current.

Direct current devices will probably play an even greater role in the future as solid state rectifiers become more fully developed.

Energy. Electric energy is usually expressed in kilowatt-hours (kWhr). One thousand W used for 1 hr, or 500 W used for 2 hr equals 1000 Whr or 1 kWhr. Note that power is expressed in watts, but that energy is expressed in watt-hours.

One kWhr of electric energy may cost anywhere from ½ cent to 4 cents, depending upon the size of the load and, of course, the location and conditions of the contract for purchasing power. Certainly, one of the primary responsibilities of an operator is to keep the use of power at a minimum because of its cost. For example, by allowing ten 100-W lamps (1000 W) to burn continuously for one month (720 hr), 720,000 Whr or 720 kW of electric energy are used. If it is assumed that energy costs one cent per kWhr, this would amount to $7.20. In a plant where there are hundreds of lamp fixtures and devices, the needless use of electricity becomes an unnecessary operating expense.

It is also interesting to compare the costs of operating the same ten 100-W lamp fixtures with the cost for operating a 1000-hp motor for 1 hr. One-thousand hp equals 746 kW. Therefore, this load is 746 kWhr, which at one cent per kWhr would amount to $7.46. Note that it only cost 26 cents more to operate the 1000-hp motor for 1 hr than it did to leave the ten lamp fixtures on for a month. This example helps to show why there is usually a demand charge in addition to the charge for energy, since the electrical system must be many times greater both in size and capacity to start and run a 1000-hp motor than it needs to be to energize ten 100-W lamps.

Power bills usually have both a demand and an energy charge. These charges are based on how big and how much—a concept that every operator should keep in mind. See definition of "demand" given previously.

Frequency. Frequency applies to alternating current and refers to the number of cycles that occur each second. The frequency of most power systems in water-distribution facilities will be 60 Hz. Some power systems in the US are 25 Hz. Hertz is the new term for cycles-per-second.

In the past there has been considerable argument about choosing a standard frequency. As frequency increases, the impedance to the flow of current in a circuit often also increases, which makes the line losses and voltage drops on electric transmission and distribution systems greater; however, as frequency increases, the physical size of equipment becomes smaller. At 25 Hz, many people could sense a light bulb dimming and brightening 25 times/sec. This, perhaps more than anything else, caused the industry finally to standardize on 60-Hz power. Single-phase and three-phase are most common.

On airplanes it is common to use a power system with a 400-Hz frequency. This allows equipment to be smaller and, of course, transmission lines don't have to be any longer than the length of the plane.

Tone-transmitting equipment used for telemetering and supervisory control usually covers a range of 300-3000 Hz.

Radio equipment operates at very much higher frequencies and can use very small components.

The power utilities regulate frequency very closely, making it possible for all of our electric clocks to remain synchronized. If frequency were lowered, our clocks would lose time and all motors would slow down. If frequency were raised, our clocks would gain time and all motors would speed up. Varying the frequency can be useful. Equipment is presently available to use frequency-variable motor controllers for variable-speed pump drives.

Frequency can be measured with a frequency meter.

Ground. Ground is an expression representing an electrical connection to earth or to a large conductor that is known to be at earth's potential. Properly grounded equipment and circuits minimize the hazard of accidental shocks to persons operating electrical equipment. Grounding also serves to facilitate the operation of protective devices used on the equipment. A good ground will generally have a resistance of not more than 10 ohms and preferably less. Motor frames and electrical-equipment enclosures should all be connected together (called bonded) and to ground. This whole subject is often referred to as "grounding."

Impedance. Impedance is the total opposition offered by a circuit to the passage of electrons—that is, the flow of current. Impedance is made up of three parts —resistance, inductance, and capacitance. It is interesting to note that once a certain flow of direct current is established, the inductive reactance and the capacitive reactance are ineffective—only the resistance of the circuit limits the flow of current because inductance and capacitance act only to oppose changes in electrical flow.

Impedance is measured in ohms and is made up of so many ohms of inductive reactance, so many ohms of capacitive reactance, and so many ohms of resistance.

A series circuit with 10 ohms inductive reactance, 10 ohms capacitive reactance, and 10 ohms resistance would not, however, have an impedance of 30 ohms. Remember that capacitive reactance and inductive reactance oppose one another; in this case they are equal, so the impedance is only 10 ohms, the same as if there were no inductance and capacitance in the circuit. This would be an example of a "tuned" circuit. Neither would a circuit of 10 ohms inductance and 10 ohms resistance have an impedance of 20 ohms. You do not add them directly as one might expect. For those who might be wondering how to add them, it is done by taking the square root of the sum of the squares. For example, 10 ohms resistance plus 10 ohms inductance is the square root of 10 × 10 + 10 × 10, or the square root of 100 + 100; that is, the square root of 200, which is a little over 14 ohms. This would be an inductive circuit having an impedance of 14 ohms.

Inductance. Inductance is part of the total impedance of an electrical circuit's tending to resist the flow of changing current. Inductance also is quite similar to inertia in that it tries to prevent a sudden change of current. If a current is flowing in a circuit and a switch is opened, it is the inductance of the circuit that tries to keep the current from instantaneously falling to zero. You have no doubt noticed sparks as you opened a switch or removed a jumper from a car battery. These sparks are simply the electric current, jumping across the air gap as the switch is being opened and trying to maintain itself. Similarly, when a switch is closed, the inductance of the circuit prevents the current from rising from zero to its maximum amount instantaneously. It takes some time for the current to rise to its full value. This can be seen by looking at an ammeter when the switch is opened and closed. Inductance and capacitance may be thought of generally as causing similar but opposite effects on a circuit. Inductance can be added to cancel the effect of capacitance, and capacitance can be added to cancel the effect of inductance.

Inductive reactance is measured in ohms.

Interrupting current. This is a term applying to the fault current that a fuse or circuit breaker is designed to interrupt or clear without becoming destructive. A fuse or circuit breaker that interrupts a short circuit current greater than its designed interrupting current will act as a bomb.

Kilo (k). A prefix meaning one thousand.

Kilowatt (kW). A measure of electric power. Since a unit of power equal to 1 W is so small in many instances in the water utility, it is more convenient to express power in units of 1000 W, or 1 kW.

To get a concept of the size of this unit of measurement, remember that 1 hp equals 746 W. Power is the capacity to do work in a given time.

A kilowatt is abbreviated kW and means 1000 W. One hp equals 0.746 kW.

Kilowatt-hour (kWhr). A measure of electric energy. Since a unit of energy equal to 1 Whr is so small, it is convenient to express units of electric energy in units of 1000 Whr.

A kWhr of electric energy is a measure of the amount of work done. The use of 1 W for 1000 hr, or 1000 W for 1 hr, in each case, would be the use of 1 kWhr of electric energy.

A kilowatt-hour is abbreviated kWhr. One hp-hr equals 0.746 kWhr.

Kilovar (kvar). This is an expression meaning thousand reactive volt-amperes and is a measurement of unreal power. It will not be used very often; however, one should have some concept of its meaning. It is used to express the size of power capacitors. You may find some induction motors with capacitors connected in parallel with the motor. These capacitors will have a nameplate similar to a transformer nameplate, stating size in kvar.

A capacitor connected to a motor terminal may be used to correct the power factor. The capacitor is added to make a reduction in the amount of unreal power (kvar) used by the motor circuit. One reason that unreal power is objectionable is that it consumes additional current flow, and current flow causes voltage drop, which results in low voltage at the equipment.

Kilovolt (kV). The kilovolt is an expression meaning 1000 V and is abbreviated kV.

Common voltages used by water utilities are 2.4 kV (2400 V), 4.16 kV (4160 V), 15 kV (15,000 V), 33 kV (33,000 V), and 69 kV (69,000 V). Many still higher voltage classes are standard transmission voltages throughout the US. Usually voltages below 1000 V are expressed directly, such as 6 V, 24 V, 120 V, 460 V, or 600 V.

kVA. This is an expression meaning thousand volt-amperes, but usually it is never referred to in any other way than the three letters, *kVA*. The term kVA is used in expressing the size of transformers, generators, and various other pieces of equipment. It is also used to express the amount of power that might be released from an electrical system during a fault such as a short circuit. Such expressions may also be used to describe the size of a system load. Saying the plant-connected load is approximately so many kVA is essentially the same as saying the plant-connected load is approximately so many hp. It is useful to recognize that, for a plant which contains mostly squirrel-cage-type induction motors, kVA is approximately equal to 1 hp. Common sizes of lighting transformers are 50, 75, 100, 225, 500, 750, and 1000 kVA.

The quantity kVA is actually a measurement of apparent power which is made up of two parts—real power and unreal power. Real power is expressed in kW and unreal (wattless) power is expressed in kvar. Apparent power equals real power plus unreal power, or kVA = kW + kvar. It is important to note here, however, that this is not a direct arithmetical addition. The sum is the square root of the sum of the squares of the two parts, similar to the addition described under the discussion of impedance.

Megohm. The term megohm is used to express 1 million ohms because the ohm is a very small quantity. Ohm is a measure of the impedance a circuit offers to the flow of current. Megohms is particularly useful when speaking of the insulation of high-voltage components in a particular type of equipment such as transformers, switchgear, and motors or in wiring materials such as cable or insulators on open-wire power lines. The insulation of a high-voltage motor during winding will, for example, have an impedance of several thousand megohms.

Maximum demand. The maximum demand is the maximum kilowatt load that occurs and persists for a full demand interval during any billing period, usually a month. It is not uncommon for a maximum demand established during one month to be used as a factor in figuring power bills for the succeeding eleven months, unless of course a still higher demand is established. Demand intervals are usually 15 min or 30 min to permit starting large loads for a few minutes during testing and maintenance without establishing a demand used for billing purposes.

Ohm. The ohm is a measure of the ability of a path to resist, hinder, or impede flow of electricity. Obviously, if you want electricity to flow, you must choose a material having little resistance to flow. Likewise, if you do not want electricity to flow, you must choose a material having great resistance to its flow. The material offering little resistance is called a conductor. The material offering great resistance is called an insulator. All materials are really conductors of electricity, but the poorest conductors may make the best insulators.

The practical unit of electric resistance is an ohm. By definition, it is the resistance in which 1 V will maintain a current flow of 1 A. The ohm is used in expressing resistance, reactance, and impedance, all of which are characteristics of a circuit that tend to restrict the flow of current.

Phase. The term phase identifies the type of electrical service either supplied or required. From a practical standpoint, there are only two types of phase—one-phase and three-phase, more commonly referred to as single-phase and three-phase. It is possible to develop two-phase or other number-phase systems, but they would be special designs not generally found in water utilities.

Single-phase power simply refers to a system having a single-winding generator or transformer requiring two conductors, one from each end of the winding, to supply power. This single winding may be center-tapped so that three conductors are used. Such a system is called a three-wire, single-phase source. A single, 120-V winding, two-wire system may have one end of the winding connected to ground; the other end then would be the hot lead. Such a system would be called a two-wire, grounded, 120-V power supply.

Similarly, a 240-V winding may have a center-tap connected to ground; both ends of the winding would then be hot leads; that is, 120 V to ground. This system would be a three-wire, 240/120-V, single-phase power source. In this sytem you could connect 240-V equipment across the two outer leads and 120-V equipment from either outer lead to ground. One of these two systems—the 120-V, two-wire, single-phase or the 240-V, three-wire, single-phase—is normally used for residential small loads. A single-phase motor greater than 3-5 hp is considered a large single-phase motor. Single-phase circuits are most adaptable for lighting and convenience outlets, small fractional-horsepower motors, and appliances. Three-phase motors are frequently referred to as polyphase motors. The term polyphase simply means several phases and may be used in conjunction with different types of equipment, such as a polyphase switchboard.

Three-phase power simply refers to an electric system having a three-winding generator or transformer requiring at least three conductors, one from each winding, to supply the power. Three-phase power systems do not have any practical limitation on the size of services of equipment, so all larger load applications, like those found in most industrial plants, treatment plants, and pumping stations, will have three-phase power systems. Some systems will be three-phase, three-wire and some will be three-phase, four-wire, depending upon how the three windings are connected inside the generator or transformer. The three windings can be connected in either of two configurations—as a delta or as a wye. The delta-connected transformer only has three corner connections and can only be a three-phase, three-wire system, since a lead or wire is connected at each corner. The wye-connected transformer has one end of each of the three windings connected in common so that the other ends are available for a three-phase, three-wire system. It is possible to connect a lead to the common junction (called the neutral) of the three windings, making a fourth wire connection. In such cases, the system is then a three-phase, four-wire system. If the neutral wire, the one connected to the common junction, is also connected to ground, this system is then referred to as a three-phase, four-wire, grounded system.

The term phase is also used frequently when paralleling (connecting together) two or more power supplies. Such systems may be referred to as in-phase or out-of-phase. They definitely must be in-phase when interconnected. The word phase, when used in this respect, actually refers to the sequence in which the voltage reaches its maximum value on each of the three power leads. For example, three power leads are connected to a three-phase induction motor and the motor turns in one direction; this is an in-phase system. If you interchange the connection of any two of these three motor leads, the motor will then run in the opposite direction; this is an out-of-phase system. Actually, what you have done is reverse the phase sequence. If you had two power supplies to your plant and these two supplies were out-of-phase, all of the motors would run in one direction when served by one power feeder and in the opposite direction when served by the second power feeder. When the power supplies are in-phase, they are said to be synchronized. The instrument used to show that two systems are in-phase is called a syncroscope.

Sometimes you will hear the expression that a three-phase motor is single-phasing. This means that one of the three leads (phases) to the motor has been disconnected by some means and the motor is trying to run with only two leads energized. Obviously, the motor will be overloaded. Loss of phase and reverse-phase can be detected by protective devices, called relays, that will stop equipment when such a condition develops.

Single-phase circuits can easily be served from three-phase systems by using the voltage between the two leads of a three-phase, three-wire system, or between the neutral and any of the three hot leads of a three-phase, four-wire system.

Three-phase circuits can also be served from a single-phase system but only in very small loads, by using special converter equipment.

The term phase is also used when referring to the relationship between AC voltage and the current in a particular circuit. If a circuit has only resistance, the voltage and the current are said to be in-phase; that is, the voltage and the current reach their maximum values in both their forward and reverse directions and return to 0 at the same instant. If a circuit has both resistance and a predominate inductive reactance, the current will then lag behind the voltage. The current is then spoken of as being phase-shifted behind or lagging the voltage. If the circuit has both resistance and a predominate capacitive reactance, the current leads the voltage. Most systems are made up of motor loads that are inductive, so it will be usual that the current lags the voltage in any water-utility facility. This is also expressed as a load having a lagging power factor.

Pole. This term is descriptive of one end of a magnet (or electromagnet) or one electrode of a battery. Magnets are said to have north and south poles; batteries have positive and negative poles.

Power factor (pf). The term power factor is used in many ways somewhat similar to the term phase, and both are somewhat related. Power factor refers to the amount that the current actually lags or leads the voltage in an AC circuit. It is a most useful figure for expressing the characteristics of a particular load, such as a motor or a complete electrical system.

During starting, an induction motor may have a very low lagging power factor, perhaps 0.3-0.4; then when it is running at full speed, it will have a lagging power factor of approximately 0.8. A synchronous motor may have a power factor of 1.0 when running; in this case, it is usually referred to as unity power factor. Some synchronous motors may be designed for operations at 0.8 leading power factor.

The power factor is also the ratio of the useful or real power to the apparent power in a circuit. Real power is usually expressed in kW (kilowatts), whereas apparent power is expressed in kVA. Apparent power multiplied by power factor equals real power ($kVA \times pf = kW$).

Power factor can be measured by a power-factor meter.

It may be found that a plant is operating at a very low, lagging power factor. This low power factor can be corrected by adding capacitors or by adding leading, power-factor, synchronous motors.

A low power factor may be the cause of low voltage.

The power factor of a system is the result of the design and selection of equipment. Usually any adjustment of power factor by an operator will be performed in one of two ways—by operating switches to add or remove capacitors or by manually adjusting the field rheostat on synchronous-motor controllers.

In some cases, a power company may have a power factor clause in its published rate structure that will require the purchaser of power to pay extra if the power factor of the plant load is below some expressed limit. Usually the power factor of a plant load is approximately 0.8-0.85, which is generally high enough to escape any billing penalty. Generally, power-factor correction by means of modification of the plant design is not undertaken unless there is either a low-voltage condition or there is an apparent savings in the cost of power.

The operator should understand this concept of real and unreal power. For example, a 100-V, 100-A load at 1.0 pf will furnish 10 000 volt-amperes (10 kVA) or apparent power. When multiplied by the 1.0 pf, this equals 10 kW of real power. Similarly, a 100-V, 100-A load at 0 pf will furnish 10 kVA of apparent power but no real power. Likewise, a 100-V, 100-A load at 0.8 pf will furnish 10 kVA of apparent power and 8 kW of real power. Notice in every case that the current in these circuits is still 100 A. The amount of unreal power is expressed in terms of reactive volt-amperes (kvar).

To summarize, kVA is apparent power; kW is real power; kvar is unreal power; and pf is the power factor, or the ratio of kW per kVA.

It is not usually economical to design plant loads to have a unity power factor. Power factor is usually less than 1.0, but may be lagging or leading.

Power. Power is the ability to do a certain amount of work within a certain amount of time. The unit of measurement of electric power is the watt (W). Since the watt is quite small, insofar as the loads on water-utility systems are concerned, it is usually more convenient to express power in terms of 1000-W units—that is, the kilowatt (kW). When we speak of power, we are always referring to real power (kW). It is necessary to recognize that there are other expressions for power, such as horsepower (hp). One hp equals 0.746 kW, and 1 kW equals 1.341 hp.

There are three types of power in an electrical circuit—apparent power (kVA), real power (kW), and unreal power (kvar).

A motor rated 10 hp has the ability to do 7.46 kW of real work. If a 10-hp induction motor operates at 0.8 pf, it will require 9.3 kVA of apparent power from the electrical system (7.46 kW divided by 0.8 pf equals 9.3 kVA). This would require approximately 10 kVA of transformer capacity. As a rule of thumb on equal plant loads, it may be said that 1 hp equals approximately 1 kVA.

On DC systems, volts × amperes = watts. On AC systems, volts × amperes × power factor = watts.

Primary. The term primary is often used in referring to the high-voltage side of a transformer. The voltage of a power line that serves a plant site is referred to as the primary voltage. A substation transformer will usually be provided to step the voltage down to a lower voltage, called the secondary voltage, for the customer's use.

It is not uncommon to find that power companies will sell power at either primary voltage or secondary voltage. Usually, a customer will furnish the transformer and substation if power is purchased at primary voltage; likewise, the power company will furnish the transformer and substation if power is furnished at secondary voltage. The rates are generally referred to as primary and secondary rates. Secondary voltage does not necessarily mean low voltage (600 V or lower); it simply means some voltage lower than the power-line distribution voltage. Obviously, power is sold at a somewhat cheaper rate at primary voltage since the user furnishes his own substation.

Resistance. Resistance is a part of the total impedance of a circuit's tending to resist or restrict the flow of current. The resistance of a circuit causes power losses in the form of heat produced by the current in the resistor. Inductive reactance and capacitance, the other two components of a circuit, do not consume any real power.

Reactance. Reactance is the general name given to the other two parts of a circuit besides resistance which tend to prevent the flow of current. The two reactances are inductive reactance and capacitive reactance. Any piece of electrical equipment will have an impedance of so many ohms, a resistance of so many ohms, an inductive reactance of so many ohms, and a capacitive reactance of so many ohms. The inductive and capacitive reactances have opposite effects, so that they tend to cancel each other. If they are equal, the circuit has a net reactance of zero and is considered a purely resistive circuit. If they are not equal, the circuit's net reactance is either inductive or capacitive. A circuit that has a predominate inductive reactance has a lagging power factor, and the current lags the voltage. A circuit that has a predominate capacitive reactance has a leading power factor, and the current leads the voltage.

Rating. Rating is a term used to designate the limit of output power, current, or voltage at which a material, a machine, a device, or an apparatus is designed to operate under specific normal conditions. Electrical materials and equipment operated in excess of this rating will have a shortened life. Excessive heating or vibration is frequently an indication of operation beyond rated output.

Real power. Real power is expressed in watts (W) or kilowatts (kW) and is the product of kVA and pf.

Reactive power. Reactive power is expressed in vars or kilovars (kvar). Reactive power is also called unreal power.

Safety factor. The safety factor (or service factor) is the percentage above which a rated device cannot be handled without damage or shortened life. This factor is usually found above the nameplate rating. In some cases the safety factor implies that the device can be overloaded at some stipulated expense to life of the device. Present-day motors, for example, may have a safety factor of 1.15, which means that they may be operated at 15 per cent overload. However, the motor manufacturer may expect the user to recognize that at 15 per cent overload, the motor may operate $10°$ F hotter than the winding insulation is designed for. A general rule-of-thumb is that for every $10°$ F above rated temperature, the motor will have its life expectancy cut in half. The safety factor of equipment should be used only for infrequent and abnormal conditions; never should it be used during a continuous or normal operating condition.

Secondary. The term is commonly applied to the low-voltage winding of a transformer. It is also used frequently in reference to the voltage of an incoming power circuit. Power can be purchased at either primary of secondary voltage. See the discussion under primary.

The term secondary is also applied to the rotor windings and controls of a wound-rotor induction motor. A secondary controller of a wound-rotor motor is the equipment used for varying its speed.

Volt. This is the practical unit of voltage, potential, or electromotive force. One volt (V) will send a current of 1 ampere (A) through a resistance of 1 ohm. The term voltage is more often used in place of electromotive force, potential, potential difference, potential gradient, or voltage drop to designate the electric pressure that exists between two points, which is capable of producing a current flow when a closed circuit is connected between the two points. Usually on a three-phase system, the voltage is expressed as the voltage from line to line, whereas on a single-phase system, the voltage is expressed as the voltage from line to ground. "High Voltage" is a danger sign and should be respected.

Voltage is measured by a voltmeter.

Watt. A watt (W) is a practical unit of electric power. In a DC circuit, watts are equal to volts times amperes. In an AC circuit, watts are equal to volts times amperes times power factor. Power is the capability of the time-rate of doing work. Watts can be measured by a wattmeter.

Watt-hour. A watt-hour (Whr) is a practical unit of electric energy equal to the power of 1 W being used continuously for 1 hr. The rate of doing work determines the number of watts, whereas the amount of work done, regardless of how long it takes, determines the number of watt-hours. One thousand Whr are the equivalent of 1 kilowatt-hour (kWhr). The price of a kilowatt-hour may vary from approximately ½ cent to 4 or 5 cents, depending upon the size of the load and the conditions of the contract for purchase of power.

Each operator will find it most helpful to continue this list of terms by describing new terms in his own words as he encounters them in his work.

Review Questions

1. Current in a system starts at zero voltage, rises to a maximum, returns to zero, increases to a maximum in the opposite direction, and returns again to zero; it repeats this 60 times per second. What is the term applied to this current?

2. How much current would flow through a resistance of 1 ohm when the voltage across the resistance is 1 volt?

3. A battery can deliver 25 amperes for 6 hours; what is the battery's rating?

4. How is apparent power expressed in units?

5. What device is sometimes used to prevent a sudden rise in voltage?

6. Define *demand.*

7. How is electric energy expressed?

8. What are the two types of charges usually shown on power bills?

9. What is the frequency of the alternating current used by most US distribution systems?

10. An electrical connection to earth or to a large conductor known to be at the same potential as earth is called _____.

11. Define *impedance* and list the three parts that make up impedance.

12. The prefix *kilo* means?

13. What is measured in kilovars?

14. What is measured in ohms? Define *ohm* in terms of volts and amperes.

15. Are larger load applications usually single-phase or three-phase?

16. How many wires are required for a single-phase system? How many for a three-phase system?

17. An 18 kVA motor is running at a 0.8 power factor. How many kilowatts of real power is it drawing?

18. How is *power* defined? How is it measured in electrical systems?

19. The high-voltage side of a step-down transformer is called the _____. The low voltage side is called the _____.

20. Resistance causes power losses in the form of _____ produced by the current in the resistor.

21. Reactance is composed of two parts: _____ and _____.

22. Reactive power is another word for _____.

23. The safety factor of a motor allows for a slight overload; what will happen if the motor is operated continuously at this slight overload?

24. The electric potential needed to send a current of 1 ampere through a resistance of 1 ohm is _____.

Detailed Answers

1. 60-Hz AC; that is, 60 cycle-per-second alternating current.

2. 1 ampere.

3. 150 amp-hr, calculated as follows: 25 amp $\times$ 6 hr = 150 amp-hr.

4. Apparent power is the total power flowing in a circuit—it is made up of real power and unreal power, and is expressed in VA or kVA.

5. Capacitor.

6. Demand is the amount of electric power expressed in watts (W) or kilowatts (kW) that may be required during a certain time interval.

7. Electric energy is expressed in kilowatt-hours (kWhr).

8. Demand and energy charges.

9. 60 Hz.

10. Ground.

11. Impedance is the total opposition offered by a circuit to the passage of electrons. It is made up of resistance, inductance, and capacitance.

12. Kilo means one thousand.

13. Unreal power.

14. Ohms are a measure of the ability of a path to resist, hinder, or impede flow of electricity. An ohm is defined as the resistance through which 1 V will maintain a current of 1 A.

15. Three-phase.

16. Single-phase: two or three wires; three-phase: three or four wires.

17. 14.4 kW, calculated as follows:

$$kVA \times pf = kW$$
$$18 \times 0.8 = 14.4kW$$

18. Power is the ability to do a certain amount of work within a certain amount of time. Electric power is measured in watts (W) or kilowatts (kW).

19. Primary; secondary.

20. Heat.

21. Inductive reactance and capacitive reactance.

22. Unreal power.

23. The life of the motor will be considerably shortened. Equipment should be overloaded to the safety factor only for infrequent and abnormal conditions.

24. 1 volt.

Electricity 3

Functions and Ratings
of Electrical Equipment

E3-1. Functions of Electrical Equipment

Because there are so many kinds of electrical equipment and materials, it is helpful to think of equipment in broad categories or classes based upon function or purpose. For example, a reference made to "household appliances" immediately causes one to think of such items as radios, television sets, and toasters, but obviously wouldn't direct one's thoughts or attention to X-ray machines or substation equipment.

A complete electric power system may also be thought of in general categories such as the following:

1. Equipment to generate electricity
2. Equipment to store electricity
3. Equipment to change electricity from one form to another
4. Equipment to transport electricity from one location to another
5. Equipment to distribute electricity throughout a plant area
6. Equipment to measure electricity
7. Equipment to convert electricity into other forms of energy
8. Equipment to protect other electrical equipment
9. Equipment to operate and control other electrical equipment
10. Equipment to convert some condition or occurrence into an electric signal (sensors)
11. Equipment to convert some measured variable to a representative electrical signal (transmitter and transducers)

Any assembly of electrical devices that makes up a certain piece of electrical equipment will usually contain elements from several of these categories.

The discussion in this book covers the first eight categories listed above. A detailed discussion of certain items in the last three categories is contained in *Introduction to Water Distribution,* Pumps and Prime Movers Module. The equipment covered in this volume pertains to the power systems commonly used in water supply. There are many more specialized types of equipment used in other fields such as medicine, photography, the military, communications, and exploration.

The operation of a water-utility distribution system will generally be performed by an employee whose title is "operator." The operator of a water-distribution system will generally be responsible for one or perhaps several electrical systems. In effect, the operator operates an electrical system that in turn operates the distribution system. Some of the systems interposed between the operator and the actual service he is controlling are power systems, metering systems, control systems, and communications systems, all of which are in some respect specialized electrical systems. Remote control via a leased telephone circuit of a pumping station by means of supervisory control equipment and telemetering would be an example. So would the control of a pumping station from a control room using direct wire for control-and-metering circuits.

To be a pump station operator, one need not be an electrician. Neither must one be an automotive mechanic to operate a car or a truck. But some general knowledge about a vehicle is necessary to drive and maintain it properly.

This volume is not designed to make you an electrician. But it will introduce some general concepts and ideas about electricity and about the electrical equipment and materials that are found in water-distribution systems. Although the installation and repair of electrical equipment should be left to a qualified electrician, the operator should be able to maintain the equipment.

One should make note of the difference in the two words—repair and maintain. Repair means to put back in good condition after damage. Maintain means to keep in good condition. Although the operator is not required to repair electric failures, he is expected to perform necessary preventive maintenance.

Preventive maintenance may be defined as the periodic inspection of equipment as necessary to uncover conditions that may lead to breakdown or excessive wear, and the upkeep as necessary to prevent or correct such conditions.

Properly applied and correctly installed equipment is obviously one of the first requirements for a well maintained plant. No one can do a good maintenance job on equipment that is either not appropriate for its task or is installed haphazardly.

The second requirement for good maintenance is an operator who has a general knowledge of what the equipment is for and how it should work.

To operate and maintain an electrical system, one must become acquainted and be familiar with all of its parts. The only way to do this is to pay a great deal of attention to the equipment. Find out what the equipment is suposed to do; then become familiar with how it does it.

How does the equipment normally sound? You use this method to judge the condition of your car engine. The sound of certain electrical equipment under operating conditions is characteristic. There is usually a distinctive hum from coils and transformers. Motor starters and circuit breakers make quite a noise when operated, but this noise is usually the same each time. Motors and rotating equipment make an air noise during operation; this air noise is nearly always the same.

How does the equipment normally feel? Knowledge of the normal operating temperature of a piece of equipment is always helpful in judging its performance.

How does the equipment normally look? Generally, a visual inspection of equipment will show points of unusual wear, corrosion, or overheating. Is it clean and dry? Remember that a piece of equipment's two worst enemies are dirt and moisture.

Most of the maintenance of electrical equipment will be of a mechanical nature (cleaning, lubricating, tightening connections, and recognizing abnormal performance such as excessive vibration, noise, or heating).

Equipment to Generate Electricity

From a practical standpoint, any large block of electricity is generated as alternating current by a rotating-type piece of equipment called a generator. Most generators found in a water utility are AC generators. DC generators, however, are used frequently for battery chargers, synchronous motor field current, and special control systems. Even in DC generators, the initial form of the generated electric current is alternating, but this is changed to DC at the output of the machine by an arrangement of current collectors, or brushes, called the commutator. DC generators are frequently found on AC synchronous motors to supply the motor field with direct current; these DC generators are called exciters. Thus a synchronous motor with a direct-connected exciter is actually an AC motor with a direct-connected DC generator to supply the motor field current. The motor field current is adjusted to vary the power factor of the synchronous motor.

Power is generally purchased by water utilities, and any on-site generation is provided only for standby power. Generally, it is necessary to reduce the operating load and operate only essential equipment and essential station auxiliaries when using the standby power generating equipment. Generators used for standby power will normally be the gasoline, diesel, or gas-turbine types, because these can be started immediately. It would be unusual to find steam-driven equipment for standby service since the steam equipment would have to remain in continuous operation to be immediately available.

Engine-driven generators may be rated in any one of several ways—that is, for continuous duty, primary-power duty, continuous standby duty, intermittent duty, and standby duty. Those units rated for intermittent or standby duty should not be operated continuously at full-load nameplate rating. Following are the definitions of various ratings:

Continuous rating. The rating "continuous" as applied to an engine generator set indicates that the equipment can deliver its continuous kilowatt rating for the duration of any power outage and is capable of 10 per cent overload for a period of 2 hr out of each 24 hr of operation.

Standby rating. The rating "standby" as applied to an engine generator set indicates the 2-hr rating at which the unit can operate. For the balance of the time, it would have to meet the continuous rating previously defined.

Continuous standby rating. The rating "continuous standby" as applied to an engine generator set indicates that the equipment can deliver its continuous standby kilowatt for the duration of any power outage. No overload is permitted on this rating.

Primary power rating. The "primary power" rating is stated as some percentage (always less than 100 per cent) of the "continuous standby" rating.

As the reader can see, the practice of rating engine generator sets is somewhat confusing, if not actually misleading.

Generating equipment is generally classified by the source of power used to drive the generator's rotating machinery. This machine is usually called the prime mover. There are a number of different prime movers used in such generators.

Hydro-generators. Hydro-generators are generators driven by water power through the use of water wheels or hydraulic turbines. Some of the earliest uses of water wheels and hydraulic turbines were for the direct drive of pumps. In such cases, the speed of the drive was set by local conditions, so systems of belts and pulleys or gearing were connected to the driven equipment to obtain the required speeds. The conversion of water power to electric power is a good example of the conversion of raw-energy sources to electric energy. Hydro-generators may be of practically any size.

Steam-turbine generators. Both the steam engine and the steam turbine have been used for the conversion to electricity of raw energy found in fossil fuels such as coal, oil, gas, and other combustible substances. In such cases, the heat is used to make steam, which in turn is used to drive the engine or turbine. Steam turbines operate at much higher speeds than engines, so are used almost exclusively for driving generating equipment. The steam turbine now completely dominates the field of prime movers on large generators in both fossil-fueled and nuclear-fueled power plants. Nuclear power plants simply convert the nuclear energy to heat, which is in turn used to produce steam to drive steam turbine generators. Steam turbine generators are usually any size above 2500 kVA.

Engine-driven generators. Gasoline motors and diesel motors are both used to drive generators. Motor-driven generators are generally referred to as engine-driven units. The smaller engines are usually similar in nearly all respects to car and truck motors, whereas larger units are diesel motors of the large stationary type. The diesel engines can use natural gas or oil fuels. Engine-driven generators are usually any size below 2500 kVA.

Gas-turbine generators. Developed by the aircraft industry, gas turbines are becoming more popular as prime movers because of their light weight, compactness, high speed, and minimum auxiliary requirements. Gas-turbine generators are generally sized from approximately 250 to 2500 kVA.

Wind-driven generators. Wind motors can be used in isolated areas to generate small amounts of electricity. Usually batteries are an adjunct of wind generators so that some power is available during periods when there is a lack of wind. In such cases, the wind generator usually operates a battery charger, which keeps a storage battery charged.

Equipment to Store Electricity

The storage of electricity has been man's dream for years, but no practical means has yet been developed to store large amounts of electricity. The nearest approach to this end has been the installation of large pump-storage systems wherein water is pumped to a high-level reservoir, then allowed to drain out through a hydroelectric generator. In such cases, a pump and motor combination is designed to be run backward so that the pump acts as a hydraulic turbine and the pump motor acts as a generator. Obviously, this pump-storage concept is a means of storing electricity but is not direct storage of electricity.

All stored electricity is direct current, and only comparatively small amounts of electricity can be stored. Such amounts are so small, in fact, that the measure for stored electricity is in terms of ampere-hours. One thousand amp-hr of stored electricity may "sound" like a large amount but the quantity would only power an ordinary lighting circuit (approximately 1500 W) for approximately 75 hr or run a 5-hp motor approximately 1 day.

The two basic types of electrical equipment that store electricity are the battery and the capacitor.

Battery. An electric battery is a device for the direct transformation of chemical energy into electric energy. There are two types of batteries—the primary battery and the storage battery.

Primary battery. A primary battery is one in which the chemical action is irreversible. In a primary battery the parts that react chemically are destroyed. A flashlight battery is a primary battery. There are two types of primary batteries, the wet cell and the dry cell. The flashlight battery is a dry-cell battery and has a capacity of approximately 2-3 amp-hr. Wet-cell batteries have a liquid electrolyte and are therefore not practical for usual industrial use, but they are used frequently as standard cells in laboratory work and in various analytical measuring devices.

Storage battery. A storage battery is one in which the chemical reactions are almost completely reversible. That is, by returning current to the battery, a chemical action takes place to recondition the parts to their original state. This process is called "charging" the battery. Storage batteries will always be accompanied by battery-charging equipment. Storage batteries are made up of individual cells and can be connected in series and in parallel to develop any practical desired voltage and capacity. Usual voltage ratings vary from 6 to 250 V DC, and capacity ratings generally vary from approximately 100 to 1000 amp-hr.

It is interesting to note that the storage battery really doesn't store electricity. It converts electricity to chemical energy during charge and converts chemical energy to electricity during discharge. This is analogous to the pump storage system for storage of water power mentioned earlier.

It is common to find two types of battery-charging equipment used in conjunction with a large-station battery. Automatic chargers generally are applied to maintain the battery at normal charge; manual chargers are provided for a quick charge after an excessive drain, as might occur during an emergency.

Storage batteries should be well ventilated because some will give off hydrogen gas, which if allowed to accumulate can cause an explosion. Battery rooms obviously must be well ventilated.

Condenser. A capacitor, also referred to as a condenser, consists of a pair of metallic plates separated by a piece of insulating material called a dielectric. If a capacitor is connected across a voltage source (a potential) and is then disconnected, it will be charged and will hold this voltage for a considerable period. If its terminals are connected together, there will be an immediate discharge causing a current to flow through the connection. The actual amount of electricity stored in a condenser is very small compared with that from even a small storage battery. This amount of electricity, however, can be very dangerous.

The charging of a condenser or capacitor is the only true means of actually storing electricity. A condenser will allow alternating current to flow through it, but it will block the flow of direct current. Capacitors vary in size from extremely small, like those used in radio and TV sets, to large power capacitors used in substations as switched capacitors for power-factor correction. They are also used to suppress voltage surges that may be caused by lightning strokes or by switching large power circuits. In such applications, these condensers are called surge capacitors or surge suppressors. Capacitors are also used frequently in parallel with motors to correct, or raise, the power factor of an induction motor load.

Equipment to Change Electricity From One Kind to Another

There are many electrical devices that simply change electricity from one type or characteristic to another. This is done for various reasons to make it more suitable for some particular use or specific requirement. Some of the more common types of electrical equipment that perform this function are as follows:

Transformer. A transformer is a device for transferring energy in an AC system from one circuit, called the primary, to a second circuit, called the secondary. The primary winding and the secondary winding are essentially two completely independent circuits linked together with a common magnetic circuit. One of the most common uses of the transformer is to change voltage from one level to another. A transformer that raises the voltage in the secondary winding above the voltage in the primary winding is called a step-up transformer. Similarly, a transformer that lowers the voltage in the secondary winding is called a step-down transformer. Transformers can have more than two windings, and in such cases they are referred to as three-winding or four-winding transformers.

Transformers are probably one of the most essential devices for effecting the economical distribution of electric power. They allow for adjustment of the voltage and current levels to practical values. For example, a circuit of 120 VA of apparent power may have a voltage of 120 V with a current of 1 A; or it could have a voltage of 480 V with a current flow of ¼A. When one considers that transmission losses are proportional to the square of the current and that voltage drop in the transmission line is directly proportional to the current, one can see that reducing the current flow also reduces the line losses and the voltage drop. Not only does the lower current reduce losses and voltage drop, but it also allows the use of smaller wire size for the circuit, since wire size is selected according to the number of amperes to be carried.

Transformers are made in all sizes from small control power transformers that can be carried in one's hand, to medium sizes that are pole-mounted, to large power transformers like those seen in substations. Transformers can also be constructed as single-phase, three-phase, multi-phase, or in any combination. Transformers found at water utilities will generally be single-phase and three-phase.

Substation. The term substation is included here because it is used so frequently in place of or as having the same meaning as transformer. A substation more specifically describes any power-switching station that may be and generally is accompanied by power transformer installations. Substations may contain high-voltage-line terminal towers, high-voltage power circuit breakers, switches, fuses, power transformers, regulators, lightning arrestors, grounding reactors, couplings for carrier-type communications, batteries, and various other auxiliaries.

The expressions "package substation" and "unit substation" are frequently used to describe substations that are prefabricated to include all of the switchgear, transformer, and auxiliary devices built into one integral assembly.

Voltage regulator. A regulator can be similar in many respects to a transformer, with the exception that it is designed to maintain automatically a certain voltage on the secondary winding. A voltage regulator may be thought of as analogous to an automatic pressure-regulating valve on a water line. Power voltage regulators are not found frequently in water utilities, but they are used extensively by power utilities on long transmission or distribution circuits.

Small-size voltage regulators are used frequently in the water utility for instrumentation circuits, telemetering, supervisory control, data-handling equipment, and computers.

Converters. Several types of equipment are called converters. Generally, the term converter alone refers to an AC motor-driven, DC generator for converting alternating current to direct current.

A frequency converter can be an AC motor-driven AC generator in which the frequency of the alternating current used to power the motor is different from that which is generated by the generator. Such equipment, for example, might be served by a 60-Hz system to generate power at 400 Hz.

The term static converter refers to a non-rotating-type of converter. Static converters may convert AC to DC or they may be frequency converters.

Rectifier. A rectifier is a device that changes AC to DC.

Inverter. An inverter is a device that changes DC to AC.

Current transformer. A current transformer, commonly referred to as a CT, is a special transformer designed to have a 5-A secondary circuit when the primary winding carries full-load current. A 500-A to 5-A current transformer (500-A/5-A CT), for example, would have 5 A in the secondary when the primary current is 500 A, or 4 A in the secondary when the primary current is 400 A. The purpose of the CT is to allow a standard 5-A instrument to be used as, for example, an ammeter suitable for 0-5 A with a scale calibrated at 0-500 A.

The advantage of using a standard CT is that all of the instruments requiring current, such as ammeters, wattmeters, watt-hour meters, demand meters, power factor meters, var meters, or protective relays can be standardized for operation over a 0-5-A range. Another advantage is that the CT isolates the high-voltage primary circuit being measured from the secondary metering circuit so that all metering and instrumentation is at low voltage and low current.

Potential transformers. A potential transformer, commonly referred to as a PT, is a special transformer designed to have a 120-V maximum secondary voltage when the primary winding carries full voltage. A 2400/120-V PT, for example, would have 120 V on the secondary winding when the primary winding voltage is 2400 V, 100 V on the secondary when the primary voltage is 2000 V, or 60 V on the secondary when the primary voltage is 1200 V. The purpose of the PT is to allow a standard 120-V instrument to be used as, for example, a voltmeter suitable for 0-120 V with a scale calibrated at 0-2400 or 0-5000 V.

The advantage of using a standard PT is all instruments requiring voltage, such as voltmeters, wattmeters, watt-hour meters, demand meters, power-factor meters, var meters, or protective relays can be standardized for operation over a 0-120-V range. Another advantage is that the PT isolates the high-voltage primary circuit being measured from the secondary metering circuits so that all metering and instrumentation is at low voltage and low current.

Instrument transformers. Current transformers and potential transformers are referred to as instrument transformers. CTs and PTs are designed for instrument and metering use and can be obtained with various accuracy classifications.

Control power transformers. Control power transformers are designed to give a secondary voltage, usually of either 120 V or 240 V, when the primary circuit voltage is at full voltage. This allows the use of all control items such as pushbutton stations, selector switches, control relays, timers, and indicating lights to be rated at a standard low voltage. A 4160-V motor controller, for example, would have a 4160/240-V control power transformer; similarly, a 2400-V motor controller would have a 2400/240-V control power transformer. In both cases, the control items could be identical in style, type, and voltage rating. It is therefore not uncommon to find control power transformers designed for two secondary winding connections, 4160/240-120, for example, so that both 240-V and 120-V control items may be used in the control wiring as may be required.

Current regulator. A current regulator is a device that regulates current automatically to stay within certain limits. The current regulator is similar in many respects to the voltage regulator. Current regulators are referred to sometimes as constant-current power sources.

Equipment to Transport Electricity From One Place to Another

Electricity can be transmitted from one place to another by means of controlled or uncontrolled paths. An uncontrolled path would be hazardous since it could cause a flashover or an arc to ground. It is essential to design controlled paths for electric energy to flow from one point to another. The controlled path for electric current is always a conductor, which may be a wire, a bus, or a conducting material of any configuration that is sufficiently insulated to prevent the escape of electricity.

The controlled paths used for the movement of large blocks of power are frequently referred to as transmission systems and distributions systems. These terms are used almost synonymously but they do carry specific connotations.

Power utilities frequently refer to high-voltage lines, those 69,000 V and greater, as transmission systems and to lines 33,000 V and less as distribution systems. The term *transmission system* can also be interpreted to include components of the high-voltage system such as transformers, substations, regulators, switching stations, steel towers, or poles associated with that particular voltage class. Similarly, reference to the distribution system by a power utility will generally be interpreted to include all of the same components associated with that lower-voltage level.

In the water utility, these terms are used with a somewhat different meaning. A transmission system is generally thought of as that system which supplies power to the water-utility site, regardless of the voltage level. For example, an incoming power line with voltage as low as 2400 V is referred to as a transmission line. Generally, the expression transmission system as used by the water utility seldom relates to any on-site part of the water-utility system—but it may.

The term *distribution system* as used by the water utility is generally interpreted to mean the on-site electrical system of the water-utility facilities, regardless of whether it is overhead, underground, and regardless of the voltage class.

Most generally, from the water-utility standpoint, the term transmission system will refer to overhead power lines, whereas the term distribution system will refer to on-site underground or in-plant power cables.

The water-utility operator will be more concerned and more closely associated with distribution systems, which will be discussed under the heading "Equipment to Distribute Electricity Over a Plant Site."

Overhead transmission. From an electrical standpoint, the design of an overhead power line must take into account voltage selection, conductor size, line regulation, line losses, lightning protection, and grounding.

From a mechanical standpoint, the design must take into account conductor composition, conductor spacing, type of insulators, amount of sag, wind loading, ice loading, and selection of hardware.

From a structural standpoint, the design must take into account the size and type of structures, foundations, guys, and anchoring.

Other features of an overhead line that must be taken into account are line location, acquisition of right-of-way, road crossings, joint construction with other utilities, and access for line maintenance.

From an operator's standpoint, perhaps the most important aspect of overhead lines is service and maintenance. To maintain an overhead line, it is essential to have all hardware bolts tightened, all grounding connections tightened, cracked insulators replaced, poles inspected and treated, lightning arrestors in order, tree limbs trimmed away from the conductors, and guy wires tightened and protected. Any line work should be done by an electrician.

Weakened lines will generally fail during storms. Loose hardware and cracked insulators can cause radio and television interference. Damaged lightning arrestors can allow lightning to damage equipment served by the power lines. Poor ground connections or opened ground wires can be hazardous to anyone adjacent to poles and guy wires.

Underground transmission. The trend in some localities is to use underground construction for distribution circuits, particularly in residential areas. The use of high-voltage cable, however, for underground transmission of power, has been practiced for years with highly reliable results. An underground cable is less likely to be damaged than an overhead line that is exposed to trafficways or storms. However, it should be noted that should damage occur, an overhead line can be serviced and put back into operation much sooner than an underground power cable can be pulled and replaced. Frequently, it is possible to arrange for two transmission lines to serve a water utility, one overhead and the other underground; then the utility can incorporate the advantages of each.

Underground transmission construction is much more expensive than conventional overhead construction, but in some densely built sections of cities, it is virtually a necessity since there is no space for overhead lines.

Underground construction must take into account many of the same features as overhead construction, such as location and routing and acquisition of right-of-way, in addition to location of vaults and manholes, access for services, and flooding. Other features to take into account to design underground transmissions are voltage selection, conductor size, regulation, losses, lightning protection, and grounding—the same as for overhead construction.

Unlike overhead construction, there is nothing that an operator can do normally to service and maintain underground lines, with the exception of a periodic high-potential test for suspected deterioration of insulation. Some very high-voltage cables are gas- or oil-filled and require service and attendance from an operator; these types of underground cables, however, are not generally found in water utilities.

Equipment to Distribute Electricity Over a Plant Site

The electrical distribution system of any water utility begins with a source of electric energy which must be distributed to each and every electrical device on the site. This source may be either on-site generation or an interconnection with a power utility.

Sometimes this power source consists of only one incoming power feeder from the power utility; sometimes there are two incoming power feeders, a preferred incoming feeder and a standby. Or it may be that both incoming power feeders are normally in use with the load arranged to be served with half the load on each feeder, but with switching arrangements so that all of the plant load can be served from either feeder. In other cases, the source of power may be a combination of incoming power services from the power utility and on-site generation for emergency standby power. Obviously, there can be various arrangements for any number of service entrances and any number of on-site generators. The following discussion will review some of the types of distribution systems and their major components:

Power utility circuits. The power utility circuit can be called the incoming feeder, power service, service, service entrance, or power supply. In each case, its design will be adapted to meet certain requirements of the water utility. Each utility will have specific standards with which the service entrance must comply. These standards will vary widely, but in most cases will cover such items as voltage, physical arrangement, metering facilities, accessibility, grounding provisions, testing facilities, and type of circuit.

These circuits must, for billing purposes, include the metering equipment, which may consist of watt-hour meters for measuring energy used, demand meters for measuring the maximum amount of power used during any given time interval (usually 15- or 30-min periods), power-factor meters, kvar meters, and any associated current transformers and potential transformers. Such metering equipment is normally the property of the power utility but is generally housed in compartments furnished by the customer. These compartments, suitable for padlocking or sealing, are designed and located to meet the power utility's requirements. Frequently, these compartments are located within the customer's incoming service switchgear and must be compatible with the switchgear installation.

In addition to the metering facilities, an incoming feeder will have an automatic disconnect that will open in case of an excessive load as a result of a short circuit or fault in the customer's equipment. It is not uncommon for an incoming feeder circuit breaker to have associated relaying equipment that must be coordinated with the power utility's transmission-system protection. The power utilities aim to provide reliable service to all customers. To accomplish this, the power utility usually requires that whenever a fault occurs on the user's system, it will be disconnected automatically from the utility's transmission system. On small services, this automatic disconnect may be a fuse or a plastic-case circuit breaker that is usually a part of a distribution panel or switchboard. On large systems, this disconnect device consists of a power circuit with associated relays that are usually a part of the switchgear.

Relays are used frequently to open incoming service breakers under conditions such as loss of power, low voltage, or reverse phase. In such instances, the user is without power until the service can be automatically or manually restored. Frequently, causes of power outages are of very short duration and the problem is self-restoring. In such cases, it is to the advantage of the user to have additional relay equipment to cause the incoming feeder breaker to reclose automatically. Power utilities also use automatic reclosing equipment to improve the customer's service.

Continuity of power service is extremely important to water utilities in many instances and particularly in cases where pumping equipment is supplying a closed water-distribution system. In such cases, an "automatic throw-over" is employed, wherein two incoming services are used. In this instance, in order to ensure the greatest continuity of power service, it is absolutely necessary that the two incoming feeders be completely independent. Completely independent power feeders could, of course, be supplied from independent transmission systems. But since it usually isn't feasible to acquire services from completely independent transmission systems, generally the two services are obtained from separate power-utility substations. The whole purpose of the two services is to ensure continual power at the site; it is presumed that when power is not available from one source, it will be available from the other. Even the reliability of service provided by two incoming feeders is not considered satisfactory in many instances; it is then necessary to arrange for three or even more services.

The Kansas City, Mo., high-service pumping station, for example, has two 15-kV underground power feeders and two 69-kV overhead power feeders to ensure reliability of power. Naturally, each incoming service must be equipped with the necessary metering, relaying, and disconnect means, so it is not uncommon in such instances to find a good portion of the main switchgear made up of only those items concerning incoming power circuit facilities.

The arrangement of incoming services can take several forms. Generally, if two services are available, it is common practice to designate one as a preferred and the other as a standby incoming service. Sometimes these are referred to as the normal and alternate, or emergency, sources. For any particular site, the power utility may or may not allow the two incoming feeders to be parallel. A parallel connection means that both incoming feeders may be connected to the same load at the same time.

If the two incoming feeders may be connected in parallel, and if either incoming feeder can handle the full load, then the load may be transferred at will from one service to another. This type of switching is called a *closed transition.*

If the two incoming feeders may not be connected in parallel, and if either incoming feeder can handle the full load, then it is necessary to disconnect one service before switching on, or closing, the second service. This type of switching is called an *open transition.*

Generally, a closed transition is not permitted if the two incoming services are independent. Obviously, an open-transition switching arrangement results in a period during which the station load is completely without a power service connection to the power utility. If this period of transfer time is very long, it is obvious that all of the load will be turned off; the motors will stop and lights will go out. It is possible, however, to have this transfer occur automatically and so quickly that it will be hardly noticeable; only a flicker of the lights is seen, and the motors are kept running. This fast-switching operation is the function of the automatic throw-over equipment. This equipment and necessary accessories monitor the loss of power, the condition of the standby source, the preference of sources, the transfer time, and control power provisions for operation of the switches. Hopefully, an automatic throw-over scheme will function within a transfer time of only a few electric-energy cycles, a few one-sixtieths of second, so that for all practical purposes, the station continues to operate as if there had been no interruption of power supply.

But what happens when power is restored from the normal source? Sometimes it really doesn't make any difference to the power utility which service is supplying the load; in other cases, the power utility desires that service is taken normally from the preferred source and that the standby source is used only when necessary. There are several different types of automatic throw-over schemes—the non-preferential type, the fixed-preferential type, and the selective-preferential type—to accommodate this possibility.

The *non-preferential*-type automatic throw-over scheme allows the load to remain in operation on whichever power source it was last connected. In this case, either incoming power source can serve as normal or as standby, so they are usually designated at "alternate source #1" and "alternate source #2." The automatic throw-over equipment does not switch the load back to the original incoming source when that source has recovered. This non-preferential-type scheme is the most desirable since it reduces the switching operations by eliminating the need to switch back to a preferred source.

The *fixed-preferential*-type automatic throw-over scheme allows the load to remain on the standby supply only until power is restored on the preferred source; then it transfers the load back to the preferred source. After a power failure, when the condition of the preferred source has become normal, the automatic return transfer of the load to the preferred source may be immediate, or may be delayed until after some adjustable time delay, or it may be only partially automatic, being delayed indefinitely until released manually.

The *selective-preferential*-type scheme is similar in all respects to the fixed-preferential scheme, except that the choice of the feeder designated as preferred may be selected by means of a manually operated control-transfer switch. This scheme is usually used whenever the incoming services are essentially the same in all respects and the selection is simply a matter of flexibility or convenience that can be easily incorporated in the control scheme. Selective-preference controls are used when the utility wishes to use the two sources alternately for a reasonably long time on the selected preference to equalize the work load and maintain the electrical equipment in good condition.

In some cases, throw-over equipment is not designed to perform the return to normal switching automatically; return switching is done manually. The return-to-normal switching is never urgent; hence a prolonged time may be reasonable and allows the operator time to make sure that the disturbance which caused the normal source to fail has been cleared up.

Sources of power that can be parallel are referred to as synchronous sources. Local generation would never be considered a synchronous source with respect to a source from a power utility. Most utilities will require that local generating equipment be both mechanically and electrically interlocked so that the sources can never become parallel. Likewise, two incoming circuits from the power utility that are not synchronous sources must be interlocked so that they cannot be inadvertently connected in parallel.

It is desirable in some instances to use both of the incoming services continuously, and this can be arranged. To do so, the station load is designed to be split essentially in half, with each half of the station normally operating from its respective incoming feeder. A load tie breaker is then employed, so that when one of the incoming feeders is disconnected, its load is connected to the remaining incoming feeder, which then will serve the entire load until power is restored to the original feeder. One great advantage of this circuit arrangement is that only one half of the station load is interrupted by the failure of an incoming service. The switchgear in this case not only has the two incoming feeder circuit breakers, but also a third breaker usually called a tie breaker or bus tie breaker. Under normal conditions, the two incoming feeder breakers are closed and the bus tie breaker is open. During outage of either incoming feeder, the bus tie breaker and the incoming feeder breaker for the "live" circuit are closed. Under any condition, normal or emergency, this scheme allows only two of the three breakers to be closed. This is a very effective scheme to keep part of the station in service since the automatic throw-over facilities need to transfer only half instead of all of the load, thus causing considerably less shock to the transmission system. Usually the switchgear is designed with two main power buses, sometimes referred to as a sectionalized bus, and these busses are connected by the bus tie breaker, each bus being supplied normally by its respective incoming feeder. Return switching is supplied in one of the regular ways, such as immediately upon restoration of power, after a fixed or adjustable delay, or manually.

It must be clear that the electrical equipment required to meter, protect, and switch two incoming power circuits can require a considerable amount of space and be quite complex and expensive. It is important to note that although it is the power-utility circuits that are involved, the equipment discussed herein must usually be provided and maintained by the water utility—the user. The use of additional incoming power sources does increase the reliability of the station at the expense of more equipment and controls. Although no attempt is made at this time to discuss the requirement of control power for operation of these breakers, it must be understood that some form of reliable energy must be available for this service. Usually the control power is derived from some mechanical stored-energy mechanism or battery.

Particular care in design, installation, and maintenance must be put into these incoming power-circuit facilities because their emergency function will be called upon at an unexpected moment, usually after a very long period of idleness.

User's circuits. Beyond the power utility's metering equipment, the distribution system belongs to the user. Where the user wishes to duplicate the power utility's metering equipment, it must be done independently at the user's expense. Usually the water utility does not duplicate such metering facilities, but accepts their accuracy and reliability since they can be verified by calibration.

The plant electrical system can accommodate many arrangements for the distribution of electric power to the various pieces of equipment of the plant site. Immediately after passing through the metering facilities, electricity is connected to a main bus. If there are two main power supplies, there may be two main busses, and if the two main busses can be interconnected, the main bus is referred to as a sectionalized main bus.

The usual and most simple electrical distribution system consists of a main bus with various branch circuits routed from the main to various loads. This type of distribution system is called a simple radial system. The name refers to the point source from which paths radiate outward. The radial system can be used with either the single or sectionalized main bus.

Another type of distribution system consists of a main bus with branch circuits that are connected at the outer ends so that a loop is formed. This is called a loop system and allows a particular load to be served from either of the two branch circuits. Usually a loop system has a number of sectionalizing switches arranged so that the loop may be opened or so that particular loads may be bypassed or taken out of service.

Other types of distribution systems are combinations or radial feeders and loops arranged to form a network. In such systems there are various paths through which electric energy can flow to a particular load. Generally, networks are not found in water utilities.

The user's circuits can take many forms including any of the following:

Insulated single conductor cables in conduit
Insulated multiconductor cables in conduit
Insulated single conductor cables in trays or open raceways
Insulated multiconductor cables in trays or open raceways
Armored, insulated multiconductor cables either exposed or carried in trays
Underground cables of various types in duct banks
Direct buried underground cables of various types
Bundled single-conductor cables supported overhead
Preformed cables supported overhead
Open, uninsulated, bare-wire overhead construction.

User's equipment. In addition to the circuits involved, equipment will be provided for protecting and disconnecting each circuit. Since each path that electric energy must flow through to get from the main bus to any particular piece of equipment is a power circuit, there must be many protective and disconnecting devices on the distribution system. A fuse, for example, is a protective device, and a switch is a disconnect device. The fuse will cause the circuit to be disconnected if the current flow becomes greater than normal, whereas the switch is simply a manual means of opening the circuit.

It is natural to assume that in place of a switch and a fuse, a single device might be designed to accomplish both functions. The circuit breaker is such a device.

Circuit breakers have two current ratings—the normal current rating and an interrupting rating. The normal rating of a circuit breaker depends entirely upon the size of the normal load that is connected to the circuit. The interrupting rating, however, depends upon the size of the electrical system to which it is connected. In one case, for example, a circuit breaker may be rated at 15 A continuous at 240 V and require an interrupting rating of 10,000 A. Yet, for an identical load at another location, the breaker would be rated at 15 A continuous at 240 V, but would require an interrupting rating of 75,000 A. Obviously, these two circuit breakers would not be identical, the latter being larger and more expensive. Similarly, a different shutoff valve would be used on a 1000-psi constant-pressure system than on a 50-psi system.

The voltage of a distribution system is generally selected to allow the most economical size of circuit conductors. For example, a 37.5-kVA transformer at 208 V could sustain a continuous current flow of 104 A, and a 750-kVA transformer at 4160 V could also have a current flow of 104 A. In both cases, the size of the cable could be the same—approximately a No. 2 AWG (American Wire Gauge) conductor. A 750-kVA transformer at 208 V could sustain a continuous current flow of 2080 A, requiring several large cables. Similarly, 1000 gpm of water can be pumped through a 6-in. pipe under any configuration, but the power consumed would vary with the head in the system.

Distribution voltages in water utilities will generally be any of the following, depending upon the size of loads and amount of space covered by the plant site—15,000 V, 4160 V, 2400 V, 460 V, 240 V, 208 V, or 120 V.

The physical size of the equipment will also vary with the voltage—the higher the voltage, the larger the equipment. The combination of high-voltage circuits having high-current-interrupting requirements results in the necessary use of very large circuit breakers.

Generally, since a number of circuits are required, it is customary to purchase assemblies of circuit protective and disconnect equipment arranged for a number of circuits. Such assemblies are generally referred to as load centers, fuse panels, lighting panels, or distribution panels on systems of 600 V or less. Such assemblies for systems of more than 600 V are generally referred to as *switchboards* and *switchgear*. The term switchboard is seldom used anymore since most high-voltage equipment falls within present switchgear classifications.

It is important to recognize that one of the main purposes of load-center, distribution-type equipment is to disconnect automatically any faulted branch circuits as rapidly as possible in order to leave the rest of the branch circuits in operation. From the operator's standpoint, it is generally sufficient to have some means by which to monitor the condition of the branch circuit protective and disconnect devices. Usually on switchgear applications, red and green indicating lamps are used to give an indication of the last operating position assumed by the device.

For monitoring a distribution panel or a lighting panel, it is generally sufficient to glance at the position of the operating handles of the individual breakers. There is usually a clearly noticeable position assumed by the handle of a breaker that has tripped automatically. This position is easily distinguished from the handle position of a breaker that has been turned on or off manually. (See under "Distribution panel," which follows.) To monitor a fuse box or panel, it is necessary to actually examine each fuse to determine its condition.

The following general classes of equipment are, for the most part, necessary items for the development of a plant distribution system.

Fuse box. On small loads, such as single-family residences, for example, an incoming 120- or 240-V, single-phase power service is connected directly from the utility distribution system to a fuse box. This fuse box may have a main switch that, when opened, disconnects the power service and, when closed, connects the power service to the fuse-holder assembly. The fuse-holder assembly has several branch circuit fuses, each going to certain appliances or groups of appliances or to groups of lights. The fuse box, in its simplest form contains fuses sized to prevent each branch circuit from becoming overloaded and provides a means of disconnecting the branch circuit by removing the fuse. Other types of fuse boxes not only have the fuse for each branch circuit but also a knife-blade-type switch for use in connecting or disconnecting the branch circuit. When a fuse is blown as a result of an excessive load on its branch circuit, it is necessary to replace the fuse. The fuse box may be referred to as the fuse panel, or lighting panel, although it may serve loads other than lights.

Lighting panel. Lighting panels are similar in function to fuse boxes and fuse panels; however, the term lighting panel most generally implies the use of molded-case circuit breakers instead of fuses. The lighting panel for a single-family residence would connect the single-phase branch circuits to the main 120-V or 240-V incoming power service. Generally, there is a main breaker for disconnecting the branch circuit-breaker assembly from the power service, and the branch circuits will each serve an appliance, a group of appliances, or a group of lights.

There are mainly two advantages to the use of molded-case breakers as compared with fuses. Rather than one's having to replace a fuse, the circuit breaker can be reset. The other advantage of the breaker is that it is calibrated to trip at a certain overload that cannot be changed readily, as can a fuse, to a higher rating, thus relieving the danger from overloading branch-circuit wiring and devices.

Distribution panel. The distribution panel performs a function similar to that of the fuse panel or the lighting panel. The main difference is that a distribution panel is designed to handle larger branch-circuit loads than normally found on lighting panels, and the distribution panel can supply three-phase circuits. The distribution panel may consist of fuses or molded-case circuit breakers and may supply either three-phase or single-phase branch loads. Distribution panels are generally rated to handle 240-V, 480-V, and up to 600-V incoming service and branch circuits.

The molded-case circuit-breaker-type distribution panel has the advantages described for the molded-case circuit-breaker-type lighting panel. The molded-case circuit breakers have a handle that assumes three positions: one when the breaker is opened manually, another when the breaker is closed manually, and a third intermediate position when the breaker trips automatically. The current rating of the breaker normally is printed on the side of the operating handle.

Distribution panels are generally constructed in rectangular wall-mounted boxes with hinged front covers. The branch circuit breakers are generally arranged in two rows, with breakers stacked one above the other. Each breaker is generally provided with a nameplate or a number. Where breakers are numbered, a directory is usually mounted inside the front cover. The directory describes the branch circuit load for each breaker by reference to the breaker number.

Switchboard. The switchboard performs the same general functions as the distribution panel. The main difference is in the type of construction. Generally, the switchboard is a free-standing panel with switches, fuses, or circuit breakers mounted on the panel. Old-style switchboards frequently had the open-type knife switches and fuses mounted on the front of the panel. Newer-type switchboards generally are constructed with a "dead-front," using molded-case circuit breakers mounted on the back of the panel with only their operating handles extending through the panel to the front. Switchboards are generally employed only at less than 600 V. The term switchboard has been used very broadly and in some cases, such as in power plants, refers to a complete control and instrument panel. Generally this term does not imply such broad coverage when used with reference to water utilities.

Switchgear. Switchgear is a general term covering an assembly of large switching equipment. Generally, the term switchgear applies to switching equipment of higher voltage and current ratings than that used in the assembly of a distribution panel. Switchgear usually applies to equipment rated from 600 V through 15,000 V and for current ratings of several hundred to thousands of amperes. Power-circuit breakers are used for connecting and disconnecting the main branch circuits, with relays, instruments, and control devices used for sensing conditions and directing the operations of the breakers. Old-style switchgear in many cases consisted of lever-operated mechanisms arranged for manual operation. Modern switchgear is electrically controlled and arranged for manual operation by means of a push button or control switch.

Equipment to Measure Electricity

Devices for measuring electricity include a wide range of equipment and apparatus for the specific purpose of metering and control. The sizes of measurements can vary from the detection of a slight charge and millionths of an ampere to extremely high voltages and currents.

Voltmeter. A voltmeter is an instrument for measuring voltage. Its scale may be graduated in microvolts, millivolts, volts, or kilovolts. A microvolt is one one-millionth of a volt, a millivolt is one one-thousandth of a volt, and a kilovolt is 1000 V. On three-phase power systems voltmeters are usually used in combination with a voltmeter switch. The voltmeter switch switches the voltmeter to read the voltage between the conductors of phase 1 and phase 2, phase 2 and phase 3, or phase 3 and phase 1. Voltmeters generally are not constructed to measure voltages above 1000 V. When it is desired to measure voltages above 1000 V, it is necessary to use a voltage transformer, commonly called a potential transformer, or PT.

Ammeter. An ammeter is an instrument for measuring the amount of current. Its scale is generally graduated in microamperes, milliamperes, or amperes. A microampere is one one-millionth of an ampere. A milliampere is one-thousandth of an ampere. On a three-phase system an ammeter switch is frequently used with an ammeter. The ammeter switch switches the ammeter to read the current in phase 1, phase 2, or phase 3. Ammeters are generally constructed to measure current only on low-voltage circuits, not more than 600-1000 V. When it is desired to measure the current on high-voltage circuits, it is necessary to use current transformers, commonly called CTs.

Wattmeter. A wattmeter is an instrument for measuring the amount of real power in watts. Its scale is generally graduated in watts, kilowatts, or megawatts. A kilowatt is 1000 W. A megawatt is 1,000,000 W. Wattmeters are not designed to measure power on high-voltage circuits. They are designed to operate at a maximum voltage of 120 V and a maximum current of 5 A. On high-voltage circuits it is therefore necessary to use potential transformer and current transformers with 120-V and 5-A secondary windings, respectively. It should be remembered that how much work is done must take time into account. The wattmeter reading on a power service circuit at any given instant is the instantaneous demand.

Demand meter. A demand meter is an instrument that measures the average power (rate of doing work) of a load during some specific time interval, such as a 5-, 15-, 30-, or 60-min period. Generally, contracts with electric utilities for the purchase of electric power have demand charges based on 15- or 30-min time intervals. The demand is usually spoken of, for example, as a 15-min demand or a 30-min demand. Both the wattmeter and the demand meter are graduated to indicate watts or kilowatts.

An *indicating wattmeter* shows the immediate and instantaneous demand. A *recording wattmeter* records the instantaneous demand. The average of the instantaneous demands over a particular time interval would be equivalent to a demand-meter indication for that same time interval. The maximum demand established during a month is the maximum average load developed over the specified time interval, not the maximum instantaneous demand. Power bills usually are submitted each month showing, in addition to the charges, the kilowatt-hours of energy consumed and the maximum demand applicable for that month. One word of caution: the demand shown on a power bill for a particular month does not necessarily mean that that demand occurred that month because some contracts are based upon the premise that a new maximum demand one month will apply to each of the following eleven months.

Demand meters may be indicating-type with a pointer and scale arranged so that the pointer remains at its maximum reading until reset to zero when the meter is read. Other types of demand meters may be printing-type or recording-type. Generally, demand meters are furnished only on the incoming power service feeders to a plant site.

Watt-hour meters. A watt-hour meter is used for measuring watt-hours or kilowatt-hours of electric energy. A kilowatt-hour is 1000 watt-hours. Watt-hour meters are generally provided only on the incoming power feeder; however, it is not uncommon to find them used on large and more important plant motors and loads.

The kilowatt-hour meter generally has a register consisting of four dials, each dial graduated from 0 to 9. Examining the dials from right to left, the first or right-hand dial reads clockwise, the second reads counterclockwise, the third reads clockwise, and the fourth or left-hand dial reads counterclockwise. Adjacent pointers move in opposite directions. Dials are read from left to right, using the figure over which the pointer has passed last. The meter will have a dial register constant, ten, for example, which is multiplied by the meter reading to obtain kilowatt-hours.

On most kilowatt-hour meters a rotating disk can be observed through the glass front or through a small window in the front of the meter case. Revolutions of the disk can be counted by means of a black mark on the disk. The speed of the disk is at any instant proportional to the kilowatt load; it provides a quick and accurate means of measuring kilowatts over short periods of time. The meter will also have a disk constant. This constant expresses the watt-hours (or watt-seconds) per revolution of the meter disk. To measure kilowatt load, determine with a stopwatch the number of seconds required for the disk to make five or ten revolutions, depending upon the disk speed. Then, knowing the disk constant, use the following formula to calculate the kilowatt load:

$$\text{kilowatts} = \frac{\text{disk watt-hours constant} \times \text{revolutions} \times 3600}{\text{seconds} \times 1000}$$

When the disc constant is in watt-seconds, it must be converted to watt-hours by dividing it by 3600 for use in the above formula.

A great amount of information can be obtained about a plant load by observing the kilowatt-hour meter. If all the plant motors and lighting circuits are turned off, any rotation of the disk would most likely represent the power required to magnetize the lighting and auxiliary-power transformers. The kilowatt load determined under this condition would be the magnetizing power requirement of the system. Then by one's turning on all of the plant lights and determining the kilowatt load again, the amount of lighting load could be determined by subtracting the magnetizing power load from this latter kilowatt load.

Similarly, by adding various increments of the plant load in succeeding steps, the kilowatt load imposed by the various plant auxiliaries could be determined.

Varmeter. A varmeter is an instrument for measuring the amount of unreal or wattless power in vars. Its scale is generally graduated in vars or kilovars. A kilovar is 1000 vars. Varmeters are not designed to measure power on high-voltage circuits. Varmeters, like wattmeters, are designed to operate on potential and current transformers at maximum voltages of 120 V and a maximum current of 5 A. Generally, varmeters are not found in water utilities unless there are some large synchronous motors at the site.

Power-factor meter. Power-factor is the ratio of the amount of real power to the total apparent power. Apparent power includes both real and unreal power. A power-factor meter is an instrument that measures this ratio directly. The power-factor meter is generally graduated with a unity center scale, from lagging 0.8, to 1.0, to 0.8 leading power factor.

Generally, a water treatment plant with the usual complement of induction motors will have a lagging power factor of approximately 0.85. There is not much that a water-utility plant operator can do to change the power factor or the var flow at a station. These values are quite well determined by the design of the station.

Frequency meter. A frequency meter is an instrument for measuring the frequency of an AC system. Frequency meters require potential transformers, PTs, for connection to circuits operating above 1000 V. The frequency of a system depends upon the speed of the generators. If the plant operates on an engine generator set, the frequency will be adjusted by adjusting the governor on the engine-generator set.

Equipment to Convert Electricity Into Other Forms of Energy

Electric energy can be converted into many more purposeful types of energy to provide particular services.

Mechanical energy. The rotating shaft, as developed by various types of motors, is perhaps one of the most common forms of electric-to-mechanical energy conversion. Another widely used form of energy is the linear thrust, the pull or push as developed by a solenoid thruster, commonly used for operating a solenoid valve or a doorbell.

Heat energy. Electric heaters, together with their thermostatic controls, are employed for practically any heating requirement. Electric heaters are in demand because of their cleanliness, convenience, flexibility, accuracy, safety, and in many cases their economy. Other special equipment such as welders and industrial equipment makes use of the direct conversion of electric energy to heat.

Light energy. Perhaps no other form of conversion is as well known as the conversion to light. The introduction of electric lighting generally preceded all other appliances and equipment.

Chemical energy. Electrical equipment is used for corrosion-protection equipment. Such equipment, referred to as cathodic-protection equipment, influences the degree and rate of the chemical reactions that take place in the oxidation, rusting, or decomposition that occurs on and within metal structures.

Radio energy. Electromagnetic waves derived from electrical equipment provide the carrier, commonly referred to as the radio, over which the well-known forms of communication presently take place.

Equipment to Protect Other Electrical Equipment

Certain types of electrical equipment are designed and used only for protecting other types of electrical equipment. Even though this type equipment is actually non-productive, the operator should know what it is for, and it should be serviced and maintained for proper operation.

Fuse. A fuse is an overcurrent protective device containing a special metal link that melts when the current through it exceeds some rated value for a definite period of time. Fuse links are made of a special alloy that melts at a relatively low temperature. A fuse is inserted in series with the circuit so that it opens the circuit automatically during an overload and thereby prevents excessive current from damaging other parts of the circuit. Fuses are rated according to voltage, continuous current, interrupting capacity, and speed of response.

The *voltage rating* of a fuse should be equal to or greater than the voltage of the circuit on which it is applied. This voltage is not a measure of its ability to withstand the voltage while carrying current. Rather, it is the ability of the fuse to prevent the open-circuit voltage from restriking and establishing an arc once the fuse link has melted.

The *current rating* of the fuse should be equal to or slightly larger than the current rating of the circuit or device that it protects. The interrupting capacity of a fuse must be such that it can interrupt the inrush of current available during a fault as determined not by the load but by the nature of the source of power being supplied through the fuse.

The *speed of response* simply defines how fast the fuse link will melt. Obviously, the faster the fuse link will melt the lower will be the limit that the fault current can reach before the circuit is opened. Fuses expressly designed for exceedingly fast response are referred to as current-limiting fuses.

Molded-case circuit breaker. A circuit breaker is an electro-mechanical overcurrent device that opens a circuit automatically when the current rises in excess of a predetermined value. It can be reset by operating a lever to its original position and used over and over again. There are thermal types, magnetic types, and thermal-magnetic types of circuit breakers.

The *thermal circuit breaker* responds only to temperature change in a bimetallic element. This element is made of two strips of different metals, bonded together. The lengths of the strips increase with temperature—but not equally—so that the composite element bends more and more as its temperature rises. Current flows through the bimetallic element and generates heat. The greater the current, the higher the temperature of the element. The mechanism is adjusted so that the bimetallic element bends just enough to open its contacts at a predetermined value of current. Because this type breaker must respond to heating, its speed of response may be quite slow.

The *magnetic circuit breaker* responds only to current, and then only if the current is sufficiently great to attract a movable pole-piece by magnetic force. Movement of the pole piece causes a contact to open. In the fully magnetic breaker, there is only one actuating element, the magnetic coil assembly. Its operation is independent of temperature. Because this type of breaker responds only to the magnitude of current, its speed of response may be very fast. The thermal-magnetic breaker operates in exactly the same way as the thermal breaker at low values of current, and in exactly the same way as the magnetic circuit breaker at high values of current. Thus it essentially combines the good features of both. Slight overloads that persist for a long time will be disconnected by the thermal element, whereas short circuits will be disconnected immediately by the magnetic element.

Power-circuit breaker. A power-circuit breaker differs greatly from the molded-case circuit breaker in that it is much larger and does not have the means for determining when it should operate or what it should do to carry out its protective assignment. Power-circuit breakers generally depend upon relays to sense abnormalities and to initiate their operation. The advantage of the power-circuit breaker is its ability to interrupt exceedingly high values of fault current on both low- and high-voltage power circuits. Power-circuit breakers are electromechanical devices used for switching (opening and closing) power circuits under either normal or fault conditions. Power-circuit breakers may be either oil-filled or air-insulated mechanisms. Generally, a row of power-circuit breakers enclosed in a metal housing is referred to as metal-clad switchgear.

Protective relays. A relay is a device activated by either an electrical or physical condition to cause the operation of some other device in an electrical circuit. Although there are all sorts of relays, the protective relay's principal function is to protect power service from interruption and to prevent or limit damage to equipment.

Lightning arrestors. Lightning can produce the most destructive of all types of overvoltages. Overhead power lines, exposed to lightning, can be elevated to a potential of several million volts by direct stroke and possibly one-half million volts by induction from a near strike in a time interval of only a few millionths of a second. This impressed voltage, called a voltage surge, tends to travel over the complete circuit in the form of a steep-front voltage wave. Since power lines and equipment cannot, from a practical standpoint, be insulated to withstand such overvoltages, it is obvious that a breakdown of insulation will occur between the phase conductors and ground. The lightning arrester is designed to break down, thus allowing the other circuit elements to remain in service.

A lightning arrester is a protective device for limiting surge voltages on equipment by discharging a surge current to ground; however, it is designed to prevent a continued flow of follow current after the voltage has returned to normal, and it is capable of restoring itself and repeating the same protective function. The types and ratings of arresters are selected for the particular voltage class and degree of protection desired. Arresters may be classified as follows:

- Station class—for 60,000-276,000-V systems, pad-mounted
- Intermediate class —for 3000-121,000-V systems, pad-mounted
- Distribution class —for 3000-18,000-V systems, pole-mounted
- Pellet type —for 1000-15,000-V systems, pole-mounted
- AC rotating machine—capacitor and arrestor combinations for 2400-13,000-V equipment
- Secondary type —for ratings from 0 to 600 V.

E3-2. Rating Electrical Materials and Equipment

To ensure safe and efficient operation of an electrical system, materials and equipment must be chosen with the proper rating. This is usually the function of the designing engineer, but the operator should be familiar with the factors taken into consideration.

Materials

All materials fall within two general classifications regarding their properties and abilities to allow electric current to flow through them. They are either conductors or insulators. A conductor is a substance which permits the flow of electricity, especially one which conducts electricity with ease. An insulator is a substance which offers very great resistance or hindrance to the flow of electric current.

The path through which current flows has to be a carefully designed circuit from start to finish. This path is made up of a combination of two classes of materials—conductors and insulators.

The windings in a generator are made of copper conductors, but they are wrapped in insulating material such as dry paper, mica, glass, or rubber so that electricity can only flow in a controlled path. Commercial power is usually generated at 15,000 V and is then transformed to still higher voltage levels for transmission and distribution. The transmission lines usually consist of overhead pole construction using bare copper or aluminum cable for conductance and glass or porcelain and air for insulation. The separation of the conductors is great enough that a flash-over between the lines in open air will not occur. At a user's substation, it may be found that the overhead bare wires connect directly to the insulated bushings of a power transformer. The output of the transformer will probably be at lower voltage and connected directly to power cables. Power cables are made up of conductors covered by insulating material such as rubber, varnished cambric, or paper with a lead cover. At each and every point in the electrical system, conductors are provided to keep the flow of current on its intended path, and insulating material is provided everywhere along this path to contain the current.

A good conductor has low resistance and is said to have high conductance. Conductance is just the inverse of resistance and is measured in mhos, which is ohm(s) spelled backward.

A good insulator has high resistance and is measured in ohms. Insulation is also called dielectric. An insulated cable with high dielectric strength is another way of saying the cable is very well insulated.

Following is a list of materials categorized by their ability to conduct electricity.

Good Conductors	Fair Conductors	Poor Conductors	Good Insulators
Silver	Charcoal	Water	Slate
Copper	Carbon	The body	Oils
Aluminum	Acid solutions	Flame	Porcelain
Zinc	Sea water	Linen	Dry paper
Brass	Saline solutions	Cotton	Rubber
Iron	Metallic ores	Wood	Mica
Tin	Vegetation	Fibers	Glass
Lead	Moist earth	Marble	Dry air

Usually, any fault found on an electrical system or a piece of electrical equipment will be either a broken conductor or a breakdown of the insulation.

Conductors usually break because of mechanical failure as a result of excessive vibration or movement.

Insulation usually fails because of deterioration as a result of excessive heating and aging. Overheating is probably the most common cause of insulation breakdown.

Insulation also fails whenever it is subjected to abnormally high voltage as will occur when it is struck by lightning or subjected to switching surges. Whenever a power switch is opened or closed, surges in system voltage will occur, just as water-hammer occurs when a valve is operated quickly.

Equipment

Electrical systems and equipment can be designed for any voltage, any frequency, and any size, but most are standardized. Imagine the deplorable condition that would exist if each of the states within the United States had its own standard voltages and frequencies and its own unique equipment sizes. For example, your 120-V, 60-Hz power tools would be useful only in your state, whereas in a neighboring state the power tools might be rated 300-V, 40-Hz, or 80-V, 100-Hz. Moreover, motors might be manufactured in 5-, 7½-, 10-, 15-, and 20-hp sizes by one company, and 4-, 8-, 12-, 18-, and 24-hp by another company. There was a period when such lack of standardization did exist; during that time equipment was designed at various ratings. There are still systems in operation which use 25 Hz instead of 60 Hz, and other-than-standard voltage classes of equipment are used in some cases.

Standardization of voltage ratings, system frequency, and sizes of equipment has permitted the mass production of equipment at competitive prices, so necessary for the rapid industrial development and expansion that has occurred within the past 50 years. This is not to say that the evolved standards were necessarily the best but that they have become recognized standards by general usage. A knowledge of some of the standard voltage classifications, equipment ratings, and sizes makes it easier to become familiar with a particular system or specific equipment.

Voltage standards. There are two terms that are generally used when expressing voltage—rated voltage and nominal voltage. Rated voltage is used with reference to the operating characteristics of equipment. Nominal voltage is used in referring to an electrical system that distributes electricity for ultimate use by other equipment. Rated voltage is not applicable here since various pieces of equipment that make up a given system may have different voltage ratings.

For example, a standard system is one having a nominal voltage of 480 V, but the motor used on that system will have a standard rated voltage of 460 V, the motor controller or motor starter may be rated 460 V, and a disconnect switch used on that motor circuit may be rated 600 V.

The following table shows the standard nominal system voltages and the generally accepted ranges considered as favorable and tolerable within which power may be delivered at each of these nominal voltage levels.

Table 1. Standard System Voltages

Minimum Tolerable	Minimum Favorable	Nominal System	Maximum Favorable	Maximum Tolerable	Type (Phase) of System
107	110	120	125	127	10
200	210	240	240	250	30
214/428	220/440	240/480	250/500	254/508	10
244/422	250/434	265/460	227/480	288/500	30
400	420	480	480	500	30
2 100	2 200	2 400	2 450	2 540	30
3 630	3 810	4 160	4 240	4 400	30
6 040	6 320	6 900	7 050	7 300	30
12 100	12 600	13 200	13 800	14 300	30
12 600	13 000	14 400	14 500	15 000	30
30 000		34 500		38 000	30
60 000		69 000		72 500	30
100 000		115 000		121 000	30
120 000		138 000		145 000	30
140 000		161 000		169 000	30

Obviously, one may be quite confused when he hears in the discussion of a particular electrical system the reference to so many voltages, as for example the reference to a 236-V fluorescent lamp ballast, a 220-V motor, a 230-V capacitor, and 240-V switchgear on a specific 240-V system.

Table 2 lists only some of the standard nominal voltages and rated voltages for AC equipment, and clearly shows why so many expressions of voltage level may be used.

Reference to any one of the above voltage levels for a particular type of equipment naturally implies the use of the other related voltages for the various associated items of equipment used on that same electrical system.

Current standards. Because the amperes of electric current vary with the size of electric load and the voltage, there is no standard classification of system nominal currents as there is of system nominal voltage levels. Only the rated current of the equipment is standardized.

There is, however, for each type of equipment at rated voltage level, a set of standard current ratings for equipment available. Examples of some standard

Table 2.   Standard Nominal Voltages

Nom-inal System	Gener-ator Rated	Trans-former Secon-dary	Switch-gear Rated	Capac-itor Rated	Motor Rated	Starter Rated	Ballast Rated
Single-Phase Systems							
120	120	120	120	—	115	115	118
120/240	120/240	120/240	240	230	230	230	236
208/120	208/120	208/240	240	230	115	115	118
Three-Phase Systems							
240	240	240	240	230	240*	220	236
480/277	480/277	480/277	480	460	460*	440	460
480	480/277	480/277	480	480	460*	440	460
2 400	2 400/ 1 388	2 400	2 400	2 400	2 300	2 300	—
4 160	4 160/ 2 400	4 160/ 2 400	4 160	4 160	4 000	4 000	—
6 900	6 900/ 3 980	6 900/ 3 980	7 200	6 640	6 600	6 600	—
7 200	6 900/ 3 980	7 200/ 4 160	13 800	7 200	7 200	7 200	—
12 000	12 500/ 7 210	12 000/ 6 920	13 800	12 470	11 000	11 000	—
13 200	13 800/ 7 970	13 800/ 7 610	13 800	13 200	13 200	13 200	—
14 400	14 000/ 8 320	13 800/ 7 970	14 400	14 400	13 200	13 200	—

current ratings of equipment are shown below; this does not mean, however, that other ratings are not available but only that they will probably be more expensive than the standards.

The following descriptions are of several types of apparatus that are used to protect systems from electric-current overload.

Safety switches. A safety switch is a disconnect switch. It will have a voltage rating, which means it can be used on systems up to and including its rated voltage. Its current rating means it will carry up to its rated current, regardless of whether or not the voltage of the system is at rated voltage or some lower voltage. From a practical standpoint, the electrical industry has standardized the following eight current ratings for low-voltage switches.

Table 3.   Current Ratings for Low-Voltage Switches, in Amperes

120/240 V	230 V	240 V	600 V
30	30	30	30
60	60	60	60
100	100	100	100
200	200	200	200
	400	400	400
	600	600	600
	800	800	800
	1200	1200	1200

Fuses. Fused disconnect switches are essentially safety switches having a fuse holder and fuse built in series with the switch. Standard, fused, disconnect-switch current ratings are therefore the same as the safety-switch ratings listed previously. A fuse holder can accommodate any size fuse below the fuse-holder rating. Standard current ratings of low-voltage fuses are 15, 20, 25, 30, 35, 40, 45, 50, 60, 70, 80, 90, 100, 110, 125, 150, 200, 300, 400, 500, and 600 A. Where fuses larger than 600 A are required, two or more of these fuses can be used in parallel, or a high-capacity fuse, in ratings from 650 to 5000 A, may be more appropriate. The standard fuses rated 0-600 A have an interrupting rating of only 13,000 A, which is barely equal to the short circuit or fault current that can be let through a 250-kVA, 220-V transformer. Obviously, whenever selecting a fuse, not only the load current rating, but also the interrupting current rating of the fuse must be taken into account. The high-capacity-type fuses have current interrupting ratings in the range of 100,000-200,000 A.

Circuit breakers. These can do two things—open a circuit automatically when an overload occurs and open a circuit when operated manually. Circuit breakers can be arranged to be operated intentionally, either manually or electrically. The size of a circuit breaker depends upon the voltage of the circuit and upon the amount of current it must handle under both normal and abnormal conditions. Under normal conditions, the circuit breaker must carry the rated current continuously without overheating. Under abnormal conditions, that is, when the breaker carries the current during a fault or short circuit, the circuit breaker must be able to withstand that amount of current until it has successfully interrupted that current flow.

Molded-case circuit breakers. The molded-case circuit breaker is used on low-voltage sysems, 600 V and less, in locations that would otherwise be held by a safety switch, disconnect switch, fused disconnect switch, or fuse. Single-pole circuit breakers are commonly used in lighting panels. Three-pole circuit breakers are commonly used in distribution panels. Two-pole circuit breakers may be used in either lighting panels or distribution panels. Circuit breakers have a voltage rating, a frame-size current rating, a continuous current rating, and a fault current ampere interrupting rating.

Functionally, circuit breakers serve as disconnecting means or as protective devices. Compared with switches, they have a much higher interrupting capacity. This means that they are capable of interrupting the flow of considerably higher current than could be interrupted by a switch. The size of a circuit breaker is based not only on the maximum continuous current it must carry but also on its interrupting ability. The interrupting capacity of the breaker must be at least as great as the largest fault current that the power system can cause to flow.

Molded-case circuit breakers are available in certain frame sizes that basically describe the maximum continuous current rating of that breaker for that frame size. A breaker of a certain frame size may be calibrated to carry a continuous current that is only a fraction of the maximum continuous current that the breaker can carry. These continuous-current calibrations are referred to as the rated-current setting of the breaker. A 100-A frame-size breaker may be calibrated to a rating of 10 A. This means that the breaker will carry up to 10 A continuously. Above 10 A the breaker would eventually open; how soon would depend upon how much the current exceeds the rated current. Currents up to approximately ten times the rated current will cause the breaker's thermal element to open the breaker. Currents exceeding ten times the rated current will cause the breaker's magnetic element to open the breaker.

Typical of some of the more common ratings of molded-case circuit breakers are the following.

Table 4. Molded-Case Circuit-Breaker Design

Frame Size A	Interrupting Capacity A	Continuous Current Rating A
(240 V AC)		
400	50 000	70, 90, 100, 125, 150, 175, 200, 225, 250, 300, 350, 400
800	50 000	125, 150, 175, 200, 225, 250, 300, 350, 400, 500, 600, 700, 800
1000	75 000	125, 150, 175, 200, 225, 250, 300, 350, 400, 500, 600, 700, 800, 900, 1000
(480 V AC)		
100	15 000	5, 8, 10, 12, 15, 20, 25, 30, 35, 40, 50, 70, 90, 100
225	25 000	70, 100, 125, 150, 175, 200, 225
400	35 000	70, 90, 100, 125, 150, 175, 200, 225, 250, 300, 350, 400
800	35 000	125, 150, 175, 200, 225, 250, 300, 350, 400, 500, 600, 700, 800
1000	40 000	125, 150, 175, 200, 225, 250, 300, 350, 400, 500, 600, 700, 800, 900, 1000

Low-voltage air circuit breakers. When breakers with continuous- or interrupting-current ratings greater than those available for molded-case circuit breakers are required, low-voltage air circuit breakers are used. These breakers are mechanical devices, completely metal-enclosed, and frequently are of drawout-type construction for convenience of service and maintenance. An assembly of low-voltage circuit breakers is generally called a line-up of low-voltage switchgear. Low-voltage air circuit breakers have voltage ratings, frame-size current ratings, continuous-current ratings, and fault-current interrupting ratings similar to those of the molded-case circuit breakers.

Typical of some of the more common ratings of low-voltage air circuit breakers are the following.

Table 5. Air Circuit-Breaker Design

Frame Size A	Interrrupting Capacity A	Continuous Current Rating
(240 V AC)		
225	30 000	15, 29, 30, 40, 50, 70, 90, 100, 125, 150, 175, 200, 225
600	50 000	40, 50, 70, 90, 100, 125, 150, 175, 200, 225, 250, 300, 350, 400, 500, 600
1600	75 000	200, 225, 250, 300, 350, 400, 500, 600, 800, 1000, 1200, 1600
(480 V AC)		
225	25 000	15, 20, 30, 40, 50, 70, 90, 100, 125, 150, 175, 200, 225
600	35 000	40, 50, 70, 90, 100, 125, 150, 175, 200, 225, 250, 300, 350, 400, 500, 600
1600	60 000	200, 225, 250, 300, 350, 400, 500, 600, 800, 1000, 1200, 1600

High-voltage circuit breakers. High-voltage circuit breakers are used on electrical systems operating above 600 V. These breakers may be oil-filled or air circuit breakers. An assembly of high-voltage circuit breakers is generally referred to as a line-up of high-voltage switchgear. Modern high-voltage switchgear for use up to 15 000 V is generally composed of air circuit breakers of the draw out type, completely metal enclosed. It is generally an accepted practice to refer to the interrupting capacity of high-voltage switchgear in terms of the apparent power (MVA) that can be interrupted during a fault instead of the fault current in amperes. The abbreviation MVA stands for million volt-amperes. For example, a fault current of 1000 A on a 2400-V, single-phase system would represent 2,400,000 VA or 2.4 MVA. High-voltage switchgear circuit breakers have a voltage rating, a frame-size current rating, a fault-current interrupting rating, and a fault MVA rating. Typical ratings follow:

Table 6. High-Voltage Switchgear Design

System Nominal Voltage	Frame Size A	Momentary Interrupting Capacity A	Interrupting Capacity MVA
2 400	1 200	20 000	30
4 160	2 000	40 000	150
4 160	2 000	60 000	250
4 160	1 200	80 000	350
4 160	2 000	80 000	350
7 200	1 200	70 000	500
7 200	2 000	70 000	500
13 800	1 200	20 000	150
13 800	2 000	35 000	250
13 800	1 200	40 000	500
13 800	2 000	40 000	500
13 800	1 200	60 000	750
13 800	2 000	60 000	750
13 800	1 200	80 000	1000

Other breakers at higher voltages are available through all voltage classes; however, water utility operators will seldom be required to operate breakers above the 15-kV (15,000-V) class of equipment.

It should be noted that continuous (tripping) current ratings are not listed for the high-voltage breakers. These breakers, in contrast to the molded-case breakers and low-voltage air current breakers, do not have any built-in means of sensing current magnitudes and therefore cannot be individually set at specific current ratings. Auxiliary current measuring transformers (current transformers) and relays can, however, be set to cause the breakers to operate at any desired magnitude of current.

Review Questions

1. Define *preventative maintenance.*

2. Large amounts of alternating current are usually produced by _____.

3. Generally, on-site generation of electric power at water utilities is used only for _____.

4. List at least three types of prime movers used with generators.

5. What are the two basic electrical devices used to store electricity?

6. In a storage battery, the chemical reactions that produce electricity are almost completely reversible. What practical meaning does this have?

7. What is an important safety consideration when charging batteries?

8. What is one of the most common uses of a transformer?

9. Define (a) *rectifier* and (b) *inverter.*

10. What are two types of instrument transformers?

11. Identify factors involved in the maintenance of overhead transmission lines. Who should be authorized to perform line work?

12. What are two types of equipment found in every incoming feeder installation?

13. Wherever more than one incoming feeder is used, equipment is needed to automatically switch from one feeder to another. This switching equipment is referred to as _____ equipment.

14. What is a radial distribution system?

15. What are two devices used to protect circuits?

16. Name the instruments used to measure the following:

   (a) Electrical potential
   (b) Current
   (c) Real power
   (d) Unreal power

(e) Average power of a load during some specific time interval

(f) Electric energy

(g) Ratio of real power to apparent power

(h) Frequency

17. Electric energy is usually converted to another form of energy for use in industry or home. List at least three types of energy that it may be converted to, and name the device involved in the conversion.

18. What are the four ratings that describe the applicability of a fuse?

19. In general, is a molded-case circuit breaker or a power-circuit breaker larger?

20. What is the term for a substance that allows the passage of electricity with great ease? What is the term for a substance that offers very great resistance to the flow of electricity?

21. In what units is resistance measured? In what units is conductance measured?

22. Identify two causes of insulation failure.

23. Can a fuse holder be used with a fuse of a higher or a lower rating than the rating of the holder?

24. Identify two functions of a circuit breaker.

25. What is the maximum voltage rating in which molded-case type circuit breakers are available?

26. What measure is used for the interrupting capacity of high-voltage switchgear?

Detailed Answers

1. Preventative maintenance is the periodic inspection of equipment as necessary to uncover conditions that may lead to breakdown or excessive wear, and the upkeep as necessary to prevent or correct such conditions.

2. A generator.

3. Standby power.

4. Water wheels or hydraulic turbines, steam turbines or steam engines, gasoline and diesel motors, gas turbines, wind motors.

5. Battery and capacitor.

6. A storage battery can be recharged and reused many times.

7. Batteries must be well ventilated during charging because some give off hydrogen gas, which can accumulate and cause an explosion.

8. A transformer is commonly used to change one voltage to another.

9. (a) A rectifier changes AC current to DC current. (b) An inverter changes DC current to AC.

10. Current transformers and potential transformers.

11. All hardware bolts should be tightened, all grounding connections tightened, cracked insulators replaced, posts inspected and treated, lightning arrestors in order, tree limbs trimmed away from the conductors, and guy wires tightened and protected. Any line work should be done by an electrician.

12. Metering equipment and an automatic disconnect.

13. Automatic throw-over.

14. A radial distribution system consists of a main bus with various branch circuits routed from the main to various loads.

15. Fuses and circuit breakers.

16. (a) Voltmeter, (b) ammeter, (c) wattmeter, (d) varmeter, (e) demand meter, (f) watt-hour meter, (g) power-factor meter, (h) frequency meter.

17. Mechanical energy, from a motor or solenoid; heat energy, from a heater

or a welder; light energy, from electric lights; chemical energy, used in cathodic protection; radio energy, used to transmit information over radio and television.

18. Voltage, continuous current, interrupting capacity, speed of response.

19. A molded-case circuit breaker is much smaller than a power-circuit breaker.

20. Conductor; insulator.

21. Resistance: ohms; conductance: mhos.

22. Overheating and age.

23. The holder can be used with a lower rated fuse but not with a higher rated one.

24. The circuit breaker can open the circuit automatically when an overload occurs, and it can open it when operated manually.

25. 600 V.

26. Apparent power in MVA (million volt-amperes).

**Basic Science
Concepts and
Applications**

Appendices

Appendix A
Conversion Tables

Table A1   Conversion Factors

Conversions		Procedure			Approximations
From	To	Multiply number of	by	To get number of	(Actual answer will be within 25% of approximate answer.)
acres	hectares (ha)	acres	0.4047	ha	1 acre ≈ 0.4 ha
acres	square feet (ft2)	acres	43,560	ft2	1 acre ≈ 40,000 ft2
acres	square kilometres (km2)	acres	0.004047	km2	1 acre ≈ 0.004 km2
acres	square metres (m2)	acres	4047	m2	1 acre ≈ 4000 m2
acres	square miles (mi2)	acres	0.001563	mi2	1 acre ≈ 0.0015 mi2
acres	square yards (yd2)	acres	4840	yd2	1 acre ≈ 5000 yd2
acre-feet (acre-ft)	cubic feet (ft3)	acre-ft	43,560	ft3	1 acre-ft ≈ 40,000 ft3
acre-feet (acre-ft)	cubic metres (m3)	acre-ft	1233	m3	1 acre-ft ≈ 1000 m3
acre-feet (acre-ft)	gallons (gal)	acre-ft	325,851	gal	1 acre-ft ≈ 300,000 gal
centimetres (cm)	feet (ft)	cm	0.03281	ft	1 cm ≈ 0.03 ft
centimetres (cm)	inches (in.)	cm	0.3937	in.	1 cm ≈ 0.4 in.
centimetres (cm)	metres (m)	cm	0.01	m	—
centimetres (cm)	millimetres (mm)	cm	10	mm	—
centimetres per second (cm/sec)	metres per minute (m/min)	cm/sec	0.6	m/min	—
cubic centimetres (cm3)	cubic feet (ft3)	cm3	0.00003531	ft3	1 cm3 ≈ 0.00004 ft3
cubic centimetres (cm3)	cubic inches (in.3)	cm3	0.06102	in.3	1 cm3 ≈ 0.06 in.3
cubic centimetres (cm3)	cubic metres (m3)	cm3	0.000001	m3	—
cubic centimetres (cm3)	cubic yards (yd3)	cm3	0.000001308	yd3	1 cm3 ≈ 0.0000015 yd3
cubic centimetres (cm3)	gallons (gal)	cm3	0.0002642	gal	1 cm3 ≈ 0.0003 gal
cubic centimetres (cm3)	litres (L)	cm3	0.001	L	—
cubic feet (ft3)	acre-feet (acre-ft)	ft3	0.00002296	acre-ft	1 ft3 ≈ 0.00002 acre-ft
cubic feet (ft3)	cubic centimetres (cm3)	ft3	28,320	cm3	1 ft3 ≈ 30
cubic feet (ft3)	cubic inches (in.3)	ft3	1728	in.3	1 ft3 ≈ 1500 in.3
cubic feet (ft3)	cubic metres (m3)	ft3	0.02832	m3	1 ft3 ≈ 0.03 m3
cubic feet (ft3)	cubic yards (yd3)	ft3	0.03704	yd3	1 ft3 ≈ 0.04 yd3
cubic feet (ft3)	gallons (gal)	ft3	7.481	gal	1 ft3 ≈ 7 gal
cubic feet (ft3)	kilolitres (kL)	ft3	0.02832	kL	1 ft3 ≈ 0.03 kL
cubic feet (ft3)	litres (L)	ft3	28.32	L	1 ft3 ≈ 30 L
cubic feet (ft3)	pounds (lb) of water	ft3	62.4	lb of water	1 ft3 ≈ 60 lb of water
cubic ft per second (cfs)	cubic metres per second (m3/sec)	cfs	0.02832	m3/sec	1 cfs ≈ 0.03 m3/sec

Table A1   Conversion Factors (continued)

Conversions		Procedure			Approximations
From	To	Multiply number of	by	To get number of	(Actual answer will be within 25% of approximate answer.)
cubic feet per second (cfs)	million gallons per day (mgd)	cfs	0.6463	mgd	1 cfs ≈ 0.6 mgd
cubic feet per second (cfs)	gallons per minute (gpm)	cfs	448.8	gpm	1 cfs ≈ 400 gpm
cubic feet per minute (cfm)	gallons per second (gps)	cfm	0.1247	gps	1 cfm ≈ 0.1 gps
cubic feet per minute (cfm)	litres per second (L/sec)	cfm	0.4720	L/sec	1 cfm ≈ 0.5 L/sec
cubic inches (in.^{3})	cubic centimetres (cm^3)	in.^{3}	16.39	cm^3	1 in.^{3} ≈ 15cm^3
cubic inches (in.^{3})	cubic feet (ft^3)	in.^{3}	0.0005787	ft^3	1 in.^{3} ≈ 0.0006 ft^3
cubic inches (in.^{3})	cubic metres (m^3)	in.^{3}	0.00001639	m^3	1 in.^{3} ≈ 0.000015 m^3
cubic inches (in.^{3})	cubic millimetres (mm^3)	in.^{3}	16,390	mm^3	1 in.^{3} ≈ 15,000 mm^3
cubic inches (in.^{3})	cubic yards (yd^3)	in.^{3}	0.00002143	yd^3	1 in.^{3} ≈ 0.00002 yd^3
cubic inches (in.^{3})	gallons (gal)	in.^{3}	0.004329	gal	1 in.^{3} ≈ 0.004 gal
cubic inches (in.^{3})	litres (L)	in.^{3}	0.01639	L	1 in.^{3} ≈ 0.015 L
cubic metres (m^3)	acre-feet (acre-ft)	m^3	0.0008107	acre-ft	1 m^3 ≈ 0.0008 acre-ft
cubic metres (m^3)	cubic centimetres (cm^3)	m^3	1,000,000	cm^3	—
cubic metres (m^3)	cubic feet (ft^3)	m^3	35.31	ft^3	1 m^3 ≈ 40 ft^3
cubic metres (m^3)	cubic inches (in.^{3})	m^3	61,020	in.^{3}	1 m^3 ≈ 60,000 in.^{3}
cubic metres (m^3)	cubic yards (yd^3)	m^3	1.308	yd^3	1 m^3 ≈ 1.5 yd^3
cubic metres (m^3)	gallons (gal)	m^3	264.2	gal	1 m^3 ≈ 300 gal
cubic metres (m^3)	kilolitres (kL)	m^3	1.0	kL	—
cubic metres (m^3)	litres (L)	m^3	1000	L	—
cubic metres per day (m^3/day)	gallons per day (gpd)	m^3/day	264.2	gpd	1 m^3/day ≈ 300 gpd
cubic metres per second (m^3/sec)	cubic feet per second (cfs)	m^3/sec	35.31	cfs	1 m^3/sec ≈ 40 cfs
cubic millimetres (mm^3)	cubic inches (in.^{3})	mm^3	0.00006102	in.^{3}	1 mm^3 ≈ 0.00006 in.^{3}
cubic yards (yd^3)	cubic centimetres (cm^3)	yd^3	764,600	cm^3	1 yd^3 ≈ 800,000 cm^3
cubic yards (yd^3)	cubic feet (ft^3)	yd^3	27	ft^3	1 yd^3 ≈ 30 ft^3
cubic yards (yd^3)	cubic inches (in.^{3})	yd^3	46,660	in.^{3}	1 yd^3 ≈ 50,000 in.^{3}
cubic yards (yd^3)	cubic metres (m^3)	yd^3	0.7646	m^3	1 yd^3 ≈ 0.8 m^3
cubic yards (yd^3)	gallons (gal)	yd^3	202.0	gal	1 yd^3 ≈ 200 gal
cubic yards (yd^3)	litres (L)	yd^3	764.6	L	1 yd^3 ≈ 800 L
feet (ft)	centimetres (cm)	ft	30.48	cm	1 ft ≈ 30 cm
feet (ft)	inches (in.)	ft	12	in.	—

Table A1   Conversion Factors (continued)

Conversions		Procedure		Approximations
From	To	Multiply number of	by	To get number of	(Actual answer will be within 25% of approximate answer.)
feet (ft)	kilometres (km)	ft	0.0003048	km	1 ft ≈ 0.0003 km
feet (ft)	metres (m)	ft	0.3048	m	1 ft ≈ 0.3 m
feet (ft)	miles (mi)	ft	0.0001894	mi	1 ft ≈ 0.0002 mi
feet (ft)	millimetres (mm)	ft	304.8	mm	1 ft ≈ 300 mm
feet (ft)	yards (yd)	ft	0.3333	yd	1 ft ≈ 0.3 yd
feet (ft) of hydraulic head	kilopascals (kPa)	ft of head	2.989	kPa	1 ft of head ≈ 3 kPa
feet (ft) of hydraulic head	metres (m) of hydraulic head	ft of head	0.3048	m of head	1 ft of head ≈ 0.3 m of head
feet (ft) of hydraulic head	pascals (Pa)	ft of head	2989	Pa	1 ft of head ≈ 3000 Pa
feet (ft) of water	inches of mercury (in. Hg)	ft of water	0.8826	in. Hg	1 ft of water ≈ 0.9 in. Hg
feet (ft) of water	pounds per square foot (psf)	ft of water	62.4	psf	1 ft of water ≈ 60 psf
feet (ft) of water	pounds per square inch gage (psig)	ft of water	0.4332	psig	1 ft of water ≈ 0.4 psig
feet per hour (fph)	metres per second (m/sec)	fph	0.00008467	m/sec	1 fph ≈ 0.00008 m/sec
feet per minute (fpm)	feet per second (fps)	fpm	0.01667	fps	1 fpm ≈ 0.015 fps
feet per minute (fpm)	kilometres per hour (km/hr)	fpm	0.01829	km/hr	1 fpm ≈ 0.02 km/hr
feet per minute (fpm)	metres per minute (m/min)	fpm	0.3048	m/min	1 fpm ≈ 0.3 m/min
feet per minute (fpm)	metres per second (m/sec)	fpm	0.005080	m/sec	1 fpm ≈ 0.005 m/sec
feet per minute (fpm)	miles per hour (mph)	fpm	0.01136	mph	1 fpm ≈ 0.01 mph
feet per second (fps)	feet per minute (fpm)	fps	60	fpm	—
feet per second (fps)	kilometres per hour (km/hr)	fps	1.097	km/hr	1 fps ≈ 1 km/hr
feet per second (fps)	metres per minute (m/min)	fps	18.29	m/min	1 fps ≈ 20 m/min
feet per second (fps)	metres per second (m/sec)	fps	0.3048	m/sec	1 fps ≈ 0.3 m/sec
feet per second (fps)	miles per hour (mph)	fps	0.6818	mph	1 fps ≈ 0.7 mph
foot-pounds per minute (ft-lb/min)	horsepower (hp)	ft-lb/min	0.00003030	hp	1 ft-lb/min ≈ 0.00003 hp
foot-pounds per minute (ft-lb/min)	kilowatts (kW)	ft-lb/min	0.00002260	kW	1 ft-lb/min ≈ 0.00002 kW
foot-pounds per minute (ft-lb/min)	watts (W)	ft-lb/min	0.02260	W	1 ft-lb/min ≈ 0.02 W
gallons (gal)	acre-feet (acre-ft)	gal	0.000003069	acre-ft	1 gal ≈ 0.000003 acre-ft
gallons (gal)	cubic centimetres (cm³)	gal	3785	cm³	1 gal ≈ 4000 cm³
gallons (gal)	cubic feet (ft³)	gal	0.1337	ft³	1 gal ≈ 0.15 ft³
gallons (gal)	cubic inches (in.³)	gal	231.0	in.³	1 gal ≈ 200 in.³
gallons (gal)	cubic metres (m³)	gal	0.003785	m³	1 gal ≈ 0.004 m³

Table A1   Conversion Factors (continued)

Conversions		Procedure		Approximations
From	To	Multiply number of	by	To get number of	(Actual answer will be within 25% of approximate answer.)
gallons (gal)	cubic yards (yd³)	gal	0.004951	yd³	1 gal ≈ 0.005 yd³
gallons (gal)	kilolitres (kL)	gal	0.003785	kL	1 gal ≈ 0.004 kL
gallons (gal)	litres (L)	gal	3.785	L	1 gal ≈ 4 L
gallons (gal)	pounds (lb) of water	gal	8.34	lb of water	1 gal ≈ 8 lb of water
gallons (gal)	quarts (qt)	gal	4	qt	
gallons per capita per day (gpcd)	litres per capita per day (Lpcd)	gpcd	3.785	Lpcd	1 gpcd ≈ 4 Lpcd
gallons per day (gpd)	cubic metres per day (m³/day)	gpd	0.003785	m³/day	1 gpd ≈ 0.004 m³/day
gallons per day (gpd)	litres per day (L/day)	gpd	3.785	L/day	1 gpd ≈ 4 L/day
gallons per day per foot (gpd/ft)	square metres per day (m²/day)	gpd/ft	0.01242	m²/day	1 gpd/ft ≈ 0.01 m²/day
gallons per day per foot (gpd/ft)	square millimetres per second (mm²/sec)	gpd/ft	0.1437	mm²/sec	1 gpd/ft ≈ 0.15 mm²/sec
gallons per day per square foot (gpd/ft²)	millimetres per second (mm/sec)	gpd/ft²	0.0004716	mm/sec	1 gpd/ft² ≈ 0.0005 mm/sec
gallons per hour (gph)	litres per second (L/sec)	gph	0.001052	L/sec	1 gph ≈ 0.001 L/sec
gallons per minute (gpm)	cubic feet per second (cfs)	gpm	0.002228	cfs	1 gpm ≈ 0.002 cfs
gallons per minute (gpm)	litres per second (L/sec)	gpm	0.06309	L/sec	1 gpm ≈ 0.06 L/sec
gallons per minute per square foot (gpm/ft²)	millimetres per second (mm/sec)	gpm/ft²	0.6790	mm/sec	1 gpm/ft² ≈ 0.7 mm/sec
gallons per second (gps)	cubic feet per minute (cfm)	gps	8.021	cfm	1 gpm ≈ 8 cfm
gallons per second (gps)	litres per minute (L/min)	gps	227.1	L/min	1 gps ≈ 200 L/min
grains (gr)	grams (g)	gr	0.06480	g	1 gr ≈ 0.06 g
grains (gr)	pounds (lb)	gr	0.0001428	lb	1 gr ≈ 0.00015 lb
grams (g)	grains (gr)	g	15.43	gr	1 g ≈ 15 gr
grams (g)	kilograms (kg)	g	0.001	kg	— —
grams (g)	milligrams (mg)	g	1000	mg	— —
grams (g)	ounces (oz), avoirdupois	g	0.03527	oz	1 g ≈ 0.04 oz
grams (g)	pounds (lb)	g	0.002205	lb	1 g ≈ 0.002 lb
hectares (ha)	acres	ha	2.471	acres	1 ha ≈ 2 acres
hectares (ha)	square metres (m²)	ha	10,000	m²	1 ha ≈ 0.004 mi²
hectares (ha)	square miles (mi²)	ha	0.003861	mi²	
horsepower (hp)	foot-pounds per minute (ft-lb/min)	hp	33,000	ft-lb/min	1 hp ≈ 30,000 ft-lb/min
horsepower (hp)	kilowatts (kW)	hp	0.7457	kW	1 hp ≈ 0.7 kW
horsepower (hp)	watts (W)	hp	745.7	W	1 hp ≈ 700 W

Table A1   Conversion Factors (continued)

From	To	Multiply number of	by	To get number of	Approximations (Actual answer will be within 25% of approximate answer.)
inches (in.)	centimetres (cm)	in.	2.540	cm	1 in. ≈ 3 cm
inches (in.)	feet (ft)	in.	0.08333	ft	1 in. ≈ 0.08 ft
inches (in.)	metres (m)	in.	0.02540	m	1 in. ≈ 0.03 m
inches (in.)	millimetres (mm)	in.	25.40	mm	1 in. ≈ 30 mm
inches (in.)	yards (yd)	in.	0.02778	yd	1 in. ≈ 0.03 yd
inches of mercury (in. Hg)	feet (ft) of water	in. Hg	1.133	ft of water	1 in. Hg ≈ 1 ft of water
inches of mercury (in. Hg)	inches (in.) of water	in. Hg	13.60	in. of water	1 in. Hg ≈ 15 in. of water
inches of mercury (in. Hg)	pounds per square foot (psf)	in. Hg	70.73	psf	1 in. Hg ≈ 70 psf
inches of mercury (in. Hg)	pounds per square inch (psi)	in. Hg	0.4912	psi	1 in. Hg ≈ 0.5 psi
inches per minute (in./min)	millimetres per second (mm/sec)	in./min	0.4233	mm/sec	1 in./min ≈ 0.4 mm/sec
inches (in.) of water	inches of mercury (in. Hg)	in. of water	0.07355	in. Hg	1 in. of water ≈ 0.07 in. Hg
inches (in.) of water	pounds per square foot (psf)	in. of water	5.198	psf	1 in. of water ≈ 5 psf
inches (in.) of water	pounds per square inch gage (psig)	in. of water	0.03610	psig	1 in. of water ≈ 0.04 psig
kilograms (kg)	grams (g)	kg	1000	g	—
kilograms (kg)	pounds (lb)	kg	2.205	lb	1 kg ≈ 2 lb
kilolitres (kL)	cubic feet (ft3)	kL	35.31	ft3	1 kL ≈ 40 ft3
kilolitres (kL)	cubic metres (m3)	kL	1.0	m3	—
kilolitres (kL)	gallons (gal)	kL	264.2	gal	1 kL ≈ 300 gal
kilolitres (kL)	litres (L)	kL	1000	L	—
kilometres (km)	feet (ft)	km	3281	ft	1 km ≈ 3000 ft
kilometres (km)	metres (m)	km	1000	m	—
kilometres (km)	miles (mi)	km	0.6214	mi	1 km ≈ 0.6 mi
kilometres (km)	yards (yd)	km	1094	yd	1 km ≈ 1000 yd
kilometres per hour (km/hr)	feet per minute (fpm)	km/hr	54.68	fpm	1 km/hr ≈ 50 fpm
kilometres per hour (km/hr)	feet per second (fps)	km/hr	0.9113	fps	1 km/hr ≈ 1 fps
kilometres per hour (km/hr)	metres per minute (m/min)	km/hr	16.67	m/min	1 km/hr ≈ 15 m/min
kilometres per hour (km/hr)	metres per second (m/sec)	km/hr	0.2778	m/sec	1 km/hr ≈ 0.3 m/sec
kilometres per hour (km/hr)	miles per hour (mph)	km/hr	0.6214	mph	1 km/hr ≈ 0.6 mph
kilopascals (kPa)	feet (ft) of hydraulic head	kPa	0.3346	ft of head	1 kPa ≈ 0.3 ft of head
kilowatts (kW)	foot-pounds per minute (ft-lb/min)	kW	44,250	ft-lb/min	1 kW ≈ 40,000 ft-lb/min

Table A1 Conversion Factors (continued)

From	Conversions To	Multiply number of	by	To get number of	Approximations (Actual answer will be within 25% of approximate answer.)
kilowatts (kW)	horsepower (hp)	kW	1.341	hp	1 kW ≈ 1.5 hp
kilowatts (kW)	watts (W)	kW	1000	W	—
litres (L)	cubic centimetres (cm³)	L	1000	cm³	—
litres (L)	cubic feet (ft³)	L	0.03531	ft³	1 L ≈ 0.04 ft³
litres (L)	cubic inches (in.³)	L	61.03	in.³	1 L ≈ 60 in.³
litres (L)	cubic metres (m³)	L	0.001	m³	—
litres (L)	cubic yards (yd³)	L	0.001308	yd³	1 L ≈ 0.0015 yd³
litres (L)	gallons (gal)	L	0.2642	gal	1 L ≈ 0.3 gal
litres (L)	kilolitres (kL)	L	0.001	kL	—
litres (L)	millilitres (mL)	L	1000	mL	—
litres (L)	ounces (oz), fluid	L	33.81	oz (fluid)	1 L ≈ 30 oz (fluid)
litres (L)	quarts (qt), fluid	L	1.057	qt (fluid)	1 L ≈ 1 qt (fluid)
litres per capita per day (Lpcd)	gallons per capita per day (gpcd)	Lpcd	0.2642	gpcd	1 Lpcd ≈ 0.3 gpcd
litres per day (L/day)	gallons per day (gpd)	L/day	0.2642	gpd	1 L/day ≈ 0.3 gpd
litres per minute (L/min)	gallons per second (gps)	L/min	0.004403	gps	1 L/min ≈ 0.004 gps
litres per second (L/sec)	cubic feet per minute (cfm)	L/sec	2.119	cfm	1 L/sec ≈ 2 cfm
litres per second (L/sec)	gallons per hour (gph)	L/sec	951.0	gph	1 L/sec ≈ 1000 gph
litres per second (L/sec)	gallons per minute (gpm)	L/sec	15.85	gpm	1 L/sec ≈ 15 gpm
megalitres per day (ML/day)	million gallons per day (mgd)	ML/day	0.2642	mgd	1 ML/day ≈ 0.3 mgd
metres (m)	centimetres (cm)	m	100	cm	—
metres (m)	feet (ft)	m	3.281	ft	1 m ≈ 3 ft
metres (m)	inches (in.)	m	39.37	in.	1 m ≈ 40 in.
metres (m)	kilometres (km)	m	0.001	km	—
metres (m)	miles (mi)	m	0.0006214	mi	1 m ≈ 0.0006 mi
metres (m)	millimetres (mm)	m	1000	mm	—
metres (m)	yards (yd)	m	1.094	yd	1 m ≈ 1 yd
metres (m) of hydraulic head	feet (ft) of hydraulic head	m of head	3.281	ft of head	1 m of head ≈ 3 ft of head
metres (m) of hydraulic head	pounds per square inch gage (psig)	m of head	1.422	psig	1 m of head ≈ 1.5 psig
metres per minute (m/min)	centimetres per second (cm/sec)	m/min	1.667	cm/sec	1 m/min ≈ 1.5 cm/sec
metres per minute (m/min)	feet per minute (fpm)	m/min	3.281	fpm	1 m/min ≈ 3 fpm

Table A1 Conversion Factors (continued)

Conversions		Procedure		Approximations
From	To	Multiply number of	by	To get number of	(Actual answer will be within 25% of approximate answer.)
metres per minute (m/min)	feet per second (fps)	m/min	0.05468	fps	1 m/min ≈ 0.05 fps
metres per minute (m/min)	kilometres per hour (km/hr)	m/min	0.06	km/hr	—
metres per minute (m/min)	miles per hour (mph)	m/min	0.03728	mph	1 m/min ≈ 0.04 mph
metres per second (m/sec)	feet per hour (fph)	m/sec	11,810	fph	1 m/sec ≈ 10,000 fph
metres per second (m/sec)	feet per minute (fpm)	m/sec	196.8	fpm	1 m/sec ≈ 200 fpm
metres per second (m/sec)	feet per second (fps)	m/sec	3.281	fps	1 m/sec ≈ 3 fps
metres per second (m/sec)	kilometres per hour (km/hr)	m/sec	3.6	km/hr	1 m/sec ≈ 4 km/hr
metres per second (m/sec)	miles per hour (mph)	m/sec	2.237	mph	1 m/sec ≈ 2 mph
miles (mi)	feet (ft)	mi	5280	ft	1 mi ≈ 5000 ft
miles (mi)	kilometres (km)	mi	1.609	km	1 mi ≈ 1.5 km
miles (mi)	metres (m)	mi	1609	m	1 mi ≈ 1500 m
miles (mi)	yards (yd)	mi	1760	yd	1 mi ≈ 2000 yd
miles per hour (mph)	feet per minute (fpm)	mph	88	fpm	1 mph ≈ 90 fpm
miles per hour (mph)	feet per second (fps)	mph	1.467	fps	1 mph ≈ 1.5 fps
miles per hour (mph)	kilometres per hour (km/hr)	mph	1.609	km/hr	1 mph ≈ 1.5 km/hr
miles per hour (mph)	metres per minute (m/min)	mph	26.82	m/min	1 mph ≈ 30 m/min
miles per hour (mph)	metres per second (m/sec)	mph	0.4470	m/sec	1 mph ≈ 0.4 m/sec
milligrams (mg)	grams (g)	mg	0.001	g	—
millilitres (mL)	litres (L)	mL	0.001	L	—
millimetres (mm)	centimetres (cm)	mm	0.1	cm	—
millimetres (mm)	feet (ft)	mm	0.003281	ft	1 mm ≈ 0.003 ft
millimetres (mm)	inches (in.)	mm	0.03937	in.	1 mm ≈ 0.04 in.
millimetres (mm)	metres (m)	mm	0.001	m	—
millimetres (mm)	yards (yd)	mm	0.001094	yd	1 mm ≈ 0.001 yd
millimetres per second (mm/sec)	gallons per day per square foot (gpd/ft2)	mm/sec	2121	gpd/ft2	1 mm/sec ≈ 2000 gpd/ft2
millimetres per second (mm/sec)	gallons per minute per square foot (gpm/ft2)	mm/sec	1.473	gpm/ft2	1 mm/sec ≈ 1.5 gpm/ft2
millimetres per second (mm/sec)	inches per minute (in./min)	mm/sec	2.362	in./min	1 mm/sec ≈ 2 in./min
million gallons per day (mgd)	cubic feet per second (cfs)	mgd	1.547	cfs	1 mgd ≈ 1.5 cfs
million gallons per day (mgd)	megalitres per day (ML/day)	mgd	3.785	ML/day	1 mgd ≈ 4 ML/day
ounces (oz), avoirdupois	grams (g)	oz	28.35	g	1 oz ≈ 30 g

Table A1   Conversion Factors (continued)

Conversions		Procedure		Approximations
From	To	Multiply number of	by	To get number of	(Actual answer will be within 25% of approximate answer.)
ounces (oz), avoirdupois	pounds (lb)	oz	0.0625	lb	1 oz ≈ 0.06 lb
ounces (oz), fluid	litres (L)	oz	0.02957	L	1 oz ≈ 0.03 L
pascals (Pa)	feet (ft) of hydraulic head	Pa	0.0003346	ft of head	1 Pa ≈ 0.0003 ft of head
pascals (Pa)	pounds per square inch (psi)	Pa	0.0001450	psi	1 Pa ≈ 0.00015 psi
pounds (lb)	grains (gr)	lb	7000	gr	
pounds (lb)	grams (g)	lb	453.6	g	1 lb ≈ 500 g
pounds (lb)	kilograms (kg)	lb	0.4536	kg	1 lb ≈ 0.5 kg
pounds (lb)	ounces (oz), avoirdupois	lb	16	oz	
pounds (lb) of water	cubic feet (ft³)	lb of water	0.01603	ft³	1 lb of water ≈ 0.015 ft³
pounds (lb) of water	gallons (gal)	lb of water	0.1199	gal	1 lb of water ≈ 0.1 gal
pounds per square foot (psf)	feet (ft) of water	psf	0.01603	ft of water	1 psf ≈ 0.015 ft of water
pounds per square foot (psf)	inches of mercury (in. Hg)	psf	0.01414	in. Hg	1 psf ≈ 0.015 in. Hg
pounds per square foot (psf)	inches (in.) of water	psf	0.1924	in. of water	1 psf ≈ 0.2 in. of water
pounds per square inch gage (psig)	feet (ft) of water	psig	2.31	ft of water	1 psig ≈ 2 ft of water
pounds per square inch (psi)	inches of mercury (in. Hg)	psi	2.036	in. Hg	1 psi ≈ 2 in. Hg
pounds per square inch gage (psig)	inches (in.) of water	psig	27.70	in. of water	1 psig ≈ 30 in. of water
pounds per square inch gage (psig)	metres (m) of hydraulic head	psig	0.7034	m of head	1 psig ≈ 0.7 m of head
pounds per square inch (psi)	pascals (Pa)	psi	6895	Pa	1 psi ≈ 7000 Pa
quarts (qt)	gallons (gal)	qt	0.25	gal	
quarts (qt)	litres (L)	qt	0.9464	L	1 qt ≈ 0.9 L
square centimetres (cm²)	square inches (in.²)	cm²	0.1550	in.²	1 cm² ≈ 0.15 in.²
square centimetres (cm²)	square millimetres (mm²)	cm²	100	mm²	
square feet (ft²)	acres	ft²	0.00002296	acres	1 ft² ≈ 0.00002 acre
square feet (ft²)	square inches (in.²)	ft²	144	in.²	1 ft² ≈ 150 in.²
square feet (ft²)	square metres (m²)	ft²	0.09290	m²	1 ft² ≈ 0.09 m²
square feet (ft²)	square millimetres (mm²)	ft²	92,900	mm²	1 ft² ≈ 90,000 mm²
square feet (ft²)	square yards (yd²)	ft²	0.1111	yd²	1 ft² ≈ 0.1 yd²
square inches (in.²)	square centimetres (cm²)	in.²	6.452	cm²	1 in.² ≈ 6 cm²
square inches (in.²)	square feet (ft²)	in.²	0.006944	ft²	1 in.² ≈ 0.007 ft²
square inches (in.²)	square metres (m²)	in.²	0.0006452	m²	1 in.² ≈ 0.0006 m²

Table A1 Conversion Factors (continued)

From	To	Multiply number of	by	To get number of	Approximations (Actual answer will be within 25% of approximate answer.)
square inches (in.²)	square millimetres (mm²)	in.²	645.2	mm²	1 in.² ≈ 600 mm²
square inches (in.²)	square yards (yd²)	in.²	0.0007716	yd²	1 in.² ≈ 0.0008 yd²
square kilometres (km²)	acres	km²	247.1	acres	1 km² ≈ 200 acres
square kilometres (km²)	square miles (mi²)	km²	0.3861	mi²	1 km² ≈ 0.4 mi²
square metres (m²)	acres	m²	0.0002471	acres	1 m² ≈ 0.0002 acre
square metres (m²)	hectares (ha)	m²	0.0001	ha	——
square metres (m²)	square feet (ft²)	m²	10.76	ft²	1 m² ≈ 10 ft²
square meters (m²)	square inches (in.²)	m²	1550	in.²	1 m² ≈ 1500 in.²
square metres (m²)	square miles (mi²)	m²	0.0000003861	mi²	1 m² ≈ 0.0000004 mi²
square metres (m²)	square yards (yd²)	m²	1.196	yd²	1 m² ≈ 1 yd²
square metres per day (m²/day)	gallons per day per foot (gpd/ft)	m²/day	80.53	gpd/ft	1 m³/day ≈ 80 gpd/ft
square miles (mi²)	acres	mi²	640	acres	1 mi² ≈ 600 acres
square miles (mi²)	hectares (ha)	mi²	259.0	ha	1 mi² ≈ 300 ha
square miles (mi²)	square kilometres (km²)	mi²	2.590	km²	1 mi² ≈ 3 km²
square miles (mi²)	square metres (m²)	mi²	2,590,000	m²	1 mi² ≈ 3,000,000 m²
square millimetres (mm²)	square centimetres (cm²)	mm²	0.01	cm²	——
square millimetres (mm²)	square feet (ft²)	mm²	0.00001076	ft²	1 mm² ≈ 0.00001 ft²
square millimetres (mm²)	square inches (in.²)	mm²	0.001550	in.²	1 mm² ≈ 0.0015 in.²
square millimetres per second (mm²/sec)	gallons per day per foot (gpd/ft)	mm²/sec	6.958	gpd/ft	1 mm²/sec ≈ 7 gpd/ft
square yards (yd²)	acres	yd²	0.0002066	acres	1 yd² ≈ 0.0002 acre
square yards (yd²)	square feet (ft²)	yd²	9	ft²	
square yards (yd²)	square inches (in.²)	yd²	1296	in.²	1 yd² ≈ 1500 in.²
square yards (yd²)	square metres (m²)	yd²	0.8361	m²	1 yd² ≈ 0.8 m²
watts (W)	foot-pounds per minute (ft-lb/min)	W	44.25	ft-lb/min	1 W ≈ 40 ft-lb/min
watts (W)	horsepower (hp)	W	0.001341	hp	1 W ≈ 0.0015 hp
watts (W)	kilowatts (kW)	W	0.001	kW	——
yards (yd)	feet (ft)	yd	3	ft	——
yards (yd)	inches (in.)	yd	36	in.	1 yd ≈ 40 in.
yards (yd)	kilometres (km)	yd	0.0009144	km	1 yd ≈ 0.0009 km
yards (yd)	metres (m)	yd	0.9144	m	1 yd ≈ 0.9 m
yards (yd)	miles (mi)	yd	0.0005681	mi	1 yd ≈ 0.0006 mi
yards (yd)	millimetres (mm)	yd	914.4	mm	1 yd ≈ 900 mm

Table A2.   Temperature Conversions, Celsius to Fahrenheit: °F = 9/5 (°C) + 32

°C to °F		°C to °F		°C to °F		°C to °F		°C to °F		°C to °F	
-29	-20.2	1	33.8	31	87.8	61	141.8	91	195.8	121	249.8
-28	-18.4	2	35.6	32	89.6	62	143.6	92	197.6	122	251.6
-27	-16.6	3	37.4	33	91.4	63	145.4	93	199.4	123	253.4
-26	-14.8	4	39.2	34	93.2	64	147.2	94	201.2	124	255.2
-25	-13.0	5	41.0	35	95.0	65	149.0	95	203.0	125	257.0
-24	-11.2	6	42.8	36	96.8	66	150.8	96	204.8	126	258.8
-23	-9.4	7	44.6	37	98.6	67	152.6	97	206.6	127	260.6
-22	-7.6	8	46.4	38	100.4	68	154.4	98	208.4	128	262.4
-21	-5.8	9	48.2	39	102.2	69	156.2	99	210.2	129	264.2
-20	-4.0	10	50.0	40	104.0	70	158.0	100	212.0	130	266.0
-19	-2.2	11	51.8	41	105.8	71	159.8	101	213.8	131	267.8
-18	-0.4	12	53.6	42	107.6	72	161.6	102	215.6	132	296.6
-17	+1.4	13	55.4	43	109.4	73	163.4	103	217.4	133	271.4
-16	3.2	14	57.2	44	111.2	74	165.2	104	219.2	134	273.2
-15	5.0	15	59.0	45	113.0	75	167.0	105	221.0	135	275.0
-14	6.8	16	60.8	46	114.8	76	168.8	106	222.8	136	276.8
-13	8.6	17	62.6	47	116.6	77	170.6	107	224.6	137	278.6
-12	10.4	18	64.4	48	118.4	78	172.4	108	226.4	138	280.4
-11	12.2	19	66.2	49	120.2	79	174.2	109	228.2	139	282.2
-10	14.0	20	68.0	50	122.0	80	176.0	110	230.0	140	284.0
-9	15.8	21	69.8	51	123.8	81	177.8	111	231.8	141	285.8
-8	17.6	22	71.6	52	125.6	82	179.6	112	233.6	142	287.6
-7	19.4	23	73.4	53	127.4	83	181.4	113	235.4	143	289.4
-6	21.2	24	75.2	54	129.2	84	183.2	114	237.2	144	291.2
-5	23.0	25	77.0	55	131.0	85	185.0	115	239.0	145	293.0
-4	24.8	26	78.8	56	132.8	86	186.8	116	240.8	146	294.8
-3	26.6	27	80.6	57	134.6	87	188.6	117	242.6	147	296.6
-2	28.4	28	82.4	58	136.4	88	190.4	118	244.4	148	298.4
-1	30.2	29	84.2	59	138.2	89	192.2	119	246.2	149	300.2
0	32.0	30	86.0	60	140.0	90	194.0	120	248.0	150	302.0

Table A3. Temperature Conversions, Fahrenheit to Celsius: °C = 5/9 (°F − 32)

°F	to °C	°F	to °C	°F	to °C	°F	to °C	°F	to °C	°F	to °C
−19	−28.3	11	−11.7	41	5.0	71	21.7	101	38.3	131	55.0
−18	−27.8	12	−11.1	42	5.6	72	22.2	102	38.9	132	55.6
−17	−27.2	13	−10.6	43	6.1	73	22.8	103	39.4	133	56.1
−16	−26.7	14	−10.0	44	6.7	74	23.3	104	40.0	134	56.7
−15	−26.1	15	−9.4	45	7.2	75	23.9	105	40.6	135	57.2
−14	−25.6	16	−8.9	46	7.8	76	24.4	106	41.1	136	57.8
−13	−25.0	17	−8.3	47	8.3	77	25.0	107	41.7	137	58.3
−12	−24.4	18	−7.8	48	8.9	78	25.6	108	42.2	138	58.9
−11	−23.9	19	−7.2	49	9.4	79	26.1	109	42.8	139	59.4
−10	−23.3	20	−6.7	50	10.0	80	26.7	110	43.3	140	60.0
−9	−22.8	21	−6.1	51	10.6	81	27.2	111	43.9	141	60.6
−8	−22.2	22	−5.6	52	11.1	82	27.8	112	44.4	142	61.1
−7	−21.7	23	−5.0	53	11.7	83	28.3	113	45.0	143	61.7
−6	−21.1	24	−4.4	54	12.2	84	28.9	114	45.6	144	62.2
−5	−20.6	25	−3.9	55	12.8	85	29.4	115	46.1	145	62.8
−4	−20.0	26	−3.3	56	13.3	86	30.0	116	46.7	146	63.3
−3	−19.4	27	−2.8	57	13.9	87	30.6	117	47.2	147	63.9
−2	−18.9	28	−2.2	58	14.4	88	31.1	118	47.8	148	64.4
−1	−18.3	29	−1.7	59	15.0	89	31.7	119	48.3	149	65.0
0	−17.8	30	−1.1	60	15.6	90	32.2	120	48.9	150	65.6
1	−17.2	31	−0.6	61	16.1	91	32.8	121	49.4	151	66.1
2	−16.7	32	0.0	62	16.7	92	33.3	122	50.0	152	66.7
3	−16.1	33	+0.6	63	17.2	93	33.9	123	50.6	153	67.2
4	−15.6	34	1.1	64	17.8	94	34.4	124	51.1	154	67.8
5	−15.0	35	1.7	65	18.3	95	35.0	125	51.7	155	68.3
6	−14.4	36	2.2	66	18.9	96	35.6	126	52.2	156	68.9
7	−13.9	37	2.8	67	19.4	97	36.1	127	52.8	157	69.4
8	−13.3	38	3.3	68	20.0	98	36.7	128	53.3	158	70.0
9	−12.8	39	3.9	69	20.6	99	37.2	129	53.9	159	70.6
10	−12.2	40	4.4	70	21.1	100	37.8	130	54.4	160	71.1

Table A3. Temperature Conversions, Fahrenheit to Celsius (continued)

°F	to °C	°F	to °C	°F	to °C	°F	to °C	°F	to °C	°F	to °C
161	71.7	191	88.3	221	105.0	251	121.7	281	138.3	311	155.0
162	72.2	192	88.9	222	105.6	252	122.2	282	138.9	312	155.6
163	72.8	193	89.4	223	106.1	253	122.8	283	139.4	313	156.1
164	73.3	194	90.0	224	106.7	254	123.3	284	140.0	314	156.7
165	73.9	195	90.6	225	107.2	255	123.9	285	140.6	315	157.2
166	74.4	196	91.1	226	107.8	256	124.4	286	141.1	316	157.8
167	75.0	197	91.7	227	108.3	257	125.0	287	141.7	317	158.3
168	75.6	198	92.2	228	108.9	258	125.6	288	142.2	318	158.9
169	76.1	199	92.8	229	109.4	259	126.1	289	142.8	319	159.4
170	76.7	200	93.3	230	110.0	260	126.7	290	143.3	320	160.0
171	77.2	201	93.9	231	110.6	261	127.2	291	143.9	321	160.6
172	77.8	202	94.4	232	111.1	262	127.8	292	144.4	322	161.1
173	78.3	203	95.0	233	111.7	263	128.3	293	145.0	323	161.7
174	78.9	204	95.6	234	112.2	264	128.9	294	145.6	324	162.2
175	79.4	205	96.1	235	112.8	265	129.4	295	146.1	325	162.8
176	80.0	206	96.7	236	113.3	266	130.0	296	146.7	326	163.3
177	80.6	207	97.2	237	113.9	267	130.6	297	147.2	327	163.9
178	81.1	208	97.8	238	114.4	268	131.1	298	147.8	328	164.4
179	81.7	209	98.3	239	115.0	269	131.7	299	148.3	329	165.0
180	82.2	210	98.9	240	115.6	270	132.2	300	148.9	330	165.6
181	82.8	211	99.4	241	116.1	271	132.8	301	149.4	331	166.1
182	83.3	212	100.0	242	116.7	272	133.3	302	150.0	332	166.7
183	83.9	213	100.6	243	117.2	273	133.9	303	150.5	333	167.2
184	84.4	214	101.1	244	117.8	274	134.4	304	151.1	334	167.8
185	85.0	215	101.7	245	118.3	275	135.0	305	151.7	335	168.3
186	85.6	216	102.2	246	118.9	276	135.6	306	152.2	336	168.9
187	86.1	217	102.8	247	119.4	277	136.1	307	152.8	337	169.4
188	86.7	218	103.3	248	120.0	278	136.7	308	153.3	338	170.0
189	87.2	219	103.9	249	120.6	279	137.2	309	153.9	339	170.6
190	87.8	220	104.4	250	121.1	280	137.8	310	154.4	340	171.1

Appendix B

Periodic Table and List of Elements

Periodic Table of the Elements

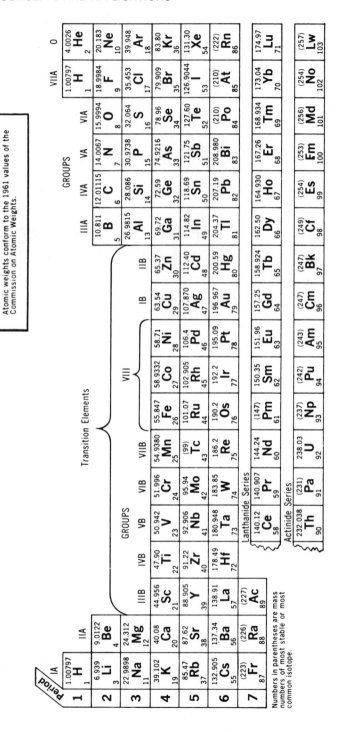

List of Elements

Name	Symbol	Atomic Number	Atomic Weight
Actinium	Ac	89	227*
Aluminum	Al	13	26.98
Americium	Am	95	243*
Antimony	Sb	51	121.75
Argon	Ar	18	39.95
Arsenic	As	33	74.92
Astatine	At	85	210*
Barium	Ba	56	137.34
Berkelium	Bk	97	247*
Beryllium	Be	4	9.01
Bismuth	Bi	83	208.98
Boron	B	5	10.81
Bromine	Br	35	79.90
Cadmium	Cd	48	112.40
Calcium	Ca	20	40.08
Californium	Cf	98	249*
Carbon	C	6	12.01
Cerium	Ce	58	140.12
Cesium	Cs	55	132.91
Chlorine	Cl	17	35.45
Chromium	Cr	24	52.00
Cobalt	Co	27	58.93
Copper	Cu	29	63.55
Curium	Cm	96	247*
Dysprosium	Dy	66	162.50
Einsteinium	Es	99	254*
Erbium	Er	68	167.26
Europium	Eu	63	151.96
Fermium	Fm	100	253*
Fluorine	F	9	19.00
Francium	Fr	87	223*
Gadolinium	Gd	64	157.25
Gallium	Ga	31	69.72
Germanium	Ge	32	72.59
Gold	Au	79	196.97
Hafnium	Hf	72	178.49
Helium	He	2	4.00
Holmium	Ho	67	164.93

* Mass number of most stable or best known isotope.
† Mass of most commonly available, long-lived isotope.

List of Elements (continued)

Name	Symbol	Atomic Number	Atomic Weight
Hydrogen	H	1	1.01
Indium	In	49	114.82
Iodine	I	53	126.90
Iridium	Ir	77	192.22
Iron	Fe	26	55.85
Krypton	Kr	36	83.80
Lanthanum	La	57	138.91
Lawrencium	Lr	103	257*
Lead	Pb	82	207.2
Lithium	Li	3	6.94
Lutetium	Lu	71	174.97
Magnesium	Mg	12	24.31
Manganese	Mn	25	54.94
Mendelevium	Md	101	256*
Mercury	Hg	80	200.59
Molybdenum	Mo	42	95.94
Neodymium	Nd	60	144.24
Neon	Ne	10	20.18
Neptunium	Np	93	237.05†
Nickel	Ni	28	58.71
Niobium	Nb	41	92.91
Nitrogen	N	7	14.01
Nobelium	No	102	254*
Osmium	Os	76	190.2
Oxygen	O	8	16.00
Palladium	Pd	46	106.4
Phosphorus	P	15	30.97
Platinum	Pt	78	195.09
Plutonium	Pu	94	242*
Polonium	Po	84	210*
Potassium	K	19	39.10
Praseodymium	Pr	59	140.91
Promethium	Pm	61	147*
Protactinium	Pa	91	231.04†
Radium	Ra	88	226.03†
Radon	Rn	86	222*
Rhenium	Re	75	186.2
Rhodium	Rh	45	102.91
Rubidium	Rb	37	85.47
Ruthenium	Ru	44	101.07

* Mass number of most stable or best known isotope.
† Mass of most commonly available, long-lived isotope.

List of Elements (continued)

Name	Symbol	Atomic Number	Atomic Weight
Samarium	Sm	62	150.4
Scandium	Sc	21	44.96
Selenium	Se	34	78.96
Silicon	Si	14	28.09
Silver	Ag	47	107.87
Sodium	Na	11	22.99
Strontium	Sr	38	87.62
Sulfur	S	16	32.06
Tantalum	Ta	73	180.95
Technetium	Tc	43	98.91†
Tellurium	Te	52	127.60
Terbium	Tb	65	158.93
Thallium	Tl	81	204.37
Thorium	Th	90	232.04†
Thulium	Tm	69	168.93
Tin	Sn	50	118.69
Titanium	Ti	22	47.90
Tungsten	W	74	183.85
Uranium	U	92	238.03
Vanadium	V	23	50.94
Xenon	Xe	54	131.30
Ytterbium	Yb	70	173.04
Yttrium	Y	39	88.91
Zinc	Zn	30	65.38
Zirconium	Zr	40	91.22

* Mass number of most stable or best known isotope.
† Mass of most commonly available, long-lived isotope.

Appendix C

Chemical Equations and Compounds Common in Water Treatment

Chemical Equations Common in Water Treatment

Taste, Odor, and Color Removal

$4Fe(OH)_2 + O_2\uparrow + 2H_2O \rightarrow 4Fe(OH)_3\downarrow$

$2MnSO_4 + O_2\uparrow + 4NaOH \rightarrow 2MnO_2 + 2Na_2SO_4 + 2H_2O$

$2H_2S + O_2\uparrow \rightarrow 2H_2O + 2S\downarrow$

$2NaClO_2 + Cl_2\uparrow \rightarrow 2ClO_2 + 2NaCl$

Alum Coagulation

$Al_2(SO_4)_3 + 3Ca(HCO_3)_2 \rightarrow 2Al(OH)_3\downarrow + 3CaSO_4 + 6CO_2$

$Al_2(SO_4)_3 + 3Na_2CO_3 + 3H_2O \rightarrow 2Al(OH)_3\downarrow + 3Na_2SO_4 + 3CO_2$

$Al_2(SO_4)_3 + 3Ca(OH)_2 \rightarrow 2Al(OH)_3\downarrow + 3CaSO_4$

Ferric Sulfate Coagulation

$Fe_2(SO_4)_3 + 3Ca(HCO_3)_2 \rightarrow 2Fe(OH)_3\downarrow + 3CaSO_4 + 6CO_2$

Iron and Manganese Removal

$2Fe(HCO_3)_2 + Cl_2 + Ca(HCO_3)_2 \rightarrow 2Fe(OH)_3\downarrow + CaCl_2 + 6CO_2$

$MnSO_4 + Cl_2 + 4NaOH \rightarrow (MnO_2) + 2NaCl + Na_2SO_4 + 2H_2O$

$4Fe(HCO_3)_2 + O_2 + 2H_2O \rightarrow 4Fe(OH)_3\downarrow + 8CO_2$

$2MnSO_4 + O_2 + 4NaOH \rightarrow 2MnO_2\downarrow + 2Na_2SO_4 + 2H_2O$

$3Fe(HCO_3)_2 + KMnO_4 + 7H_2O \rightarrow MnO_2\downarrow + 3Fe(OH)_3 + KHCO_3 + 5H_2CO_3$

$3Mn(HCO_3)_2 + 2KMnO_4 + 2H_2O \rightarrow 5MnO_2\downarrow + 2KHCO_3 + 4H_2CO_3$

$3MnSO_4 + 2KMnO_4 + 2H_2O \rightarrow 5MnO_2\downarrow + K_2SO_4 + 2H_2SO_4$

Hardness Removal

$Ca(HCO_3)_2 + Ca(OH)_2 \rightarrow 2CaCO_3\downarrow + 2H_2O$

$Mg(HCO_3)_2 + Ca(OH)_2 \rightarrow CaCO_3\downarrow + MgCO_3 + 2H_2O$

$MgCO_3 + Ca(OH)_2 \rightarrow CaCO_3\downarrow + Mg(OH)_2$

$CaSO_4 + Na_2CO_3 \rightarrow CaCO_3\downarrow + Na_2SO_4$

$CaCl_2 + Na_2CO_3 \rightarrow CaCO_3\downarrow + 2NaCl$

$MgCl_2 + Ca(OH)_2 \rightarrow Mg(OH)_2\downarrow + CaCl_2$

$MgSO_4 + Ca(OH)_2 \rightarrow Mg(OH)_2\downarrow + CaSO_4$

$CO_2 + Ca(OH)_2 \rightarrow CaCO_3\downarrow + H_2O$

Corrosion Control

$$Fe(OH)_2 + 2H_2CO_3 \rightarrow Fe(HCO_3)_2 + 2H_2O$$
$$4Fe(HCO_3)_2 + 10H_2O + O_2 \rightarrow 4Fe(OH)_3\downarrow + 8H_2CO_3$$
$$4Fe(OH)_2 + 2H_2O + O_2 \rightarrow 4Fe(OH)_3\downarrow$$

Chlorination

$$Cl_2 + H_2O \rightarrow HOCl + HCl$$
$$NH_3 + HOCl \rightarrow NH_2Cl + H_2O$$
$$NH_2Cl + HOCl \rightarrow NHCl_2 + H_2O$$
$$NHCl_2 + HOCl \rightarrow NCl_2 + H_2O$$
$$Ca(OCl)_2 + 2H_2O \rightarrow 2HOCl + Ca(OH)_2$$
$$NaOCl + H_2O \rightarrow HOCl + NaOH$$

Compounds Common in Water Treatment

Chemical Name	Common Name	Chemical Formula
Aluminum hydroxide	Alum floc	$Al(OH)_3$
Aluminum sulfate	Filter alum	$Al_2(SO_4)_3 \cdot 14H_2O$
Ammonia	Ammonia	NH_3 (Ammonia gas)
Calcium bicarbonate	—	$Ca(HCO_3)_2$
Calcium carbonate	Limestone	$CaCO_3$
Calcium chloride	—	$CaCl_2$
Calcium hydroxide	Hydrated lime (slaked lime)	$Ca(OH)_2$
Calcium hypochlorite	HTH	$Ca(OCl)_2$
Calcium oxide	Unslaked lime (quick lime)	CaO
Calcium sulfate	—	$CaSO_4$
Carbon	Activated carbon	C
Carbon dioxide	—	CO_2
Carbonic acid	—	H_2CO_3
Chlorine	—	Cl_2
Chlorine dioxide	—	ClO_2
Copper sulfate	Blue vitriol	$CuSO_4 \cdot 5H_2O$
Dichloramine	—	$NHCl_2$
Ferric chloride	—	$FeCL_3 \cdot 6H_2O$
Ferric hydroxide	Ferric hydroxide floc	$Fe(OH)_3$
Ferric sulfate	—	$Fe_2(SO_4)_3 \cdot 3H_2O$

Compounds Common in Water Treatment (continued)

Chemical Name	Common Name	Chemical Formula
Ferrous bicarbonate	—	$Fe(HCO_3)_2$
Ferrous hydroxide	—	$Fe(OH)_2$
Fluosilicic acid (hydrofluosilicic acid)	—	H_2SiF_6
Hydrochloric acid	Muriatic acid	HCl
Hydrofluosilicic acid (fluosilicic acid)	—	H_2SiF_6
Hydrogen sulfide	—	H_2S
Hypochlorous acid	—	$HOCl$
Magnesium bicarbonate	—	$Mg(HCO_3)_2$
Magnesium carbonate	—	$MgCO_3$
Magnesium chloride	—	$MgCl_2$
Magnesium hydroxide	—	$Mg(OH)_2$
Manganese dioxide	—	MnO_2
Manganous bicarbonate	—	$Mn(HCO_3)_2$
Manganous sulfate	—	$MnSO_4$
Monochloramine	—	NH_2Cl
Potassium bicarbonate	—	$KHCO_3$
Potassium permanganate	—	$KMnO_4$
Sodium bicarbonate	Soda	$NaHCO_3$
Sodium carbonate	Soda ash	Na_2CO_3
Nitrogen trichloride (trichloramine)	—	NCl_3
Sodium chloride	Salt	$NaCl$
Sodium chlorite	—	$NaClO_2$
Sodium fluoride	—	NaF
Sodium fluosilicate (sodium silicofluoride)	—	Na_2SiF_6
Sodium hydroxide	Lye	$NaOH$
Sodium hypochlorite	—	$NaOCl$
Sodium phosphate	—	$Na_3PO_4 \cdot 12H_2O$
Sodium silicofluoride (sodium fluosilicate)	—	Na_2SiF_6
Sodium sulfate	—	Na_2SO_4
Sulfuric acid	Oil of vitriol	H_2SO_4
Trichloramine (nitrogen trichloride)	—	NCl_3

Appendix D
Algae Color Plates

The six color plates on the following pages illustrate how some of the more common types of algae encountered in water treatment would appear under a microscope. For positive identification, a biologist familiar with algae should be consulted.

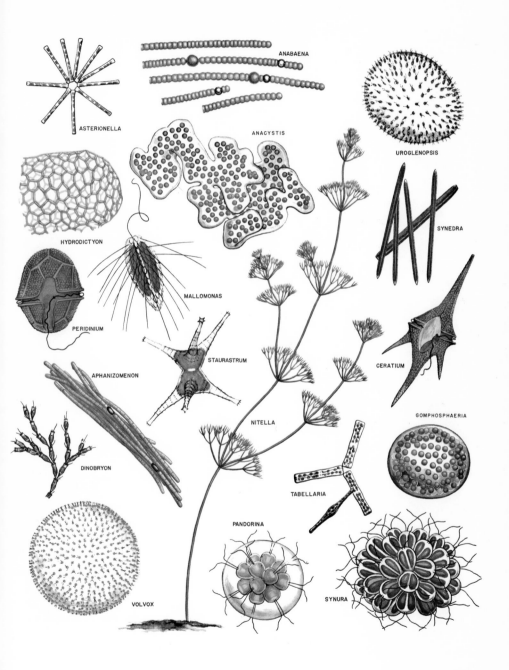

ASTERIONELLA

ANABAENA

UROGLENOPSIS

HYDRODICTYON

ANACYSTIS

SYNEDRA

PERIDINIUM

MALLOMONAS

APHANIZOMENON

STAURASTRUM

CERATIUM

NITELLA

GOMPHOSPHAERIA

DINOBRYON

TABELLARIA

VOLVOX

PANDORINA

SYNURA

Plate A.   Taste and Odor Algae

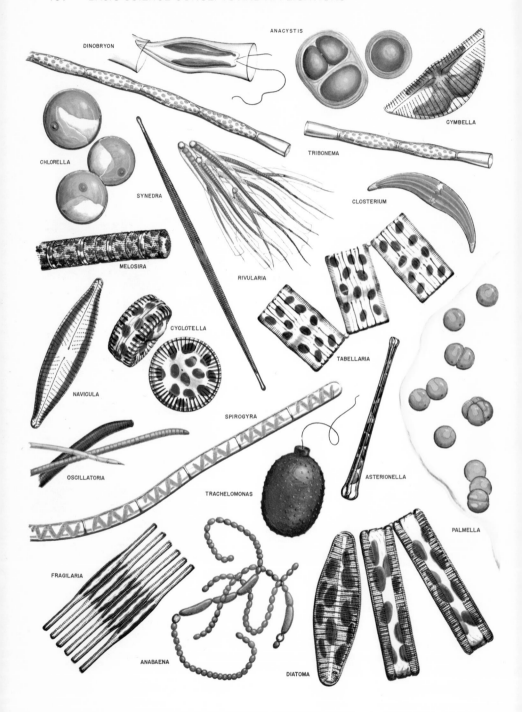

Plate B.   Filter Clogging Algae

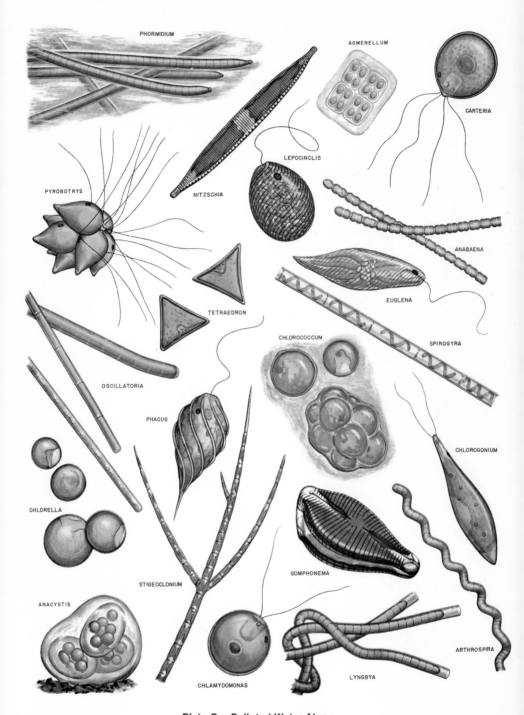

Plate C. Polluted Water Algae

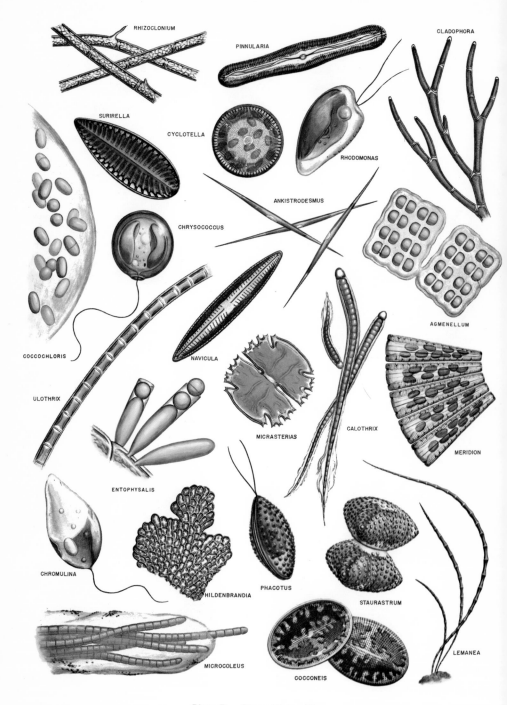

Plate D.   Clean Water Algae

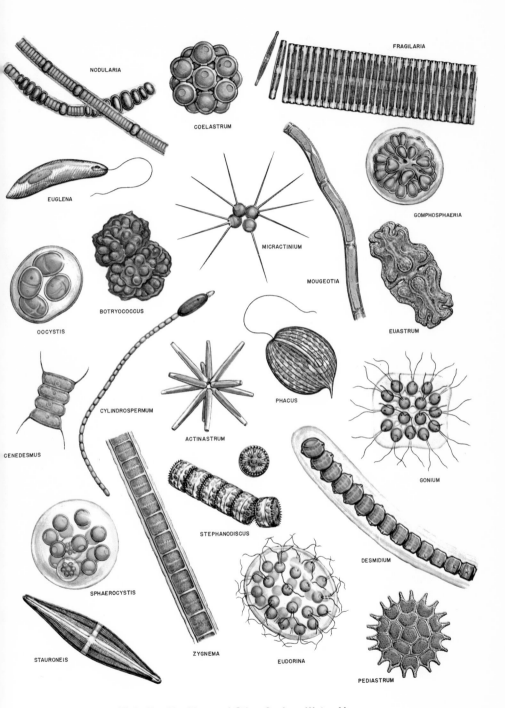

Plate E.   Plankton and Other Surface Water Algae

Plate F.   Algae Growing on Reservoir Walls

Glossary/Index

Glossary/Index

Words set in SMALL CAPITAL LETTERS in the text are defined in the following glossary. The page number following each word is generally the page where that word first appears in the text. However, where a word is introduced on one page but not explained in detail until several pages later, the page number where the more complete explanation can be found is given. There are no words set in small capitals in the Electricity Section, but all important terms in the section are included in the glossary.

Arithmetic mean, 71   A measurement of average value, calculated by summing all terms and dividing by the number of terms.

Arithmetic scale, 166   A scale is a series of intervals (marks or lines), usually marked along the side and bottom of a graph, that represents the range of values of the data. When the marks or lines are equally spaced, it is called an arithmetic scale. Compare LOGARITHMIC SCALE.

Atom, 479   The smallest particle of an ELEMENT that still retains the characteristics of that element.

Atomic number, 480   The number of PROTONS in the NUCLEUS of an ATOM.

Atomic weight, 480   The sum of the number of PROTONS and the number of NEUTRONS in the NUCLEUS of an atom.

Average daily flow (ADF), 233   A measurement of the amount of water treated by a plant each day. It is the average of the actual DAILY FLOWS that occur within a period of time, such as a week, a month, or a year. Mathematically, it is the sum of all daily flows divided by the total number of daily flows used.

Average flow rate, 419   The average of the INSTANTANEOUS FLOW RATES over a given period of time, such as a day.

Balanced, 501   A CHEMICAL EQUATION is balanced when, for each element in the equation, as many atoms are shown on the right side of the equation as are shown on the left side.

Base, 546   Any substance that releases hydroxyl IONS (OH⁻) when it dissociates in water.

Basic solution, 547   A solution that contains significant numbers of OH⁻ IONS.

Battery, 675   A device for producing DC electric current from a CHEMICAL REACTION. In a storage battery, the process may be reversed, with current flowing into the battery, thus reversing the chemical reaction and recharging the battery.

Bicarbonate alkalinity, 548   Alkalinity caused by bicarbonate IONS (HCO_3^-).

Bond   See CHEMICAL BOND.

Brake horsepower, 379   The power supplied to a pump by a motor. Compare WATER HORSEPOWER and MOTOR HORSEPOWER.

Buffer, 548   In chemistry, the ability of a solution to resist a reduction in pH as acid is added is known as the solution's "acid buffering capacity."

Bulk density, 288   The weight per standard volume (usually pounds per cubic foot) of material as it would be shipped from the supplier to the treatment plant.

Calcium carbonate saturation index   See LANGLIER INDEX.

Capacitance, 656   A part of the total IMPEDANCE of an electrical circuit tending to resist the flow of current. Capacitance can be added to cancel the effect of INDUCTANCE.

Capacitor, 656   Two conductive plates separated by a nonconductor, used to create CAPACITANCE in a circuit. Also called a CONDENSOR

Capacity, 387   The FLOW RATE that a pump is capable of producing.

Carbonate alkalinity, 548   ALKALINITY caused by carbonate ions (CO_3^{-2}).

Carbonate hardness, 561   HARDNESS caused primarily by bicarbonate. Compare NONCARBONATE HARDNESS.

Cation, 481   A positive ION.

Cation exchange materials, 564   Materials that release nontroublesome IONS into water in exchange for hardness causing ions.

Chelation, 570   A chemical process used to control scale formation, in which a chelating agent "captures" scale-causing IONS and holds them in solution, thus preventing them from precipitating out and forming scale.

Chemical bond, 473   The force that holds atoms together within MOLECULES. A chemical bond is formed when a CHEMICAL REACTION takes place. Two types of chemical bond are IONIC BONDS and COVALENT BONDS.

Chemical equation, 500   A shorthand way, using CHEMICAL FORMULAS, of writing the reaction that takes place when chemicals are brought together. The left side of the equation indicates the chemicals brought together (the REACTANTS); the arrow indicates which direction the reaction occurs; and the right side of the equation indicates the results (the PRODUCTS) of the chemical reaction.

Chemical formula   See FORMULA.

Chemical reaction, 489   A process that occurs when ATOMS of certain elements are brought together and combine to form MOLECULES, or when molecules are broken down into individual atoms.

Circuit breaker, 691, 697   A device that functions both as a CURRENT overload protective device and as a SWITCH.

Circumference, 90   The distance measured around the outside edge of a circle.

Compounds, 485   Two or more ELEMENTS bonded together by a CHEMICAL REACTION.

Concentration, 517   In chemistry, a measurement of how much SOLUTE is contained in a given amount of SOLUTION. Concentrations are commonly measured in PARTS PER MILLION (ppm), MILLIGRAMS PER LITRE (mg/L), or GRAINS PER GALLON (gpg).

Condensor   See CAPACITOR.

Conductor, 693   A substance that permits the flow of electricity, especially one that conducts electricity with ease.

Converter, 677   Generally, a DC GENERATOR driven by an AC motor.

Covalent bond, 473   A type of CHEMICAL BOND in which electrons are shared. Compare IONIC BOND.

Cross multiplication, 59   A method used to determine if two RATIOS are in PROPORTION. Using this method, the NUMERATOR of the first ratio is multiplied by the DEMONINATOR of the second ratio. Similarly, the denominator of the first ratio is multiplied by the numerator of the second ratio. If the products of both multiplications are the same, the two ratios are in proportion to each other.

Current, 651   The "flow rate" of electricity, measured in AMPERES. Compare POTENTIAL.

Current regulator, 678   A device that automatically holds electric CURRENT within certain limits.

Cycle, 656   In ALTERNATING CURRENT electrical systems, one complete change in direction of current flow from zero to maximum to zero in one direction, then to maximum and back to zero in the opposite direction.

Daily flow, 233   The volume of water that passes through a plant in one day (24 hr). More precisely called "daily flow volume."

DC   See DIRECT CURRENT.

Demand, 657   The amount of electric POWER expressed in watts or kilowatts that may be required during a certain time interval.

Demand meter, 688   An instrument that measures the average POWER of a load during some specific interval.

Denominator   The part of a fraction below the line. A fraction indicates division of the NUMERATOR by the denominator.

Density, 287   The weight of a substance per a unit of its volume; for example pounds per cubic foot or pounds per gallon.

Design point, 389   The mark on the H–Q (head/capacity) curve of a PUMP CHARACTERISTICS CURVE that indicates the head and capacity at which the pump is intended to operate for best efficiency in a particular installation.

Detention time, 271   The average length of time a drop of water or a suspended particle remains in a tank or chamber. Mathematically, it is the volume of water in the tank divided by the FLOW RATE through the tank. The units of flow rate used in the calculation (gpm, gph, or gpd) is dependent on whether the detention time is to be calculated in minutes, hours, or days.

Diameter, 90   The length of a straight line measured through the center of a circle from one side to the other.

Digit, 24   Any one of the ten arabic numerals (zero through nine) by which all numbers may be expressed.

Direct current, 657   Electric CURRENT that flows continuously in one direction.

Drawdown, 279   The amount the water level in a well drops once pumping begins. Drawdown equals STATIC WATER LEVEL minus PUMPING WATER LEVEL.

Dynamic discharge head, 369   The difference in height measured from the PUMP CENTERLINE at the discharge of the pump to the point on the HGL directly above it.

Dynamic head, 367   See TOTAL DYNAMIC HEAD.

Dynamic suction head, 369   The distance from the PUMP CENTERLINE at the suction of the pump to the point of the HGL directly above it. Dynamic suction head exists only when the pump is below the PIEZOMETRIC SURFACE of the water at the pump suction. When the pump is above the piezometric surface, the equivalent measurement is DYNAMIC SUCTION LIFT.

Dynamic suction lift, 369   The distance from the PUMP CENTERLINE at the suction of the pump to the point on the HGL directly below it. Dynamic suction lift exists only when the pump is above the PIEZOMETRIC SURFACE of the water at the pump suction. When the pump is below the piezometric surface, the equivalent measurement is called DYNAMIC SUCTION HEAD.

Dynamic water system, 314   The condition of a water system when water is moving through the system.

Effective height, 373   The total feet of HEAD against which a pump must work.

Efficiency, 379   The ratio of the total energy output to the total energy input, expressed as percent.

Electromagnetics, 649   The study of the combined effects of electricity and magnetism.

Electron, 479, 649   One of the three elementary particles of an ATOM (along with PROTONS and NEUTRONS). An electron is a tiny, negatively charged particle which orbits around the NUCLEUS of an atom. The number of electrons in the outermost shell is one of the most important characteristics of an atom in determining how chemically active an ELEMENT will be and with what other elements or COMPOUNDS it will react.

Element, 479   Any of more than 100 fundamental substances that consist of ATOMS of only one kind and that constitute all matter.

Elevation head, 328   The energy possessed per unit weight of a fluid because of its elevation above some reference point (called "reference datum"). Elevation head is also called "position head" or "potential head."

Energy, 658   As pertains to electricity, energy is usually expressed as KILOWATT-HOURS. It is a measure of how much electricity is used. Compare POWER, the measure of the rate at which electricity is used.

Energy grade line (EGL), 329   (Sometimes called "energy-gradient line" or "energy line.") A line joining the elevations of the energy heads; a line drawn above the HYDRAULIC GRADE LINE by a distance equivalent to the velocity head of the flowing water at each section along a stream, channel, or conduit.

Equivalent weight, 528   The weight of an ELEMENT or COMPOUND which, in a given CHEMICAL REACTION, has the same combining capacity as 8 g of oxygen or as 1 g of hydrogen. The equivalent weight for an element or compound may vary with the reaction being considered.

Exponent, 3   An exponent indicates the number of times a base number is to be multiplied together. For example, given a base number of three with an exponent of five, written: 3^5. This indicates that the base number is to be multiplied together five times: $3^5 = 3 \times 3 \times 3 \times 3 \times 3$.

Filter backwash rate, 261   A measurement of the number of gallons each minute flowing upward (backwards) through a square foot of filter surface area. Mathematically, it is the gallons-per-minute backwash flow rate divided by the total square feet of filter area.

Filter loading rate, 253   A measurement of the number of gallons of water applied to each square foot of filter surface area. Mathematically, it is the gallons-per-minute flow rate into the filter divided by the total square feet of filter area.

Flow rate, 419   A measure of the volume of water moving past a given point in a given period of time. Compare INSTANTANEOUS FLOW RATE and AVERAGE FLOW RATE.

Formula, 492   Using the chemical symbols for each element, a formula is a shorthand way of writing what elements are present in a molecule and how many atoms of each element are present in each of the molecules. Also called a chemical formula.

Formula weight, 495   See MOLECULAR WEIGHT.

Free water surface, 309   The surface of water that is in contact with the atmosphere.

Frequency, 658   For ALTERNATING CURRENT, electricity, the number of CYCLES that occur each second. Frequency is measured in HERZ (Hz), which are defined as cycles per second (cps).

Friction head loss, 341   The HEAD lost by water flowing in a stream or conduit as the result of (1) the disturbance set up by the contact between the moving water and its containing conduit and (2) intermolecular friction.

Fuse, 691, 696   A protective device that disconnects equipment from the power source when CURRENT exceeds a specified value.

Gage pressure, 298   The water pressure as measured by a gage. Gage pressure is not the total pressure. Total water pressure (ABSOLUTE PRESSURE) also includes the atmospheric pressure (about 14.7 psi at sea level) exerted upon the water. However, because atmospheric pressure is exerted everywhere (against the outside of the main as well as the inside, for example), it is generally not written in to water system calculations. Gage pressure in pounds per square inch is expressed as "psig."

Gallons per capita per day (GPCD), 211   A measurement of the average number of gallons of water used by the average person each day in a water system. The calculation is made by dividing the total gallons of water used each day by the total number of people using the water system.

Generator, 673   A piece of equipment used to transform rotary motion (for example, the output of a diesel engine) to electric current.

Grains per gallon (gpg), 517   A measure of the CONCENTRATION of SOLUTIONS. 1 gpg 17.12 mg/L.

Gram mole   See MOLE.

Ground, 659   An electrical connection to earth or to a large conductor that is known to be at the same potential as earth.

Groups, 482   The vertical columns of elements in the PERIODIC TABLE.

Hardness, 530, 561   A characteristic of water, caused primarily by the salts of calcium and magnesium. Causes deposition of scale in boilers, damage in some industrial processes, and sometimes objectionable taste. May also decrease the effectiveness of soap.

Head, 327   (1) A measure of the energy possessed by water at a given location in the water system, expressed in feet. (2) A measure of the pressure or force exerted by water, expressed in feet.

Head loss, 341   The amount of energy used by water in moving from one location to another.

Herz (Hz), 655   A measurement of FREQUENCY, equal to cycles per second.

Homogenous, 517   A substance with a uniform structure or composition throughout is said to be homogeneous.

Horsepower, 374   A standard unit of power equal in the US to 746 watts, approximately equal to 33,000 ft-lb per min.

Hydraulic grade line (HGL), 313   A line (hydraulic profile) indicating the piezometric level of water at all points along a conduit, open channel, or stream. In an open channel, the HGL is the free water surface.

Hydroxyl alkalinity, 548   ALKALINITY caused by hydroxyl IONS (OH^-).

Impedance, 659   The total opposition offered by a circuit to the flow of electric current.

Inductance, 659   A part of the total IMPEDANCE of an electrical circuit that tends to resist the flow of changing CURRENT. Inductance can be added to cancel the effect of CAPACITANCE.

Instantaneous flow rate, 419   A FLOW RATE of water measured at one particular instant, such as by a metering device or by a calculation ($Q = A V$) involving the cross sectional area of the channel or pipe and the velocity of the water at one instant.

Insulator, 693   A substance that offers very great RESISTANCE or hindrance to the flow of electric CURRENT.

Interpolation, 166   A technique used to determine values that fall between the marked intervals on a scale.

Interrupting current, 660   The fault current that a FUSE or CIRCUIT BREAKER is designed to interrupt without becoming destructive.

Inverter, 677   A device that changes DC electric current to AC.

Ion, 481   An atom which is electrically unstable because it has more or less electrons than protons. A positive ion is called a CATION. A negative ion is called an ANION.

Ionic bond, 473   A type of CHEMICAL BOND in which ELECTRONS are transferred. Compare COVALENT BOND.

Isotopes, 480   Atoms of the same ELEMENT, but containing varying numbers of NEUTRONS in the NUCLEUS, are called isotopes of that element. For each element, the most common naturally occurring isotope is called the PRINCIPAL ISOTOPE of that element.

Kilo (k), 660   A prefix meaning one thousand.

Kilovolt, 660   One thousand VOLTS.

Kilowatt (kW), 660   A measure of electric POWER equal to 1000 WATTS. 1 hp   0.746 kW. See WATT.

Kilowatt-hour (kWhr), 660   A measure of electric ENERGY.

kVA, 660   Thousand volt-amperes, a measurement of APPARENT POWER.

kvar, 660   Thousand reactive volt-amperes, a measurement of UNREAL POWER.

Langlier index, 569   A method for reporting the tendency of water to scale or corrode pipes, based on the pH and alkalinity of the water. Full name is "Langlier's calcium carbonate saturation index." Also called "saturation index."

Logarithmic scale (Log scale), 166   A scale is a series of intervals (marks or lines), usually marked along the side and bottom of a graph, that represents the range of values of the data. When the marks or lines are varied logarithmically (and are therefore not equally spaced), the scale is called a logarithmic or log scale.

Maximum demand, 661   In electricity, the maximum kilowatt load that occurs and persists for a full demand interval during any billing period, usually a month.

Megohm, 661 One million OHMS.

Milligrams per litre (mg/L), 517 A measure of the CONCENTRATION of a SOLUTION. One milligram weight of SOLUTE in every litre volume of solution. Generally interchangable with PARTS PER MILLION in water treatment calculations.

Minor head loss, 349 The energy losses that result from the resistance to flow as water passes through valves, fittings, inlets, and outlets of a piping system.

Mixture, 486 Two or more elements, compounds, or both, mixed together with no CHEMICAL REACTION (BONDING) occurring.

Molality A measure of CONCENTRATION defined as the number of MOLES of SOLUTE per litre of *SOLVENT*. Not commonly used in water treatment. Compare MOLARITY.

Molarity, 525 A measure of CONCENTRATION defined as the number of MOLES of SOLUTE per litre of SOLUTION.

Mole, 504, 525 Used in this text and generally as an abbreviation for GRAM-MOLE. A mole is the quantity of a COMPOUND or ELEMENT that has a weight in grams equal to the substance's MOLECULAR or ATOMIC WEIGHT.

Molecular weight, 495 The sum of the ATOMIC WEIGHTS of all the ATOMS in the COMPOUND. Also called formula weight.

Molecule, 486 Two or more ATOMS joined together by a CHEMICAL BOND.

Motor horsepower, 379 The horsepower equivalent to the watts of electric power supplied to a motor. Compare BRAKE HORSEPOWER and WATER HORSEPOWER.

Neutralization, 547 The process of mixing an ACID and a BASE to form a SALT and water.

Neutralize See NEUTRALIZATION

Neutron, 479 An uncharged elementary particle that has a mass approximately equal to that of the PROTON. Neutrons are present in all known atomic NUCLEI except the lightest hydrogen nucleus.

Nomograph, 184 A graph in which three or more scales are used to solve mathematical problems.

Noncarbonate hardness, 530 HARDNESS caused by the salts of calcium and magnesium.

Normality, 528 A method of expressing the CONCENTRATION of a SOLUTION. It is the number of EQUIVALENT WEIGHTS of SOLUTE per litre of SOLUTION.

Nucleus (plural: nuclei), 479 The center of an ATOM, made up of positively charged particles called PROTONS and uncharged particles called NEUTRONS.

Numerator The part of a fraction above the line. A fraction indicates division of the numerator by the DENOMINATOR.

Ohm, 661 A measure of the ability of a path to resist or impede the flow of electric CURRENT.

Ohm's Law, 653 An equation expressing the relationship between the POTENTIAL (E) in volts, the RESISTANCE (R) in ohms, and the CURRENT (I) in amperes for electricity passing through a metallic conductor.
Ohm's law is: $E = I \times R$.

Organic compounds, 555 Generally, COMPOUNDS containing carbon.

Organics See ORGANIC COMPOUNDS.

Parts per million (ppm) A measure of the CONCENTRATION of a SOLUTION. One part of SOLUTE in every million parts of solution. Generally interchangable with MILLIGRAMS PER LITRE in water treatment calculations.

Pascal (Pa) A unit of pressure in the metric system. To convert from Pascals to psi, multiply the number of Pascals by 0.000145.

Percent (%), 77 The fraction of the whole expressed as parts per one hundred.

Perimeter, 89   The distance around the outer edge of a shape.

Periodic table, 481   A chart showing all ELEMENTS arranged according to similarities of chemical properties.

Periods, 482   The horizontal rows of elements in a PERIODIC TABLE.

pH, 547   A measurement of how ACIDIC or BASIC a substance is. The pH scale runs from 0 (most acidic) to 14 (most basic). The center of the range (7) indicates the substance is neutral, neither acidic or basic.

Phase, 661   The type of electrical service either supplied or required; generally, it will be single-phase or three-phase. Phase is also used to refer to the relationship between AC CURRENT and AC VOLTAGE in a circuit; phase in such a circuit is determined by the interaction between RESISTANCE, INDUCTANCE, and CAPACITANCE.

Pi (π), 90   The RATIO of the circumference of a circle to the diameter of that circle, approximately equal to 3.14159, or about 22/7.

Piezometer, 309   An instrument for measuring PRESSURE HEAD in a conduit, tank, or soil, by determining the location of the FREE WATER SURFACE.

Piezometric surface, 309   An imaginary surface that coincides with the level of the water in an aquifer, or the level to which water in a system would rise in a piezometer.

Pole, 663   One end of a magnet: the north or south pole.

Potential, 651   The "pressure" of electricity, measured in VOLTS. Compare CURRENT.

Pounds per square inch, 295   A measurement of pressure. The comparable metric unit is the PASCAL (Pa) 1 psi = 6895 Pa.

Pounds per square inch absolute (psia), 298   The sum of GAGE PRESSURE (psig) and atmospheric pressure. See ABSOLUTE PRESSURE. Compare POUNDS PER SQUARE INCH GAGE.

Pounds per square inch gage (psig), 298   PRESSURE measured by a gage and expressed in terms of pounds per square inch. See GAGE PRESSURE. Compare POUNDS PER SQUARE INCH ABSOLUTE.

Power (in hydraulics or electricity), 374, 664   The measure of the amount of WORK done in a given period of time. The rate of doing WORK. Measured in watts or horsepower.

Power (in mathematics)   See EXPONENT.

Power factor (pf), 663   The ratio of useful or REAL POWER to the APPARENT POWER in a circuit (related to PHASE).

Pressure, 295   The force pushing on a unit area. Normally pressure can be measured in POUNDS PER SQUARE INCH (psi), feet of HEAD, or PASCALS (Pa).

Pressure head, 328   A measurement of the amount of energy in water due to water pressure.

Primary, 665   The high-voltage side of a step-down TRANSFORMER.

Principal isotopes   See ISOTOPES.

Products, 500   The results of a CHEMICAL REACTION. The products of a reaction are shown on the right side of a CHEMICAL EQUATION.

Proportion (proportionate), 59   When the relationship between two numbers in a RATIO is the same as that between two other numbers in another ratio, the two ratios are said to be in proportion, or proportionate.

Proton, 479   One of the three elementary particles of an ATOM (along with NEUTRONS and ELECTRONS). The proton is a positively-charged particle located in the NUCLEUS of an atom. The number of protons in the nucleus of an atom determines the ATOMIC NUMBER of that element.

psi   See POUNDS PER SQUARE INCH.

psig   See POUNDS PER SQUARE INCH GAGE.

Pump centerline, 364   An imaginary line through the center of a pump.

Pump characteristic curve, 387   A curve or curves showing the interrelation of speed,

DYNAMIC HEAD, CAPACITY, BRAKE HORSEPOWER, and EFFICIENCY of a pump.

Pumping water level (PWL), 279 The water level measured when the pump is in operation.

Radicals, 491 Groups of elements CHEMICALLY BONDED together and acting like single ATOMS or IONS in their ability to form other compounds.

Radius, 102 The distance from the center of a circle to its edge. One half of the DIAMETER.

Rating, 665 In electricity, the limit of power, current, or voltage at which a material or piece of equipment is designed to operate under specific conditions.

Ratio, 59 A relationship between two numbers. A ratio may be expressed using colons (for example, 1:2 or 3:7), or it may be expressed as a fraction (for example, 1/2 or 3/7).

Reactance, 665 The combined effect of CAPACITANCE and INDUCTANCE.

Reactants, 500 The chemicals brought together in a CHEMICAL REACTION. The chemical reactants are shown on the left side of a CHEMICAL EQUATION.

Reactive power See UNREAL POWER.

Real power, 665 The part of the power supplied to a device that is actually available to do work; the product of APPARENT POWER and the POWER FACTOR. See APPARENT POWER, UNREAL POWER.

Recarbonation, 563 The reintroduction of carbon dioxide into the water, either during or after lime-soda ash softening.

Rectifier, 677 A device that changes AC electric current to DC.

Regeneration, 566 The process of reversing the ion exchange softening reaction of ION EXCHANGE MATERIALS, removing the hardness IONS from the used materials and replacing them with nontroublesome ions, thus rendering the materials fit for reuse in the softening process.

Relay, 692 A device activated by an electrical or physical condition to cause the operation of some other device in an electrical circuit.

Resistance, 665 A part of the total IMPEDANCE of a circuit tending to restrict the flow of current, measured in OHMS. Resistance consumes REAL POWER in the form of heat produced as current flows through the resistor; the other types of impedance, CAPACITANCE and INDUCTANCE, do not consume real power.

Rule of continuity, 427 The rule states that the flow (Q) that enters a system must also be the flow that leaves the system. Mathematically, this rule is generally stated as $Q_1 = Q_2$ or (since $Q = AV$), $A_1 V_1 = A_2 V_2$.

Safety factor, 665 The percentage above which a rated electrical device cannot be operated without damage or shortened life.

Salts, 547 Compounds resulting from ACID BASE mixtures.

Saturation Index (S.I.) See LANGLIER INDEX.

Scientific notation, 5 A method by which any number can be expressed as a number between 1 and 9 multiplied by a POWER of 10.

Secondary, 666 The low-voltage side of a step-down TRANSFORMER.

Sequestration, 571 A chemical process used to control scale formation, in which a sequestering agent holds scale-causing ions in solution, preventing them from precipitating out and forming scale.

Side water depth (SWD), 111 The depth of water measured along a vertical interior wall.

Solute, 517 The substance dissolved in a SOLUTION. Compare SOLVENT.

Solution, 517 A liquid containing a dissolved substance. The liquid alone is called the SOLVENT, the dissolved substance is called the SOLUTE. Together they are called a solution.

Solvent, 517   The liquid used to dissolve a substance. See SOLUTION.

Specific capacity, 280   A measurement of the WELL YIELD per unit (usually per foot) of DRAWDOWN. Mathematically, it is the gallons-per-minute well yield divided by the total feet of drawdown.

Specific gravity, 290   The ratio of the DENSITY of a substance to a standard density. For solids and liquids, the density is compared to the density of water (62.4 lb/cu ft). For gases the density is compared to the density of air (0.075 lb/cu ft).

Standard solution, 534   A solution with an accurately known CONCENTRATION, used in the lab to determine the properties of unknown solutions.

Static discharge head, 364   The difference in height (in feet) between the PUMP CENTERLINE and the level of the discharge FREE WATER SURFACE.

Static suction head, 364   The difference in elevation between the PUMP CENTERLINE and the FREE WATER SURFACE of the reservoir feeding the pump. In the measurement of static suction head, the PIEZOMETRIC SURFACE of the water at the suction side of the pump is higher than the pump; otherwise, STATIC SUCTION LIFT is measured.

Static suction lift, 364   The difference in elevation between the PUMP CENTERLINE of a pump and the FREE WATER SURFACE of the liquid being pumped. In a static suction lift measurement, the PIEZOMETRIC SURFACE of the water at the suction side of the pump is lower than the pump, otherwise STATIC SUCTION HEAD is measured.

Static water level (SWL), 279   The water level in a well measured when no water is being taken from the aquifer, either by pumping or by free flow.

Static water system, 313   The condition in a water system when water is not moving through the system.

Substation, 677   An electric power switching station, generally accompanied by a TRANSFORMER.

Surface overflow rate, 241   A measurement of the amount of water leaving a sedimentation tank per square foot of tank surface area. Mathematically, it is the gallons-per-day flow rate from the tank divided by the square feet of tank surface.

Switch   A device to manually disconnect electrical equipment from the power source.

Thrust, 465   A force resulting from water under pressure and in motion. Thrust pushes against fittings, valves, and hydrants, and can cause couplings to leak or to pull apart entirely.

Thrust anchor, 466   A block of concrete, often a roughly shaped cube, cast in place below a fitting to be anchored against vertical THRUST, and tied to the fitting with anchor rods.

Thrust block, 466   A mass of concrete, cast in place between a fitting to be anchored against THRUST and the undisturbed soil at the side or bottom of the pipe trench.

Total alkalinity, 548   The combined effect of HYDROXYL ALKALINITY (OH⁻), CARBONATE ALKALINITY (CO_3^-) and BICARBONATE ALKALINITY (HCO_3).

Total dynamic head, 369   The difference in height between the HGL on the discharge side of the pump and the HGL on the suction side of the pump. This head is a measure of the total energy that a pump must impart to the water to move it from one point to another.

Total static head, 365   The total height that the pump must lift the water when moving it from one point to another. The vertical distance from the suction free water surface to the discharge free water surface.

Transformer, 650   A device used with AC to increase CURRENT while decreasing VOLTAGE, or to decrease CURRENT while increasing VOLTAGE.

Trihalomethanes (THMs), 555   Certain organic COMPOUNDS, sometimes formed when water containing natural organics is chlorinated. Some THMs, in large enough concentrations, may be carcinogenic.

Valence, 490   One or more numbers assigned to each ELEMENT indicating the ability of the element to enter into CHEMICAL REACTIONS with other elements.

Valence electrons, 489   The electrons in the outermost electron shells. These electrons are one of the most important factors in determining which atoms will combine with which other atoms.

Velocity head, 328   A measurement of the amount of energy in water due to its velocity, or motion.

Volt (V), 666   The practical unit of electric POTENTIAL. One volt will send a CURRENT of one AMPERE through a RESISTANCE of one OHM.

Voltage regulator, 677   A device to automatically maintain a certain voltage.

Voltmeter, 688   An instrument for measuring VOLTS.

Water hammer   The potentially damaging slam, bang, or shudder that occurs in a pipe when a sudden change in water velocity (usually as a result of too-rapidly starting a pump or operating a valve) creates a great increase in water pressure.

Water horsepower (WHP), 374, 379   The portion of the power delivered to a pump that is actually used to lift water. Compare BRAKE HORSEPOWER and MOTOR HORSEPOWER.

Watt, 666   The practical unit of electric POWER. In DC, WATTS equals VOLTS times AMPERES. In AC, WATTS equals VOLTS times AMPERES times POWER FACTOR.

Watt-hour, 666   A measure of electric ENERGY.

Wattmeter, 688   An instrument for measuring REAL POWER in WATTS.

Weir overflow rate, 247   A measurement of the number of gallons per day of water flowing over each foot of weir in a sedimentation tank or circular clarifier. Mathematically, it is the gallons-per-day flow over the weir divided by the total length of the weir in feet.

Well yield, 277   The volume of water that is discharged from a well during a specified time period (usually minutes). Mathematically, it is the total number of gallons discharged, divided by the minutes during which the discharge was monitored.

Whole numbers, 24   Any of the natural numbers, such as 1, 2, 3, etc.; the negative of these numbers, such as -1, -2, -3, etc.; and zero. Also called "integers" or "counting numbers."

Wire-to-water efficiency, 379   The ratio of the total power input (electric current expressed as MOTOR HORSEPOWER) to a motor and pump assembly to the total power output (WATER HORSEPOWER), expressed as a percent.

Work, 373   The operation of a force over a specific distance.

NOTES

NOTES

NOTES

NOTES

NOTES